Crystal Indentation Hardness

Special Issue Editors

Ronald W. Armstrong
Stephen M. Walley
Wayne L. Elban

MDPI • Basel • Beijing • Wuhan • Barcelona • Belgrade

MDPI

Special Issue Editors
Ronald W. Armstrong
University of Maryland, College Park
USA

Stephen M. Walley
University of Cambridge
UK

Wayne L. Elban
Loyola University Maryland
USA

Editorial Office
MDPI
St. Alban-Anlage 66
Basel, Switzerland

This edition is a reprint of the Special Issue published online in the open access journal *Crystals* (ISSN 2073-4352) in 2017–2018 (available at: http://www.mdpi.com/journal/crystals/special_issues/cryst_indentation_hardness).

For citation purposes, cite each article independently as indicated on the article page online and as indicated below:

Lastname, F.M.; Lastname, F.M. Article title. *Journal Name* **Year**, *Article number*, page range.

First Editon 2018

Cover image courtesy of Ronald W. Armstrong and Carl Cm. Wu.

ISBN 978-3-03842-967-8 (Pbk)
ISBN 978-3-03842-968-5 (PDF)

Table of Contents

About the Special Issue Editors

Ronald W. Armstrong, Professor Emeritus Ronald Armstrong obtained a B.E.S. degree from Johns Hopkins University in 1955 and a Ph.D. degree from Carnegie Institute of Technology in 1958. A postdoctoral year was spent at Leeds University, U.K., in 1959, followed by appointments at Westinghouse Research Laboratories during 1959–1964, then at Brown University, 1965–1968, and the University of Maryland, 1968–1999. From 2000–2003, he was a senior scientist in the Munitions Directorate, Eglin Air Force Base, FL. His research experience has involved mainly the dislocation mechanics of plasticity and fracturing.

Stephen M. Walley, Dr Stephen Walley graduated with a PhD from the University of Cambridge in 1983. He then worked as a Research Associate at the Cavendish Laboratory. Although he retired in 2014, he is still professionally active in SMF Fracture and Shock Physics at the Cavendish. Over the years, he has been involved in a number of projects, including ballistic impact on glass/polymer laminates, ignition mechanisms of propellants, and high strain rate mechanical properties of polymers, metals and energetic materials. He is secretary of the DYMAT Association (and on its Governing Board).

Wayne L. Elban, Professor Wayne Elban received a BChE with distinction in 1969 and a PhD in Applied Sciences: Metallurgy in 1977 from the University of Delaware and a MS in Engineering Materials in 1972 from the University of Maryland, College Park. From 1969–1985, he was a research engineer at the Naval Surface Warfare Center, White Oak Laboratory, Silver Spring, Maryland. Since 1985, he has been in the engineering department at Loyola College (now Loyola University Maryland). He has worked on a variety of projects, including the development and characterization of energetic materials, metallographic examination and hardness testing of historic wrought iron, and the synthesis and characterization of organic-clay hybrids.

Preface to "Crystal Indentation Hardness"

The first known test machines for measuring the indentation hardness of engineering materials were developed in the middle of the nineteenth century, but it was not until the start of the twentieth century that indentation testing became widespread. This was because quality control of certain manufactured goods required that the strength of every item mass-produced in a factory had to be determined, and this could only be achieved by using a nondestructive testing technique. Also, an emphasis on the science aspects of the topic started in the mid-twentieth century. The subject has received expanded interest in modern times, especially with the advent of nanoindentation hardness testing. Now the complete elastic, plastic and, when appropriate, cracking behaviors of crystals are readily determined using suitable indentation testing procedures. Contact mechanics, dislocation observation and modeling, and indentation fracture mechanics analyses are available for the interpretation of measurements.

The present editorial and collection of review and research articles are intended to provide an impression of the current status of hardness testing methods and the types of analyses applied to the interpretation of the measurements. Diverse crystal types have been investigated, including results for both the softest (organic) and hardest (covalently bonded) structures. Point loading and continuous force-displacement measurements are reported for the range from nano- to macro-dimensional scales. Atomic modeling calculations are presented.

We editors hope that the current sampling of articles will spur the interest of readers, and that they will be successful in their own research efforts to achieve further progress on the topic.

The editors express appreciation to the current authors for the valuable contribution of their diverse articles and express the same appreciation to the reviewers of the articles who have done a great service in more than several instances to help with the explication of the reported hardness measurements and their analyses. Very importantly, we co-editors express our appreciation to Ms. Xin Guo and her Crystals journal team for their considerable support and guidance provided all through the duration of our bringing the present Special Issue effort to fruition as a Crystals e-book.

Ronald W. Armstrong, Stephen M. Walley and Wayne L. Elban
Special Issue Editors

crystals

MDPI

Editorial

Crystal Indentation Hardness

Ronald W. Armstrong [1,*], Stephen M. Walley [2,†] and Wayne L. Elban [3,†]

[1]   Department of Mechanical Engineering, University of Maryland, College Park, MD 20742, USA
[2]   Cavendish Laboratory, Madingley Road, Cambridge CB3 0HE, UK; smw14ster@googlemail.com
[3]   Department of Engineering, Loyola University Maryland, Baltimore, MD 21210, USA; welban@loyola.edu
*   Correspondence: rona@umd.edu; Tel.: +1-410-723-4616
†   These authors contributed equally to this work.

Academic Editor: Sławomir Grabowski
Received: 5 January 2017; Accepted: 6 January 2017; Published: 12 January 2017

Abstract: There is expanded interest in the long-standing subject of the hardness properties of materials. A major part of such interest is due to the advent of nanoindentation hardness testing systems which have made available orders of magnitude increases in load and displacement measuring capabilities achieved in a continuously recorded test procedure. The new results have been smoothly merged with other advances in conventional hardness testing and with parallel developments in improved model descriptions of both elastic contact mechanics and dislocation mechanisms operative in the understanding of crystal plasticity and fracturing behaviors. No crystal is either too soft or too hard to prevent the determination of its elastic, plastic and cracking properties under a suitable probing indenter. A sampling of the wealth of measurements and reported analyses associated with the topic on a wide variety of materials are presented in the current Special Issue.

Keywords: crystal hardness; nanoindentations; dislocations; contact mechanics; indentation plasticity; plastic anisotropy; stress–strain characterizations; indentation fracture mechanics

1. Introduction

The concept of material hardness has been tracked historically by Walley starting from biblical time and proceeding until the 1950s, with pictorial emphasis given to the earliest 19th century design of testing machines and accumulated measurements [1]. Walley's review leads up to David Tabor setting a new course via science connection with his seminal 1951 book *The Hardness of Metals* [2] and further leading, for example, to a later 1973 conference proceedings on *The Science of Hardness Testing and Its Research Applications* [3]. The subject has gained increased importance with the relatively recent advent of orders of magnitude greater force and displacement measuring capabilities available with modern nano-indentation test systems. A review was presented in 2013 of the complete elastic, plastic and, when appropriate, cracking behaviors that can be monitored for crystals, polycrystals, composites and amorphous materials under suitable probing indentation [4]. A substantial reference list was included in the review of conferences and books produced until that time. As will be seen in the current updated collection of research and review articles, no crystal can be too soft or too hard within its environment to escape measurement with a suitable probing indenter applied using appropriate test conditions.

2. Continuous Indentation Testing

The original application of the hardness test corresponded to the obtainment of a single reference point for measurement of an applied load and resultant plastic deformation. Much has been made, and continues to be made, of correlating the measurement with some aspect of the same material unidirectional stress–strain behavior beginning from historical association with the ultimate tensile stress. In like manner, current achievements of continuous indentation load–deformation measurements, particularly obtained with nanoindentation test instruments, are being developed to describe the full material indentation-based stress–strain behavior. Figure 1 provides an example in which a number of measurements are compared. The hardness stress is load divided by contact area; and, the hardness strain is based on a spherical-tipped indenter and evaluated in terms of the surface-projected indentation diameter, *d*, divided by the actual or effective ball diameter, *D* [5]. The terminal open circle points on the dashed and solid elastic loading lines are computed for indicated *D* values on the basis of an indentation fracture mechanics description, as will be described. Pathak and Kalidindi have given an updated description of such nanoindentation stress–strain curves [6]. Here, a brief preview is given of information currently available from point-by-point measurements made conventionally or from continuous load/penetration measurements.

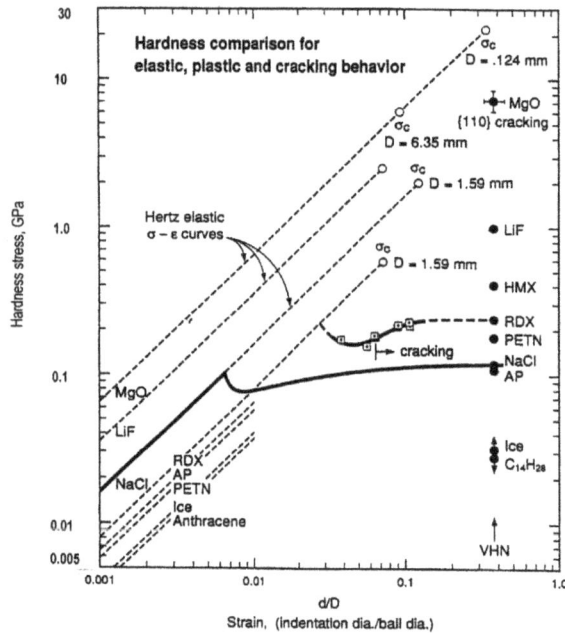

Figure 1. Hardness-based stress–strain description of indentation test measurements. The hardness stress is load divided by the projected area of contact; and, the hardness strain is the projected contact diameter divided by the actual or effective ball diameter of the indenter tip [5]. The solid experimental (ball test) and dashed linear dependencies are computed in accordance with a Hertzian description; and, the Vickers hardness numbers are plotted on the abscissa scale at a position corresponding to the indenter diagonal, *d* = 0.375*D*.

2.1. Continuous Load–Deformation Measurements

The NaCl crystal (solid curve) hardness stress–strain measurement shown in Figure 1 was obtained on indenting an {001} crystal surface with a 6.35 mm ball in a standard compression test. Such measurments are made much easier with the naturally-rounded tips of nanometer- or micrometer-scale indenter tips in nano-test indentation systems. Figure 2 shows an example elastic loading/unloading behavior recorded for nanoindentation of a {0001} sapphire crystal surface [7]. The nonlinear dependence on penetration depth, *h*, may be employed to determine the effective elastic modulus for a known tip diameter in accordance with the Hertzian relationship:

$$E^* = [\{(1 - \nu_B)/E_B\} + \{(1 - \nu_S)/E_S\}]^{-1} = (3\sqrt{2}/4)(P/D^{1/2})h^{-3/2} \qquad (1)$$

In Equation (1), ν_B, E_B and ν_S, E_S are Poisson's ratio and Young's modulus for ball and specimen, respectively, P is load, and again, D is (effective) ball diameter. As indicated in Figure 2, the value of E^* can be determined from fit to the indentation loading curve if D for the rounded indenter tip is known. Dub, Brazhkin, Novikov, Tolmachova et al. have reported similar measurements on sapphire and stishovite single crystals [8]. Elastic modulus determinations for aluminum have been reported for both micro- and nano-scale test systems [9,10]. Solhjoo and Vakis have provided a molecular dynamics assessment of surface roughness on E^* determinations [11]. The initial loading method compares with an alternative method of determining E^* from an unloading curve after small plastic deformation [12].

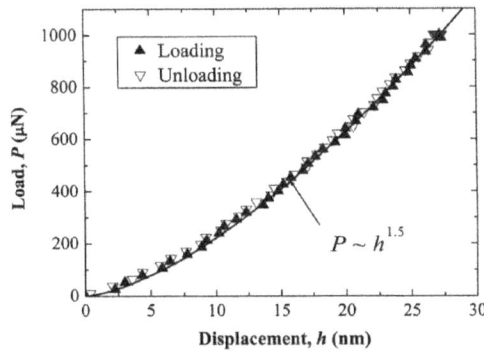

Figure 2. Continuous load/unload curve for nanoindentation of a (0001) sapphire crystal surface [7].

2.2. Crystal Plasticity

The omnipresent elastic loading is eventually interrupted moreso sooner than later by the so-called "pop-in" initiation of crystal plasticity. An example is shown in Figure 3 for the ambient temperature nanoindentation of several different body-centered cubic (bcc) tantalum crystal surfaces [13]. Beyond the higher hardness of the (001) crystal surface, one might note the rapid hardening at the end of the pop-in displacement and the indication of greater strain hardening within the hardened (001) indentation region.

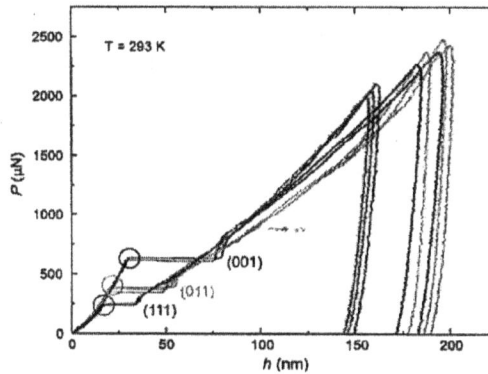

Figure 3. Nanoindentation load–penetration curves obtained with impressing a (rounded point) Berkovich tri-pyramidal indenter into various tantalum (111), (011) and (001) crystal surfaces [13].

Such measurements as shown in Figure 3 have provided valuable information for dislocation mechanics modeling of the crystal deformation behavior, particularly in connection with determining the theoretical shear stress required for dislocation nucleation within a zone below the indentation. Ruestes, Stukowski, Tang, Tramontina et al. have reported on atomic simulation of dislocation creation at nanoindentations in tantalum crystals, including the occurrence of deformation twinning [14]. Alhafez, Ruestes, Gao and Urbassek have investigated nanoindentations made in hexagonal close-packed (hcp) crystal surfaces [15]. A standard (elongated) Knoop indenter system is often employed to evaluate the plastic anisotropy of both ductile and brittle crystal materials [16].

2.3. Crystal Cracking

There are two main aspects of indentation-induced (cleavage) cracking determinations that are of interest: (1) investigation of dislocation mechanism(s) for crack formation; and (2) specification of the indentation fracture mechanics stress intensity for crack propagation.

2.3.1. Crack Formation

Figure 4 provides an example of crack formation at a crystallographically-aligned diamond pyramid indentation impressed into an ammonium perchlorate (AP) {210} crystal surface [17]. The top-side cleavage crack has been generated by sessile dislocations reacted at the intersections of the indicated juxtaposed {111} slip planes, in the same manner reported for cracking at similar indentations put into {001} MgO crystal surfaces and originally proposed for cleavage crack formation in α-iron and other bcc metals [4]. Close examination of the reflected optical image gives an indication that the crack has formed in the valley between otherwise raised surface regions on either side of the indentation. Such "piling up" results from secondary slip occurring to accommodate the primary indentation-forming displacement [4].

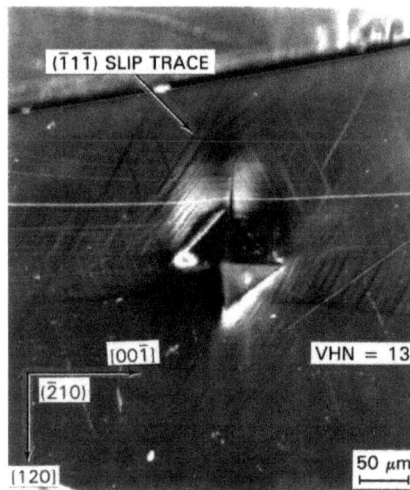

Figure 4. {001} cleavage crack produced by Cottrell-type dislocation reaction at a diamond pyramid indentation impressed into a {210} ammonium perchlorate (AP), NH_4ClO_4, crystal surface [17].

2.3.2. Indentation Fracture Mechanics

An example of elastic, plastic and cracking hardness measurements spanning load and size characterizations for nano- to micro-scale indentations and cracking is shown in Figure 5 [18]. The left-side linear solid and dashed line is the computed Hertzian dependence for elastic behavior, following a $P \sim d^3$ dependence. The next broad band containing filled circle, triangle and square points applies for the determination of the hardness dependence, with d_i being the indentation diagonal length and is shown to approximate, at higher P value, to a $P \sim d_i^2$ dependence. The open symbols correspond to the shift of the load and diagonal measurements to an effective ball diameter result. The furthest right-side measurements are for an indentation fracture mechanics assessment applied to crack extensions following a theoretical $P \sim d_C^{3/2}$ dependence. The figure indicates that plasticity precedes cracking, in line with other results shown in Figure 1 that also show that cracking requires a higher load at smaller effective ball sizes. Wan, H.; Shen, Y.; Chen, Q. and Chen, Y. have elaborated on the plastic 'damage' preceding such cracking produced in silicon crystals with different type indenters [19]. Vodenitcharova, Borrero-López and Hofffman described the related cracking associated with scratching along different directions on {100} silicon crystal wafers [20]. The hardness literature shows such Griffith-based crack analyses to be widely employed for characterizing the cracking behavior of brittle ceramic and glass materials.

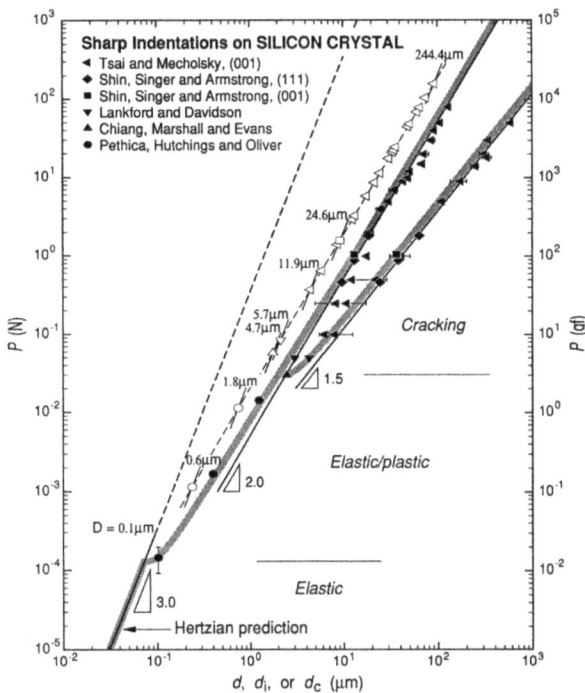

Figure 5. The load, P, versus elastic, d, plastic, d_i, or crack extent, d_c, dependence for cracking of silicon crystal surfaces indented with various sharp indenters [18].

3. Applications

Particular attention was given to example hardness aspects of surface films and coatings in the previous review [4]. Here, we mention, first, an important hardness connection with material compaction behaviors that arise for softer organic crystals relating to pharmaceutical tableting [21] and formulated energetic material processing [5]. Research effort on the former topic has been carried on to investigating temperature [22] and strain rate [23] influences on molecular crystal hardness. In the latter energetic crystal case, a recent report on hardness and molecular dynamics aspects of RDX (cyclotrimethylenetrinitramine) crystals has dealt with concern for their mechanical and tribological properties [24].

Beyond the referenced-above consideration of hardness-determined elastic deformation behavior of sapphire, its hardness properties have been investigated to relate with abrasive wear resistance [25]. Tribological concern has been with chemomechanical aspects of sapphire crystal, polycrystal ZnO coating and other ceramic material hardness properties [26]. In addition, there are interesting optical (cathodoluminescence) properties associated with dislocation zones at nanoindentations in ZnO crystals [27].

3.1. Polycrystals, Polyphases and Amorphous Phases

Particular attention was also directed in [4] to relating hardness properties between individual crystals and their polycrystalline counterparts. The connection was reasonably shown to be well-established in the so-called Hall–Petch (H–P) relationship for hardness:

$$H = H_0 + k_H \ell^{-1/2} \qquad (2)$$

In Equation (2), H is Meyers hardness, $P/(\pi d^2/4)$; H_0 is the single crystal hardness; k_H is a microstructural stress intensity; and ℓ is average (crystal) grain diameter, generally measured on a line intercept basis. Other hardness values are easily related to the Meyer hardness.

Equation (2) is related to the same type equation for the true compressive (or tensile) stress (σ_ε)–true strain (ε) behavior for which $k_H \approx 3k_\varepsilon$. Recent estimations have been reported for H–P k values determined for steel materials through nanoindentation measurements [28]. In line with the H–P model description, the obstacle presented to slip penetrations by grain boundaries has been investigated in the bcc metals case by Soer, Aifantis and De Hosson [29] and more recently in (hcp) α-titanium via high resolution X-ray and electron microscope methods [30]. Related hardness and grain size dependent scratch results have been reported for polycrystalline copper [31]. The influence of H–P strengthening of polycrystalline diamond has been described [32] and also for cubic boron nitride via nanotwinned (boundary) strengthening [33].

An important crystal size-dependent connection has been made with composite WC–Co cermet material for which indentations made on the individual components were known to follow separate H–P dependencies [34]. A recent report on the system has been made by Roa, Jimenez–Pique, Verge, Tarragó et al. [35]; see Figure 6. In related work, Roa et al. have determined an intermediate k_H for the combined deformation of phases [36]. Zhang, Wang and Dai have employed nanoindentation testing to investigate the rate dependence of plastic flow in metallic glass materials [37].

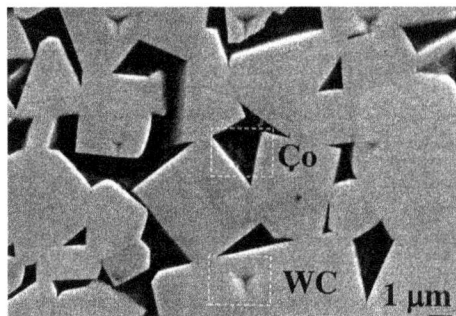

Figure 6. Berkovich nanoindentations made within the WC particle and Co binder phases of the composite cermet material [35].

3.2. Hardness as a Test Probe

In this concluding section, we return to reference [3] in which Gilman pointed to the important use of local hardness testing as a strength microprobe [38] and was able to correlate hardness with a number of material properties such as yield strength, elastic modulus, glide activation energy and energy gap density for a variety of crystals. The advent of nanoindentation testing has given greater meaning to such probe capability. Emphasis was given previously in [4] to connection with probe aspects of atomic force microscopy for which a new tip fabrication procedure has recently been reported [39]. Another recent application has been to investigate the strain hardening surrounding larger Rockwell indentations made in electrodeposited nanocrystalline nickel material with different grain sizes [40]. Additional examples include (1) investigating shear banding in metallic glass materials [41]; and, nano-probing of diffusion-controlled deformation on a copper crystal surface [42] and on different surfaces of ZnO crystals [43].

4. Summary

An editorial introduction to the Special Issue on crystal indentation hardness has been presented by the authors while building upon their previous review entitled *Elastic, Plastic and Cracking Aspects of the Hardness of Materials* [4]. The wide variation in currently referenced journals gives an indication of the broad interest in the subject. Hopefully, the more recent referenced reports will be viewed as indicating an expanded interest in the topic of crystal hardness testing, particularly as provided in this sampling of important measurements and analyses stemming from current attention being given to all aspects of nanoindentation hardness testing.

Conflicts of Interest: The authors declare no conflict of interest.

References

1. Walley, S.M. Review: Historical origins of indentation hardness testing. *Mater. Sci. Technol.* **2012**, *28*, 1028–1044. [CrossRef]
2. Tabor, D. *The Hardness of Metals*; Clarendon Press: Oxford, UK, 1951.
3. Westbrook, J.H.; Conrad, H. (Eds.) *The Science of Hardness Testing and Its Research Applications*; American Society for Metals: Metals Park, OH, USA, 1973.
4. Armstrong, R.W.; Elban, W.L.; Walley, S.M. Elastic, plastic and cracking aspects of the hardness of materials. *Int. J. Mod. Phys. B* **2013**, *28*, 1330004. [CrossRef]
5. Armstrong, R.W.; Elban, W.L. Review: Hardness properties across multi-scales of applied loads and material structures. *Mater. Sci. Technol.* **2012**, *28*, 1060–1071. [CrossRef]
6. Pathak, S.; Kalidindi, S.R. Spherical nanoindentation stress-strain curves. *Mater. Sci. Eng. R* **2015**, *91*, 1–36. [CrossRef]
7. Lu, C.; Mai, Y.W.; Tam, P.L.; Shen, Y.G. Nanoindentation-induced elastic-plastic transition and size effect in α-Al$_2$O$_3$ (0001). *Philos. Mag. Letts.* **2007**, *87*, 409–415. [CrossRef]
8. Dub, S.N.; Braxhkin, V.V.; Novikov, N.V.; Tolmachova, G.N.; Litvin, P.M.; Litagina, L.M.; Dyuzheva, T.I. Comparative Studies of Mechanical Properties of Stishovite and Sapphire Single Crystals by Nanoindentation. *(RU) J. Superhard Mater.* **2010**, *32*, 406–414. [CrossRef]
9. Ferrante, L., Jr.; Armstrong, R.W.; Thadhani, N.N. Elastic/plastic deformation behavior in a continuous ball indentation test. *Mater. Sci. Eng. A* **2004**, *371*, 251–255. [CrossRef]
10. Yoshida, M.; Sumomogi, T.; Endo, T.; Maeta, H.; Kino, T. Nanoscale Evaluation of Strength and Deformation Properties of Ultrahigh-Purity Aluminum. *(JPN) Mater. Trans.* **2007**, *48*, 1–5. [CrossRef]
11. Solhjoo, S.; Vakis, A.I. Continuum mechanics at the atomic scale: Insights into non-adhesive contacts using molecular dynamics simulations. *J. Appl. Phys.* **2016**, *120*, 215102. [CrossRef]
12. Oliver, W.C.; Pharr, G.M. An improved technique for determining hardness and elastic modulus using load and displacement sensing indentation experiments. *J. Mater. Res.* **1992**, *7*, 1564–1583. [CrossRef]
13. Alcala, J.; Dalman, R.; Franke, O.; Biener, M.; Biener, J.; Hodge, A. Planar defect nucleation and annihilation mechanisms in nanocontact plasticity of metal surfaces. *Phys. Rev. Lett.* **2010**, *109*, 075502. [CrossRef] [PubMed]

14. Ruestes, C.J.; Sukowski, A.; Tang, Y.; Tramontina, D.R.; Erhart, P.; Remington, B.A.; Urbassek, H.M.; Meyers, M.A.; Bringa, E.M. Atomistic simulation of tantalum nanoindentation: Effects of indenter diameter, penetration velocity, and interatomic potentials on defect mechanisms and evolution. *Mater. Sci. Eng. A* **2014**, *613*, 390–403. [CrossRef]

15. Alhafez, I.A.; Ruestes, C.J.; Gao, Y.; Urbassek, H.M. Nanoindentation of hcp metals: A comparative simulation study of the evolution of dislocation networks. *Nanotechnology* **2016**, *27*, 045706. [CrossRef] [PubMed]

16. Gao, F. Theoretical model of hardness anisotropy in brittle materials. *J. Appl. Phys.* **2012**, *112*, 023506. [CrossRef]

17. Elban, W.L.; Armstrong, R.W. Plastic anisotropy and cracking at hardness impressions in single crystal ammonium perchlorate. *Acta Mater.* **1998**, *46*, 6041–6052. [CrossRef]

18. Armstrong, R.W.; Ruff, A.W.; Shin, H. Elastic, plastic and cracking indentation behavior of silicon crystals. *Mater. Sci. Eng. A* **1996**, *209*, 91–96. [CrossRef]

19. Wan, H.; Shen, Y.; Chen, Q.; Chen, Y. A plastic damage model for finite element analysis of cracking of silicon under indentation. *J. Mater. Res.* **2010**, *25*, 2224–2237. [CrossRef]

20. Vodenitcharaova, T.; Borrero-López, O.; Hoffman, M. Mechanics prediction of the fracture pattern on scratching wafers of single crystal silicon. *Acta Mater.* **2012**, *60*, 4448–4460. [CrossRef]

21. Egart, M.; Janković, B.; Srčič, S. Application of instrumented nanoindentation in preformulation studies of pharmaceutical active ingredients and excipients. *Acta Pharm.* **2016**, *66*, 303–330. [CrossRef] [PubMed]

22. Mohamed, R.M.; Mishra, M.K.; Al-Harbi, L.M.; Al-Ghamdi, M.S.; Asiri, A.M.; Reddy, C.M.; Ramamurty, U. Temperature Dependence of Mechanical Properties in Molecular Crystals. *Cryst. Growth Des.* **2015**, *15*, 2474–2479. [CrossRef]

23. Raut, D.; Kiran, M.S.R.N.; Mishra, M.K.; Asiri, A.M.; Ramamurty, U. On the loading rate sensitivity of plastic deformation in molecular crystals. *CrystEngComm* **2016**, *18*, 3551–3555. [CrossRef]

24. Weingarten, N.S.; Sausa, R.C. Nanomechanics of RDX Single Crystals by Force-Displacement Measurements and Molecular Dynamics Simulations. *J. Phys. Chem. A* **2015**, *119*, 9338–9351. [CrossRef] [PubMed]

25. Voloshin, A.V.; Dolzhenkova, E.F.; Litvinov, L.A. Anisotropy of Deformation and Fracture Processes in Sapphire Surface. *(RU) J. Superhard Mater.* **2015**, *37*, 341–345. [CrossRef]

26. Bull, S.J.; Moharrami, N.; Hainsworth, S.V.; Page, T.F. The origins of chemomechanical effects in the low-load indentation hardness and tribology of ceramic materials. *J. Mater. Sci.* **2016**, *51*, 107–125. [CrossRef]

27. Juday, R.; Silva, E.M.; Huang, J.Y.; Caldas, P.G.; Prioli, R.; Ponce, F.A. Strain-related optical properties of ZnO crystals due to nanoindentation on various surface orientations. *J. Appl. Phys.* **2013**, *113*, 183511. [CrossRef]

28. Seok, M.-Y.; Choi, I.-C.; Moon, J.; Kim, S.; Ramamurty, U.; Jang, J.-I. Estimation of the Hall–Petch strengthening coefficient of steels through nanoindentation. *Scr. Mater.* **2014**, *87*, 49–52. [CrossRef]

29. Soer, W.A.; Aifantis, K.E.; de Hosson, J.T.M. Incipient plasticity during nanoindentation at grain boundaries in body-centered cubic metals. *Acta Mater.* **2005**, *53*, 4665–4676. [CrossRef]

30. Guo, Y.; Collins, D.M.; Tarleton, E.; Hofmann, F.; Tischler, J.; Liu, W.; Xu, R.; Wilkinson, A.J.; Britton, T.B. Measurements of stress fields near a grain boundary: Exploring blocked arrays of dislocations in 3D. *Acta Mater.* **2015**, *96*, 229–236. [CrossRef]

31. Kareer, A.; Hou, X.D.; Jennett, N.M.; Hainsworth, S.V. The interaction between lateral size effect and grain size when scratching polycrystalline copper using a Berkovich indenter. *Philos. Mag.* **2016**, *96*, 3414–3429. [CrossRef]

32. Xu, B.; Tian, Y. Ultrahardness: Measurement and Enhancement. *J. Phys. Chem.* **2015**, *119*, 5633–5638. [CrossRef]

33. Li, B.; Sun, H.; Chen, C. Large indentation strain stiffening in nanotwinned cubic boron nitride. *Nat. Commun.* **2014**, *5*, 4965–4971. [CrossRef] [PubMed]

34. Lee, H.C.; Gurland, J. Hardness and deformation of cemented tungsten carbide. *Mater. Sci. Eng.* **1978**, *33*, 125–133. [CrossRef]

35. Roa, J.J.; Jiménez-Pique, E.; Verge, C.; Tarragó, J.M.; Mateo, A.; Llanes, L. Intrinsic hardness of constitutive phase in WC–Co composites: Nanoindentation testing, statistical analysis, WC crystal orientation effects and flow stress for the constrained metallic binder. *J. Eur. Ceram. Soc.* **2015**, *35*, 3419–3425. [CrossRef]

36. Roa, J.J.; Jiménez-Pique, E.; Tarragó, J.M.; Sandoval, D.A.; Mateo, A.; Fair, J.; Llanes, L. Hall–Petch strengthening of the constrained metallic binder in WC–Co cemented carbides: Experimental assessment by means of massive nanoindentation and statistical analysis. *Mater. Sci. Eng. A* **2016**, *676*, 487–491. [CrossRef]
37. Zhang, M.; Wang, Y.J.; Dai, L.H. Correlation between strain rate sensitivity and α relaxation of metallic glasses. *AIP Adv.* **2016**, *6*, 075022. [CrossRef]
38. Gilman, J.J. Hardness—A Strength Microprobe. In *The Science of Hardness Testing and Its Research Applications*; Westbrook, J.H., Conrad, H., Eds.; American Society for Metals: Metals Park, OH, USA, 1973; pp. 51–74.
39. Göring, G.; Dietrich, P.-I.; Blaicher, M.; Sharma, S.; Korvink, J.G.; Schimmel, T.; Koos, C.; Hölscher, H. Tailored probes for atomic force microscopy fabricated by two-photon polymerization. *Appl. Phys. Lett.* **2016**, *109*, 063101. [CrossRef]
40. Tang, B.T.F.; Zhou, Y.; Zabev, T. Reduced hardening of nanocrystalline nickel under multiaxial indentation loading. *J. Mater. Res.* **2015**, *30*, 3528–3541. [CrossRef]
41. Wang, J.Q.; Perepezko, J.H. Focus: Nucleation kinetics of shear bands in metallic glass. *J. Chem. Phys.* **2016**, *145*, 211803. [CrossRef]
42. Samanta, A.; Weinan, E. Interfacial diffusion aided deformation during nanoindentation. *AIP Adv.* **2016**, *6*, 075002. [CrossRef]
43. Lin, P.H.; Du, X.H.; Chen, Y.H.; Chen, H.C.; Huang, J.C. Nano-scaled diffusional or dislocation creep analysis of single-crystal ZnO. *AIP Adv.* **2016**, *6*, 095125. [CrossRef]

crystals

MDPI

Review

Progress in Indentation Study of Materials via Both Experimental and Numerical Methods

Mao Liu [1,2,*], Jhe-yu Lin [1], Cheng Lu [2], Kiet Anh Tieu [2], Kun Zhou [3,*] and Toshihiko Koseki [1]

[1]   Department of Materials Engineering, The University of Tokyo, 7-3-1 Hongo, Bunkyo-ku,
     Tokyo 113-8656, Japan; 3jg275@gmail.com (J.L.); koseki@material.t.u-tokyo.ac.jp (T.K.)
[2]   School of Mechanical, Materials and Mechatronic Engineering, University of Wollongong,
     Wollongong, NSW 2522, Australia; chenglu@uow.edu.au (C.L.); ktieu@uow.edu.au (K.A.T.)
[3]   School of Mechanical & Aerospace Engineering, Nanyang Technological University, 50 Nanyang Avenue,
     Singapore 639798, Singapore
*   Correspondence: ml818@uowmail.edu.au (M.L.); kzhou@ntu.edu.sg (K.Z.)

Academic Editor: Ronald W. Armstrong
Received: 19 June 2017; Accepted: 9 August 2017; Published: 13 October 2017

Abstract: Indentation as a method to characterize materials has a history of more than 117 years. However, to date, it is still the most popular way to measure the mechanical properties of various materials at microscale and nanoscale. This review summarizes the background and the basic principle of processing by indentation. It is demonstrated that indentation is an effective and efficient method to identify mechanical properties, such as hardness, Young's modulus, etc., of materials at smaller scale, when the traditional tensile tests could not be applied. The review also describes indentation process via both experimental tests and numerical modelling in recent studies.

Keywords: indentation; mechanical properties; hardness; nanoscale; experiment; modelling

---

1. Introduction

Indentation tests were first performed by a Swedish iron mill's technical manager Brinell who used spherical balls from hardened steel ball bearings or made of cemented tungsten carbide as indenters to measure the plastic properties of materials in 1900 [1–4]. Brinell's testing approach is schematically shown in Figure 1a. Brinell also proposed the following formula to determine the hardness

$$H_B = \frac{2P}{\pi D\left(D - \sqrt{D^2 - d^2}\right)} \tag{1}$$

here H_B is the Brinell hardness; P is the load; D is the diameter of the ball indenter and d is the diameter of residual area of the impression.

Brinell's work was then followed and improved by Meyer in 1908 [5]. In his work, the hardness (H) is calculated by the load (P) divided by the projected area (A), namely

$$H = \frac{P}{A} \tag{2}$$

In 1922, the Vickers test was carried out and commercialized by the Firth-Vickers company [6]. A square-based pyramid diamond indenter with a 136° semi-angle was used instead of a ball indenter, as shown in Figure 1b. The Vickers hardness (H_V) is defined as the load divided by the surface area of the impression, namely

$$H_V = 1.8544 \frac{P}{d_V^2} \tag{3}$$

where d_V is the length of diagonal of the surface area.

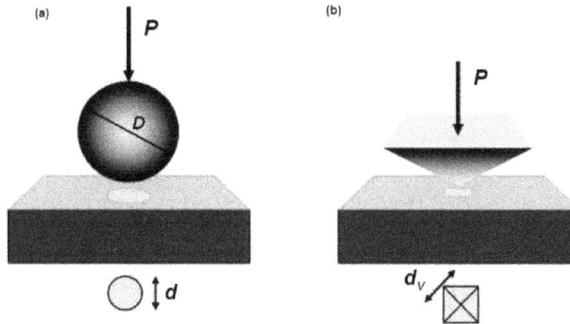

Figure 1. Commonly used methods of indentation hardness tests: (**a**) spherical indenter (Brinell and Meyer); (**b**) diamond pyramid (Vickers) [7]. Reproduced with permission from Ian M. Hutchings, Journal of Materials Research; published by Cambridge University Press, 2009.

The Brinell, Meyer and Vickers methods were then widely used in the metallurgical and engineering industries in the early 20th century since indentation tests offered simplicity, low-cost and high speed compared with conventional tensile testing. No special shape or extra fabrication of sample is required for indentation tests except for a simple sample with a flat surface. In addition, several indentation tests could be quickly performed on a small area without destroying the whole sample.

Another significant finding is that the hardness is load-dependent, which was proposed by Meyer through ball indentation experiments on a wide range of metals [7]. For a given ball size, the diameter of the impression after unloading was found to be related to the applied load by the following empirical relationship:

$$P = Cd^n \tag{4}$$

here P is the applied load; C is a constant of proportionality; the exponent n is the well-known Meyer index and d is the diameter of the residual area of the impression after unloading. When the n-value is less than 2, according to Equations (2) and (4), the hardness increases with decreasing the load. While the n-value is larger than 2, the hardness decreases with decreasing the load. If the Meyer index equals to 2, the hardness is a constant, namely load-independent. For most metals that can be work hardened by the indentation process, $n > 2$ [7]. Therefore, the Meyer index has been found to be strongly dependent on the work hardening of the tested material and to be independent of the size of the ball indenter [5]. Meyer [5] also found that the same hardness using balls of different diameters could be obtained only if the indentations were geometrically similar, namely with the same ratio d/D.

Tabor's work [8] performed in 1948 represents a landmark in the understanding of the indentation process. He qualitatively described the procedure how an indentation by a ball initially led to elastic deformation, then to plastic flow associating with work hardening, and final on removal of the load to elastic recovery. In 1951, Tabor [1] had proven that the indentation hardness (H) could be related to the yield stress (σ) of the material by an equation based on the theory of indentation of a rigid perfectly plastic solid, namely

$$H = C\sigma \tag{5}$$

where C is a constant, which depends on the geometry of the indenters. For the strain-hardened materials and the materials which consequently have no definite yield stress, the stress measured at a representative strain ε_r can be used as σ [9]. The representative strain denotes the strain at where the corresponding stress can be regarded as the yield stress during tensile deformation for the materials without definite yield stress. The value of the representative strain is relevant to the geometry of the indenter. For instance, $\varepsilon_r \approx 0.08$ for a Vickers diamond indenter.

Tabor [7] started taking an interest in the indentation response of polymers and of macroscopically brittle materials in the studies of their frictions. In the following couple of years, Tabor and

King [10] reported the method of estimating the yield pressure on polyethylene, PMMA (Polymethyl methacrylate), PTFE (Polytetrafluoroethylene), and halocarbon polymer via Vickers hardness measurements. Subsequently, Pascoe and Tabor [11] reported a range of polymers obeyed Meyer's laws. The single-crystal rock salt was also investigated by King and Tabor [12] with Vickers indentations, by which, they found the values of yield stress from the indentation matched well with the compression experiments.

In 1960s, indentation at high temperature was investigated by Atkins and Tabor [13]. They studied the mechanical properties of single crystals of MgO at temperatures of 600 °C to 1700 °C via the mutual indentation hardness technique. It was found that the short-time hardness decreased when the temperature increased.

Conventional indentation tests have the length scale of the penetration in microns or millimetres. In the mid-1970s, the indentation technique was applied to measure the hardness of small volumes of material, such as thin film. The length scale of the penetration is usually in nanometres. Therefore, this new technique is called nanoindentation. Apart from the penetration length scale, the distinguishing feature of nanoindentation testing is the indirect measurement of the contact area [14]. In conventional indentation tests, the contact area is directly measured from the residual impression area. In nanoindentation test, the residual impression area is too small to be directly measured. Therefore, the contact area is determined by the measured penetration depth in nanoindentation.

Since 1980s, especially after 1990s, extensive experimental studies of nanoindentation have been performed on many different types of materials. This review will provide the systematic description of indentation study of materials via both experimental and numerical methods. The contents of this article can be summarized as follows. The first section provides the background necessary for understanding the mechanics of indentation and hence for understanding its applications in materials science. The second part reviews methods of mechanical properties characterization and measurement and follows with application to experimental measurement reported in the literature. Next, we discuss the numerical study of deformation mechanism of materials induced by indentation. We follow with a comprehensive review of the models involved in texture and indentation size effect prediction, ranging in length scale from the meso- and micro- to the nanoscale, and a critical assessment of their performance.

2. Nanoindentation Facilities

The most famous manufacturers of nanoindentation equipment include Keysight Technologies (Santa Rosa, CA, USA), Micro Materials Ltd. (Wrexham, UK), Fischer-Cripps Laboratories Pty Ltd. (Killarney Heights, Australia), Hysitron Inc. (Eden Prairie, MN, USA), and Anton Paar (Ashland, VA, USA). The iconic products of these manufacturers are shown in Figure 2. The working theories of different representative instruments are given in Figure 3. All of these instruments include three principal parts, namely indenter, load application, and capacitive sensor for measuring the displacements of the indenter. For instance, in Figure 3a, the load is applied by an electromagnetic coil which is connected to the indenter shaft by a series of leaf springs. The deflection of the springs is a measure of the load applied to the indenter, and the displacement usually can be measured by a capacitive sensor.

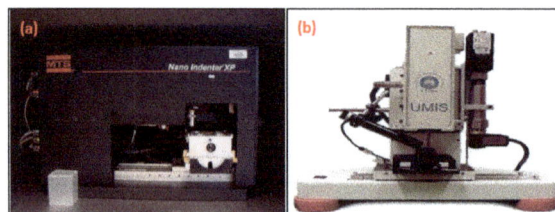

Figure 2. *Cont.*

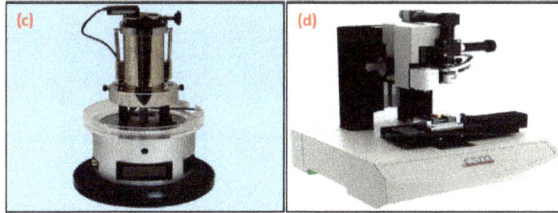

Figure 2. Commercial representative indentation instruments: (**a**) Nano-Indenter XP (Keysight Technologies); (**b**) Ultra-Micro-Indentation System (IBIS); (**c**) TS-75-TriboScope (Hysitron); (**d**) Table-Topnanoindentation (Anton Paar).

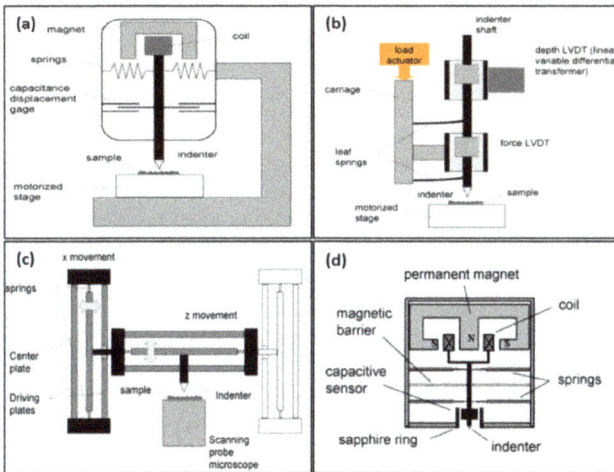

Figure 3. Sketches of the representative indentation instruments: (**a**) Nano-Indenter; (**b**) The Ultra-Micro-Indentation System (UMIS); (**c**) TriboScope; (**d**) The Nano-Hardness Tester (NHT) [15].

The indenter is conventionally made of diamond which has been ground to shape and sintered in a stainless steel chuck. The frequently used shapes of indenter are shown in Figure 4a–f [16]. The conical indenter has a sharp, self-similar geometry. Normally, the cone angle is either $60°$ or $90°$, and the tip radii are 0.7, 1, 2, 5, 10, 20, 50, 100 and 200 μm [16]. The applications of conical indenter are extensive, including scratch testing, wear testing, nano-scale 3D imaging capturing and tensile, and compression tests on MEMS (Microelectromechanical systems). Berkovich indenter is the most frequently used indenter for indentation tests. The most noticeable feature of Berkovich indenter is that it is a three-sided pyramid which can be ground to a point, making it easy to maintain a self-similar geometry to micro-scale or nano-scale. The radius is about 150 nm when it is new and will become 250 nm 12 months later. The applications of Berkovich indenter are much more extensive, such as bulk materials tests, thin films tests, polymers tests, scratch testing, wear testing, MEMS tests and in situ imaging. The Vickers indenter is a four-sided pyramid and suitable for measuring mechanical properties on the very small scale, such as nano-scale as the line of conjunction at the tip limits the sharpness of tip for determination of hardness for very shallow indentation. The recommended applications include bulk materials tests, films and foils tests, scratch testing and wear testing.

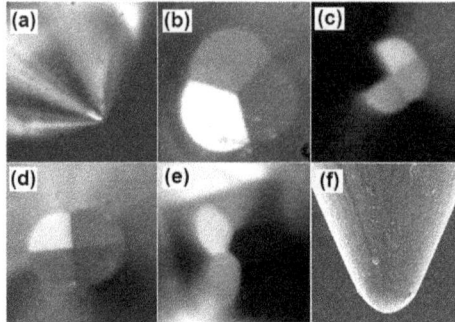

Figure 4. Shapes of different indenters: (**a**) Conical indenter; (**b**) Berkovich indenter; (**c**) Vickers indenter; (**d**) Knoop indenter; (**e**) Cube-corner indenter; (**f**) Spherical indenter [16].

The Knoop indenter is originally designed for hard metals. It is also can be used to probe anisotropy in sample surface. The Cube-corner indenter is a three-sided pyramid with mutually perpendicular faces, which is like the corner of a cube. The sharpness of the cube corner produces much higher stresses and strains in the area in contact with the indenter, thus makes it capable of producing very small and well-defined cracks around imprint in brittle materials. The toughness at microscale or nanoscale can be investigated via these induced fine cracks. Meanwhile, the Cube-corner indenter is fairly fragile and easily broken. The spherical indenter can be used to examine yielding and work hardening theoretically. Moreover, the elastic-plastic transition can also be investigated. The reason is, the contact stresses with spherical indenter are initially small and produce only elastic deformation, and with increasing indentation depth, a transition from elastic to plastic deformation can be captured. The parameters of all these indenters are listed in Table 1. Here, A denotes the projected area; R represents the contact radius of the indent pressed by spherical indenter; h_p is the contact indentation depth; θ represents the semi angle of indenters; a is the effective cone angle and β denotes the geometry correction factor.

Table 1. Parameters of different indenters [17].

Indenter Type	Projected Area	Semi Angle (θ)	Effective Cone Angle (a)	Intercept Factor	Geometry Correction Factor (β)
Sphere	$A \approx \pi 2Rh_p$	N/A	N/A	0.75	1
Berkovich	$A = 3h_p^2\tan^2\theta$	65.3°	70.2996°	0.75	1.034
Vickers	$A = 4h_p^2\tan^2\theta$	68°	70.32°	0.75	1.012
Knoop	$A = 2h_p^2\tan\theta_1\tan\theta_2$	$\theta_1 = 86.25°\ \theta_2 = 65°$	77.64°	0.75	1.012
Cube-corner	$A = 3h_p^2\tan^2\theta$	35.26°	42.28°	0.75	1.034
Conical	$A = \pi h_p^2\tan^2 a$	a	a	0.72	1

3. Application of Nanoindentation in Materials

In 1992, Oliver and Pharr [18] proposed a method to measure the Young's modulus based on the indentation load-displacement curve, which was frequently used until now. Subsequently, they [19] provided a update of how to implement the method to make the most accurate measurents as well as the discussion of its limitations. Kucharski and Mroz [20] presented a new procedure for determining the plastic stress—strain curve by means of a cyclic spherical indentation test in 2007, which constitutes an ground-breaking improvement with respect to the traditional method. Kruzic et al. [21] improved the indentation techniques of evaluating the fracture toughness of biomaterials and hard human bones although accurately measuring the fracture toughness of brittle materials can be quite challenging. The fracture toughness can be obtained directly from indent crack length measurements, as shown in Figure 5.

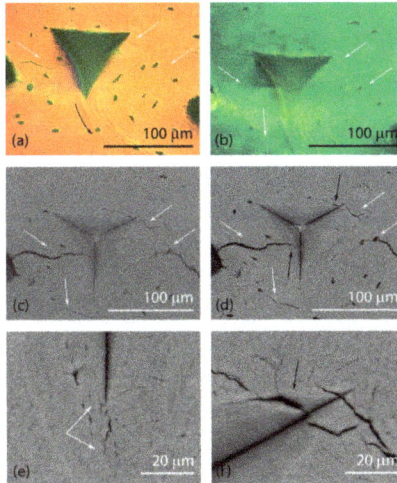

Figure 5. Indentation site on the transverse section of human cortical bone. Images are of (**a**) as-indented; (**b**) dehydrated; (**c**) ESEM (environmental scanning electron microscope) and (**d**) SEM. The bottom corner of the indentation; (**e**) shows a crack with 1 = 10 μm, which is one of the two cracks emanated from indent corners among 30 corners. High-magnification imaging of the top right corner; (**f**) revealed that no cracks were generated from this indent corner [21]. Reproduced with permission from J.J. Kruzic et al., Journal of the Mechanical Behavior of Biomedical Materials; published by Elsevier, 2009.

Huber and Tsakmakis [22] proposed that nanoindentation tests can be used to identify effects of kinematic hardening on the material response. They also indicated that the identification may rely on the measurement of the opening of the hysteresis loop produced in the indentation load-displacement curve as shown in Figure 6. It is obvious that each material has a unique opening diagram indicating the corresponding effect of kinematic hardening.

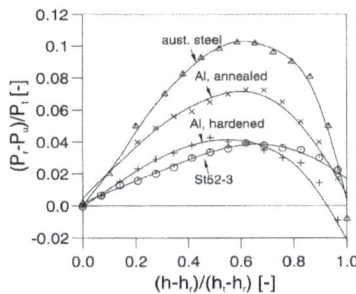

Figure 6. Opening of experimental measured hysteresis loops [22]. Reproduced with permission from N. Huber et al., Mechanics of Materials; published by Elsevier, 1998.

Durban and Masri [23] in 2007 found that the nanoindentation test data with conical indenter over a range of cone angles can be used to reconstruct the axial stress-strain curve. Mata et al. [24] modified the previous hardness formulation within the elastic-plastic transition derived for solids by Hill [25] in 1950 and Marsh [26] in 1964 as it did not exhibit strain hardening.

Application of nanoindentation tests to determine the mechanical properties of surface coatings is another milestone. Normally, the investigations of the mechanical properties of the coatings are difficult as the traditional compression and tensile tests are unable to apply well at very small scales. Rodriguez et al. [27] performed depth sensing indentation in plasma sprayed Al_2O_3–13% TiO_2 nano-coatings in order to determine the Young's modulus and hardness. It was found the mechanical properties were dramatically enhanced in the nanostructured coating compared to the conventional one, which is shown in Figure 7. Swain, Menčík and their coworkers [28,29] investigated the application of five approximation functions (linear, exponential, reciprocal exponential, Gao's, and the Doerner and Nix functions) for determining the Elastic modulus of thin homogeneous films. By conducting various experimental testes, they found that generally the Gao analytical function is able to predict the indentation response of film/substrate composites. For determining the thin film modulus from experimental data, satisfactory results can also be obtained with the exponential function, while linear function can only be used for thick films where the relative depths of penetration are small. For determing the hardness of thin films, Jönsson and Hogmark [30] built a physical model which was verified for chromium films on four different substrate materials.

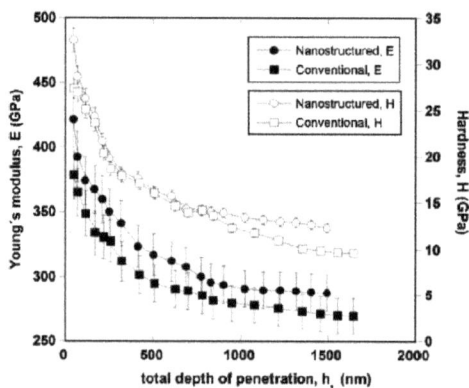

Figure 7. Young's modulus and hardness vs. total penetration depth for both coatings [27]. Reproduced with permission from J. Rodriguez et al., Acta Materialia; published by Elsevier, 2009.

Thin water film was studied by Opitz et al. [31] in 2003. It was believed that the thin water films which covered most of the micro-scale and nano-scale surfaces could be particularly important for microelectromechanical systems (MEMS) and the upcoming nanoelectromechanical systems (NEMS), as they played a critical role in defining the micro- and nano-tribological properties of a system. Kim et al. [32] presented a nanoindentation method to measure the Poisson's ratio of thin films for MEMS applications. In their test, a double-ring-shaped sample was designed to conduct the measurement of the Poisson's ratio as shown in Figure 8. The load-deflection data of the double ring sample after nano-indenter loading was analysed to obtain the Poisson's ratio. Lou et al. [33] investigated the mechanical behaviour of LIGA (an acronym of the German words "lithographie, galvanoformung, abformung") nickel MEMS structures which were developed for applications in micro-switches and accelerometers via the nanoindentaion method. Both Berkovich and Cube-corner indenter were used to conduct the nanoindentation tests and study the effects of residual indentation depth on the hardness of LIGA Ni MEMS structures between the mico-scale and nano-scales. Almost no apparent size dependence has been found for LIGA Ni films indented by a Berkovich indenter. In contrary, the hardness dramatically increased with decreasing residual indent depth for films indented by the Cube-corner indenter.

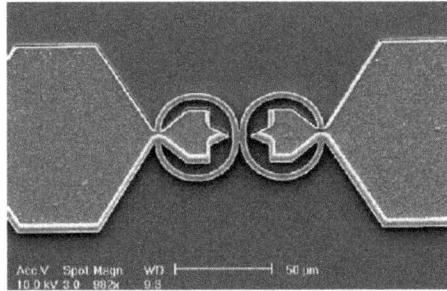

Figure 8. Top view of the double ring sample [32]. Reproduced with permission from Jong-Hoon Kim et al., Sensors and Actuators A: Physical; published by Elsevier, 2003.

Indentation tests were also used to evaluate adhesion strength of the thermal barrier coatings (TBCs) which were used to improve the performance and efficiency of advanced gas turbines [34–38]. Interfacial strength was one of the most important properties in TBCs, and traditional tensile method was found to be restricted due to the size dependence. Yamazaki et al. [39] used the indentation method to investigate the interfacial strength of TBC which was subjected to thermal cycle fatigue. An indent was made directly at the interface on the polished surface of the sample as shown in Figure 9. The crack length and the diagonal length of the indentation were measured using SEM right after indentation as shown in Figure 10. The typical interfacial crack initiated by the indentation test in Figure 10a showed the crack mainly propagated in the ceramic top coat near the interface. Even after 500 thermal cycles with the formation and growth of TGO (thermally grown oxide) layer shown in Figure 10b, the crack propagated only in the ceramic top coat too.

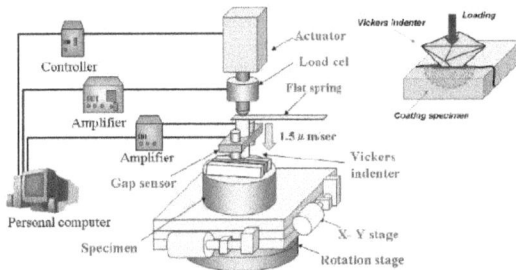

Figure 9. Schematic illustration of the instrumented indentation test equipment [39]. Reproduced with permission from S. Kuga et al., Procedia Engineering; published by Elsevier, 2011.

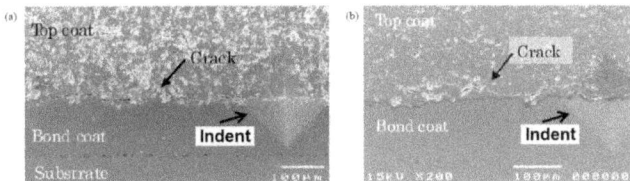

Figure 10. Typical interfacial cracks initiated by the indentation test: (**a**) for 0 thermal cycles (As-sprayed); (**b**) after 500 cycles [39]. Reproduced with permission from S. Kuga et al., Procedia Engineering; published by Elsevier, 2011.

The dependence of nanoindentation piling-up patterns and of micro-textures on the crystallographic orientation was studied by Wang et al. [40] using high purity copper single crystal with three different initial orientations. The indentation tests were performed on a Hysitron nanoindentation setup using a conical indenter to avoid symmetries. The results are shown in Figure 11. Four-, two-, and sixfold symmetrical piling-up patterns were captured on the surface of (001), (011) and (111) initial oriented single crystal, individually, which could be explained in terms of the strong crystallographic anisotropy of the out-of-plane displacements around the indenter.

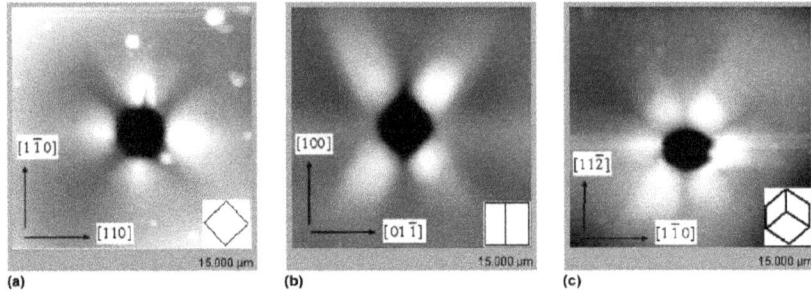

Figure 11. Experimentally observed contour plots of the pile-up patterns on (a) (001), (b) (011) and (c) (111) oriented copper single crystal surfaces after indentation with a conical indenter [40]. Reproduced with permission from Y. Wang et al., Acta Materialia; published by Elsevier, 2004.

With the development of focused ion beam milling of site-specific electron transparent foils, the investigation of cross-sections of nanoindentations with the transmission electron microscope (TEM) or electron backscatter diffraction has recently become feasible [41]. Lloyd [42] and his colleagues combined nanoindentation and TEM to survey the deformation behaviour in a range of single crystal materials with different resistances to dislocation flow as shown in Figure 12. The principal deformation models included phase transformation (silicon and germanium), twinning (gallium arsenide and germanium at 400 °C), lattice rotations (spinel), shear (spinel), lattice rotations (copper) and lattice rotations and densification (TiN/NbN multilayers). Generally, the residual impresses were sectioned through the tip of the indent with the thin foil normal approximately parallel to either [110] or [100] zone axis.

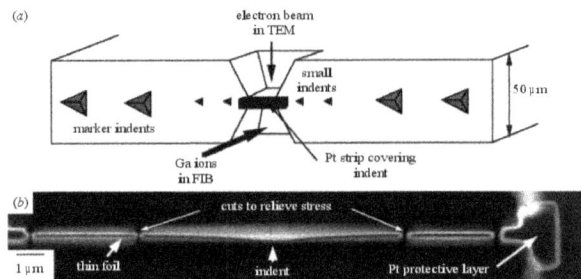

Figure 12. (a) Schematic illustrating how the indents were sectioned in the FIB to allow examination with a TEM; (b) Secondary electron FIB image of the final stages of preparation of a thin foil through a 50 mN indent in silicon [42]. Reproduced with permission from S.J. Lloyd et al., Proceedings of the Royal Society A; published by The Royal Society, 2005.

The indents in (001) silicon at loads of 30 mN and 60 mN are shown in Figure 13. It was found a transformed region was under the area of the Berkovich indenter in both cases, and a crack was originating at the base of the transformed region in the case of 60 mN.

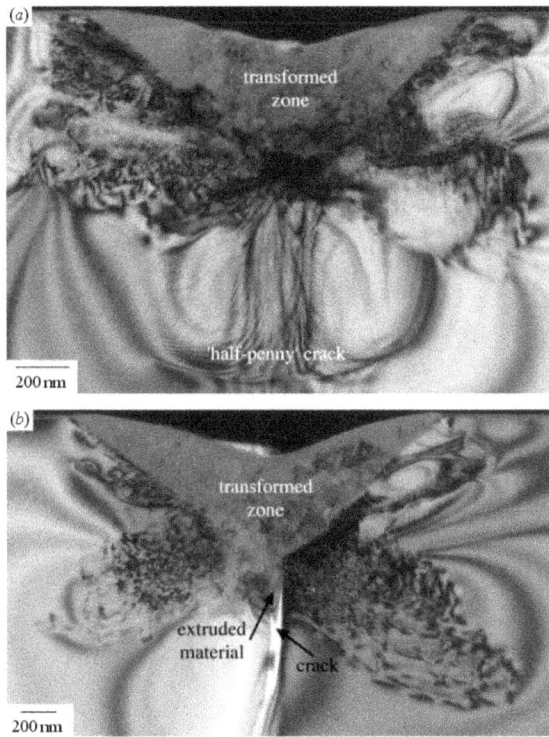

Figure 13. Bright field images of indents in silicon formed with loads of (**a**) 30 mN and (**b**) 60 mN. The structure is observed here after the load has been removed, allowing the high-pressure phase to transform back to a mixture of other structures [43]. Reproduced with permission from S.J. Lloyd et al., *Philosophical Magazine A*; published by Taylor & Francis, 2002.

Shear bands were captured in spinel crystals, which are shown in Figures 14 and 15. It was found the shear band spacing increased with increasing distance from the indent tip, and the spacing on the steep side of the indent was a little smaller for the large load. Lloyd [43] concluded that the increase of the shear band spacing with distance far away from the indenter tip indicated there was a limit to the amount of displacement occurring through any shear band due to strain hardening. Consequently, a high concentration of shear bands was close to the indenter tip where had the largest vertical displacements, while a low density of slip bands was sufficient to accommodate the relatively small vertical displacement in the region far away from the indenter tip.

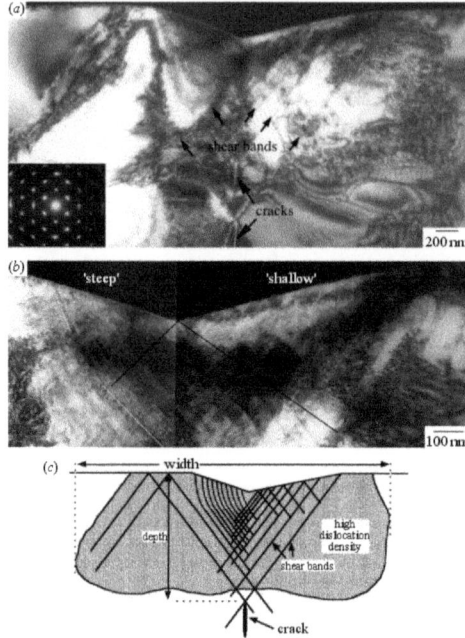

Figure 14. (a) Bright field image of an 80 mN indent in spinel with an inset diffraction pattern; (b) Bright field image of a 60 mN indent in spinel showing the region of shear bands under the indent more clearly; (c) Schematic illustrating the principle components of the deformation pattern in spinel [42]. Reproduced with permission from S.J. Lloyd et al., Proceedings of the Royal Society A; published by The Royal Society, 2005.

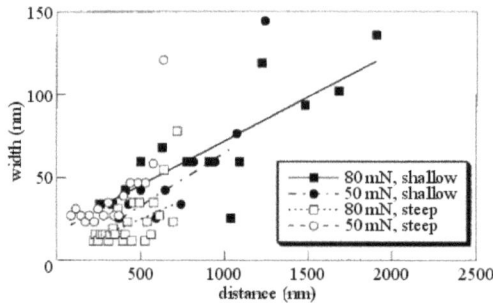

Figure 15. Slip band spacing in spinel as a function of distance from the indenter tip (see Figure 14b) for the 50 mN and 80 mN indents. The error in measurement of the shear band spacing is approximately 10 nm [42]. Reproduced with permission from S.J. Lloyd et al., Proceedings of the Royal Society A; published by The Royal Society, 2005.

Lattice rotation angles around an axis perpendicular to the [110] zone axis were investigated [42] and are shown in Figure 16. It was found the rotations only occurred in the region immediately below the indent impression. The greatest rotations were quite near the indent tip and the magnitude of rotation angles decreased significantly with the increasing distance from the indent tip along the surface on the shallow side.

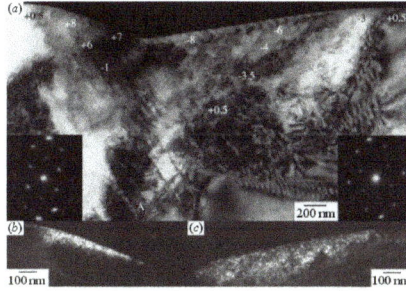

Figure 16. (a) Bright field image of a 50 mN indent in spinel with lattice rotations (in degrees) at selected positions indicated; (b,c) are dark field images of the surface of the same indent taken using streaks from either side of the 004 reflection [42]. Reproduced with permission from S.J. Lloyd et al., Proceedings of the Royal Society A; published by The Royal Society, 2005.

The indentation-induced plastic zones below indentations in copper single crystals, with depths ranging from 250 nm to 250 μm, were examined by Rester et al. [44] via the implementation of focussed ion beam (FIB) and electron backscatter diffraction (EBSD) techniques. Two or three distinguishable regimes shown in Figure 17 were captured by analysing scanned orientation micrographs. Meanwhile, the changes in the evolution of the microstructure were reflected in the hardness curve shown in Figure 18. Regime α described the impression which is smaller than 300 nm, in which no significant orientation changes are observed by EBSD (Figure 17a). Regime β describes the indentation depth between 300 nm and 30 μm, which is characterized by regions having noticeable changes of the orientation (Figure 17c,d). It was also found that the orientation differences increase with growing indentation depth in this regime. Regime γ is associated with indentation depth larger than 30 μm exhibiting a typical substructure of FCC (face centered cubic) single crystal of pure metals during indentation. Therefore, Rester et al. [44] concluded that the hardness of a material varies with the size of the indent impression and the source size becomes the dominant effect only for very small impressions (i.e., in regime α).

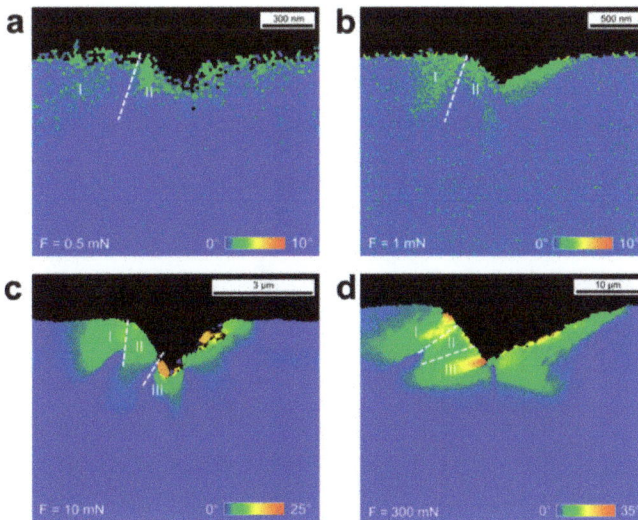

Figure 17. *Cont.*

21

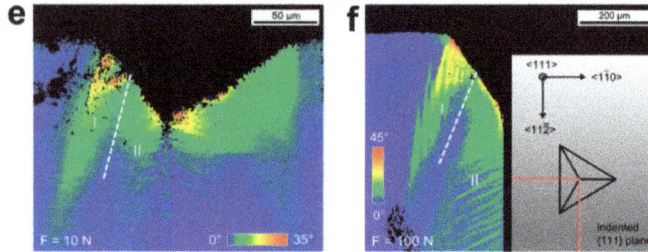

Figure 17. Misorientation maps of indentations in copper for loads of (**a–d**) 0.5, 1, 10, 300 mN; (**e**) 10 N and (**f**) 100 N [44]. Reproduced with permission from M. Rester et al., Scripta Materialia; published by Elsevier, 2008.

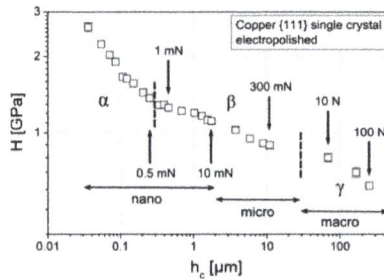

Figure 18. Logarithmic plot of the hardness versus the indentation depth for loads ranging from 40 μN to 100 N. The arrows mark the hardness values of imprints investigated in course of this work [44]. Reproduced with permission from M. Rester et al., Scripta Materialia; published by Elsevier, 2008.

Zaafarani et al. [45] investigated texture and microstructure below a conical nano-indent in a (111) oriented Cu single crystal using 3D EBSD. The tests were performed using a joint high-resolution field emission scanning electron microscopy/electron backscatter diffraction (EBSD) set-up coupled with serial sectioning in a focused ion beam system in the form of a cross-beam 3D crystal orientation microscope (3D EBSD) as shown in Figure 19.

The EBSD tests conducted in sets of subsequent $(11\bar{2})$ cross-section planes exhibited a pronounced deformation-induced 3D patterning of the lattice rotations below and around the indent, which are shown in Figure 20.

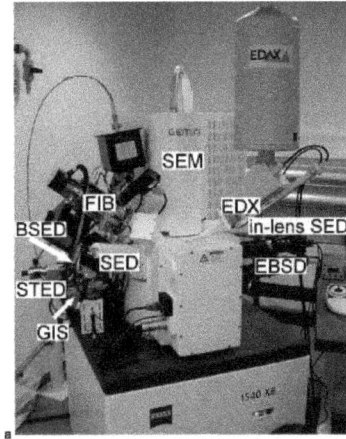

Figure 19. *Cont.*

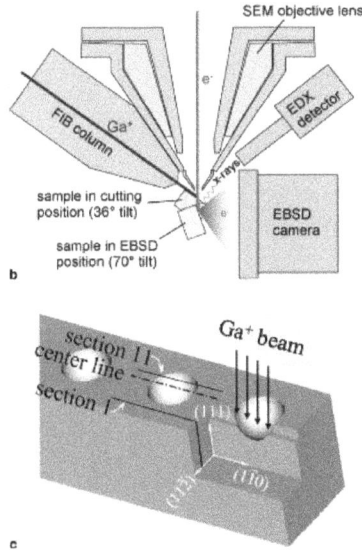

Figure 19. (a) Joint high-resolution field emission SEM/electron backscatter diffraction (EBSD) set-up together with a focussed ion beam (FIB) system in the form of a cross-beam 3D crystal orientation microscope for conducting 3D EBSD measurements by serial sectioning (Zeiss); (b) Schematic of the joint FIB/EBSD set-up; (c) Schematic of the FIB sectioning geometry [45]. Reproduced with permission from N. Zaafarani et al., Acta Materialia; published by Elsevier, 2006.

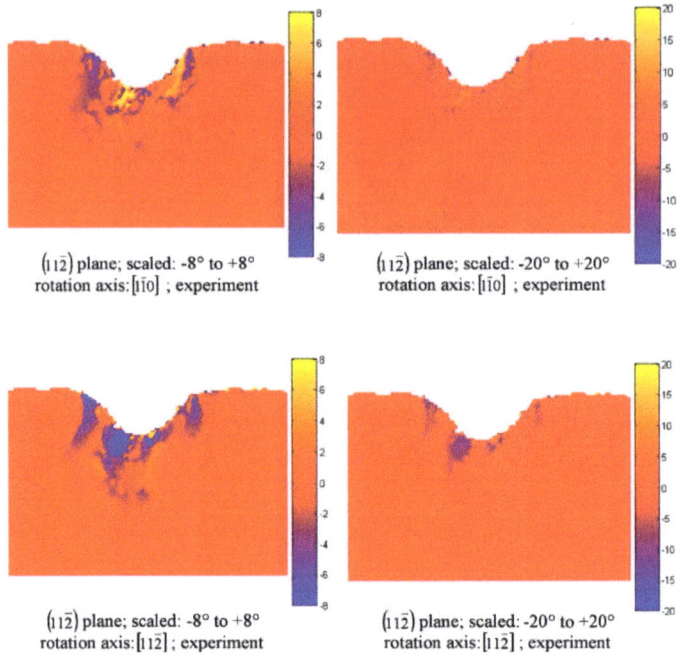

$(1\bar{1}2)$ plane; scaled: -8° to +8°
rotation axis:$[1\bar{1}0]$; experiment

$(1\bar{1}2)$ plane; scaled: -20° to +20°
rotation axis:$[1\bar{1}0]$; experiment

$(1\bar{1}2)$ plane; scaled: -8° to +8°
rotation axis:$[11\bar{2}]$; experiment

$(1\bar{1}2)$ plane; scaled: -20° to +20°
rotation axis:$[11\bar{2}]$; experiment

Figure 20. Rotation angles and rotation directions in the $(11\bar{2})$ plane [45]. Reproduced with permission from N. Zaafarani et al., Acta Materialia; published by Elsevier, 2006.

4. Nanoindentation Size Effect of Materials

Over the past several decades, with the development of new technologies, people felt that it was very important to understand how materials perform at small scales because the mechanical properties are significantly different from those at macro-scales. Therefore, a great number of researchers became interested in micro- and nano-scale deformation phenomena which also made them try to look for a new method to examine these physical phenomena at ever-decreasing length scales. In the field of the mechanical behaviour of materials, one of the more interesting small-scale phenomena is an increase in yield or flow strength that is often observed when the size of the test sample is reduced to micro-meter and sub-micrometer dimensions [46]. Pharr et al. [46] believed such size-dependent increases in strength were due to unique deformation phenomena which could be observed only when the sample dimensions approached the average dislocation spacing and when plastic deformation was controlled by a limited number of defects.

In the metal micro-forming process, grain size, grain orientation and material's dimension are the important influence factors for the material's deformation. It is reported that when the dimension of the material is downscaled to a much smaller scale, the mechanical properties are significantly different from those on the macro-scale, which is called 'size effect' [47]. Size effect is an interesting and important topic for the development of micro-forming technologies. Recently, numerous experiments have shown that metallic materials display noticeable size effects once the size of non-uniform plastic deformation zone associated their characteristic length size are on the order of microns [47]. Fleck et al. [48] found a dramatic increase of plastic work hardening when the wire's diameter decreased from 170 μm to 12 μm via doing thin copper wire's torsion experiments. Stolken and Evans [49] observed that the plastic work hardening increased significantly when the nickel beam thickness was decreased from 50 μm to 12.5 μm while doing a micro-bend test. Gau et al. [50] found that the conventional concept of spring-back cannot be applied on brass sheet metal when its thickness is less than 350 μm by studying the spring-back behaviour of brass in micro-sheet forming. Geiger et al. [51] and Kals and Eckstein [52] have conducted compression tests, tension tests, and bending tests, respectively, in order to study how the material properties change due to size effect. Saotome et al. [53] carried out investigations in micro-deep drawing, and found that the relative punch diameter (punch diameter related to the sheet thickness) has a significant influence on the limit draw ratios (LDR).

The indentation size effect (ISE) is often observed for materials that are indented with geometrically self-similar indenters like pyramidal or conical tips (see Figure 21) [46]. In general, the hardness, H, defined as the load on the indenter normalized with the projected contact area of the hardness impression (see Figure 21), should be independent of the depth of penetration, h. However, over the past 60 years, it was observed that there were significant variations of hardness with respect to penetration depth, especially when the depths decreased to less than a few micro-meters [54–60]. In addition, two types of indentation size effects have been reported. One is the normal ISE (the tip is pyramidal or conical)—the hardness increases with decreasing penetration depths, according to the expression "smaller is stronger" (see Figure 21). Another one is the reverse ISE (the tip is spherical) [55,56,61], which displayed that the hardness decreases with increasing depths. However, for the reverse ISE, it was observed that the hardness also increases with the decreasing of tip radius. The reverse ISE is thought to be derived from testing artefacts such as vibration in the testing system or problems with accurate imaging and measuring the sizes of hardness impressions at dimensions approaching the limits of optical microscopy [46].

Although classical descriptions of the ISE show a decrease in hardness for increasing indentation depth (Figure 22), recently new experiments [62] have shown that after the initial decrease, hardness increases with increasing indentation depth. After this increase, eventually the hardness decreases with increasing indentation (Figure 23). This phenomenon is very prominent for copper, but not noticeable for aluminium.

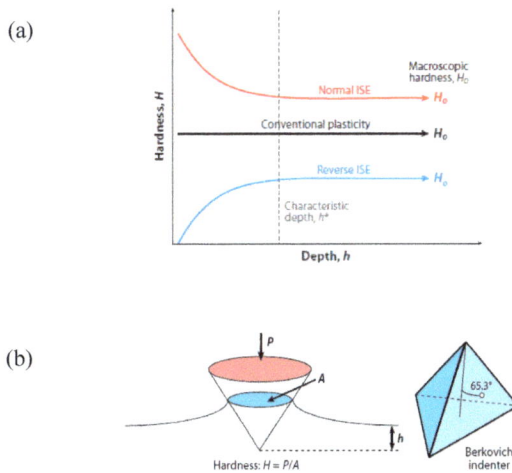

Figure 21. Indentation size effect (ISE) (**a**) for geometrically self-similar indenters such as a conical or pyramidal tip (**b**) [46]. Reproduced with permission from George M. Pharr et al., Proceedings of Annual Review of Materials Research; published by Annual Reviews, 2010.

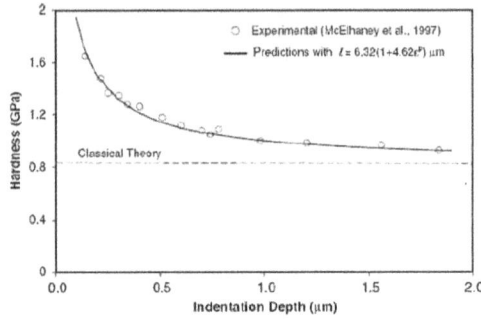

Figure 22. Visualization of the ISE: hardness decreases with increasing indentation depth for cold-worked polycrystalline copper [62]. Reproduced with permission from G.Z. Voyiadjis et al., Acta Mechanica; published by Springer, 2010.

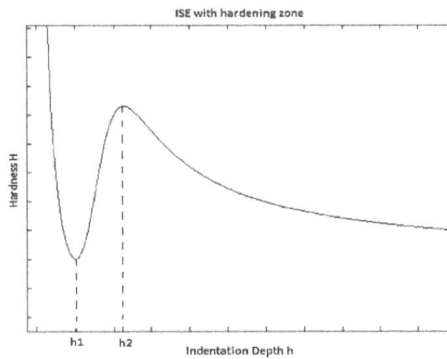

Figure 23. Visualization of the ISE with the incorporated hardening effect. For $h < h_1$, hardness decreases, for $h_1 < h < h_2$, hardness increases and for $h > h_2$, hardness decreases again [62]. Reproduced with permission from G.Z. Voyiadjis et al., Acta Mechanica; published by Springer, 2010.

There are different theories to explain ISE. The most popular theory is based on strain gradient plasticity which is a class of continuum theories aimed to bridge the gap between classical plasticity and dislocation. This theory assumes that the flow stress of metals depend on the density of statistically stored dislocations (SSD) which relate to effective strain, and the density of the geometrically necessary dislocations (GND) which relate to the strain gradient. Dislocations are generated, moved and stored during the process of plastic deformation. The process of storing dislocation is also a process of strain hardening. It is assumed that dislocations become stored because they either accumulate by randomly trapping each other or they are required for compatible deformation of various parts of the material. When they randomly trap each other, they are often known as the statistically stored dislocation [63], whereas when they are required for the compatibility purpose, they are often called geometrically necessary dislocations and related to the gradient of plastic shear strain in a material [63,64].

The most famous strain gradient plasticity model for nanoindentation was proposed by Nix and Gao [65] in 1998. It has been assumed that plastic deformation of the surface is accompanied by the generation of dislocation loops below the surface, which are contained in an approximately hemispherical volume below the region in contact, as shown schematically in Figure 24. The deformation is self-similar and the angle (θ) between the conical indentation tip and indented surface remains constant.

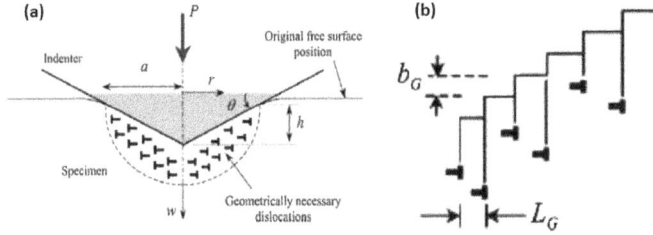

Figure 24. Model of geometrically necessary dislocation for a conical indent: (**a**) Sample being indented by a rigid conical indenter; (**b**) Deformation loops created during indentation process [62]. Reproduced with permission from G.Z. Voyiadjis et al., Acta Mechanica; published by Springer, 2010.

The angle θ can be calculated by

$$tan(\theta) = \frac{h}{a} = \frac{b_G}{L_G}, \; L_G = \frac{b_G a}{h} \tag{6}$$

where h is the residual plastic depth, and a is the contact radius, and b_G is Burger's vector. The number of geometrically necessary dislocation loops is h/b_G. S is the spacing between individual slip steps on the indentation surface, as shown in Figure 24. Assume that λ is the total length of the injected loops, thus between r and $r + dr$ we have

$$d\,\lambda = 2\pi r \frac{dr}{S} = 2\pi r \frac{h}{ba} dr \tag{7}$$

which after integration from 0 to a gives the total length of dislocation loops,

$$\lambda = \int_0^a \frac{h}{ba} 2\pi r dr = \frac{\pi h a}{b} \tag{8}$$

The model assumes that the dislocations are distributed uniformly in a hemispherical volume with the contact radius.Thus, we have $V = 2\pi a^3/3$, and therefore the density of geometrically necessary dislocation is

$$\rho_G = \frac{\lambda}{V} = \frac{3}{2bh} tan^2\theta \tag{9}$$

Taylor hardening model has been used to find the shear strength which can be used to measure the deformation resistance:

$$\tau = \alpha\mu b\sqrt{\rho_T} = \alpha\mu b\sqrt{\rho_G + \rho_S} \tag{10}$$

where τ is the resolved shear stress, μ is the shear modulus, b is the Burgers vector and α is a constant which is usually in the range 0.3–0.6 for FCC metals [66]. Here, they note that ρ_S does not depend on the depth of indentation. Rather it depends on the average strain in the indentation, which is related to the shape of the indenter ($tan(\theta)$). They also assume that the von Mises flow rule applies and that Tabor' factor of 3 can be used to convert the equivalent flow stress to hardness:

$$\sigma = \sqrt{3}\tau, H = 3\sigma \tag{11}$$

According to these relations the hardness can be expressed by as:

$$\frac{H}{H_0} = \sqrt{1 + \frac{h^*}{h}} \tag{12}$$

where

$$H_0 = 3\sqrt{3}\,\alpha\mu b\sqrt{\rho_S} \tag{13}$$

is the macroscopic hardness from the statistically stored dislocations alone, in the absence of any geometrically necessary dislocations, and

$$h^* = \frac{81}{2} b\alpha^2 tan^2\theta \left(\frac{\mu}{H_0}\right)^2 \tag{14}$$

is a length scale for the depth dependence of hardness.

From Equation (12), it can be seen that the indentation hardness H is related only to indentation depth h as h^* and H_0 are material constants which can be obtained by fitting the experimental results.

After Nix and Gao [67], many other researchers continue to observe ISE through experiments and simulations. Most of them explained the size effect via the strain gradient theory (Ma and Clarke [68], Fleck et al. [48], Nix and Gao [65], Poole et al. [69], Stelmashenko et al. [70], Gao et al. [71,72], Acharya and Bassani [73], Huang et al. [74], and Gurtin [75]). They believed that ISE must have a relationship with the strain gradient as the geometrically necessary dislocation density ρ_G is usually related to an effective strain gradient η as

$$\rho_G = \frac{2\eta}{b} \tag{15}$$

According to Equation (9), strain gradient was inversely proportional to the indentation depth h, which meant strain gradient should be larger at the shallow depth.

However, Demir, Raabe and Zaafarani [76] expressed a different theory about ISE. They thought as ρ_G is the GND density that is required to accommodate a curvature ω, the crystallographic misorientation between two neighboring points can be used as an approximate measure for the GNDs as shown in the following equation:

$$\rho_G = \frac{\omega}{b} \tag{16}$$

According to Equations (15) and (16), the size dependence of indentation hardness has been associated with strain gradients which exist in the lattice through GNDs. Thus, these researchers decided to directly measure lattice rotations below indents with the aim of quantifying the density of these defects. For this purpose they performed an experiment using a tomographic high-resolution electron backscatter diffraction orientation microscope in conjunction with a focused ion beam instrument to map the orientation distribution below four nanoindents of different depths. Unfortunately, the experimental result contradicted the commonly expected inverse relationship between the indentation depth and the density of the GNDs. In terms of GND-based strain gradient theories, larger GND densities should appear at shallow indentation depth, not at a deep one. But their experiment showed an opposite trend that the total GND density below the indents reduces with decreasing indentation depth. According to the experimental results, they concluded that the explanation size-dependent material strengthening effects by using average density measure for GNDs was not sufficient to understand the indentation size effect.

Kiener [57] investigated Nix-Gao model with the cross-sectional EBSD method and proposed that Nix-Gao model's physical basis was still under debate, and its validity cannot be addressed alone with load versus displacement characteristics. There were two assumptions with respect to the Nix-Gao model, hemispherical plastic zone and self-similarity of the evolving deformation structure. However, according to the EBSD observation shown in Figure 17, it is clear that the true plastic deformed zone deviates from the assumption of a half-sphere. Other reports indicated that there were differently shaped deformation areas, depending on the indenter geometry [45,77,78]. Regarding the self-similarity, based on the Nix-Gao model, the strain gradient induced by the indenter should be determined solely by the indenter geometry. Therefore, it should be constant for self-similar indenter shapes, which means the observed misorientations should depend only on the indenter angle and not on the indent size. However, the observations for Vickers indents in copper and tungsten displayed that the maximum misorientation was along the indent flanks for different sizes, as shown in Figure 25.

Furthermore, Nix-Gao model assumed that GNDs would accommodate the impression geometry at very low indentation depth for metallic materials with no pre-existing defects, such as single-crystal metals without any deformation. Subsequently, it was generally accepted that ISE would not exist in materials with pre-existing high dislocation densities, such as ultrafine-grained materials [79] and amorphous materials. However, recently Smedskjare [80] and Liu et al. [81] found ISE existing in oxide glasses and photonic crystals, respectively. Obviously, all of these materials are not crystalline metallic materials. Therefore, it is concluded that ISE may commonly exist in natural materials, not only in crystalline materials.

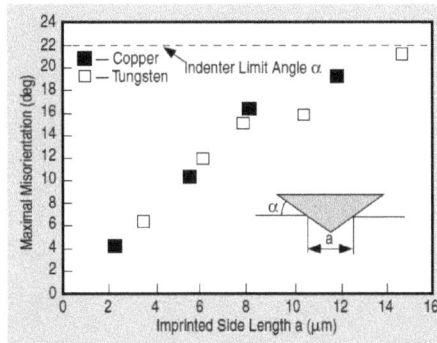

Figure 25. Maximum misorientation angle measured along the flanks of the indent for different sized Vickers indents into fcc copper (closed squares) and bcc tungsten (open squares) [82,83]. Reproduced with permission from D. Kiener et al., JOM; published by Springer, 2009.

Meyer's law is another widely used method to describe ISE. For the indenters which have ideal geometry, the relationship between the test load and the resultant indentation diagonal length curve could be obtained from [67]

$$P = C \cdot d^n \tag{17}$$

where P is the load, d is the diagonal length of impression, C is the material/indenter constant and n is the Meyer index due to the curvature of the curve. Since d is proportional to the contact depth h_c which in turn is proportional to the indentation depth h, Equation (17) could be expressed as follows:

$$P = C' \cdot h^n \tag{18}$$

where C' is constant and h is the indentation depth.

Fischer-Cripps [84] mentioned that if the plastic zone was fully developed (beyond elastic-plastic transition point), the load-displacement (P-h) curve of the loading section could be related to the square of the displacement ($P = C/h^2$). As for the loading stage of the P-h curve in the elastic-plastic field. Sakai [85] stated that the load is proportional to the square of the indentation depth. According to Equatoion (17), if $n = 2$, the materials shows no ISE. But if $n < 2$, the materials shows ISE and this case was confirmed by different materials [86,87].

Considering aforementioned facts, Ebisu and Horibe [67] analyzed the relationship between the P-h curve and ISE behaviour in their experiments by differentiating the P-h curves of the three samples. They found that for single 8Y-FSZ (8 mol %Y_2O_3–ZrO_2 single crystal) in the load range of 200–1900 mN, the index n calculated from the P-h curve was 1.873, which suggests that single 8Y-FSZ showed ISE behaviour. They also found this ISE result agreed well with the experimental results of another report [88] in which the hardness decreased as the indentation increased between the load range of 100–2000 mN. The same procedure was conducted in the 12Ce-TZP P-h curve and they found $n = 1.808$ within the load range 200–1900 mN. It also was consistent with the results (ISE behaviour) from zirconia ceramics [88,89]. However, for fused quartz, index $n = 1.961$ calculated from *P-h* curve in the load range of 200–1900 mN meant that this material showed almost no ISE behaviour [67]. This agreed with Oliver and Pharr's report [18] which stated that quartz showed very little indentation size effect.

Kolemen [90] conducted the same experiments with superconductors, which showed a apparent ISE. He then concluded that for hard materials like brittle ceramics at low indentation loads, n is significantly less than 2. According to Onitsch [82], n lied between 1 and 1.6 for hard materials and higher than 1.6 for soft materials.

However, Peng et al. [91] pointed out the correlation between n and C seemed to be of little significance for understanding the ISE, as their previous study has showed that the best-fit value of the Meyer's law coefficient C depended on the unit system used for recording the experimental data and completely different trends of n versus C may be observed in different unit systems [92] as shown in Figure 26.

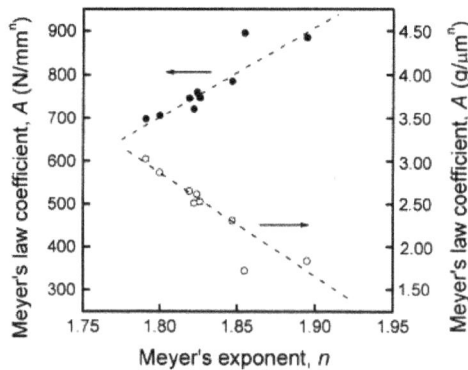

Figure 26. Variation of Meyer's law coefficient C with Meyer's exponent n. Note that completely different trends are observed when different unit systems were used for recording the experimental data [92]. Reproduced with permission from JianghongGong et al., Proceedings of Journal of the European Ceramic Society; published by Elsevier, 2000.

The first unit system used is *P* in Newton (N) and d in millimeter (mm) and the second is *P* in gram (g) and d in micrometer (mm). As can be seen from Figure 26, completely different trends of n versus *A* were observed in different unit systems. Similar conclusions were reported by Li and Bradt [93] when they analyzed the experimental data on single crystals. Therefore, they concluded that a particular care should be taken when analyzing the microstructural effects on the measured hardness based on Meyer's power law.

It should be noted that only n plays an important role in determining ISE in Meyer's law and it is not necessary to consider the variation of Meyer's law coefficient C. Although numerous researchers used Meyer's law to analyse ISE, all of them just fit the whole *P-h* curve. In fact, the *P-h* curve can be fitted separately from the ISE-boundary which is shown as in Figure 27. In the left side of ISE-boundary an n-value smaller than 2 can be obtained, while an n-value of about 2 can be achieved in the right side.

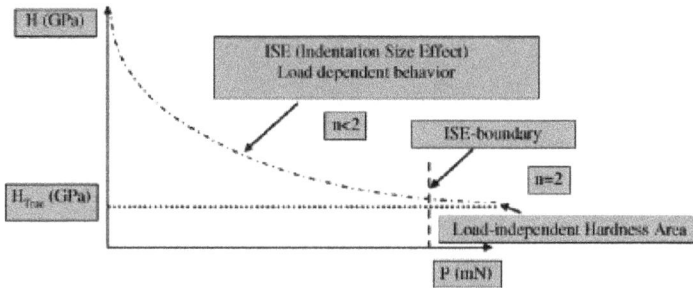

Figure 27. Schematic plot of the ISE behaviour [90]. Reproduced with permission from U. Kölemen et al., Journal of alloys and compounds; published by Elsevier, 2006.

Some other factors may influence the ISE, which include: inadequate measurement capabilities of extremely small indents [94], presence of oxides or chemical contamination on the surface [95], indenter-sample friction [96], and increased dominance of edge effects with shallow indents [68].

Elmustafa and Stone [97] proposed a vast number of hypotheses to explain the ISE, including: friction, and lack of measurement capabilities, and surface layers, oxides, chemical contamination and dislocation mechanisms. To calculate the hardness for the nanoindentation measurements, indents were imaged in calibrated optical and scanning electron microscopes. However, because of the inaccuracies inherent to "optical" method, they did not just rely on them alone. In addition, they used contact stiffness as a method to determine indirectly the projected contact area. Interestingly, the two methods agreed very well.

Elmustafa et al. [97] conducted experiments using alpha brass and aluminium as samples. It has been found that the oxide on a fresh surface of alpha brass is less than 5 nm thick, while the oxide thickness on an aluminium surface is only 1–3 nm [98]. For Elmustafa et al.'s experiment, the smallest indents were approximately 0.3 um across or 60 times larger in lateral dimensions than the oxide thickness. Therefore, the oxide layer would not greatly affect the hardness.

The indenter piling-up or sinking-in was also seen as a factor that influences ISE. McElhaney et al. [99] did numerous experiments and found that the indenter piling up and sinking in had a huge effect on the micro-indentation hardness. However, after very careful examination, they found the indentation hardness still displayed strong dependence on the penetration depth. This observation displayed that the indenter piling up and sinking in cannot explain the depth-dependent indentation hardness alone.

Loading rate was raised as another influence factor for ISE because usually materials tended to display larger a plastic work hardening at a large strain rate (loading rate) [100]. The strain rate in the indented material should be proportional to the ratio $\frac{\dot{h}}{h}$, where $\dot{h}$ is the rate of change of the indentation depth and h is the indentation depth. It is not hard to understand that for a constant rate of indentation depth ($\dot{h} = constant$), the strain rate should be very huge at the initial penetration. However, Lilleodden [101] performed indentation test at a constant ratio $\frac{\dot{h}}{h}$ and still observed the phenomenon of ISE.

Indenter tip radius was regarded as another factor which affected ISE. However, McElhaney et al. [99] and Huang et al. [102] found that the sharp indenter tip radius less than 100 nm had definitely no influence on the micron and sub-micron scale indentation, which indicated that indenter tip radius cannot be used to explain the ISE observed in the indentation test with sharp indenters.

5. Simulations of Mechanical Behaviours of Materials Undergoing Nanoindentation

5.1. Conventional Finite Element Method (FEM) Simulations

Lee and Kobayashi's work [103] was the first to conduct the finite element simulation (FEM) of indentation in 1969. Plane strain and axisymmetric flat punch indentation was simulated to study the development of the plastic zone, the load-displacement relationships, and the stress and strain distributions during continued loading, taking into account the changes of the punch friction and sample dimensions. However, problems such as the accuracy of the solutions and the efficiency of the computation still existed. Then three years later, Lee and his colleagues [83] performed another finite element simulation of indentation using ball indenter, and compared all the results with their own experiments for heat-treated SAE4340 steel. It was found that the simulated load-displacement curve, plastic zone development and indentation pressure were in good agreement with the experimental observations. In addition, by calculating the mean effective strains with FEM, the representative strains defined by Tabor were found to be equal to the mean effective strains of the plastic zone under the indenters shown in Figure 28.

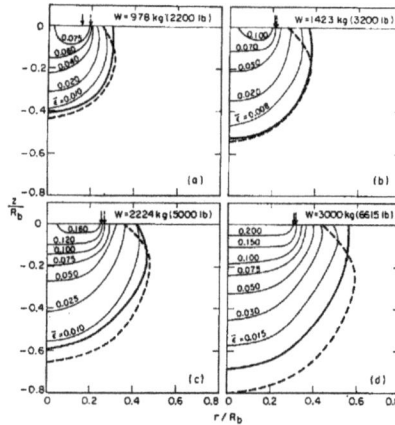

Figure 28. Strain contours and elastic-plastic boundaries under the load of (**a**) 978 kg; (**b**) 1423 kg; (**c**) 2224 kg and (**d**) 3000 kg (—FEM; —experiment) [83]. Reproduced with permission from C.H. Lee et al., International Journal of Mechanical Sciences; published by Elsevier, 1972.

Bhattacharya and Nix [104] performed elastoplastic FEM simulations of nanoindentation using conical indenter to study the elastic and plastic properties of materials on a sub-micro scale under the conditions of frictionless and completely adhesive contact. The simulated load-displacement curves for nickel and silicon were compared with experimental results as shown in Figures 29 and 30, respectively. It was concluded that the FEM is suitable to simulate nanoindentation behaviour at a sub-micro scale for different types of materials. Bhattacharya and Nix [105] then investigated the relationship between Young's modulus and yield strength, and concluded from FEM simulation that the shapes of the plastic zones for an elastic-plastic bulk material under a conical indenter depend strongly on the ratio E/σ_y (Young's modulus/yield stress) with a fixed indenter angle as shown in Figure 31.

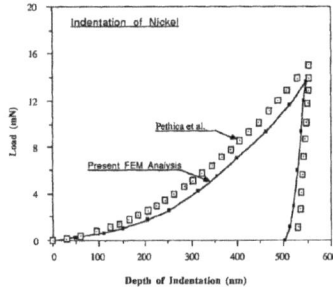

Figure 29. Comparison between the results from the present Finite Element Method (FEM) analysis and those from Pethica et al. [106] on indentation of nickel [104]. Reproduced with permission from A.K. Bhattacharya et al., International Journal of Solids and Structures; published by Elsevier, 1988.

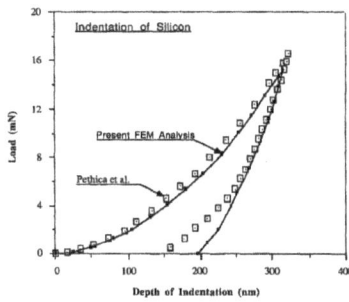

Figure 30. Comparison between the results from the present FEM analysis and those from Pethica et al. [106] on indentation of silicon [104]. Reproduced with permission from A.K. Bhattacharya et al., International Journal of Solids and Structures; published by Elsevier, 1988.

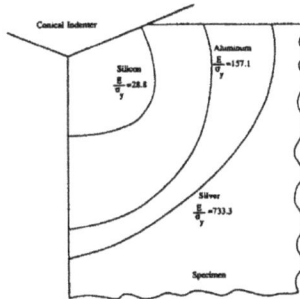

Figure 31. Comparison of the yield zones at a particular indenter angle of $136°$ for various E/σ_y, ratios for a depth of indentation of 203 nm [105]. Reproduced with permission from A.K. Bhattacharya et al., International Journal of Solids and Structures; published by Elsevier, 1991.

Subsequently, FEM nanoindentation simulation of thin films [107–110], stress distribution [111–113], hardness [114–116], friction effects [109,117], brittle cracking behaviour [118–120] and coatings [121,122] were extensively developed.

Furthermore, recently Maier et al. [123–125] proposed an inverse analysis method to identify elastic-plastic material parameters via FEM indentation simulation, and their proposed methodology was validated using "pseudo-experimental" (computer generated) data with and without noise. Chen et al. [126] proposed two alternative indentation techniques to effectively distinguish elastoplastic

33

properties of the mystical materials which have distinct elastoplastic properties yet they yield almost identical indentation behaviors, even when the indenter angle is varied in a large range

Most of the researchers thought the conventional plasticity theory cannot be used to explain ISE because its constitutive models possess no intrinsic (internal) material lengths. However, Storakers et al. [127] did observe the reversed ISE through simulation by using a parabola-shaped indenter.

5.2. Crystal Plasticity FEM Simulation

The evolution of crystallographic texture and grain lattice rotation under the indentation has not been well understood. This work must be done through the crystal plasticity based simulation.

Casals and Forest [128] investigated the anisotropy in the contact response of FCC and HCP (hexagonal closest-packed) single crystal via simulating spherical indentation experiments of bulk single crystals and thin films on hard substrates. Their simulations predicted that the plastic zone beneath the indenter preferentially grew along the slip system directions as shown in Figures 32 and 33. Consequently, in coated thin film systems, a prominent localization of plastic deformation occurred at those specific regions where slip system directions intersected the substrate. Meanwhile, these specific areas were prone to crack nucleation in terms of accumulative plastic damage. Therefore, the identification of these areas was meaningful for the prediction of potential delamination and failure of the coatings. Casals et al. [129] also used three-dimensional crystal plasticity finite element simulations to examine Vickers and Berkovich indentation experiments of strain-hardened copper. The results showed that the simulation was in a good agreement with experimental observations with respect to hardness, load-displacement curves, material piling up and sinking in development at the contact boundary.

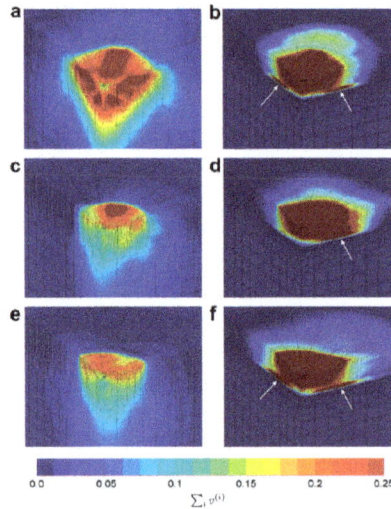

Figure 32. Details of the indentation-induced plastic zone in the simulations concerning face centered cubic (FCC) copper crystals. (**a,c,e**) correspond to the (001), (011) and (111) indented planes of bulk crystals. (**b,d,f**) correspond to their thin film counterparts. Plastic zone is assessed by considering the total acumulated plastic strain variable. White arrows point to the specific locations where high plastic strain localization occurs within the coating-substrate interface. Penetration depth in the figures is $h_s = 3.5$ μm [128]. Reproduced with permission from O. Casals et al., Computational Materials Science; published by Elsevier, 2009.

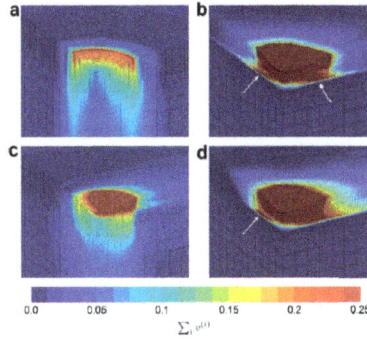

Figure 33. Details of the indentation-induced plastic zone in the simulations concerning hexagonal closest-packed (HCP) zinc crystals. (**a,c**) correspond to the basal and prismatic indented planes of bulk crystals. (**b,d**) correspond to their thin film counterparts. White arrows point to the specific locations where high plastic strain localization occurs within the coating-substrate interface. Penetration depth in the figures is $h_s = 3.5$ μm [128]. Reproduced with permission from O. Casals et al., Computational Materials Science; published by Elsevier, 2009.

Alcala et al. [130] analysed Vickers and Berkovich indentation behaviour via extensive crystal plasticity finite element simulation by recourse to the Bassani and Wu hardening model for pure FCC crystals. The simulated results have been used to illustrate the impact of the crystallographic orientation, as shown in Figure 34. It was clear that the irregular appearance of pyramidal indentations was governed by the crystallography of FCC crystals on the indented surface.

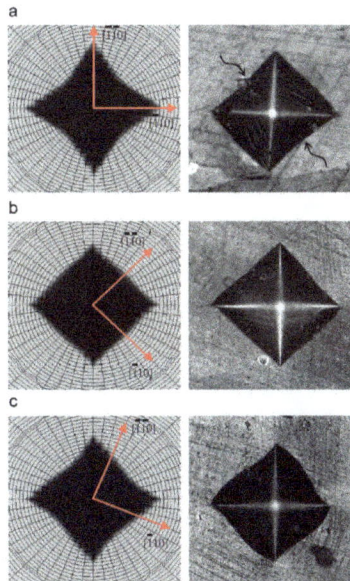

Figure 34. Imprint morphologies for Vickers indentation in the (001) plane for different orientations (rotations) of the tip. (**a**) The slip directions at the surface coincide with the edges of the indenter (arrows indicate development of pincushion effects); (**b**) The slip directions coincide with the sides of the indenter; (**c**) Intermediate orientation to those described in (**a,b**) [130]. Reproduced with permission from J. Alcala et al., Journal of the Mechanics and Physics of Solids; published by Elsevier, 2008.

Liu et al. [131] implemented crystal plasticity constitutive model initially developed by Peirce et al. [132] in a finite element code Abaqus/Explicit to study the material behaviour of nanoindentation on (001) oriented surface of single crystal copper. All of the appropriate meso-plastic parameters used in the hardening model was determined by fitting the simulated load-displacement curves to the experimental data. Their studies demonstrated that the combined nanoindentation/CPFEM simulation approach for determining meso-plasitc model parameters works reasonably well from micro level to the macro level as shown in Figure 35. They also investigated the orientation effects in nanoindentation of single crystal copper [133]. Simulated load-displacement curves were found to be in agreement with those from experimental tests as shown in Figure 36. Meanwhile, two-, three-, and four-fold symmetric piling-up patterns were observed on (011), (111), and (100) oriented surface with respect to CPFEM simulation. The anisotropic nature of the surface topographies around the imprints in different crystallographic orientations of the single crystal copper samples then were related to the active slip systems and local texture variations. Wang et al. [40] had similar observations while performing a 3D elastic-viscoplastic crystal plasticity finite element method simulation to study the dependence of nanoindentation piling-up patterns. Their simulation showed that the piling-up patterns on the surface of (001)-, (011)- and (111)-oriented single crystal copper had four-, two-, and sixfold symmetry, respectively. All the simulated piling-up patterns were in agreement with those from the experiments. The explanations of the anisotropic surface profiles were also related to the active slip systems and local texture variations.

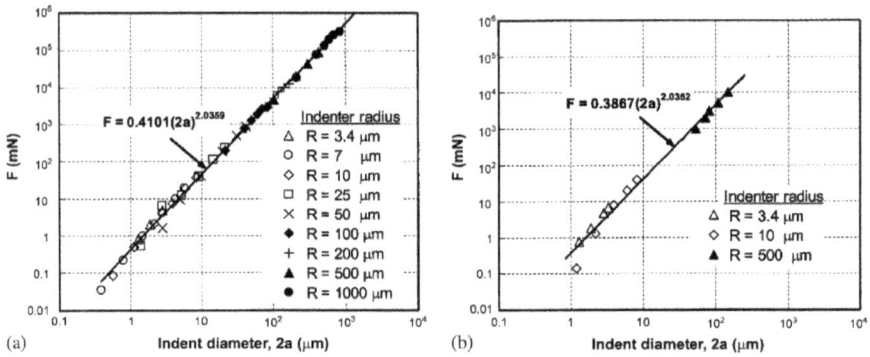

Figure 35. (**a**) Results of FEM simulations showing the variation of the indentation force, F with indent diameter, 2a for different indenter radii, R; (**b**) Results of indentation experiments showing the variation of the indentation force, F with indent diameter, 2a for different indenter radii, R [131]. Reproduced with permission from Y. Liu et al., Journal of the Mechanics and Physics of Solids; published by Elsevier, 2005.

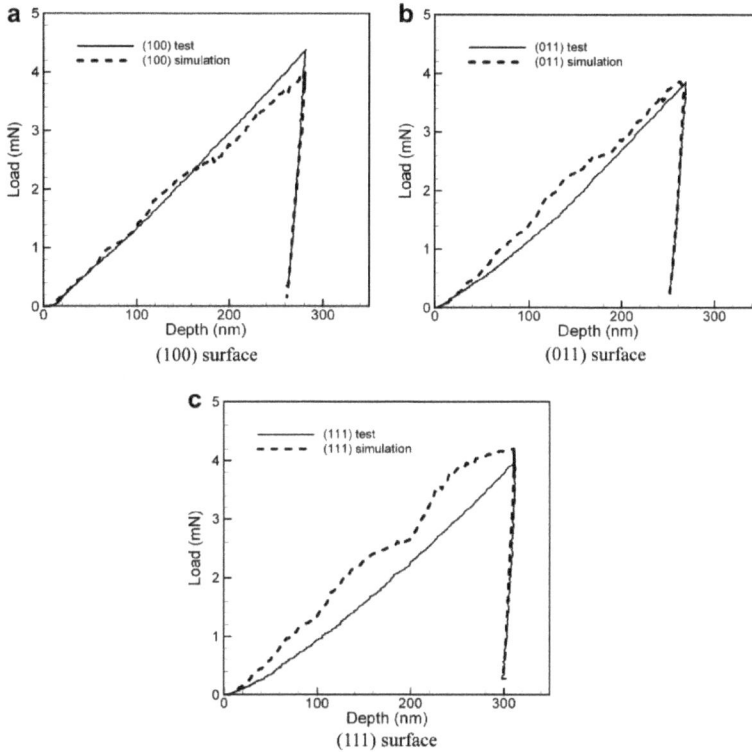

Figure 36. Comparisons between numerical and experimental load-displacement curves on copper samples of different crystallographic orientations made with a spherical indenter (tip radius 3.4 μm): (**a**) (100) plane; (**b**) (011) plane and (**c**) (111) plane [133]. Reproduced with permission from Y. Liu et al., International Journal of Plasticity; published by Elsevier, 2008.

Zaafarani et al. [45] carried out the 3D elastic-viscoplastic crystal plasticity finite element simulations with the same geometry of indenter and boundary conditions as those from experiments. Their CPFEM simulations predicted a similar pattern for the absolute orientation changes as the experiments as shown in Figure 37. However, it was found that the simulations over-emphasized the magnitude of the rotation field tangent to the indenter relative to that directly below the indenter tip. The reason was then found to be due to edge effects at the contact zone and milling-induced curvature caused by ion beam so that no complete EBSD mapping could be made up to the actual contact interface [134].

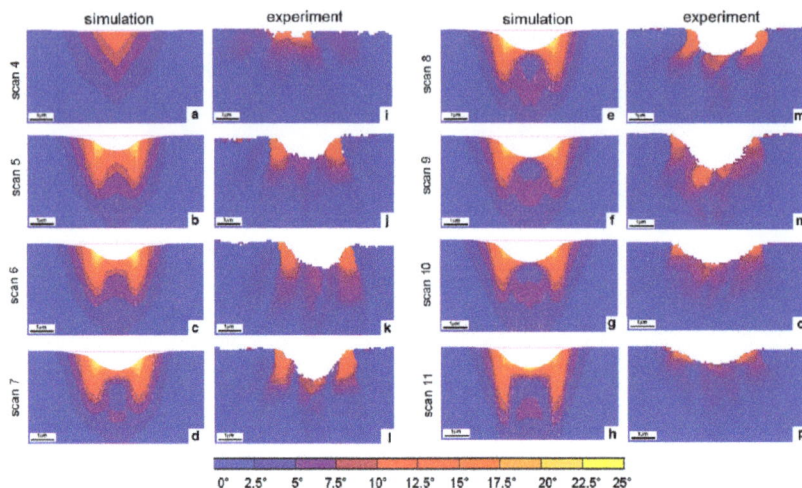

Figure 37. Rotation maps for a set of successive (11$\bar{2}$) sections perpendicular to the (111) indentation plane (surface plane perpendicular to the plane presented) with different spacing to the actual indent [45]. The images on the left-hand side (**a–h**) were obtained from viscoplastic crystal plasticity simulations. The corresponding maps on the right-hand side (**i–p**) were determined via EBSD measurements in succeeding planes prepared by serial FIB sectioning. The color code shows the magnitude of the orientation change relative to the initial crystal orientation without indicating the rotation axis or rotation direction. Scaling is identical for all diagrams. Reproduced with permission from N. Zaafarani et al., Acta Materialia; published by Elsevier, 2006.

Eidel [135] simulated pyramidal micro-indentation on the (001) surface of Ni-base superalloy single crystal with three different azimuthal orientations of the pyramidal indenter. The numerical piling-up patterns were compared with the experimental results. It was found that the resultant material piling-up was insensitive to different azimuthal orientations of the pyramidal indenter as shown in Figure 38. The reason could be due to the piling-up formation determined by crystallographic processes rather than by the stress distribution pattern, induced under the non-isotropic pyramidal indenter. He then also found that the piling-up was independent of the indenter shape (sphere or pyramid) and the elastic anisotropy, which further confirmed that only the geometry of the slip systems in the (001) oriented crystal governed piling-up, whereas stress concentrations introduced by different indenter shapes, by the azimuthal orientation of a pyramidal indenter and also by the characteristics of the elasticity law, had no significant influence.

Liu et al. [136] simulated indentation process of single-crystal aluminium with three different initial orientations via using three-sided Berkovich indenter, and all the numerical load-displacement curves and pile-up patterns from CPFEM have been validated by experimental observations, as shown in Figures 39–41. CPFEM simulation was also used to explain the anisotropic feature of pile-up patterns for different single crystals in details [137,138]. Recently, indentation on severely deformed materials [139] and bicrystalline ones [139] were also investigated via CPFEM modelling.

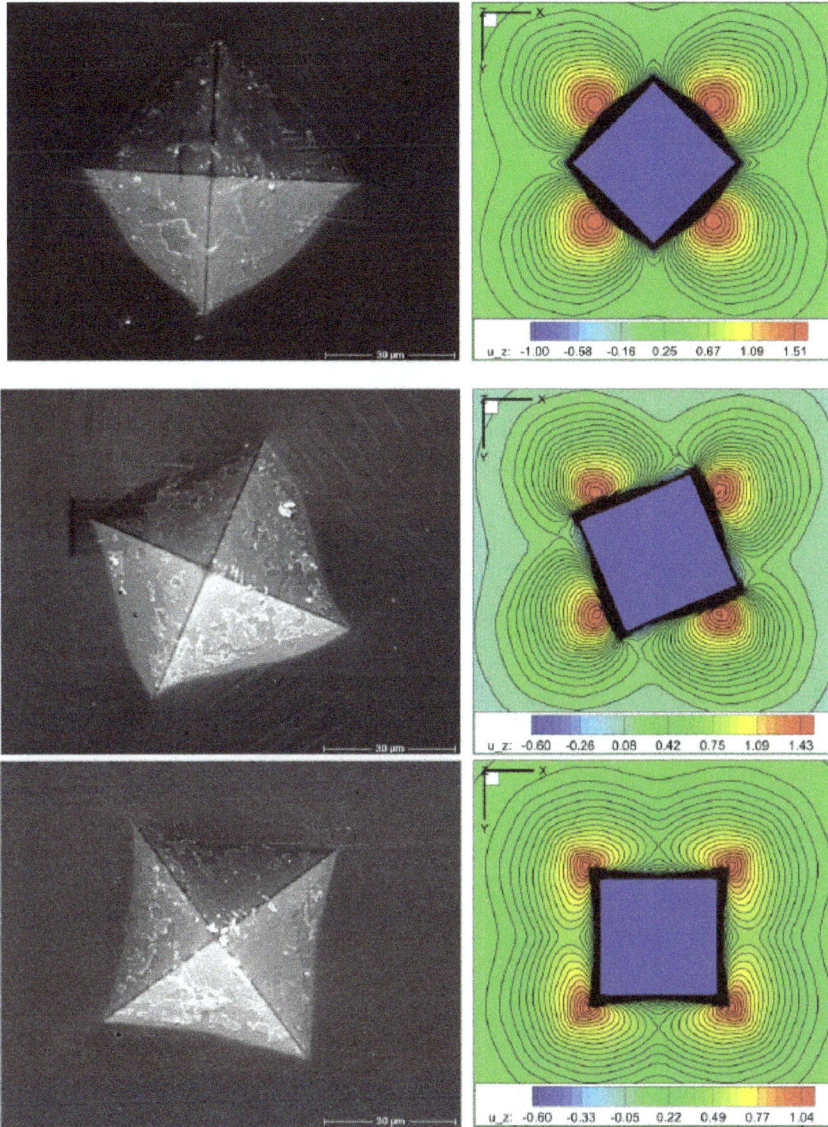

Figure 38. Pyramidal indentation experiments into (001) FCC CMSX-4. Left: experiment (SEM); right: simulation with isolines of height, uz (μm), for azimuthal orientation angle θ = 0°, 22.5°, 45° in row 1–3. In the coordinate system, *X*-, *Y*-, Z-axes each represent h001i directions. The white stains in the experimental indentation craters are debris from sputtering [135]. Reproduced with permission from Y. Liu et al., International Journal of Plasticity; published by Elsevier, 2008.

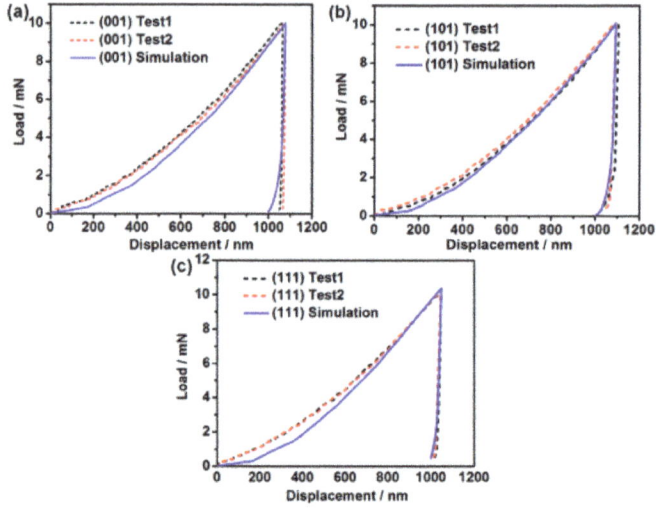

Figure 39. Comparisons between numerical and experimental load-displacement curves for single-crystal Al samples: (**a**) (001); (**b**) (101) and (**c**) (111) surfaces [136].

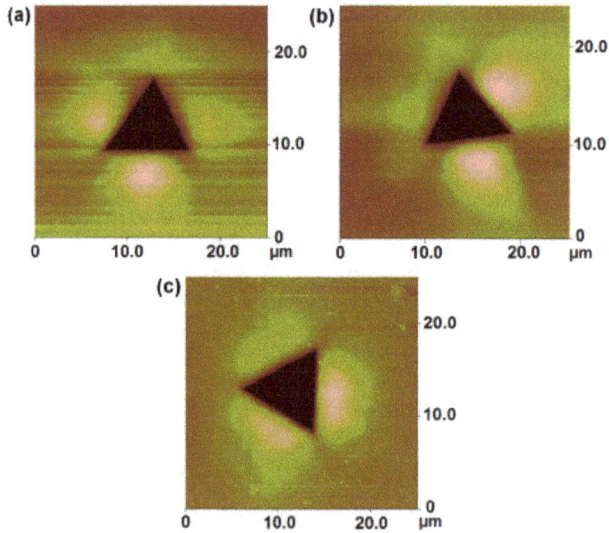

Figure 40. Atomic-force microscopy (AFM) images of the indent impressions made on a single-crystal Al workpiece with a Berkovich indenter (tip radius 200 nm) at different crystallographic orientations: (**a**) (001), (**b**) (101) and (**c**) (111) surfaces [136].

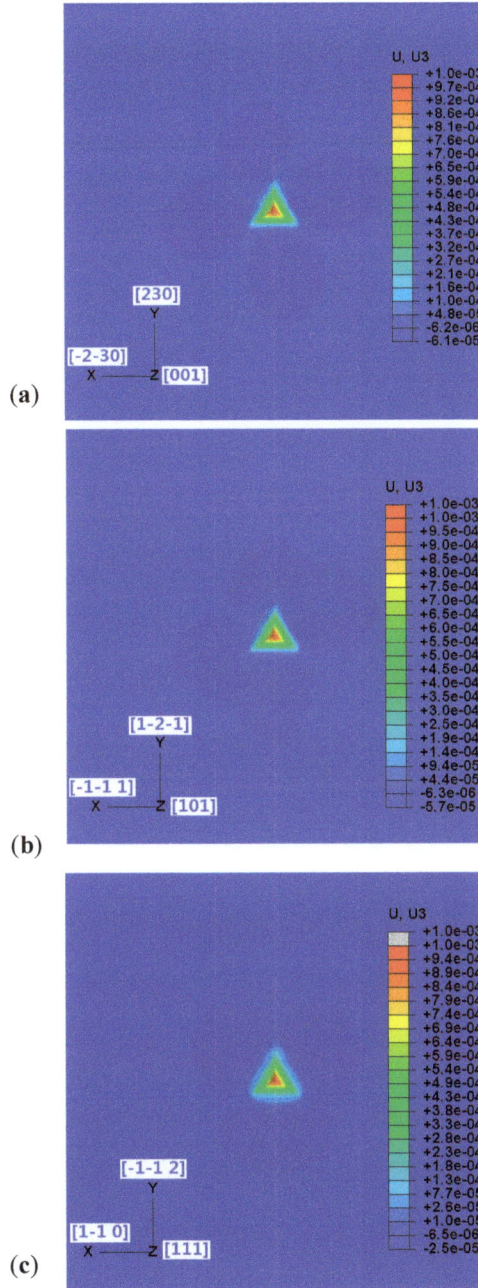

Figure 41. Simulated images of the indent impressions on a single-crystal Al workpiece with a Berkovich indenter (tip radius 200 nm) at different crystallographic orientations: (**a**) (001); (**b**) (101) and (**c**) (111) surfaces [136].

6. Conclusions

Nanoindentation is the most popular method to investigate the mechanical properties of materials at mico- and nanoscale. The key advantage of this technology is its convenience and applicability on a very small sample where normal tensile test cannot be applied. A vast number of experiments have been conducted to investigate different characters during indentation deformation, including hardness, Young's modulus, load-displacement curve, ISE, piling up and sinking in, cracks, texture evolution and lattice rotation and so on. Meanwhile, a wide range of materials were studied, such as metals, ceramics, rubbers, human bones, coatings, etc.

Indentation size effect has been extensively studied by numerous researchers. Among all of these researchers, Nix and Gao has proposed a strain gradient model which was believed to be the best way to simulate and explain ISE in the past. However, it has been found that the main assumptions for that model are in conflict with the experimental observations. Other potential influence factors, such as inadequate measurement capabilities of extremely small indents, presence of oxides or chemical contamination on the surface, indenter-sample friction, increased dominance of edge effects with shallow indents and tip radius have been eventually proven to be ineffective in determining the ISE. Therefore, the mechanism of ISE needs to be further investigated and discussed.

The conventional FEM has been used to study the indentation process on materials and to predict their hardness, stress distribution, friction effects, Young's Modulus, load-displacement curves, and brittle cracking behaviour and so on. However, the anisotropic characters of materials could not be taken into account in the conventional FEM. Therefore, Crystal plasticity FEM (CPFEM) which considers the lattice rotation and the plastic slip as the key deformation mechanism has been used to simulate the texture evolution of materials during indentation deformation. All the numerical results can be accurately validated by comparing with experimental observations. In addition, CPFEM is also the best candidate to investigate the deformation mechanism of materials at smaller scale.

Acknowledgments: The authors acknowledge the financial support from an Australian Research Council Discovery Grant (DP0773329) and from UPA and IPTA scholarships from the University of Wollongong and from Japan Society for the Promotion of Science (JSPS).

Conflicts of Interest: The authors declare no conflicts of interest.

References

1. Tabor, D. *The Hardness of Metals*; Oxford University Press: Oxford, UK, 1951.
2. Bhushan, B. *Handbook of Micro/Nanotribology*, 2nd ed.; CRC Press: Boca Raton, FL, USA, 1999.
3. Brinell, J.A. Way of determining the hardness of bodies and some applications of the same. *Teknisk Tidskrift* **1900**, *5*, 69.
4. Wahlberg, A. Brinell's method of determining hardness and other properties of iron and steel. *J. Iron. Steel Inst.* **1901**, *59*, 243.
5. Meyer, E. Investigations of hardness testing and hardness. *Phys. Z* **1908**, *9*, 66.
6. Smith, R.; Sandland, G. An accurate method of determining the hardness of metals, with particular reference to those of a high degree of hardness. *Proc. Inst. Mech. Eng.* **1922**, *102*, 623. [CrossRef]
7. Hutchings, I.M. The contributions of David Tabor to the science of indentation hardness. *J. Mater. Res.* **2009**, *24*, 581–589. [CrossRef]
8. Tabor, D. A simple theory of static and dynamic hardness. *Proc. R. Soc. Lond. Ser. A* **1948**, *192*, 247–274. [CrossRef]
9. Johnson, K.L. The correlation of indentation experiments. *J. Mech. Phys. Solids* **1970**, *18*, 115–126. [CrossRef]
10. King, R.F.; Tabor, D. The effect of temperature on the mechanical properties and the friction of plastics. *Proc. Phys. Soc. B* **1953**, *66*, 728–736. [CrossRef]
11. Pascoe, M.W.; Tabor, D. The Friction and Deformation of Polymers. *Proc. R. Soc. A* **1956**, *235*, 210–224. [CrossRef]
12. King, R.F.; Tabor, D. The strength properties and frictional behaviour of brittle solids. *Proc. R. Soc. A* **1954**, *223*, 225–238. [CrossRef]

13. Atkins, A.G.; Tabor, D. Mutual Indentation Hardness of Single-Crystal Magnesium Oxide at High Temperatures. *J. Am. Ceram. Soc.* **1967**, *50*, 195–198. [CrossRef]
14. Fischer-Cripps, A.C. A review of analysis methods for sub-micron indentation testing. *Vacuum* **2000**, *58*, 569–585. [CrossRef]
15. Available online: http://www.nanoindentation.cornell.edu/Machine/commercial_machine.htm (accessed on 1 December 2013).
16. Fischer-Cripps, A.C. *The IBIS Handbook of Nanoindentation*; Fischer-Cripps Laboratories Pty Ltd.: Sydney, Australia, 2009.
17. Kim, D.K. Nanoindentation Lecture 1 Basic Principle, KAIST: Daejeon, Korea. Available online: http://szft.elte.hu/oktat/www/mikronano/Nano_indentation.pdf (accessed on 17 August 2017).
18. Oliver, W.C.; Pharr, G.M. An Improved Technique for Determining Hardness and Elastic-Modulus Using Load and Displacement Sensing Indentation Experiments. *J. Mater. Res.* **1992**, *7*, 1564–1583. [CrossRef]
19. Oliver, W.C.; Pharr, G.M. Measurement of hardness and elastic modulus by instrumented indentation: Advances in understanding and refinements to methodology. *J. Mater. Res.* **2004**, *19*, 3–20. [CrossRef]
20. Kucharski, S.; Mroz, Z. Identification of yield stress and plastic hardening parameters from a spherical indentation test. *Int. J. Mech. Sci.* **2007**, *49*, 1238–1250. [CrossRef]
21. Kruzic, J.J.; Kim, D.K.; Koester, K.J.; Ritchie, R.O. Indentation techniques for evaluating the fracture toughness of biomaterials and hard tissues. *J. Mech. Behav. Biomed.* **2009**, *2*, 384–395. [CrossRef] [PubMed]
22. Huber, N.; Tsakmakis, C. Experimental and theoretical investigation of the effect of kinematic hardening on spherical indentation. *Mech. Mater.* **1998**, *27*, 241–248. [CrossRef]
23. Masri, R.; Durban, D. Cylindrical cavity expansion in compressible Mises and Tresca solids. *Eur. J. Mech. A-Solids* **2007**, *26*, 712–727. [CrossRef]
24. Mata, M.; Anglada, M.; Alcala, J. A hardness equation for sharp indentation of elastic-power-law strain-hardening materials. *Philos. Mag. A* **2002**, *82*, 1831–1839. [CrossRef]
25. Hill, R. *The Mathematical Theory of Plasticity*; Oxford University Press: Oxford, UK, 1950.
26. Marsh, D.M. Plastic Flow in Glass. *Proc. R. Soc. A* **1964**, *279*, 420–435. [CrossRef]
27. Rodriguez, J.; Rico, A.; Otero, E.; Rainforth, W.M. Indentation properties of plasma sprayed Al2O3-13% TiO2 nanocoatings. *Acta Mater.* **2009**, *57*, 3148–3156. [CrossRef]
28. Mencik, J.; Munz, D.; Quandt, E.; Weppelmann, E.R.; Swain, M.V. Determination of elastic modulus of thin layers using nanoindentation. *J. Mater. Res.* **1997**, *12*, 2475–2484. [CrossRef]
29. Swain, M.V.; Mencik, J. Mechanical Property Characterization of Thin-Films Using Spherical Tipped Indenters. *Thin Solid Films* **1994**, *253*, 204–211. [CrossRef]
30. Jonsson, B.; Hogmark, S. Hardness Measurements of Thin-Films. *Thin Solid Films* **1984**, *114*, 257–269. [CrossRef]
31. Opitz, A.; Ahmed, S.I.U.; Schaefer, J.A.; Scherge, M. Nanofriction of silicon oxide surfaces covered with thin water films. *Wear* **2003**, *254*, 924–929. [CrossRef]
32. Kim, J.H.; Yeon, S.C.; Jeon, Y.K.; Kim, J.G.; Kim, Y.H. Nano-indentation method for the measurement of the Poisson's ratio of MEMS thin films. *Sens. Actuator A-Phys.* **2003**, *108*, 20–27. [CrossRef]
33. Lou, J.; Shrotriya, P.; Buchheit, T.; Yang, D.; Soboyejo, W.O. Nanoindentation study of plasticity length scale effects in LIGA Ni microelectromechanical systems structures. *J. Mater. Res.* **2003**, *18*, 719–728. [CrossRef]
34. Chicot, D.; Demarecaux, P.; Lesage, J. Apparent interface toughness of substrate and coating couples from indentation tests. *Thin Solid Films* **1996**, *283*, 151–157. [CrossRef]
35. Drory, M.D.; Hutchinson, J.W. Measurement of the adhesion of a brittle film on a ductile substrate by indentation. *Proc. R. Soc. Lond. A* **1953**, *452*, 2319–2341. [CrossRef]
36. Yamazaki, Y.; Kuga, S.; Yoshida, T. Evaluation of interfacial strength by an instrumented indentation method and its application to an actual TBC vane. *Acta Metall. Sin.-Engl.* **2011**, *24*, 109–117.
37. Bartsch, M.; Mircea, I.; Suffner, J.; Baufeld, B. Interfacial fracture toughness measurement of thick ceramic coatings by indentation. *Key Eng. Mater.* **2005**, *290*, 183–190. [CrossRef]
38. Vasinonta, A.; Beuth, J.L. Measurement of interfacial toughness in thermal barrier coating systems by indentation. *Eng. Fract. Mech.* **2001**, *68*, 843–860. [CrossRef]
39. Yamazaki, Y.; Kuga, S.-I.; Jayaprakash, M. Interfacial Strength Evaluation Technique for Thermal Barrier Coated Components by Using Indentation Method. *Procedia Eng.* **2011**, *10*, 845–850. [CrossRef]

40. Wang, Y.; Raabe, D.; Kluber, C.; Roters, F. Orientation dependence of nanoindentation pile-up patterns and of nanoindentation microtextures in copper single crystals. *Acta Mater.* **2004**, *52*, 2229–2238. [CrossRef]

41. Saka, H.; Nagaya, G. Plan-View Transmission Electron-Microscopy Observation of a Crack-Tip in Silicon. *Philos. Mag. Lett.* **1995**, *72*, 251–255. [CrossRef]

42. Lloyd, S.J.; Castellero, A.; Giuliani, F.; Long, Y.; McLaughlin, K.K.; Molina-Aldareguia, J.M.; Stelmashenko, N.A.; Vandeperre, L.J.; Clegg, W.J. Observations of nanoindents via cross-sectional transmission electron microscopy: A survey of deformation mechanisms. *Proc. R. Soc. A-Math. Phys.* **2005**, *461*, 2521–2543. [CrossRef]

43. Lloyd, S.J.; Molina-Aldareguia, J.M.; Clegg, W.J. Deformation under nanoindents in sapphire, spinel and magnesia examined using transmission electron microscopy. *Philos. Mag. A* **2002**, *82*, 1963–1969. [CrossRef]

44. Rester, M.; Motz, C.; Pippan, R. Indentation across size scales—A survey of indentation-induced plastic zones in copper {111} single crystals. *Scr. Mater.* **2008**, *59*, 742–745. [CrossRef]

45. Zaafarani, N.; Raabe, D.; Singh, R.N.; Roters, F.; Zaefferer, S. Three-dimensional investigation of the texture and microstructure below a nanoindent in a Cu single crystal using 3D EBSD and crystal plasticity finite element simulations. *Acta Mater.* **2006**, *54*, 1863–1876. [CrossRef]

46. Pharr, G.M.; Herbert, E.G.; Gao, Y.F. The Indentation Size Effect: A Critical Examination of Experimental Observations and Mechanistic Interpretations. *Annu. Rev. Mater. Res.* **2010**, *40*, 271–292. [CrossRef]

47. Li, L.; Zhou, Q.; Zhou, Y.Y.; Cao, J.G. Numerical study on the size effect in the ultra-thin sheet's micro-bending forming process. *Mater. Sci. Eng. A* **2009**, *499*, 32–35. [CrossRef]

48. Fleck, N.A.; Muller, G.M.; Ashby, M.F.; Hutchinson, J.W. Strain Gradient Plasticity—Theory and Experiment. *Acta Metall. Mater.* **1994**, *42*, 475–487. [CrossRef]

49. Stolken, J.S.; Evans, A.G. A microbend test method for measuring the plasticity length scale. *Acta Mater.* **1998**, *46*, 5109–5115. [CrossRef]

50. Gau, J.T.; Principe, C.; Yu, M. Springback behavior of brass in micro sheet forming. *J. Mater. Process. Technol.* **2007**, *191*, 7–10. [CrossRef]

51. Geiger, M.; Kleiner, M.; Eckstein, R.; Tiesler, N.; Engel, U. Microforming. *CIRP Ann.-Manuf. Technol.* **2001**, *50*, 445–462. [CrossRef]

52. Kals, T.A.; Eckstein, R. Miniaturization in sheet metal working. *J. Mater. Process. Technol.* **2000**, *103*, 95–101. [CrossRef]

53. Saotome, Y.; Yasuda, K.; Kaga, H. Microdeep drawability of very thin sheet steels. *J. Mater. Process. Technol.* **2001**, *113*, 641–647. [CrossRef]

54. Mott, B.W. *Microindentation Hardness Testing*; Butterworths Scientific Publications: London, UK, 1956.

55. Bückle, H. Progress in micro-indentation hardness testing. *Metall. Rev.* **1959**, *4*, 49–100. [CrossRef]

56. Gane, N. Direct Measurement of Strength of Metals on a Sub-Micrometre Scale. *Proc. R. Soc. Lond. Ser.-A* **1970**, *317*. [CrossRef]

57. Kiener, D.; Durst, K.; Rester, M.; Minor, A.M. Revealing deformation mechanisms with nanoindentation. *JOM-US* **2009**, *61*, 14–23. [CrossRef]

58. Gerberich, W.W.; Tymiak, N.I.; Grunlan, J.C.; Horstemeyer, M.F.; Baskes, M.I. Interpretations of indentation size effects. *J. Appl. Mech.* **2002**, *69*, 433–442. [CrossRef]

59. Bull, S.J. On the origins and mechanisms of the indentation size effect. *Z. Metallkd.* **2003**, *94*, 787–792. [CrossRef]

60. Zhu, T.T.; Bushby, A.J.; Dunstan, D.J. Materials mechanical size effects: A review. *Mater. Technol.* **2008**, *23*, 193–209. [CrossRef]

61. Sangwal, K. On the reverse indentation size effect and microhardness measurement of solids. *Mater. Chem. Phys.* **2000**, *63*, 145–152. [CrossRef]

62. Voyiadjis, G.Z.; Peters, R. Size effects in nanoindentation: An experimental and analytical study. *Acta Mech.* **2010**, *211*, 131–153. [CrossRef]

63. Ashby, M.F. Deformation of Plastically Non-Homogeneous Materials. *Philos. Mag.* **1970**, *21*, 399–424. [CrossRef]

64. Nye, J.F. Some geometrical relations in dislocated crystals. *Acta Metall. Mater.* **1953**, *1*, 153–162. [CrossRef]

65. Nix, W.D.; Gao, H.J. Indentation size effects in crystalline materials: A law for strain gradient plasticity. *J. Mech. Phys. Solids* **1998**, *46*, 411–425. [CrossRef]

66. Wiedersich, H. Hardening mechanisms and the theory of deformation. *JOM* **1964**, *16*, 425–430.

67. Ebisu, T.; Horibe, S. Analysis of the indentation size effect in brittle materials from nanoindentation load-displacement curve. *J. Eur. Ceram. Soc.* **2010**, *30*, 2419–2426. [CrossRef]

68. Ma, Q.; Clarke, D.R. Size dependent hardness of silver single crystals. *J. Mater. Res.* **1995**, *10*, 853–863. [CrossRef]

69. Poole, W.J.; Ashby, M.F.; Fleck, N.A. Micro-hardness of annealed and work-hardened copper polycrystals. *Scr. Mater.* **1996**, *34*, 559–564. [CrossRef]

70. Stelmashenko, N.A.; Walls, M.G.; Brown, L.M.; Milman, Y.V. Microindentations on W and Mo Oriented Single-Crystals—An Stm Study. *Acta Metall. Mater.* **1993**, *41*, 2855–2865. [CrossRef]

71. Gao, H.; Huang, Y.; Nix, W.D. Modeling plasticity at the micrometer scale. *Naturwissenschaften* **1999**, *86*, 507–515. [CrossRef] [PubMed]

72. Gao, H.; Huang, Y.; Nix, W.D.; Hutchinson, J.W. Mechanism-based strain gradient plasticity—I. Theory. *J. Mech. Phys. Solids* **1999**, *47*, 1239–1263. [CrossRef]

73. Acharya, A.; Bassani, J.L. Lattice incompatibility and a gradient theory of crystal plasticity. *J. Mech. Phys. Solids* **2000**, *48*, 1565–1595. [CrossRef]

74. Huang, Y.; Gao, H.; Nix, W.D.; Hutchinson, J.W. Mechanism-based strain gradient plasticity—II. Analysis. *J. Mech. Phys. Solids* **2000**, *48*, 99–128. [CrossRef]

75. Gurtin, M.E. A gradient theory of single-crystal viscoplasticity that accounts for geometrically necessary dislocations. *J. Mech. Phys. Solids* **2002**, *50*, 5–32. [CrossRef]

76. Demir, E.; Raabe, D.; Zaafarani, N.; Zaefferer, S. Investigation of the indentation size effect through the measurement of the geometrically necessary dislocations beneath small indents of different depths using EBSD tomography. *Acta Mater.* **2009**, *57*, 559–569. [CrossRef]

77. Rester, M.; Motz, C.; Pippan, R. Microstructural investigation of the volume beneath nanoindentations in copper. *Acta Mater.* **2007**, *55*, 6427–6435. [CrossRef]

78. Kiener, D.; Pippan, R.; Motz, C.; Kreuzer, H. Microstructural evolution of the deformed volume beneath microindents in tungsten and copper. *Acta Mater.* **2006**, *54*, 2801–2811. [CrossRef]

79. Durst, K.; Backes, B.; Goken, M. Indentation size effect in metallic materials: Correcting for the size of the plastic zone. *Scr. Mater.* **2005**, *52*, 1093–1097. [CrossRef]

80. Smedskjaer, M.M. Indentation size effect and the plastic compressibility of glass. *Appl. Phys. Lett.* **2014**, *104*, 251906. [CrossRef]

81. Liu, M.; Xu, W.; Bai, J.; Chua, C.K.; Wei, J.; Li, Z.; Gao, Y.; Kim, D.H.; Zhou, K. Investigation of the size effect for photonic crystals. *Nanotechnology* **2016**, *27*, 405703. [CrossRef] [PubMed]

82. Onitsch, E.M. The present status of testing the hardness of materials. *Mikroskopie* **1956**, *95*, 12–14.

83. Lee, C.H.; Masaki, S.; Kobayashi, S. Analysis of ball indentation. *Int. J. Mech. Sci.* **1972**, *14*, 417–426. [CrossRef]

84. Fischer-Cripps, A.C. Critical review of analysis and interpretation of nanoindentation test data. *Surf. Coat. Technol.* **2006**, *200*, 4153–4165. [CrossRef]

85. Sakai, M. Energy principle of the indentation-induced inelastic surface deformation and hardness of brittle materials. *Acta Metall. Mater.* **1993**, *41*, 1751–1758. [CrossRef]

86. Gong, J.; Wu, J.; Guan, Z. Examination of the indentation size effect in low-load vickers hardness testing of ceramics. *J. Eur. Ceram. Soc.* **1999**, *19*, 2625–2631. [CrossRef]

87. Ren, X.J.; Hooper, R.M.; Griffiths, C.; Henshall, J.L. Indentation size effect in ceramics: Correlation with H/E. *J. Mater. Sci. Lett.* **2003**, *22*, 1105–1106. [CrossRef]

88. Gogotsi, G.A.; Dub, S.N.; Lomonova, E.E.; Ozersky, B.I. Vickers and Knoop Indentation Behavior of Cubic and Partially-Stabilized Zirconia Crystals. *J. Eur. Ceram. Soc.* **1995**, *15*, 405–413. [CrossRef]

89. Sahin, O.; Uzun, O.; Sopicka-Lizer, M.; Gocmez, H.; Kolemen, U. Analysis of load-penetration depth data using Oliver-Pharr and Cheng-Cheng methods of SiAlON-ZrO(2) ceramics. *J. Phys. D Appl. Phys.* **2008**, *41*, 035305. [CrossRef]

90. Kolemen, U. Analysis of ISE in microhardness measurements of bulk MgB2 superconductors using different models. *J. Alloy Compd.* **2006**, *425*, 429–435. [CrossRef]

91. Peng, Z.; Gong, J.; Miao, H. On the description of indentation size effect in hardness testing for ceramics: Analysis of the nanoindentation data. *J. Eur. Ceram. Soc.* **2004**, *24*, 2193–2201. [CrossRef]

92. Gong, J.H.; Zhao, Z.; Guan, Z.D.; Miao, H.Z. Load-dependence of Knoop hardness of Al2O3-TiC composites. *J. Eur. Ceram. Soc.* **2000**, *20*, 1895–1900. [CrossRef]

93. Li, H.; Bradt, R.C. The Microhardness Indentation Load Size Effect in Rutile and Cassiterite Single-Crystals. *J. Mater. Sci.* **1993**, *28*, 917–926. [CrossRef]
94. Samuels, L.E. *Microindentation in Metals*; ASTM STP: Philadelphia, PA, USA, 1986; pp. 5–25.
95. Sargent, P.M. *Use of the Indentation Size Effect on Microhardness for Materials Characterization*; ASTM STP: Philadelphia, PA, USA, 1986; pp. 160–174.
96. Li, H.; Ghosh, A.; Han, Y.H.; Bradt, R.C. The Frictional Component of the Indentation Size Effect in Low Load Microhardness Testing. *J. Mater. Res.* **1993**, *8*, 1028–1032. [CrossRef]
97. Elmustafa, A.A.; Stone, D.S. Nanoindentation and the indentation size effect: Kinetics of deformation and strain gradient plasticity. *J. Mech. Phys. Solids* **2003**, *51*, 357–381. [CrossRef]
98. Hassel, A.W.; Lohrengel, M.M. Preparation and properties of ultra thin anodic valve metal oxide films. *Passiv. Met. Semicond.* **1995**, *185*, 581–590. [CrossRef]
99. McElhaney, K.W.; Vlassak, J.J.; Nix, W.D. Determination of indenter tip geometry and indentation contact area for depth-sensing indentation experiments. *J. Mater. Res.* **1998**, *13*, 1300–1306. [CrossRef]
100. Xue, Z.; Huang, Y.; Hwang, K.C.; Li, M. The influence of indenter tip radius on the micro-indentation hardness. *J. Eng. Mater. Technol.* **2002**, *124*, 371–379. [CrossRef]
101. Lilleodden, E.T. Indentation-Induced Plasticity of Thin Metal Films. Ph.D. Thesis, Stanford University, Palo Alto, CA, USA, 2001.
102. Huang, Y.; Xue, Z.; Gao, H.; Nix, W.D.; Xia, Z.C. A Study of Microindentation Hardness Tests by Mechanism-based Strain Gradient Plasticity. *J. Mater. Res.* **2000**, *15*, 1786–1796. [CrossRef]
103. Lee, C.H.; Kobayash, S. Elastoplastic Analysis of Plane-Strain and Axisymmetric Flat Punch Indentation by Finite-Element Method. *Int. J. Mech. Sci.* **1970**, *12*, 349–370. [CrossRef]
104. Bhattacharya, A.K.; Nix, W.D. Finite-Element Simulation of Indentation Experiments. *Int. J. Solids Struct.* **1988**, *24*, 881–891. [CrossRef]
105. Bhattacharya, A.K.; Nix, W.D. Finite-Element Analysis of Cone Indentation. *Int. J. Solids Struct.* **1991**, *27*, 1047–1058. [CrossRef]
106. Pethica, J.B.; Hutchings, R.; Oliver, W.C. Hardness Measurement at Penetration Depths as Small as 20-Nm. *Philos. Mag. A* **1983**, *48*, 593–606. [CrossRef]
107. Lichinchi, M.; Lenardi, C.; Haupt, J.; Vitali, R. Simulation of Berkovich nanoindentation experiments on thin films using finite element method. *Thin Solid Films* **1998**, *312*, 240–248. [CrossRef]
108. Bressan, J.D.; Tramontin, A.; Rosa, C. Modeling of nanoindentation of bulk and thin film by finite element method. *Wear* **2005**, *258*, 115–122. [CrossRef]
109. Huang, X.Q.; Pelegri, A.A. Finite element analysis on nanoindentation with friction contact at the film/substrate interface. *Compos. Sci. Technol.* **2007**, *67*, 1311–1319. [CrossRef]
110. Pelletier, H.; Krier, J.; Cornet, A.; Mille, P. Limits of using bilinear stress-strain curve for finite element modeling of nanoindentation response on bulk materials. *Thin Solid Films* **2000**, *379*, 147–155. [CrossRef]
111. Xu, B.; Zhao, B.; Yue, Z.F. Finite element analysis of the indentation stress characteristics of the thin film/substrate systems by flat cylindrical indenters. *Mater. Werkst.* **2006**, *37*, 681–686. [CrossRef]
112. Care, G.; FischerCripps, A.C. Elastic-plastic indentation stress fields using the finite-element method. *J. Mater. Sci.* **1997**, *32*, 5653–5659. [CrossRef]
113. Zhao, M.H.; Yao, L.P.; Zhang, T.Y. Stress analysis of microwedge indentation-induced delamination. *J. Mater. Res.* **2009**, *24*, 1943–1949. [CrossRef]
114. Wang, H.F.; Yang, X.; Bangert, H.; Torzicky, P.; Wen, L. 2-Dimensional Finite-Element Method Simulation of Vickers Indentation of Hardness Measurements on Tin-Coated Steel. *Thin Solid Films* **1992**, *214*, 68–73. [CrossRef]
115. Yan, W.Y.; Sun, Q.P.; Liu, H.Y. Spherical indentation hardness of shape memory alloys. *Mater. Sci. Eng. A-Struct.* **2006**, *425*, 278–285. [CrossRef]
116. Toparli, M.; Koksal, N.S. Hardness and yield strength of dentin from simulated nano-indentation tests. *Comput. Methods Prog. Biomed.* **2005**, *77*, 253–257. [CrossRef] [PubMed]
117. Sarris, E.; Constantinides, G. Finite element modeling of nanoindentation on C-S-H: Effect of pile-up and contact friction. *Cem. Concr. Comp.* **2013**, *36*, 78–84. [CrossRef]
118. Niezgoda, T.; Malachowski, J.; Boniecki, M. Finite element simulation of vickers microindentation on alumina ceramics. *Ceram. Int.* **1998**, *24*, 359–364. [CrossRef]

119. Larsson, P.L.; Giannakopoulos, A.E. Tensile stresses and their implication to cracking at pyramid indentation of pressure-sensitive hard metals and ceramics. *Mater. Sci. Eng. A-Struct.* **1998**, *254*, 268–281. [CrossRef]

120. Wang, H.F.; Bangert, H. 3-Dimensional Finite-Element Simulation of Vickers Indentation on Coated Systems. *Mater. Sci. Eng. A-Struct.* **1993**, *163*, 43–50. [CrossRef]

121. Zhai, J.G.; Wang, Y.Q.; Kim, T.G.; Song, J.I. Finite element and experimental analysis of Vickers indentation testing on Al2O3 with diamond-like carbon coating. *J. Cent. South Univ.* **2012**, *19*, 1175–1181. [CrossRef]

122. Chen, Z.Y.; Diebels, S. Nanoindentation of hyperelastic polymer layers at finite deformation and parameter re-identification. *Arch. Appl. Mech.* **2012**, *82*, 1041–1056. [CrossRef]

123. Bolzon, G.; Maier, G.; Panico, M. Material model calibration by indentation, imprint mapping and inverse analysis. *Int. J. Solids Struct.* **2004**, *41*, 2957–2975. [CrossRef]

124. Bocciarelli, M.; Bolzon, G.; Maier, G. Parameter identification in anisotropic elastoplasticity by indentation and imprint mapping. *Mech. Mater.* **2005**, *37*, 855–868. [CrossRef]

125. Nakamura, T.; Wang, T.; Sampath, S. Determination of properties of graded materials by inverse analysis and instrumented indentation. *Acta Mater.* **2000**, *48*, 4293–4306. [CrossRef]

126. Chen, X.; Ogasawara, N.; Zhao, M.H.; Chiba, N. On the uniqueness of measuring elastoplastic properties from indentation: The indistinguishable mystical materials. *J. Mech. Phys. Solids* **2007**, *55*, 1618–1660. [CrossRef]

127. Storåkers, B.; Biwa, S.; Larsson, P.-L. Similarity analysis of inelastic contact. *Int. J. Solids Struct.* **1997**, *34*, 3061–3083. [CrossRef]

128. Casals, O.; Forest, S. Finite element crystal plasticity analysis of spherical indentation in bulk single crystals and coatings. *Comput. Mater. Sci.* **2009**, *45*, 774–782. [CrossRef]

129. Casals, O.; Ocenasek, J.; Alcala, J. Crystal plasticity finite element simulations of pyramidal indentation in copper single crystals. *Acta Mater.* **2007**, *55*, 55–68. [CrossRef]

130. Alcala, J.; Casals, O.; Ocenasek, J. Micromechanics of pyramidal indentation in fcc metals: Single crystal plasticity finite element analysis. *J. Mech. Phys. Solids* **2008**, *56*, 3277–3303. [CrossRef]

131. Liu, Y.; Wang, B.; Yoshino, M.; Roy, S.; Lu, H.; Komanduri, R. Combined numerical simulation and nanoindentation for determining mechanical properties of single crystal copper at mesoscale. *J. Mech. Phys. Solids* **2005**, *53*, 2718–2741. [CrossRef]

132. Peirce, D.; Asaro, R.J.; Needleman, A. An Analysis of Nonuniform and Localized Deformation in Ductile Single-Crystals. *Acta Metall. Mater.* **1982**, *30*, 1087–1119. [CrossRef]

133. Liu, Y.; Varghese, S.; Ma, J.; Yoshino, M.; Lu, H.; Komanduri, R. Orientation effects in nanoindentation of single crystal copper. *Int. J. Plast.* **2008**, *24*, 1990–2015. [CrossRef]

134. Zaafarani, N.; Raabe, D.; Roters, F.; Zaefferer, S. On the origin of deformation-induced rotation patterns below nanoindents. *Acta Mater.* **2008**, *56*, 31–42. [CrossRef]

135. Eidel, B. Crystal plasticity finite-element analysis versus experimental results of pyramidal indentation into (0 0 1) fcc single crystal. *Acta Mater.* **2011**, *59*, 1761–1771. [CrossRef]

136. Liu, M.; Lu, C.; Tieu, K.A.; Peng, C.T.; Kong, C. A combined experimental-numerical approach for determining mechanical properties of aluminum subjects to nanoindentation. *Sci. Rep.* **2015**, *14*, 15072. [CrossRef] [PubMed]

137. Liu, M.; Tieu, K.A.; Zhou, K.; Peng, C.T. Investigation of the Anisotropic Mechanical Behaviors of Copper Single Crystals through Nanoindentation Modeling. *Metall. Mater. Trans. A* **2016**, *47*, 2717–2725. [CrossRef]

138. Liu, M.; Tieu, A.K.; Peng, C.T.; Zhou, K. Explore the anisotropic indentation pile-up patterns of single-crystal coppers by crystal plasticity finite element modelling. *Mater. Lett.* **2015**, *161*, 227–230. [CrossRef]

139. Liu, M.; Tieu, K.A.; Zhou, K.; Peng, C.T. Indentation analysis of mechanical behaviour of torsion-processed single-crystal copper by crystal plasticity finite-element method modelling. *Philos. Mag.* **2016**, *96*, 261–273. [CrossRef]

MDPI

Article

Application of the Improved Inclusion Core Model of the Indentation Process for the Determination of Mechanical Properties of Materials

Boris A. Galanov, Yuly V. Milman *, Svetlana I. Chugunova, Irina V. Goncharova and Igor V. Voskoboinik

Institute for Problems of Materials Science of NASU, 3 Krzhizhanovky Str., 03680 Kiev, Ukraine; gbaprofil@bk.ru (B.A.G.); yuly.milman@gmail.com (S.I.C.); irina@ipms.kiev.ua (I.V.G.); igor-d23@ipms.kiev.ua (I.V.V.)
* Correspondence: milman@ipms.kiev.ua; Tel.: +38-044-424-3184

Academic Editors: Ronald W. Armstrong, Stephen M. Walley, Wayne L. Elban and Helmut Cölfen
Received: 9 February 2017; Accepted: 14 March 2017; Published: 16 March 2017

Abstract: The improved Johnson inclusion core model of indentation by conical and pyramidal indenters in which indenter is elastically deformed and a specimen is elastoplastically deformed under von Mises yield condition, was used for determination of mechanical properties of materials with different types of interatomic bond and different crystalline structures. This model enables us to determine approximately the Tabor parameter $C = HM/Y_S$ (where HM is the Meyer hardness and Y_S is the yield stress of the specimen), size of the elastoplastic zone in the specimen, effective apex angle of the indenter under load, and effective angle of the indent after unloading. It was shown that the Tabor parameter and the size of elastoplastic deformation zone increase monotonically with the increase of the plasticity characteristic δ_H, which is determined in indentation experiments using the early elaborated by the several authors of this article method. The corresponding analytical dependencies were obtained and their physical nature is discussed. For the materials studied in this work, the Tabor parameter ranges from 1 to 4. At the same time, for structural metallic alloys its value is between 2.8 and 3.1 in agreement with the results obtained by Tabor. A very simple technique developed in this article allows one to determine from the standard indentation test not only the hardness of a material but also its yield stress and plasticity. This makes the indentation test results significantly more informative.

Keywords: mechanical properties; hardness; indentation; plasticity

1. Introduction

The study of mechanical properties of materials by the method of local loading with a rigid indenter is extensively used in practice. In indentation the Meyer hardness $HM = P/S$ (where P is the load on the indenter and S is the projection area of the hardness indent on the initial surface of the specimen) has a precise physical meaning of the average pressure under indenter and is usually determined.

Indentation models which describe theoretically the indentation process with the aim to determine other mechanical properties, particularly the yield stress of material Y_S, were proposed long ago and many times [1,2]. Among the developed models, the Johnson inclusion core model is the most successful [3,4].

The details of these investigations and historical information on this problem up to 1969 are presented in [3]. Thereafter, the concept of the inclusion core model was checked and investigated in many works (see, e.g., [5–9]). In [10], executed with the participation of several authors of this article,

Johnson's model has been improved to describe the process of continuous indentation, in which not only the sample, but also the indenter undergoes elastic-plastic deformation. In this improved model the elastic compression of the inclusion core under the indenter is taken into account for the first time, as well as the change in the apex angle of the indenter in the deformation process. In [10] for the description of such indentation process the system of five equations was derived, which has been used to study the deformation of diamond during indentation by the diamond indenter. In this paper, the model [10] is simplified for the case where only the sample is deformed elastically-plastically, and the indenter is deformed elastically. The advantages of the model [10] are preserved in this paper by taking into account the compression of the core under indenter and the change of the indenter shape as a result of elastic deformation. Simplification of the model [10] reduced the number of equations from five to three (see the system of Equations (26) in [10] and the system (1) in this article). The system (1) is used in this study to analyze the deformation process during indentation of materials with different types of interatomic bonds and various crystalline structures, to establish the functional relationship between the Tabor parameter C [11] and the plasticity of the material ($C = HM/Y_S$, where HM is the Meyer hardness and Y_S is the yield stress of the specimen), as well as for development of the simple method for determination of the yield stress as a result of standard determination of hardness.

2. Theoretical Background. Scheme and Equations of the Improved Model

Figure 1 shows a scheme in a spherical coordinate system $0r\theta\psi$ of a model of contact interaction of a conical indenter and specimen, in which a hydrostatic core of radius c forms. The non-deformed indenter is shown by a dashed line, and the following notations are used: ψ is the angle between the surface of the indenter and the indenter axis x_i under load; $0 \leq r \leq c$ is the region of the core; $c \leq r \leq b_S$ is the spherical layer of the specimen where elastoplastic deformations occurred; $r \geq b_S$ is the region of elastic deformation of the specimen. Strains are assumed to be sufficiently small.

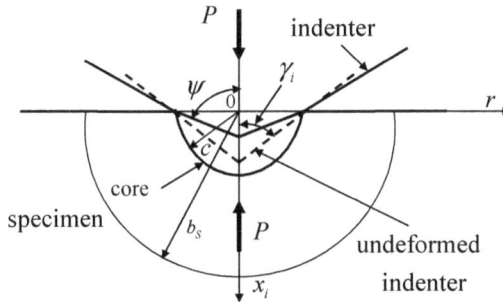

Figure 1. Scheme of interaction of an indenter and a specimen under a load P in a spherical coordinate system $0r\theta\psi$, $HM = P/(\pi c^2)$.

Dislocation approach to the mechanism of deformation during indentation is being developed intensively ([12–18], etc.). In the framework of the dislocation theory, the zone of elastoplastic deformation with the radius b_S is the zone with a sharp increase in the dislocation density around the indentation imprint with a symmetry center at the very point 0 in Figure 1. Dislocations are nucleated near the indenter and move in the radial directions to the boundaries of the elastoplastic zone under the action of shear stress, caused by the load on indenter [19]. The comparison of calculated values of b_S with the experimental data is given in the Section 3.6.

During continuous penetration of the elastic indenter, the core increases at the expense of the elastoplastic zone of the specimen. This proceeds on its boundary, where the material of this zone is compressed by the pressure of the core, which exceeds the pressure in the elastoplastic zone (in passing the boundary of the core, the jump of pressure and volume strain is observed; shear stresses, which are absent in the hydrostatic core, also change abruptly). During such penetration, the material of the elastoplastic zone is additionally densified on the boundary of the core by a pressure $\Delta p_S = 2Y_S/3$ (caused by the jump of pressure Δp_S on this boundary) and joined to the material of the core.

As mentioned above, this model has three transcendental equations for three unknown quantities: yield stress Y_S, the relative size of the elastoplastic zone $x = b_S/c$ and $z = \cot \psi$:

$$
\begin{cases}
z = \cot \psi = \cot \gamma_i - 2HM/E_i^*, & \text{(1a)} \\
(1 - \theta_S Y_S) \cdot \left(x^3 - \alpha_S\right) = z\beta_S/Y_S, & \text{(1b)} \\
(2/3 + 2\ln x) - HM/Y_S = 0, & \text{(1c)}
\end{cases}
$$

where the notation $\alpha_S = \frac{2(1-2\nu_S)}{3(1-\nu_S)}$, $\beta_S = \frac{E_S}{6(1-\nu_S)}$, $\theta_S = \frac{2(1-2\nu_S)}{E_S}$ and $E_i^* = \frac{E_i}{1-\nu_i^2}$ is used, E is the Young's modulus, ν is the Poisson's ratio, γ_i is the angle between the surface and the axis x_i of the conical non-deformed indenter. Subscripts i and s correspond to the indenter and specimen, respectively. The solution of this system for unknowns (z, x, Y_S) determines approximately the stress-strain state of the specimen in accord with the proposed model. As it is seen from Equation (1c) the Tabor constant

$$
C = HM/Y_S = 2/3 + 2\ln x, \tag{2}
$$

The system of Equations (1) takes into account the elastic compressibility during formation of the core, and, thus, the proposed model develops the model considered in [3,4]. Equation (1a) corresponds to Equation (17) of the work [10] at $\gamma_{iR} = \gamma_i$, Equation (1b) corresponds to the first equation of the system (26) of the work [10], and Equation (1c) corresponds to the fourth equation of the system (26) of the work [10].

The influence of compressibility during formation of the core, as it follows from [10] is determined by the value of $\theta_S Y_S$. This value increases with increase in the ratio Y_S/E_S and with decrease in the Poisson's ratio ν. The evaluation of $\theta_S Y_S$ shows that the ratio Y_S/E_S can attain 0.1 for covalent crystals, and $\theta_S Y_S$ becomes substantial as compared to 1. For the same crystals, ν has a minimum value, which is particularly small for diamond ($\nu = 0.07$). Diamond was not investigated in the present work because its deformation is purely elastic at room temperature. Features of the diamond deformation during indentation by diamond indenter are considered in [10]. However, we can evaluate the quantity Y_S/E_S on the basis of the Meyer hardness of diamond at room temperature $HM = 150$ GPa [20] assuming that, as for high-hardness ceramics, for diamond, $Y_S \approx HM$. For diamond and for the value $E = 1200$ GPa, we obtain $\theta_S Y_S \approx 0.23$, i.e., the compressibility of the deformation core is particularly substantial for diamond and high-hardness ceramic materials. For metals, at $Y_S/E_S \approx 0.002$, and if $\nu = 0.35$, the value of $\theta_S Y_S = 0.001$, is much smaller than 1, and taking into account the compressibility of the material during formation of the core hardly influences on the obtained results.

For the residual conical indent in the specimen, the effective angle γ_{SR} after its elastic unloading has the value ([10], Equation (16))

$$
\cot \gamma_{SR} = \cot \psi - 2HM(1 - \nu_S^2)/E_S, \tag{3}
$$

where the term $2HM(1 - \nu_S^2)/E_S$ takes account of the elastic recovery of angle ψ and elastic deflection component of the specimen surface.

The considered model was elaborated for the case of penetration of a cone with an apex angle $2\gamma_i$. The following relations between the apex angles of equivalent conical and pyramidal (trihedral and tetrahedral) indenters were proposed in [10],

$$
\cot \gamma_i = \sqrt{\pi} \cot \gamma_V/2 = \sqrt[4]{\pi^2/27} \cot \gamma_B, \tag{4}
$$

where γ_i, γ_V, γ_B are the apex angles of conical, tetrahedral (e.g., Vickers indenters, $\gamma_V = 68°$), and trihedral (e.g., Berkovich indenters, $\gamma_B = 65°$) indenters, respectively.

3. Results and Discussion

3.1. Comparative Analysis of the Deformation Process during Indentation of Materials with Different Types of Interatomic Bond and Different Crystalline Structures

In this work, results of measurement of the Vickers microhardness obtained by the authors, a substantial part of which was published [18,21–25], were used. For most presented results, the load on the indenter was close to 2 N. For the analysis of features of deformation in indentation, we chose unalloyed polycrystalline and single-crystalline metals with FCC, BCC, and HCP lattices; a number of intermetallics (Al_3Ti, $Al_{61}Cr_{12}Ti_{27}$, and $Al_{66}Mn_{11}Ti_{23}$); single-crystals of refractory carbides (WC, NbC, TiC, ZrC, and SiC), covalent crystals of Si and Ge, and partially covalent Al_2O_3 and LaB_6; amorphous alloys ($Fe_{83}B_{17}$, $Fe_{40}Ni_{38}Mo_4B_{18}$, and $Co_{50}Ni_{10}Fe_5Si_{12}B_{17}$) and quasicrystals ($Al_{63}Cu_{25}Fe_{12}$ and $Al_{70}Pd_{20}Mn_{10}$). An investigation was also performed for steel with 0.45% C and 5083 aluminum alloy.

The characteristics of the studied materials are presented in Table 1. The microhardness *HM* was calculated from the value of *HV* (*HM* = 1.08 *HV*). In calculations for the diamond indenter, E_i = 1200 GPa and ν_i = 0.07 were taken.

Table 1. Mechanical characteristics of materials (Meyer hardness *HM*, Young modulus E_S, and Poisson's ratio ν_S) and characteristics calculated according to the core indentation model (Tabor parameter *C*, yield stress Y_S, plasticity characteristic δ_H, relative size of elastoplastic zone *x*, apex angle of indenter under load ψ, and relaxed effective apex angle of a hardness indent γ_{SR}).

	Materials	HM, GPa	E_S, GPa	ν_S	C = HM/Ys	Ys, GPa	δ_H	x = b_S/c	ψ, deg.	γ_{SR}, deg.
FCC metals	Al	0.173	71	0.35	4.02	0.043	0.99	5.33	68.01	68.12
	Au	0.270	78	0.42	3.86	0.07	0.99	4.84	68.02	68.27
	Cu	0.486	130	0.343	3.74	0.13	0.98	4.47	68.04	68.32
	Ni	0.648	210	0.29	3.81	0.17	0.98	4.68	68.05	68.29
BCC metals	Cr	1.404	298	0.31	3.42	0.41	0.97	3.98	68.10	68.47
	Ta	0.972	185	0.342	3.35	0.29	0.97	3.88	68.07	68.48
	V	0.864	127	0.365	3.20	0.27	0.97	3.54	68.06	68.58
	Mo (111)	1.998	324	0.293	3.17	0.63	0.96	3.52	68.14	68.64
	Nb	0.972	104	0.397	2.94	0.33	0.96	3.16	68.07	68.76
	Fe	1.512	211	0.28	3.02	0.50	0.95	3.29	68.11	68.69
	W (001)	4.320	420	0.28	2.73	1.58	0.92	2.80	68.31	69.15
HCP metals	Ti	1.112	120	0.36	2.93	0.38	0.95	3.09	68.08	68.79
	Zr	1.156	98	0.38	2.75	0.42	0.95	2.83	68.08	68.97
	Re	3.024	466	0.26	3.09	0.63	0.95	3.38	68.22	68.75
	Mg	0.324	44.7	0.291	2.94	0.11	0.95	3.3	68.02	68.60
	Be	1.620	318	0.024	3.05	0.53	0.94	3.35	68.12	68.56
	Co	1.836	211	0.32	2.91	0.63	0.94	3.10	68.13	68.82
Intermetallics (IM)	$Al_{66}Mn_{11}Ti_{23}$ (IM$_3$)	2.203	168	0.19	2.42	0.91	0.87	2.42	68.16	69.27
	$Al_{61}Cr_{12}Ti_{27}$ (IM$_2$)	3.456	178	0.19	2.08	1.66	0.81	2.03	68.25	69.90
	Al_3Ti (IM$_1$)	5.335	156	0.30	1.67	3.19	0.76	1.65	68.38	71.16
Metallic glasses (MG)	$Fe_{40}Ni_{38}Mo_4B_{18}$ (MG$_2$)	7.992	152	0.30	1.25	6.39	0.62	1.34	68.58	72.90
	$Co_{50}Ni_{10}Fe_5Si_{12}B_{17}$ (MG$_3$)	9.288	167	0.30	1.19	7.80	0.60	1.30	68.67	73.25
	$Fe_{83}B_{17}$ (MG$_1$)	10.044	171	0.30	1.14	8.84	0.58	1.26	68.73	73.58
Quasicrystalls (QC)	$Al_{70}Pd_{20}Mn_{10}$ (QC$_2$)	7.560	200	0.28	1.55	4.88	0.71	1.55	68.54	71.67
	$Al_{63}Cu_{25}Fe_{12}$ (QC$_1$)	8.024	113	0.28	0.97	8.30	0.48	1.16	68.58	74.54
Refractory compounds	WC (0001)	18.036	700	0.31	1.89	9.56	0.81	1.84	69.31	71.40
	NbC (100)	25.920	550	0.21	1.22	21.26	0.54	1.32	69.89	74.02
	LaB_6 (001)	23.220	439	0.20	1.13	20.51	0.50	1.26	69.69	74.34
	TiC (100)	25.920	465	0.191	1.08	24.07	0.46	1.23	69.89	74.83
	ZrC (100)	23.760	410	0.196	1.06	22.48	0.46	1.22	69.73	74.85
	Al_2O_3 (0001)	22.032	323	0.23	0.94	23.40	0.41	1.15	69.60	75.56
	α-SiC (0001)	32.400	457	0.22	0.87	37.24	0.36	1.11	70.38	76.77
Covalent crystals	Ge (111)	7.776	130	0.21	1.10	7.06	0.49	1.24	68.56	73.75
	Si (111)	11.340	160	0.22	0.96	11.84	0.42	1.16	68.82	74.99
Industrial alloys	Steel 0.45%C	1.890	204	0.285	2.74	0.69	0.93	2.79	68.14	68.88
	Al alloy #5083	1.030	70.1	0.33	2.51	0.41	0.91	2.49	68.07	69.23

The analysis of the deformation process in microindentation was performed on the basis of the developed inclusion core model of indentation with the use of the system of Equations (1). The parameter *z* was calculated from Equation (1a), and then the system of Equations (1b) and

(1c) was solved to determine the yield strength Y_S and the relative size of the elastoplastic zone in the specimen $x = b_S/c$.

The apex angle of the equivalent conical indenter under load ψ was calculated by the relation $z = cot\ \psi$. The apex angle of the conical hardness indent in the specimen after unloading of the indenter γ_{SR} was calculated by Equation (3).

In accordance with [10,21], the mean plastic strain on the contact area of the indenter and specimen ε_p in the direction of the force P applied to the indenter was calculated by Equation (5), the elastic strain ε_E, corresponding to the elastic deflection component of the specimen surface, was computed by (6), and the total strain ε_t was calculated by (7)

$$\varepsilon_p = \ln\sin\gamma_{SR} = -\ln\sqrt{1 + \cot^2\gamma_{SR}} < 0, \tag{5}$$

$$\varepsilon_e = -(1 + \nu_S)(1 - 2\nu_S)HM/E_S, \tag{6}$$

$$\varepsilon_t = \varepsilon_e + \varepsilon_p. \tag{7}$$

The plasticity characteristic δ_H (introduced in [18]) was evaluated by formula (8) in Section 3.2.1. The obtained results are presented in Table 1, in which groups of materials are located in the order of decreasing plasticity characteristic δ_H. It is seen, that the Tabor parameter C decreases simultaneously with a decrease δ_H within each group of materials of Table 1, and at the comparison of values C and δ_H of the different groups.

For the most plastic materials with a FCC lattice, C = 3.8–4. For metals with BCC and HCP structures, C ≈ 3, which corresponds to the Tabor concept [11].

Among the other studied materials, intermetallic compounds have values of C ≈ 2, that are close to those for metals.

Among the studied refractory compounds, the lowest value of C, even smaller than 1, is observed for SiC and Al$_2$O$_3$. These crystals also have the smallest plasticity.

Among refractory compounds, carbide WC, as is known [24,25], is distinguished by increased plasticity $\delta_H = 0.81$, and, for it, C = 1.89, that is higher than for other refractory compounds. For covalent crystals Si and Ge, C ≈ 1. At the same time Ge has a somewhat higher plasticity and higher value of C. However, it should be taken into account that, in these crystals, indentation leads to the semiconductor–metal phase transition [26,27], which complicates the discussion of results obtained for them.

In view of the established correlation of the Tabor parameter C with the plasticity characteristic δ_H, it seems reasonable to consider the relation of these characteristics more thoroughly to elucidate the physical nature of the Tabor parameter C. The relation between C and δ_H seems to be particularly interesting because both these characteristics relate the hardness to the mechanical properties of the material, namely, to the yield strength (Tabor parameter C) and to the plasticity of the material (plasticity characteristic δ_H).

3.2. Relation between the Tabor Parameter C = HM/Y$_S$ and Plasticity Characteristic δ_H

3.2.1. Plasticity Characteristic δ_H Determined by Indentation

In modern physics plasticity is determined by the tendency of a material to undergo residual deformation under load [28,29].

The frequently used plasticity characteristics (elongation of a specimen to fracture δ and its reduction of the area to fracture Ψ) do not correspond to the physical definition of plasticity and must be considered only as convenient technological tests [18,21,30], which can be used for only metals having some elongation to fracture. For a large number of modern materials, the value $\delta = 0$ and cannot characterize their mechanical behavior. The plasticity characteristic satisfying the physical definition of plasticity was proposed in [18] in the form of the dimensionless parameter

$$\delta^* = \varepsilon_p / \varepsilon_t = 1 - \varepsilon_e / \varepsilon_t, \tag{8}$$

where ε_p, ε_e, and ε_t are, respectively, the plastic, elastic, and total strain, and $\varepsilon_t = \varepsilon_p + \varepsilon_e$.

The considered plasticity characteristic δ^* can be determined in any methods of mechanical tests (tension, compression, and bending) and, as shown in [18,21], in indentation.

It is seen from expression (8) that δ^* depends on the total strain ε_t, which follows directly from the definition of plasticity δ^* presented above.

Since the plasticity δ^* depends on the strain ε_t, a comparison of the plasticity of different materials should be performed at a representative strain $\varepsilon_t \approx const$. In tensile test, in the first stages of loading, $\varepsilon_t = \varepsilon_e$, and plastic strain is absent, i.e., the material does not retain a part of strain after unloading. For this reason representative strain ε_t must be sufficiently large (7%–10%). It is natural that, in the case of standard tensile and compression test methods, this characteristic can be determined only for sufficiently plastic metals. At the same time, the condition $\varepsilon_t \approx const$ is automatically fulfilled in indentation of materials using a pyramidal indenter, e.g., a tetrahedral Vickers pyramid or trihedral Berkovich pyramid, and the degree of total strain under these indenters lies in the interval indicated above ($\varepsilon_t \approx 7.6\%$ for a tetrahedral Vickers indenter, and $\varepsilon_t \approx 9.8\%$ for a trihedral Berkovich indenter).

During indentation, the small volume of the deformed material and a specific character of strain fields decrease the susceptibility to macroscopic fracture. This enables one to determine the hardness and plasticity characteristic for most materials even at cryogenic temperatures.

In [18,21] it was shown that, for a pyramidal indenter, the plasticity characteristic can be determined in indentation in the form

$$\delta_H = 1 - \frac{HM}{E_S \cdot \varepsilon_t} \left(1 - \nu_S - 2\nu_S^2 \right). \tag{9}$$

In particular, for a Vickers indenter, taking into account that $HV = HM \sin \gamma_i$, $\gamma_i = 68°$, and $\varepsilon_t = 7.6\%$, we have

$$\delta_H = 1 - 14,3 \cdot \left(1 - \nu_S - 2\nu_S^2 \right) HV / E_S, \tag{10}$$

The introduction of the plasticity characteristic δ_H made it possible to classify practically all (plastic and brittle materials in standard mechanical tests) on the basis of their plasticity [18,21,22]. A dependence of δ_H on the temperature, strain rate, and structural factors has been established [18,21,30]. It was possible to introduce the notion of theoretical plasticity for perfect crystals in which theoretical strength is attained [30]. It was experimentally shown that there exists a critical value of the plasticity characteristic $\delta_{H\,cr} \cong 0.9$. At smaller values of δ_H, the plasticity in tensile tests is $\delta = 0$ or has a very low value. The plasticity characteristic δ_H is fairly extensively used in works of different authors (e.g., [31–33]).

The values of the plasticity characteristic δ_H for the materials studied in the present work are presented in Table 1, which enables us to compare them with the Tabor parameter C.

Consider the theoretical relation between C and δ_H. It follows from Equation (2) that the parameter C is completely determined by the relative size of the elastoplastic zone $x = b_S/c$. This is why we first calculate the relation between x and the plasticity characteristic δ_H.

3.2.2. Relation between the Relative Size of the Elastoplastic Zone $x = b_S/c$ and the Plasticity Characteristic δ_H

As noted in Section 2, for metals, the quantity $\theta_S Y_S$ can be neglected as compared to 1 in Equation (1b). Substituting Y_S from (1c) into (1b), we find the following equation for the determination of x for metals:

$$x^3 - \alpha_S = \frac{E_S z \left(\frac{2}{3} + 2\ln x\right)}{6(1 - \nu_S)HM}, \tag{11}$$

where $\alpha_S = \frac{2(1 - 2\nu_S)}{3(1 - \nu_S)}$.

Determining HM/E_S from (10) and substituting its value into (11), for the Vickers indenter we get the following explicit dependence of δ_H on the relative size of the elastoplastic deformation zone x:

$$\delta_H = 1 - \frac{2,21z\left(\frac{2}{3} + 2\ln x\right)}{x^3 - \alpha_S}\lambda_S, \tag{12}$$

where $\lambda_S = \frac{1 - \nu_S - 2\nu_S^2}{1 - \nu_S} = 1 - 2\frac{\nu_S^2}{1 - \nu_S}$.

It follows from Equation (12) and Figure 2 that δ_H is predominantly determined by the quantity x, but the parameters z and λ_S exert some influence on the relation between δ_H and x. For metals, the parameter z is practically equal to $z \approx \cot \gamma_i$ because the angle ψ for them differs very slightly from an angle $\gamma_i = 68°$ (see Table 1). Therefore, it can be assumed that $z \approx const$. However, the parameter λ_S varies somewhat for metals having different values of Poisson's ratio ν_S, which leads to an insignificant scatter of experimental results relative to the averaged curve in Figure 2.

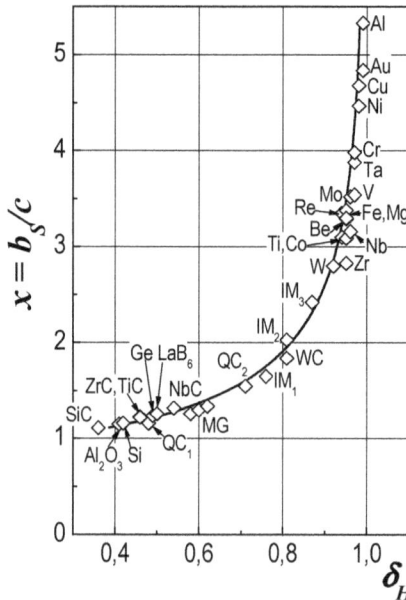

Figure 2. Relation between the plasticity characteristic δ_H and the relative size of the elastoplastic deformation zone x. Curve was constructed on the basis of Equation (12) for $z = 0.38$ and $\nu_S = 0.27$.

For metals the results of calculation of δ_H by (10) and (12) practically coincide.

Formula (12) was used for the calculation of the dependence $x(\delta_H)$ shown in Figure 2. In this case, the values of the parameters z and ν_S were varied. The smallest mean square error equal to 0.06%

was obtained for $z = 0.38$ and $\nu_S = 0.27$. Thus, it was shown that Equation (12) with the values of the parameters $z = 0.38$ and $\nu_S = 0.27$ can be used with an accuracy sufficient for practice not only for metals, but also for other materials studied in the work.

The experimental data and theoretical curve shown in Figure 2 indicate that the relative size of the elastoplastic deformation zone during indentation $x = b_S/c$ is mainly determined by the plasticity characteristic δ_H. The value of x increases monotonically with increasing δ_H. In this case, x changes from values close to 1 for ceramic materials to $x = 5.33$ for aluminum.

3.2.3. Yield Strength Y_S and Tabor Parameter HM/Y_S in the Considered Model

Figure 3 shows the relation between the Tabor parameter $C = HM/Y_S$ and plasticity characteristic δ_H. It is seen that the experimental dots for all studied materials lie on practically one curve.

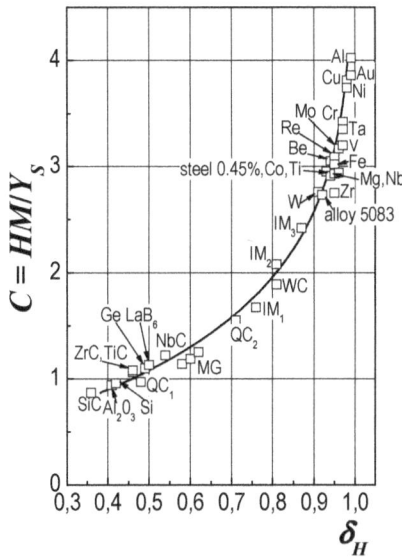

Figure 3. Relation between the Tabor parameter $C = HM/Y_S$ and the plasticity characteristic δ_H. Curve was constructed on the basis of Equation (13) for $z = 0.38$ and $\nu_S = 0.27$.

To calculate the theoretical dependence $C(\delta_H)$ for the studied materials shown in Figure 3, formulas (1c) and (12) were used. We obtained the next Equation:

$$\delta_H = 1 - \frac{2,21 z C \lambda_S}{\exp(1,5C - 1) - \alpha_S},$$ (13)

It is seen from Figure 3 that this equation satisfactorily describes the experimental results.

It should also be noted that, by analogy with $\delta_{H\,cr}$, the notion of the critical value of the Tabor parameter $C_{cr} = HM/Y_S \approx 2.6$ can be introduced. As is seen in Figure 3, this value corresponds to $\delta_{H\,cr} = 0.9$. Therefore, only for $C > 2.6$, the materials have a substantial macroscopic plasticity in tensile tests.

3.3. Physical Nature of Increase of the Tabor Parameter $C = HM/Y_S$ with Increase in the Plasticity δ_H

During indentation of low-plasticity materials, the elastoplastic deformation zone is small and its radius b_S exceeds slightly the radius of the penetrated indent c. In this case, $C \approx 1$ and $HM \approx Y_S$. However, as shown in the present work, with increase in the plasticity δ_H, the size of the elastoplastic deformation zone increases substantially, and, in most plastic materials, the value of b_S/c increases to more than 5. Therefore, during penetration of an indenter into plastic materials, deformation occurs not only under the indenter, but also in a hemisphere with a radius b_S, exceeding substantially the radius of the hardness indent c. In order for the plastic deformation to occur on a large hemisphere, the pressure $P = HM$ on the contact area of the indenter and specimen must exceed substantially the yield strength Y_S. The higher ductility of the material, the greater the size of elastic-plastic deformation zone and, hence, the pressure P and the Tabor parameter C should be higher. The mathematical relation between $C = HM/Y_S$ and the plasticity characteristic δ_H is described by Equation (13) and is shown in Figure 3.

3.4. Relaxed Effective Apex Angle of a Hardness Indent γ_{SR} and Apex Angle of an Indenter under Load ψ

It is seen from Table 1 and Figure 4 that the relaxed apex angle of the hardness indent γ_{SR} can be much larger than the corresponding angle of the indenter $\gamma_i = 68°$. As is seen in Figure 4, the value of γ_{SR} correlates with the plasticity characteristic δ_H and can be described by the linear equation $\gamma_{SR} = 80.64 - 12.55\,\delta_H$. The correlation between γ_{SR} and δ_H shows once again the fundamental character of the plasticity characteristics δ_H.

It is obvious from Table 1 that, for metals, the value of the apex angle of indenter under load ψ differs very slightly from the value of γ_i. However, for high-hardness materials ψ can exceed $70°$.

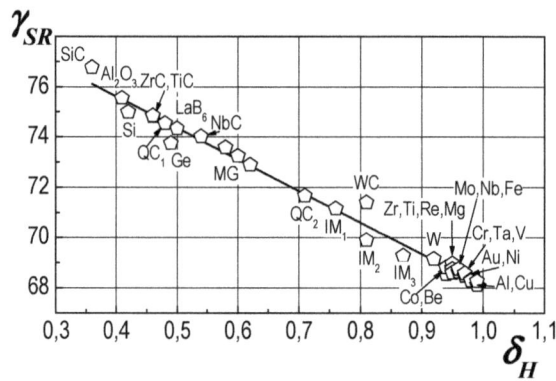

Figure 4. Dependence of the relaxed apex angle of a hardness indent γ_{SR} on the plasticity characteristic δ_H.

3.5. Simple Method of Determination of the Tabor Parameter $C = HM/Y_S$ and Yield Strength Y_S from the Hardness HM Determined with a Pyramidal Indenter

The results presented above enable us to propose a very simple method of determination of the Tabor parameter C and yield strength Y_S from the hardness HM determined with a Vickers indenter. In this method, the plasticity characteristic δ_H is calculated by the simple formula (12), the Tabor parameter C is determined from the curve shown in Figure 3 or calculated by Equation (13), and the yield strength is calculated by the formula $Y_S = HM/C$. The simplicity of the described technique makes it possible to use it extensively in indentation by the Vickers method. The authors think that the determination of the plasticity characteristic δ_H and yield strength Y_S raises significantly the informativeness and efficiency of the indentation technique. It should be noted that the simplified calculation of the Tabor parameter C and yield strength Y_S can also be carried out in the case of measuring the hardness HM by a trihedral Berkovich indenter. In this case, for the determination of the plasticity characteristic δ_H, it is necessary to use relation (9) at $\varepsilon_t \approx 9.8\%$.

3.6. Experimental Check of the Values of the Tabor Parameter $C = HM/Y_S$ and the Radius of Elastoplastic Zone b_S.

As is seen from Figure 3 and Table 1, the value of the Tabor parameter C changes quite strongly for different materials. Why did the parameter C range from 2.8 to 3.1 in the Tabor tests? This can be explained by the fact that Tabor tested structural metallic alloys. These alloys are usually hardened by alloying and heat treatment, but hardening is limited by the necessity to have good plasticity, which is measured as elongation to fracture δ, and usually $\delta \approx 10\%$–20% for these alloys. According to the data of the authors of the present paper, such values of δ corresponds to the plasticity characteristic $\delta_H = 0.93 - 0.95$. According to Figure 3, at this value of δ_H, the Tabor parameter C is actually equal to 2.8–3.1 for different materials.

In a number of earlier performed works (e.g., [3–6]), it was shown that for ceramic materials the Tabor parameter C approaches 1 as in the present work.

It follows from Figure 3 and Table 1 that materials with a plasticity characteristic lower than that for metals ($\delta_H < 0.9$: intermetallics, refractory compounds, quasicrystals, metallic glasses etc.) must also be characterized by a lower value of $C = HM/Y_S$. An experimental check of the values of C for these materials is complicated (or practically impossible) because of their insufficiently high plasticity in compression tests for the determination of Y_S at a total strain $\varepsilon_t \approx 7.6\%$. However, the values of C for these materials obtained in the present paper are fairly predictable because the values of the plasticity characteristic δ_H and the relative size of the elastoplastic deformation zone b_S/c for them are intermediate between those for metals and ceramics.

It seemed reasonable to check the high value $C \approx 4$ for pure aluminum, as a representative of the most plastic metals with a FCC lattice.

For this purpose, we prepared specimens of aluminum of 99.98% purity for uniaxial compression tests. The specimens had a diameter $d = 5$ mm and a height $h = 6$ mm. They were prepared from a commercial ingot and annealed in vacuum at a temperature of 400 °C for 1 h. The mean grain size was equal to 93 μm. The yield stress $\sigma = Y_S$ in compression to $\varepsilon_t \approx 7.6\%$ was equal to 41 MPa. As is seen from Table 1, the hardness is $HM = 173$ MPa. Therefore, $C_{exp} = HM/\sigma_{7.6\%} = 4.2$, which confirms the high value of the parameter C for aluminum, which even somewhat exceeds the value calculated using the developed model $C \approx 4.02$. In this case, for the studied aluminum, $\delta_H = 0.99$, which, according to Figure 3 and Equation (13), corresponds to $C \approx 4$–4.2.

The experimental check of the values of the Tabor parameter C by the uniaxial compression test method was also performed for 5083 aluminum alloy and carbon steel containing 0.45% C. These materials were tested in the as-delivered state. The obtained results are presented in Table 2. It is seen that the values of the yield strength Y_S and Tabor parameter C obtained by the indentation method (with calculation by Equations (1) and (2)) agree well with those obtained in mechanical tests. The values of C and δ_H for these materials are also shown in Figure 3 and coincide satisfactorily with the calculated curve $C = f(\delta_H)$.

Table 2. Results of compression mechanical tests (yield stress at tension ($\varepsilon_t = 7.6\%$) $Y_{7.6\%}$, the value of C_{exp} in tension test).

Material	$Y_{7.6\%}$, GPa	C_{exp}
Al	0.041	4.21
Al alloy #5083	0.373	2.76
Steel 0.45%C	0.64	2.95

For comparison of the actual size of the elastoplastic deformation zone with the calculated value of b_S, results of the work [34], in which dislocation rosettes around indentation were investigated for Mo (001) single crystal by etch pits method, were used. Additionally, in the present work, dislocation rosettes around indentation made at 300 °C were investigated. In Figure 5 the circles with radius b_S

are plotted on dislocation rosettes around the indentations. At the room temperature (Figure 5a) the anisotropy of the dislocation velocity in different crystallographic directions is observed, but at 300 °C such anisotropy is absent (Figure 5b). It is seen, that in both cases, the calculated values of b_S are in satisfactory agreement with the average values of the areas in which plastic deformation has occurred and dislocation density has increased.

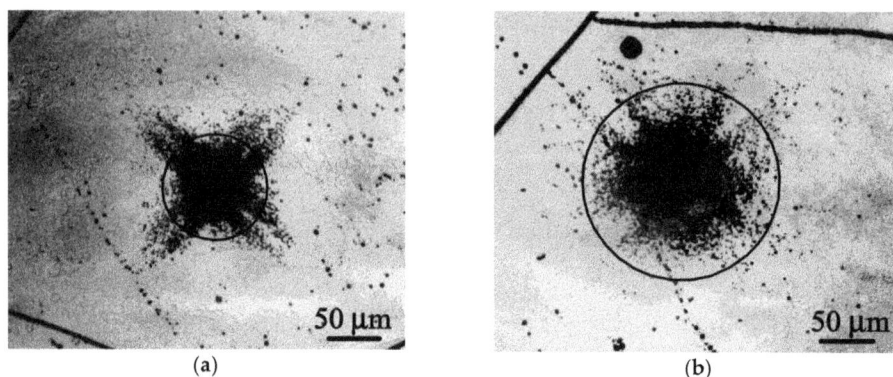

(a) (b)

Figure 5. Dislocations around indentation print for single crystal Mo (001), revealed by etch pits method. The circles with radius b_S are plotted on dislocation rosettes: (a) $t = 20$ °C, $HM = 1.998$ GPa, $b_S = 47.7$ μm [34]; (b) $t = 300$ °C, $HM = 1.026$ GPa, $b_S = 87.2$ μm, present work.

4. Conclusions

1. The developed inclusion core model of indentation by conical and pyramidal indenters makes it possible to carry out an analysis of the mechanical behavior of materials in indentation with the determination of the Tabor parameter $C = HM/Y_S$, yield strength Y_S, relative size of the elastoplastic deformation zone under an indenter b_S/c (see Figure 1), effective angle of a relaxed hardness indent γ_{SR}, and effective angle of an indenter under load ψ. In this case, for the first time, the elastic compressibility of the deformation core is taken into account. An analysis of the mechanical behavior in the indentation of materials with different types of interatomic bond and different crystalline structures has been carried out using the developed model.

2. It has been shown that the main quantities of the developed indentation model (the Tabor relation $C = HM/Y_S$ and relative size of the elastoplastic deformation zone b_S/c) correlate precisely with the determined in indentation plasticity characteristic δ_H = ***plastic strain/total strain***, which was introduced in [18]. The Tabor parameter C and the size of the elastoplastic deformation zone b_S/c increase monotonically with increasing plasticity characteristics δ_H. The Tabor parameter ranges from 1 for ceramic materials to 3.8–4.0 for the most plastic FCC metals. In structural metallic alloys, combining a high strength with an elongation at fracture $\delta = 10\%$–20% (which corresponds to $\delta_H = 0.93$–0.95), $C = 2.8$–3.1, which agrees with the results obtained by Tabor. The relative size of the elastoplastic deformation zone b_S/c changes from 1 for ceramic materials to 5.3 for aluminum. The calculated size of b_S is in the satisfactory agreement with the average values of the area in which plastic deformation under indenter is occurred and dislocation density is increased.

3. On the basis of the developed inclusion core model of indentation, analytical expressions relating C and b_S/c to the plasticity characteristic δ_H have been obtained. These expressions agree sufficiently well with the obtained experimental results and make it possible to calculate C and b_S/c from the value of the plasticity characteristic δ_H. To determine more exactly all parameters, it is necessary to solve the system (1) of three equations with three unknowns.

4. The physical nature of increase of the Tabor parameter $C = HM/Y_S$ with increasing plasticity is explained by the fact that with increase in the plasticity, the elastoplastic deformation zone b_S/c increases and b_S can substantially exceed the radius of the hardness indent c. This is why the pressure $P = HM$ on an area of radius c must provide plastic deformation not only under the indenter, but also in a hemisphere of radius b_S. Naturally, in this case, the pressure P must be substantially higher than the yield strength Y_S.

5. It has been shown that it is reasonable to introduce the notion of the critical value of the Tabor parameter $C_{cr} = 2.6$. Only at $C > 2.6$, materials have substantial macroscopic plasticity in tensile tests.

6. A very simple technique of determination of the Tabor parameter $C = HM/Y_S$ and yield strength Y_S from results of standard indentation has been proposed. In this technique, the plasticity characteristic δ_H is determined by the simple formula (10), and the Tabor parameter is determined from the calibration plot $C = f(\delta_H)$ shown in Figure 3. The yield strength Y_S is calculated by the formula $Y_S = HM/C$.

7. Thus, the inclusion core model of indentation developed in the present work and the earlier proposed technique of determination of the plasticity δ_H enable us to calculate both the yield strength and plasticity characteristic from the value of the hardness HM and elastic characteristics of the material. The authors think that the determination of the plasticity characteristic δ_H and yield strength Y_S make the indentation technique substantially more informative and efficient.

Acknowledgments: This work supported by the Program "Development of the theory and practice of determining the mechanical and trybological properties of a wide range of materials and coatings by local loading of indenter at the macro-, micro- and nano-scales" of the National Academy of Sciences of Ukraine.

Author Contributions: Boris A. Galanov—development of the improved inclusion core model of the indentation process, wrote Theoretical background, Scheme and equations of the improved model; Yuly V. Milman—relation between the Tabor parameter C = HM/Y_S and plasticity characteristic δ_H, wrote Results and Discussion, Conclusions; Svetlana I. Chugunova and Irina V. Goncharova—experimental part of the article, prepared figures and tables; Igor V. Voskoboinik—mathematical calculations.

Conflicts of Interest: The authors declare no conflict of interest.

References

1. Walley, S.M. Historical origins of indentation hardness testing. *Mater. Sci. Technol.* **2012**, *28*, 1028–1044. [CrossRef]
2. Walley, S.M. Addendum and correction to 'Historical origins of indentation hardness testing'. *Mater. Sci. Technol.* **2013**, *29*, 1148. [CrossRef]
3. Johnson, K.L. The correlation of indentation experiments. *J. Mech. Phys. Solids* **1970**, *18*, 115–126. [CrossRef]
4. Johnson, K.L. *Contact Mechanics*; Cambridge University Press: Cambridge, UK, 1985.
5. Tanaka, K. Elastic/plastic indentation hardness and indentation fracture toughness: The inclusion core model. *J. Mater. Sci.* **1987**, *22*, 1501–1508. [CrossRef]
6. Mata, M.; Anglada, M.; Alcala, J. A hardness equation for sharp indentation of elastic-power-low strain-hardening materials. *Philos. Mag. A* **2002**, *82*, 1831–1839. [CrossRef]
7. Kogut, L.; Etsion, I. Elastic-plastic contact analysis of a sphere and a rigid flat. *J. Appl. Mech.* **2002**, *69*, 657–662. [CrossRef]
8. Cheng, Y.-T.; Cheng, C.-M. Analysis of indentation loading curves obtained using conical indenters. *Philos. Mag. Lett.* **1998**, *77*, 39–47. [CrossRef]
9. Cheng, Y.-T.; Cheng, C.-M. What is indentation hardness? *Surf. Coat. Technol.* **2000**, *133–134*, 417–424. [CrossRef]
10. Galanov, B.; Milman, Yu.; Ivakhnenko, S.; Suprun, E.; Chugunova, S.; Golubenko, A.; Tkach, V.; Litvin, P.; Voskoboinik, I. Improved inclusion core model and its application for measuring the hardness of diamond. *J. Superhard Mater.* **2016**, *38*, 289–305. [CrossRef]
11. Tabor, D. *The Hardness of Metals*; Clarendon Press: Oxford, UK, 1951.
12. Chaudhri, M.M. Strain hardening around spherical indentations. *Phys. Status Solidi A* **2000**, *182*, 641–652. [CrossRef]

13. Brown, L.M. Indentation size effect and the Hall-Petch 'law'. *Mater. Sci. Forum* **2010**, *662*, 13–26. [CrossRef]
14. Milman, Y.; Chugunova, S.; Goncharova, I. Plasticity at absolute zero as a fundamental characteristic of dislocation properties. *Int. J. Mater. Sci. Appl.* **2014**, *3*, 353–362. [CrossRef]
15. Stelmashenko, N.A.; Walls, M.G.; Brown, L.M.; Milman, Y.V. Microindentation of W and Mo oriented single crystals: STM study. *Acta Metall. Mater.* **1993**, *41*, 2855–2865. [CrossRef]
16. Nix, W.P.; Gao, H. Indentation size effects in crystalline materials: A law for strain gradient plasticity. *J. Mech. Phys. Solids* **1998**, *46*, 411–425. [CrossRef]
17. McLaughlin, K.K.; Clegg, W.J. Deformation underneath low-load indentations in copper. *J. Phys. D Appl. Phys.* **2008**, *41*, 074007. [CrossRef]
18. Milman, Y.V.; Galanov, B.A.; Chugunova, S.I. Plasticity characteristic obtained through hardness measurement. *Acta Metall. Mater.* **1993**, *41*, 2523–2532. [CrossRef]
19. Gridneva, I.V.; Milman, Y.V.; Trefilov, V.I.; Chugunova, S.I. Analysis of dislocation mobility under concentrated loads at indentations of single crystals. *Phys. Status Solidi A* **1979**, *54*, 195–206. [CrossRef]
20. Trefilov, V.I.; Milman, Y.V.; Grigoriev, O.N. Deformation and rupture of crystals with covalent interatomic bonds. *Prog. Cryst. Growth. Charact. Mater.* **1988**, *16*, 225–277. [CrossRef]
21. Galanov, B.A.; Milman, Y.V.; Chugunova, S.I.; Goncharova, I.V. Investigation of mechanical properties of high-hardness materials by indentation. *J. Superhard Mater.* **1999**, *3*, 23.
22. Milman, Y.V.; Chugunova, S.I.; Goncharova, I.V. Plasticity characteristic obtained by indentation technique for crystalline and noncrystalline materials in the wide temperature range. *High Temp. Mater. Process.* **2006**, *25*, 39–46. [CrossRef]
23. Milman, Y.V.; Miracle, D.B.; Chugunova, S.I.; Voskoboinik, I.V.; Korzhova, N.P.; Legkaya, T.N.; Podrezov, Y.N. Mechanical behaviour of Al_3Ti intermetallic and $L1_2$ phases on its basis. *Intermetallics* **2001**, *9*, 839–845. [CrossRef]
24. Milman, Y.V. The effect of structural state and temperature on mechanical properties and deformation mechanisms of WC-Co hard alloy. *J. Superhard Mater.* **2014**, *36*, 65–81. [CrossRef]
25. Milman, Y.V.; Luyckx, S.; Goncharuck, A.V.; Northrop, J.T. Results from bending tests on submicron and micron WC-Co grades at elevated temperatures. *Int. J. Refract. Met. Hard Mater.* **2002**, *20*, 71–79. [CrossRef]
26. Gridneva, I.V.; Milman, Y.V.; Trefilov, V.I. Phase transition in diamond structure crystals at hardness measurement. *Phys. Status Solidi A* **1972**, *14*, 177–182. [CrossRef]
27. Kovalchenko, A.M.; Milman, Y.V. On the cracks self-healing mechanism at ductile mode cutting of silicon. *Tribol. Int.* **2014**, *80*, 166–171. [CrossRef]
28. Orlov, A.N.; Regel, V.R. Plasticity. In *Physical Encyclopedic Dictionary*; Soviet Encyclopaedia: Moscow, Russia, 1965; p. 39. (In Russian)
29. Plasticity (Physics), Wikipedia. Available online: https://en.wikipedia.org/wiki/Plasticity_%28physics%29 (accessed on 15 March 2017).
30. Milman, Y.V.; Chugunova, S.I.; Goncharova, I.V. Plasticity determined by indentation and theoretical plasticity of materials. *Bull. Russ. Acad. Sci. Phys.* **2009**, *73*, 1215–1221. [CrossRef]
31. Bozzini, B.; Boniardi, M.; Fanigliulo, A.; Bogani, F. Tribological properties of electroless Ni-P/diamond composite films. *Mater. Res. Bull.* **2001**, *36*, 1889–1902. [CrossRef]
32. Boldt, P.H.; Weatherly, G.C.; Embury, J.D. A transmission electron microscope study of hardness indentations in $MoSi_2$. *J. Mater. Res.* **2000**, *15*, 1025–1031. [CrossRef]
33. Qiang, J.B.; Zhang, W.; Xie, G.; Kimura, H.; Dong, C.; Inoue, A. An in situ bulk $Zr_{58}Al_9Ni_9Cu_{14}Nb_{10}$ quasicrystal-glass composite with superior room temperature mechanical properties. *Intermetallics* **2007**, *15*, 1197–1201. [CrossRef]
34. Lanin, A.G.; Lotsko, D.V.; Milman, Y.V.; Sibirtsev, S.A.; Fedorova, V.N.; Chugunova, S.I. The effect of temperature on the mobility of dislocations at the penetration of the indenter into a single crystal of molybdenum (001). *Phys. Met.* **1989**, *11*, 50–55. (In Russian)

![crystals logo] *crystals*

MDPI

Article

Atomistic Studies of Nanoindentation—A Review of Recent Advances

Carlos J. Ruestes [1,*], Iyad Alabd Alhafez [2] and Herbert M. Urbassek [2]

[1] Facultad de Ciencias Exactas y Naturales, Universidad Nacional de Cuyo and CONICET, Mendoza 5500, Argentina

[2] Physics Department and Research Center OPTIMAS, University Kaiserslautern, Erwin-Schrödinger-Straße, D-67663 Kaiserslautern, Germany; alhafez@rhrk.uni-kl.de (I.A.A.); urbassek@rhrk.uni-kl.de (H.M.U.)

* Correspondence: cruestes@fcen.uncu.edu.ar; Tel.: +54-261-423-6003

Academic Editors: Ronald Armstrong, Stephen M. Walley and Wayne L. Elban

Received: 12 September 2017; Accepted: 26 September 2017; Published: 29 September 2017

Abstract: This review covers areas where our understanding of the mechanisms underlying nanoindentation has been increased by atomistic studies of the nanoindentation process. While such studies have been performed now for more than 20 years, recent investigations have demonstrated that the peculiar features of nanoplasticity generated during indentation can be analyzed in considerable detail by this technique. Topics covered include: nucleation of dislocations in ideal crystals, effect of surface orientation, effect of crystallography (fcc, bcc, hcp), effect of surface and bulk damage on plasticity, nanocrystalline samples, and multiple (sequential) indentation. In addition we discuss related features, such as the influence of tip geometry on the indentation and the role of adhesive forces, and how pre-existing plasticity affects nanoindentation.

Keywords: nanoindentation; molecular dynamics; hardness

---

1. Introduction

Nanoindentation is a technique commonly used to provide information about the elastic modulus and hardness of materials [1,2]. This technique has provided insights into a broad range of material properties; as examples we mention the indentation cracking of brittle thin films on brittle substrates [3]; the fracture toughness, adhesion and mechanical properties of dielectric thin films [4]; the strain hardening and recovery in a bulk metallic glass [5]; the phase transformation of titanium dioxide thin films produced by filtered arc deposition [6]; superhard materials [7]; and even the investigation of biomaterials, such as the mechanical properties of human enamel [8]. Nanoindentation testing has become of wide-spread use when modern modern experimental testing methods were combined with the Oliver-Pharr [9] analysis. Further applications of this method are found in the investigation of the deformation mechanics of nanoparticles, micro- and nanopillars, microbeams, micro- and nanofibers, membranes, and nanofilms; this wide variety of structures are ubiquitous to the field of nanotechnology [10].

Nanoindentation is intimately related to the problem of contact between two bodies, where an indenter exerts force on a material. Contact mechanics involves all the spatial scales, from atomistic to continuum, and many temporal scales, ranging from the period of atomic vibrations to the duration of contact. It also comprises complex mechanisms such as many-body interactions, plasticity, heating and even phase transformations. Luan and Robbins [11] showed that a nanoscale contact is governed by atomistic phenomena and that it is frequent to find plastic deformation in the form of dislocation nucleation on the surface or the flattening of asperities [12]. The scale of these phenomena render their in situ experimental observation extremely difficult and that is the reason why computational tools help on the elucidation of the deformation mechanisms taking place.

The understanding of the deformation mechanisms during nanoindentation at the atomic scale has gained considerably from atomistic simulations [13]. Among the landmark contributions we mention the paper by Landman et al. [14] which helped to understand the jump-to-contact phenomenon during indentation. The contribution by Hoover et al. [15] showed that the predicted hardness was strongly influenced by the interatomic potential, temperature, and indenter speed used in the simulations. Harrison et al. [16] published simulation results of nanoindentation on the diamond (111) surface and found a fracture mode of stress relaxation under the indenter. Sinnott et al. [17] performed atomistic simulations of the nanometer-scale indentation of amorphous-carbon thin films providing qualitative insight into the mechanical deformation processes that take place during indentation, and quantitative predictions that compare well with experimental data.

Later Kelchner et al. [18] performed molecular dynamics (MD) simulations of spherical indentation in Au, and since then MD simulations have been extensively applied to study plasticity mechanisms during indentation processes. This paper reviews some of the most relevant contributions to nanoindentation that were made possible by atomistic simulations.

In a previous review [19] we gave an introduction to the methods used in atomistic simulation of nanoindentation, and in the analysis and interpretation of such simulations. A remarkable outcome of such simulations is the possibility to identify all features (peaks, load drops, etc.) of the load-depth curve with the underlying plastic changes in the material, i.e., with the generation of dislocations, their reaction, or the emission of dislocation loops. Thus in particular the nucleation of dislocations underneath the indent tip, the dissociation of prismatic loops from the network adherent to the indent pit, and the generation of pile-up surrounding the pit has been studied in detail. In the present review we do not want to repeat this analysis, but rather focus on recent advances obtained after the writing of our last review [19]. It thus exemplifies that the field of atomistic modeling of nanoindentation is both active and continues obtaining relevant results and insights.

The topics covered in this review include ideal crystals, effect of surface orientation and crystallography (Section 2), the effect of surface and bulk defects on plasticity (Section 3), multiple indentation (Section 4), the effect of the tip modeling (Section 5) and the role of adhesive forces and tip wetting (Section 6). Finally we conclude on current challenges in the field (Section 7).

2. Ideal Crystals, Effect of Surface Orientation and Crystallography

2.1. Fcc Metals

Due to the abundance of fcc metals and the technological applications of some of them, metals with this structure were the first to be studied. The homogeneous nucleation and structure of dislocations in fcc metals under indentation was first studied by Kelchner et al. [18] and later by Van Vliet et al. [20] and Lee et al. [21]. The heterogeneous nucleation of dislocations at surface steps was first studied by Zimmerman et al. [22].

Dislocation slip in fcc crystals occurs along the close-packed plane, that is a plane of type {111}. The primary glide system in this material class is the ⟨110⟩{111} system. Immediately after nucleation, shear loops are formed that tend to attach to the indenter surface. Depending on the generalized stacking fault energies (SFEs), dislocations in fcc metals dissociate and form partials that are accompanied by stacking fault planes. The reaction of dislocations can generate prismatic loops that transport material away from the surface into the substrate. Depending on the surface orientation, if glide vectors ⟨110⟩ are available that lie parallel to the surface—such as for a (111) surface—V-shaped loops are formed at the surface that are free to glide out of the high-stress indentation zone [23–26].

Li et al. [27] studied nanoindentation of Au and found good agreement of the yield strength of the single-indexed surfaces with experiment. In addition, they alloyed 5% of Zr, Cu and Ti to their Au crystal. They demonstrated that the difference of unstable and stable SFE—rather than the stable SFE itself—is a good indicator of the strength of alloys. This difference corresponds to the nucleation barrier for defects, and exhibits a strong correlation with the hardness of the alloys.

The size effect in nanoindentation was explored by Begau et al. [28]. With increasing indentation depth the dislocation density increased. Similar findings were obtained later in [29] for a single-crystalline (sc) Ni thin film, showing that indentation hardness decreases with indentation depth.

2.2. Bcc Metals

Prismatic loops are also formed in bcc metals, as was observed by Hagelaar et al. [30] during nanoindentation in tungsten. Upon indentation on a (111) surface loop generation was associated with shear stresses in their atomistic indenter. Considering tantalum as a model bcc material, Alcalá and co-workers [31] have shown that nanocontact plasticity occurs by the nucleation and propagation of twin and stacking fault bands driven by a combination of shear stresses and pressure. They suggested that dislocations appear after a thermally assisted twin annihilation and mentioned that this mechanism is common to other bcc metals.

Remington et al. [32] presented a comprehensive nanoindentation MD study of Ta single crystals along the three principal crystallographic orientations. They reported that after the formation and annihilation of planar defects similar to the ones reported by Alcalá et al. [31], shear loops form and propagate along $\langle 111 \rangle$ directions. Consistent with bcc slip systems, the shear loops grow by the advance of their edge components while the screw components undergo limited cross-slip. Eventually the screw segments may annihilate each other since they have dislocation lines with opposite signs, and a prismatic loop is pinched off [32,33].

Figure 1 presents a side view of the indentation of a (001) Ta single crystal by an 8 nm diameter spherical indenter to a penetration of 5 nm, and a detailed view of a prismatic loop produced in the process. The loop is formed of several dislocation segments in several slip planes pertaining to slip systems with the same slip direction. For methodological details, the reader is referred to reference [33].

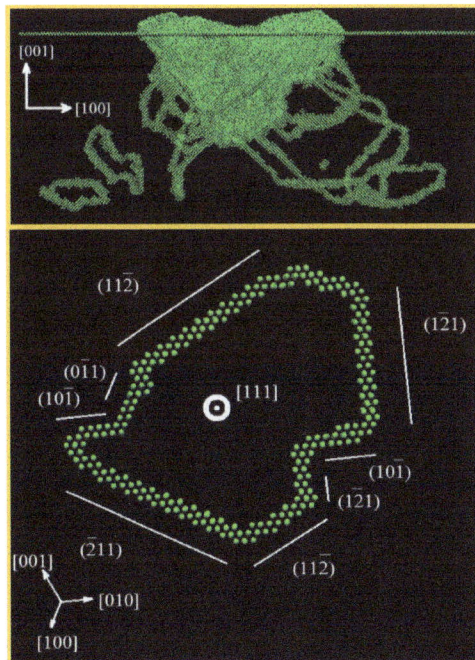

Figure 1. Prismatic loop formation in bcc Ta, see text. Original contribution of the authors.

Twinning may play a role in the plasticity of several bcc metals. Goel et al. [34] investigated the role of twinning during the nanoindentation of Ta, finding evidence for a significant twinning anisotropy.

2.3. Hcp Metals

While nanoindentation in fcc and bcc metals has been characterized fairly well using MD simulation, little work has been published on the indentation of hcp metals, and only recently. Most of these studies are concerned with Mg [35–37], while Lu et al. [38] indent into Zr. In a comparative study, Alhafez et al. [39] simulated Mg, Ti, Zr in three different orientations; they found that here the surface crystallography plays a comparatively larger role than in the fcc and bcc materials. The reason is that the anisotropy of the crystal (unequal *a* and *c* axes) makes slip in the pertinent directions differ more pronouncedly than in the cubic crystal classes. In other words, slip by basal and prismatic or pyramidal dislocations shows more variety than in the fcc and bcc crystals. Figure 2 exemplifies the dislocation network generated by indentation in an hcp metal, Ti. Dislocations are dominated by the partials $\boldsymbol{b} = \frac{1}{3}\langle\bar{1}100\rangle$, while the perfect $\boldsymbol{b} = \frac{1}{3}\langle\bar{2}110\rangle$ dislocations occur more rarely. Prismatic dislocation loops are emitted abundantly, leading to quite extended plastic zones. Figure 2 thus demonstrates the influence of surface orientation on the density of the network and the direction and intensity of loop emission.

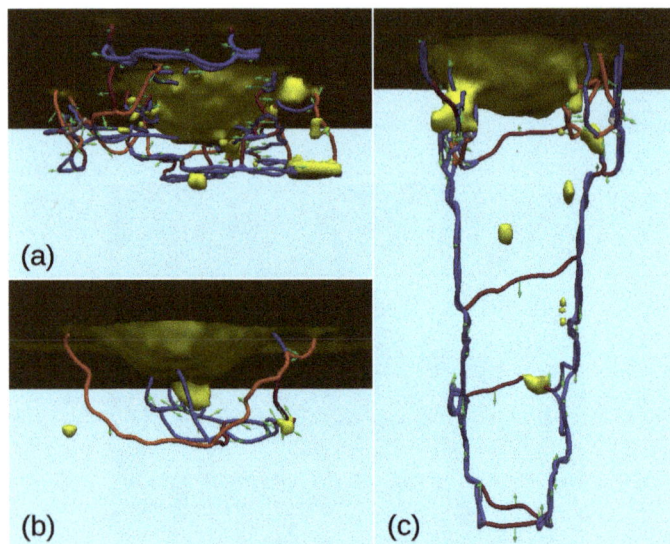

Figure 2. Dislocation networks generated for indentation into Ti. (**a**) Basal plane, (**b**) first prismatic plane, (**c**) second prismatic plane. Dislocations with Burgers vector $\frac{1}{3}\langle\bar{2}110\rangle$ are colored dark red, $\frac{1}{6}\langle\bar{2}203\rangle$ orange, $\frac{1}{2}\langle0001\rangle$ white, and $\frac{1}{3}\langle\bar{1}10\rangle$ blue. The deformed surface and other defects are colored yellow. Original contribution of the authors.

2.4. Si

Non-metals have been investigated with less systematics than metals. An exception is provided by the important material Si. Already in our previous report [19] we mentioned a number of studies on Si [40–42]. Research in this material is still very active as is evidenced by the large number of recent publications [43–48].

All studies indicate that phase transformation—in particular to the amorphous state—contributes strongly to plasticity. An important issue concerns the the question in how far dislocations take part in the plasticity. Here a comparative study of indentation into Si [46] showed that the Stillinger-Weber potential [49] produces considerably more dislocations than the Tersoff potential [50], confirming earlier studies. The exact nature of the phases created is under discussion; but it seems that—at least for the Stillinger-Weber potential—the beta-tin phase is not formed, but the bct5 phase [45]. These results were recently corroborated and extended [47]. Already previously Mylvaganam and Zhang [44] studied the effect of crystal orientation on the formation of bct-5 silicon, using the Tersoff potential.

The nature of the plastic yield in Si is still under discussion [51]. While simulations using the Stillinger-Weber potential [49] determine dislocation nucleation and amorphization are the key contributors to plasticity [52], simulations with the Tersoff potential [50] find the solid-solid transformation to the beta-tin phase as initiator of the plastic yield [41].

Abrams et al. [48] use the Tersoff potential [50] to understand the influence of crystalline and amorphous phase transitions in Si on the extrusion behavior on the surface. They find that formation of the crystalline Si-III phase can be identified by a pop-in in the force-depth curve which is absent under amorphization; both phase transformations lead to material extrusion to the surface.

Du et al. [43] report a temperature effect in nanoindentation between 10 and 300 K; with plastic indentation depth increasing and hardness decreasing when temperature increases. In addition they find an influence of the temperature on the crystalline solid-solid transformations occurring under the Tersoff potential.

A recent review of machining of Si was provided by Goel et al. [53], including information on indentation [42]; these are based on simulations using the so-called analytical bond order potential by Erhart and Albe [54]. They emphasize the high pressures obtained in the indentation zone, which may reach up to 10 GPa in their example and are responsible for the phase transformations. In comparison, the temperatures reached in the zone are appreciable only when the indentation velocity exceeds several ten m/s; this will not be relevant for indentation experiments.

It must be concluded that the mechanism of plastic yield in Si is dominated or at least strongly influenced by phase transformations.

2.5. Other Materials

Richter et al. [55] applied the molecular dynamics technique to study the nanoindentation of graphite and diamond to support their experimental studies, giving an atomistic description of the indentation process.

Szlufarska et al. [56] performed molecular dynamics simulation of indentation of nano-crystalline silicon carbide predicting a crossover from intergranular continuous deformation to intragrain discrete deformation at a critical indentation depth. Walsh et al. [57] relied on MD simulations to probe silicon nitride films reporting amorphization and cracking with a marked anisotropy.

Energetic materials can also be studied by molecular dynamics simulations, as shown by Chen et al. [58] who used MD simulations with reactive force fields to study nanoindentation of cyclotrimethylenetrintramine (RDX) by a diamond indenter. They report on significant heating of the substrate in the vicinity of the indenter, resulting in the release of molecular fragments and migration of these molecules on the indenter surfaces.

Recently also the indentation of Cu-Zr metallic glasses was attempted [59–61]. The hardness was found to increase with Cu content. Pop-in events in the load-depth curve and plastic yield were related to shear band formation.

Comparatively little work was devoted to indentation into composites. Feng et al. [62] investigated indentation into a nanocomposite formed of WC and Co layers. Special attention was paid on the action of the semi-coherent interface; it was found that it triggers dislocation generation in Co, enhancing the ductility. Indentation directly on a heterointerface, formed of Al and Si crystallites, was performed in [63]. Here enhanced dislocation mobility of the Si dislocations, mediated by the nearby interface, was reported.

Recently, also high-entropy alloys were studied [64]. For an FeCrCuAlN alloy a high hardness of 15.4 GPa is found which is claimed to be due to the low SFE and the dense atomic arrangement in the slip plane of this alloy. Further atomistic indentation studies were devoted to c-BN [65], γTi-Al alloy [66], and (001) oriented strontium titanate [67].

3. Effect of Surface and Bulk Defects on Plasticity

The response of grain boundaries (GBs) and their role in the mechanical response under indentation has attracted much attention [68,69]. Feichtinger et al. [70] performed atomistic simulations of nanoindentation on nanocrystalline (nc) Au with grain diameters of 5 and 12 nm and found GBs acting as dislocation sinks and also observed GB sliding. Ma and Yang [71] observed heterogeneous nucleation of dislocations at GBs in nc Cu. Hasnaoui et al. [72] found that for cases where the indenter size is smaller than the grain size in nc Au, GBs not only act as dislocations sinks, but that they can also reflect or emit dislocations, depending on their local structure and stress distribution. Jang and Farkas [73] studied the interaction of lattice dislocations with a grain boundary during nanoindentation of Ni and found dislocation transmission across GBs.

Liu et al. [74] explored the grain size effect in nc Ni with grain sizes ranging from 5 nm to 40 nm and found inverse Hall-Petch effect for the whole range, grain boundary absorption and that the area of the plastic zone generated is strongly dependent on the GB density. However, Huang et al. [75] only found inverse Hall-Petch effect for grain sizes below 7 nm in nc Cu. In addition they reported stress-induced grain growth as well as grain rotation as the cause for grain coarsening under indentation.

More recently Li et al. [76] studied the effect of grain size on the nanoindentation of Cu. They used both nc and nanotwinned (nt) Cu and compared to the indentation of an sc Cu specimen. They report a strong influence of dislocation interactions with GBs and with twin boundaries (TBs), respectively, which depend in size on the grain size and twin lamella thickness. In the nt Cu sample, in particular, plasticity is dominated by twinning/detwinning rather than by dislocation nucleation and motion.

Voyiadjis and Yaghoobi [77] explored the role of grain boundary on the source of size effects using bi-crystal Ni thin films and large scale MD simulations showing that the size effects mechanism influenced by GBs changes from dislocation nucleation and source exhaustion to the forest hardening mechanism as the grain size increases. Guleryuz and Mesarovic [78] studied low angle twist and asymmetric tilt boundaries in Cu and found nucleation of dislocations at GBs together with GB sliding. Talaei et al. [79] explored grain boundary effects on nanoindentation of Fe bicrystal.

Dupont and Sansoz [80] found significant softening of a nc-Al specimen under indentation which was caused by GB movement and grain rotation. In that simulation they used an indenter ($R = 15$ nm) that was larger than the average grain size (5 nm); the simulation was performed using a coupled atomistic-continuum approach.

The effect of GBs in bcc materials seems not to have been explored by MD simulation up to now, while simulations using gradient plasticity theory are available [81], where the size effects were studied.

Figure 3 shows the indentation of a (100) Fe single crystal with pre-existing defects (vacancies, divacancies, voids, and dislocations). Dislocations depicted in green correspond to a Burgers vector $\frac{1}{2}\{111\}$, while pink ones correspond to $\{100\}$ dislocations. The indentation point is just above a region where several dislocations meet. The strain gradient applied by the indenter promotes the movement of pre-existing dislocations in that zone and the nucleation of new dislocations that react with the former. Upon removal of the indenter, there is significant dislocation retraction and annihilation leading to a considerable modification of the dislocation forest in the region affected by the indenter, without apparent changes farther away.

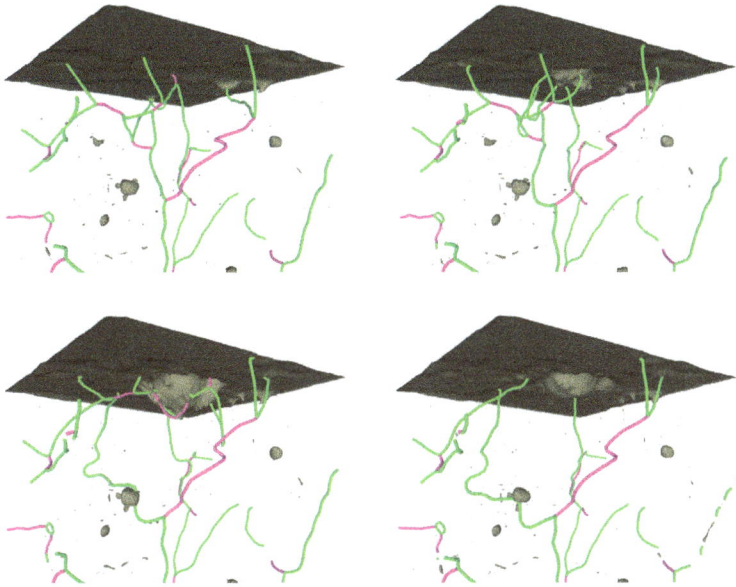

Figure 3. Sequence of indentation and release in a Fe sample with pre-existing defects, see text. Original contribution of the authors.

Esqué-de los Ojos et al. [82] studied the mechanical response under nanoscale spherical indentation employing MD simulations on single crystalline copper with an array of voids. Their simulations revealed that, for a given porosity fraction, the mechanical behavior of fcc metals with smaller pores differs more significantly from the behavior of the bulk, fully-dense counterpart. This effect is more pronounced for smaller voids than for bigger voids and is ascribed to the increase of the overall surface area as the pore size is reduced while the porosity fraction is kept constant, together with the reduced coordination number of the atoms located at the pores edges.

Ukwatta and Achuthan [83] studied the role of existing dislocations on the incipient plasticity under nanoindentation. To this end, they introduced edge dislocations into a Cu sample. They reported that the interaction between pre-existing and newly formed dislocations has a significant influence on the incipient plasticity, in particular by inducing cross-slip.

The influence of surface defects on the indentation has been investigated only rarely. Here surface roughness, surface steps, vacancy or adatom islands, or in general nanostructured surfaces may be of interest. The influence of adatom islands on the indentation process has been investigated for the example of Cu [84]; only central indent points were considered. It was found that the results are determined by the ratio of the indenter contact radius, a_c, to the radius of the adatom island, s. Small adatom islands, $s \ll a_c$, are pushed into the substrate and then transported away by prismatic loops. After the initial load drop accompanying this event, indentation proceeds as for a flat surface. If the island size matches that of the indenter, $s \cong a_c$, dislocations are generated below the island step edges and remain pinned there; dislocation activity remains localized under the island. In this size-matched case, the surface is weakest and yields first. Finally, if $s \gg a_c$, the influence of the adatom island vanishes.

4. Multiple (Sequential, Cyclic) Indentation

In 2002, Van Vliet and Suresh [85] pointed out the lack of direct and in situ studies of the evolution of damage at surfaces subjected to cyclic contact loading on the atomic level and performed simulations of cyclic indentation using the bubble-raft model [86], observing the homogeneous nucleation of dislocations beneath the indenter and showing that there is a contact-fatigue response under cyclic indentation that is different from monotonic response. Zarudi et al. [87] studied the microstructure evolution of monocrystalline Si during cyclic microindentations. Molecular dynamics simulation of repeated indentation started with the work of Komvopoulos and Yan [88], investigating the evolution of deformation and heating in an fcc model crystal with indentation cycles. Later, Cheong and Zhang [89] used MD to study the effect of repeated nano-indentations on the deformation in monocrystalline silicon. Some years later Shiari and Miller [90] performed cyclic indentation of aluminum single crystals at the nanoscale by means of a multiscale 2D approach, using an atomistic model calculated using the molecular dynamics method for the contact region and a continuum model for regions away from it.

Cordill et al. [91] performed coupled experiments and MD simulations to study the response of Ni under oscillatory dynamic nanoindentation, coining the Nano-Jackhammer effect, a combination of dislocation nucleation and strain rate sensitivity caused by indentation with a superimposed dynamic oscillation. Deng and Schuh [92] performed MD simulations of nanoindentation and cyclic loading in a Cu-Zr metallic glass, showing hardening effects and attributed this response to confined plasticity and stiffening in regions initially preferred for yielding, requiring higher applied loads for triggering secondary plasticity events. Imran et al. [93] used MD to explore the response of a Ni single crystal subjected to multiple loading-unloading nanoindentation cycles, observing that an increase in the number of loading/unloading cycles reduces the maximum load and hardness of the Ni substrate and attributed this effect to the decrease in recovery force due to defects and dislocations produced after each indentation cycle. Salehinia et al. [94] performed repeated indentation in Nb/NbC multilayers using molecular dynamics simulations showing that the damage produced by the first indentation has a significant effect on the strength and the ductility of Nb/NbC nanolaminates as measured by subsequent indentations.

Wang, Yan and Li [95] conducted a mesoscopic examination of cyclic hardening in metallic glass by combining finite-element-method simulations coupled with kinetic Monte Carlo, finding that the yield load of the metallic glass increases after cyclic indentation in the microplastic regime. More recently, Zhao et al. [61] performed an investigation on the hardening behavior of a Cu-Zr metallic glass under cyclic indentation loading via molecular dynamics simulation revealing that the cycling hardening has a dependence on the cyclic indentation amplitudes so that with higher cyclic indentation amplitudes, the hardening behavior is more pronounced.

5. Tip Geometry

On the nanoscale all tips are blunt (rounded). Available nanoindenter tips may reach nowadays radii as small as $R = 10$ nm [96,97]. Indeed most simulations have been performed for spherical tips, and occasionally for conospherical tips.

Still, it may be found useful to investigate the effect of tip shape on the indentation process. There are at least two reasons for this: (i) indentation into single crystals is governed by crystal plasticity, and the governing rules are similar in the nano- and microworld; (ii) available macroscopic laws may thus be tested in the nanoscale.

We exemplify the indentation with a Vickers indenter in Figure 4. Clearly, the imprint shape and surrounding pile-up reflect the shape of the Vickers indenter. On the other hand, the dense dislocation network developing below the surface is typical also of other indenter geometries with a large opening angle.

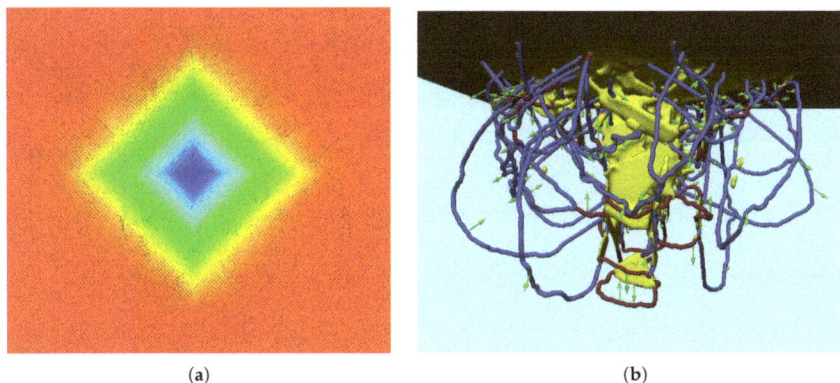

Figure 4. Indentation with a Vickers indenter into an Fe (100) surface. (**a**) Indent pit. (**b**) Dislocation network. Original contribution of the authors.

A more thorough study of the influence of the indenter shape was performed in [98] for the case of sc Fe. While this study focused on scratching, also the indentation process was included in the simulation. Systematic results could be obtained for indentation with a conical tip in dependence of the cone semi-apex angle β. The indentation hardness increased with β; this feature could be rationalized by the increasing complexity of the dislocation network beneath the indenter. For the case of the Fe (100) surface studied, the hardness increase measured 30%, if β changed from 30° to 70°.

The behavior of a Berkovich indenter agreed well with that of a cone with $\beta = 70°$; this angle agrees with the so-called equivalent cone angle of the Berkovich indenter [1,99]. The indentation behavior of a sphere shows, however, only poor agreement with the indentation of a cone with the corresponding equivalent cone angle, which in this case depends on the indentation depth. This missing agreement was attributed to the fact that cone and pyramid are self-similar structures, while the sphere is not.

6. Role of Adhesive Forces and Tip Wetting

Unlike in large-scale nanoindentation behavior, adhesion between indenter and substrate may play a significant role in nanocontact mechanics. Adhesion and tip wetting can be very pronounced at the nanoscale, with large surface area to volume ratio, clean surface and ultra high vacuum conditions. Molecular dynamics simulations of nanoindentation showed some of their potential in this area with the pioneering work of Landman et al. [14], showing metallic bonding and substrate-to-tip atom transfer (also known as the tip-wetting or jump-to-contact phenomenon) as a result of the need of optimization of the interaction energy. The high surface energies associated with clean metal surfaces can lead to strong attractive forces between surfaces close to contact, and these forces can become stronger in certain environmental conditions such as ultra-high vacuum. If the attraction is strong enough, surface atoms jump from the surface to the tip. Adhesion forces also play a role in retraction; as the tip retracts from the sample, a connective neck of atoms forms between the substrate and the tip. MD simulations also suggested that material transfer usually occurs during contact separation [30]. A similar phenomenon was found by Oliver et al. [100] when they performed one-to-one spatially matched experiment and atomistic simulations of nanometre-scale indentation; they reported that many features of the experiment were correctly reproduced by MD simulations, in some cases only when an atomically rough indenter rather than a smooth repulsive-potential indenter is used, tip wetting being one of these features. Paul et al. [101] highlighted the need to further explore the role of adhesive forces and tip wetting, ranging from the mechanisms of substrate-to-tip material transfer to electronic transport properties.

Tavazza et al. [102–104] used density-functional theory to study the details of the interaction of a diamond tip with a Ni surface. They found that the chemical interaction between the two materials leads to the formation of new ordered phases—comparable to a nickel carbide—at the contact area. This influences the substrate surface, but has also consequences for the wear of the tip, since substrate material is transferred to it. They argue that for the detailed investigation of such chemical changes quantum mechanical methods are necessary. In [102] this study is extended to oxidized and hydrogenated Ni surfaces; while O at the surface leads to similar results as a bare Ni surface, the presence of H reduces material transfer to the tip.

Figure 5 exemplifies the effect of adhesive forces between a diamond indenter (radius 10 nm) and an Fe (100) surface. A Morse potential with a well depth of 95 meV was assumed to act between C and Fe atoms [105]. We see that the indenter extracts some substrate material after retraction from the surface; this is identified by the red-colored atoms decorating the retracted tip. The pit size is smaller for the case of the adhesive interaction where the atoms move with the indenter upwards during unload.

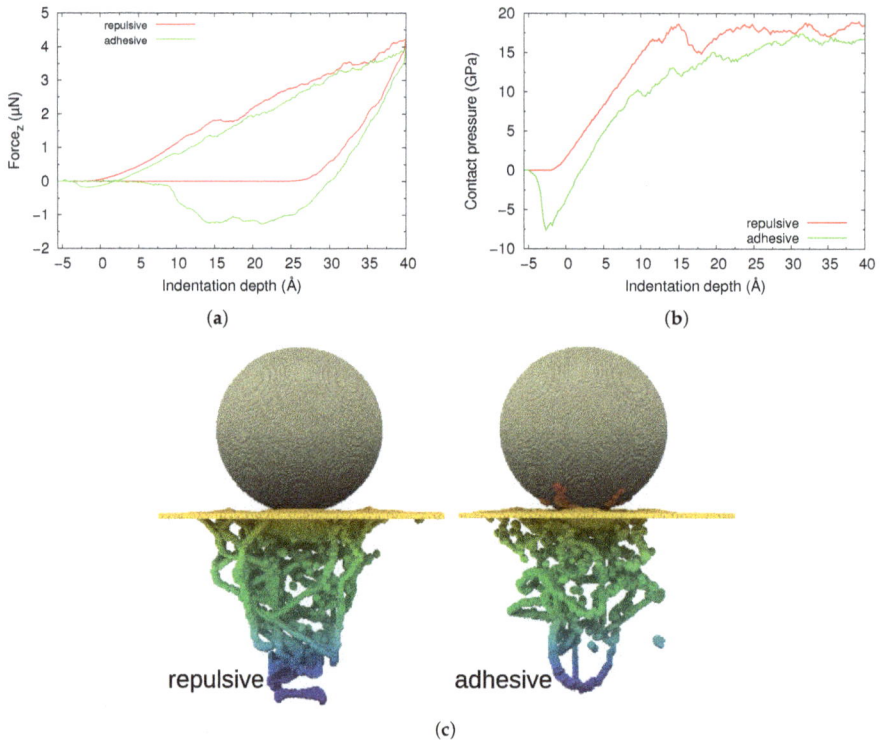

Figure 5. Comparison of indentation with a repulsive and an attractive diamond indenter into a Fe (100) surface. (**a**) Load-depth curve. (**b**) Contact pressure. (**c**) Pile-up after retraction of the indenter. Original contribution of the authors.

The load-depth curve shows a clear minimum when the indenter approaches the surface due to the mutual attraction, and forces are lower than in the purely repulsive scenario during the entire indentation process. However, adhesive effects are even larger upon pit retraction. The final hardness—that is the contact pressure after full indentation—is, however, nearly the same in the repulsive and in the attractive case.

7. Conclusions

This report on recent results in the field of MD simulation of nanoindentation demonstrates the high level of activity in this field. Seemingly a simple process—a tip is pushed in a material leading to plastic deformation—still many details are unclear, and MD simulation seems a promising technique to further the understanding of this process. Of particular interest of the nanoindentation technique is its capability of creating localized plasticity.

Our review identified the fields in which further simulations efforts are required to advance our understanding of localized plasticity even further. While the effect of GBs and TBs on dislocation activity has already been studied to some extent [106], the response of nc metals and more generally defective materials—containing preexisting vacancies, dislocations, steps, ledges, etc.—needs further clarification. Also effects of surface roughness, nanostructured surfaces or of hard inclusions appear not to have been modeled up to now and requires clarification.

Acknowledgments: Iyad Alabd Alhafez and Herbert M. Urbassek acknowledge support by the Deutsche Forschungsgemeinschaft via the Sonderforschungsbereich 926. Carlos J. Ruestes acknowledges support by ANPCyT PICT-2015-0342, SECTyP UNCuyo and high performance computing resources at Mendieta CCAD-UNC through PDC-SNCAD MinCyT initiative.

Author Contributions: The authors contributed equally to this work.

Conflicts of Interest: The authors declare no conflict of interest.

Abbreviations

The following abbreviations are used in this manuscript:

bcc	body centered cubic
fcc	face centered cubic
hcp	hexagonal close packed
nc	nanocrystalline
nt	nanotwinned
sc	single-crystalline
GB	grain boundary
MD	molecular dynamics
SFE	stacking fault energy
TB	twin boundary

References

1. Fischer-Cripps, A.C. *Nanoindentation*, 2nd ed.; Mechanical Engineering Series; Springer: New York, NY, USA, 2004.
2. Armstrong, R.W.; Elban, W.L.; Walley, S.M. Elastic, Plastic, Cracking Aspects of the Hardness of Materials. *Int. J. Mod. Phys. B* **2013**, *27*, 1330004.
3. Weppelmann, E.; Wittling, M.; Swain, M.V.; Munz, D. Indentation Cracking of Brittle Thin Films on Brittle Substrates. In *Fracture Mechanics of Ceramics*; Bradt, R.C., Hasselman, D.P.H., Munz, D., Sakai, M., Shevchenko, V.Y., Eds.; Springer: Boston, MA, USA, 1996; pp. 475–486.
4. Volinsky, A.A.; Vella, J.B.; Gerberich, W.W. Fracture toughness, adhesion and mechanical properties of low-K dielectric thin films measured by nanoindentation. *Thin Solid Films* **2003**, *429*, 201–210.
5. Yang, B.; Riester, L.; Nieh, T. Strain hardening and recovery in a bulk metallic glass under nanoindentation. *Scr. Mater.* **2006**, *54*, 1277–1280.
6. Bendavid, A.; Martin, P.; Takikawa, H. Deposition and modification of titanium dioxide thin films by filtered arc deposition. *Thin Solid Films* **2000**, *360*, 241–249.
7. Veprek, S. The search for novel, superhard materials. *J. Vac. Sci. Technol. A Vac. Surf. Films* **1999**, *17*, 2401–2420.
8. Cuy, J.L.; Mann, A.B.; Livi, K.J.; Teaford, M.F.; Weihs, T.P. Nanoindentation mapping of the mechanical properties of human molar tooth enamel. *Arch. Oral Biol.* **2002**, *47*, 281–291.

9.   Oliver, W.C.; Pharr, G.M. An improved technique for determining hardness and elastic modulus using load and displacement sensing indentation experiments. *J. Mater. Res.* **1992**, *7*, 1564–1583.

10.   Palacio, M.L.B.; Bhushan, B. Depth-sensing indentation of nanomaterials and nanostructures. *Mater. Charact.* **2013**, *78*, 1–20.

11.   Luan, B.; Robbins, M.O. The breakdown of continuum models for mechanical contacts. *Nature* **2005**, *435*, 929.

12.   Luan, B.; Robbins, M.O. Hybrid atomistic/continuum study of contact and friction between rough solids. *Tribol. Lett.* **2009**, *36*, 1–16.

13.   Sinnott, S.B.; Heo, S.J.; Brenner, D.W.; Harrison, J.A.; Irving, D.L. Computer Simulations of Nanometer-Scale Indentation and Friction. In *Springer Handbook of Nanotechnology*; Springer: Berlin/Heidelberg, Germany, 2010; pp. 955–1011.

14.   Landman, U.; Luedtke, W.D.; Burnham, N.A.; Colton, R.J. Atomistic mechanisms and dynamics of adhesion, nanoindentation, and fracture. *Science* **1990**, *248*, 454–461.

15.   Hoover, W.G.; De Groot, A.J.; Hoover, C.G.; Stowers, I.F.; Kawai, T.; Holian, B.L.; Boku, T.; Ihara, S.; Belak, J. Large-scale elastic-plastic indentation simulations via nonequilibrium molecular dynamics. *Phys. Rev. A* **1990**, *42*, 5844.

16.   Harrison, J.A.; White, C.T.; Colton, R.J.; Brenner, D.W. Nanoscale investigation of indentation, adhesion and fracture of diamond (111) surfaces. *Surf. Sci.* **1992**, *271*, 57–67.

17.   Sinnott, S.B.; Colton, R.J.; White, C.T.; Shenderova, O.A.; Brenner, D.W.; Harrison, J.A. Atomistic simulations of the nanometer-scale indentation of amorphous-carbon thin films. *J. Vac. Sci. Technol. A Vac. Surf. Films* **1997**, *15*, 936–940.

18.   Kelchner, C.; Plimpton, S.; Hamilton, J. Dislocation nucleation and defect structure during surface indentation. *Phys. Rev. B* **1998**, *58*, 11085–11088.

19.   Ruestes, C.J.; Bringa, E.M.; Gao, Y.; Urbassek, H.M. Molecular dynamics modeling of nanoindentation. In *Applied Nanoindentation in Advanced Materials*; Tiwari, A., Natarajan, S., Eds.; Wiley: Chichester, UK, 2017; Chapter 14, pp. 313–345.

20.   Van Vliet, K.; Li, J.; Zhu, T.; Yip, S.; Suresh, S. Quantifying the early stages of plasticity through nanoscale experiments and simulations. *Phys. Rev. B* **2003**, *67*, 104105.

21.   Lee, Y.; Park, J.Y.; Kim, S.Y.; Jun, S.; Im, S. Atomistic simulations of incipient plasticity under Al (111) nanoindentation. *Mech. Mater.* **2005**, *37*, 1035–1048.

22.   Zimmerman, J.A.; Kelchner, C.L.; Klein, P.A.; Hamilton, J.C.; Foiles, S.M. Surface Step Effects on Nanoindentation. *Phys. Rev. Lett.* **2001**, *87*, 165507.

23.   Ziegenhain, G.; Urbassek, H.M. Effect of material stiffness on hardness: A computational study based on model potentials. *Philos. Mag.* **2009**, *89*, 2225–2238.

24.   Ziegenhain, G.; Hartmaier, A.; Urbassek, H.M. Pair vs. many-body potentials: Influence on elastic and plastic behavior in nanoindentation of fcc metals. *J. Mech. Phys. Solids* **2009**, *57*, 1514–1526.

25.   Ziegenhain, G.; Urbassek, H.M.; Hartmaier, A. Influence of crystal anisotropy on elastic deformation and onset of plasticity in nanoindentation: A simulational study. *J. Appl. Phys.* **2010**, *107*, 061807.

26.   Gao, Y.; Ruestes, C.J.; Tramontina, D.R.; Urbassek, H.M. Comparative simulation study of the structure of the plastic zone produced by nanoindentation. *J. Mech. Phys. Solids* **2015**, *75*, 58–75.

27.   Li, Y.; Goyal, A.; Chernatynskiy, A.; Jayashankar, J.S.; Kautzky, M.C.; Sinnott, S.B.; Phillpot, S.R. Nanoindentation of gold and gold alloys by molecular dynamics simulation. *Mater. Sci. Eng. A* **2016**, *651*, 346–357.

28.   Begau, C.; Hua, J.; Hartmaier, A. A novel approach to study dislocation density tensors and lattice rotation patterns in atomistic simulations. *J. Mech. Phys. Solids* **2012**, *60*, 711–722.

29.   Yaghoobi, M.; Voyiadjis, G.Z. Atomistic simulation of size effects in single-crystalline metals of confined volumes during nanoindentation. *Comput. Mater. Sci.* **2016**, *111*, 64–73.

30.   Hagelaar, J.H.A.; Bitzek, E.; Flipse, C.F.J.; Gumbsch, P. Atomistic simulations of the formation and destruction of nanoindentation contacts in tungsten. *Phys. Rev. B* **2006**, *73*, 045425.

31.   Alcalá, J.; Dalmau, R.; Franke, O.; Biener, M.; Biener, J.; Hodge, A. Planar Defect Nucleation and Annihilation Mechanisms in Nanocontact Plasticity of Metal Surfaces. *Phys. Rev. Lett.* **2012**, *109*, 075502.

32.   Remington, T.P.; Ruestes, C.J.; Bringa, E.M.; Remington, B.A.; Lu, C.H.; Kad, B.; Meyers, M.A. Plastic deformation in nanoindentation of tantalum: A new mechanism for prismatic loop formation. *Acta Mater.* **2014**, *78*, 378–393.

33. Ruestes, C.J.; Stukowski, A.; Tang, Y.; Tramontina, D.R.; Erhart, P.; Remington, B.A.; Urbassek, H.M.; Meyers, M.A.; Bringa, E.M. Atomistic simulation of tantalum nanoindentation: Effects of indenter diameter, penetration velocity, and interatomic potentials on defect mechanisms and evolution. *Mater. Sci. Eng. A* **2014**, *613*, 390–403.
34. Goel, S.; Beake, B.; Chan, C.W.; Faisal, N.H.; Dunne, N. Twinning anisotropy of tantalum during nanoindentation. *Mater. Sci. Eng. A* **2015**, *627*, 249–261.
35. Zambaldi, C.; Zehnder, C.; Raabe, D. Orientation dependent deformation by slip and twinning in magnesium during single crystal indentation. *Acta Mater.* **2015**, *91*, 267–288.
36. Sánchez-Martín, R.; Zambaldi, C.; Pérez-Prado, M.T.; Molina-Aldareguia, J.M. High temperature deformation mechanisms in pure magnesium studied by nanoindentation. *Scr. Mater.* **2015**, *104*, 9–12.
37. Somekawa, H.; Tsuru, T.; Singh, A.; Miura, S.; Schuh, C.A. Effect of crystal orientation on incipient plasticity during nanoindentation of magnesium. *Acta Mater.* **2017**, *139*, 21–29.
38. Lu, Z.; Chernatynskiy, A.; Noordhoek, M.J.; Sinnott, S.B.; Phillpot, S.R. Nanoindentation of Zr by Molecular Dynamics Simulation. *J. Nucl. Mater.* **2015**, *467*, 742–757.
39. Alabd Alhafez, I.; Ruestes, C.J.; Gao, Y.; Urbassek, H.M. Nanoindentation of hcp metals: A comparative simulation study of the evolution of dislocation networks. *Nanotechnology* **2016**, *27*, 045706.
40. Kim, D.E.; Oh, S.I. Atomistic simulation of structural phase transformations in monocrystalline silicon induced by nanoindentation. *Nanotechnology* **2006**, *17*, 2259.
41. Mylvaganam, K.; Zhang, L.C.; Eyben, P.; Mody, J.; Vandervorst, W. Evolution of metastable phases in silicon during nanoindentation: mechanism analysis and experimental verification. *Nanotechnology* **2009**, *20*, 305705.
42. Goel, S.; Faisal, N.H.; Luo, X.; Yan, J.; Agrawal, A. Nanoindentation of polysilicon and single crystal silicon: Molecular dynamics simulation and experimental validation. *J. Phys. D* **2014**, *47*, 275304.
43. Du, X.; Zhao, H.; Zhang, L.; Yang, Y.; Xu, H.; Fu, H.; Li, L. Molecular dynamics investigations of mechanical behaviours in monocrystalline silicon due to nanoindentation at cryogenic temperatures and room temperature. *Sci. Rep.* **2015**, *5*, 16275.
44. Mylvaganam, K.; Zhang, L. Effect of crystal orientation on the formation of bct-5 silicon. *Appl. Phys. A* **2015**, *120*, 1391–1398.
45. Zhang, Z.; Stukowski, A.; Urbassek, H.M. Interplay of dislocation-based plasticity and phase transformation during Si nanoindentation. *Comput. Mater. Sci.* **2016**, *119*, 82–89.
46. Zhang, Z.; Urbassek, H.M. Comparative study of interatomic interaction potentials for describing indentation into Si using molecular dynamics simulation. *Appl. Mech. Mater.* **2017**, *869*, 3–8.
47. Zhang, J.; Zhang, J.; Wang, Z.; Hartmaier, A.; Yan, Y.; Sun, T. Interaction between phase transformations and dislocations at incipient plasticity of monocrystalline silicon under nanoindentation. *Comput. Mater. Sci.* **2017**, *131*, 55–61.
48. Abram, R.; Chrobak, D.; Nowak, R. Origin of a Nanoindentation Pop-in Event in Silicon Crystal. *Phys. Rev. Lett.* **2017**, *118*, 095502.
49. Stillinger, F.H.; Weber, T.A. Computer simulation of local order in condensed phases of Si. *Phys. Rev. B* **1985**, *31*, 5262–5271.
50. Tersoff, J. New emprirical approach for the structure and energy of covalent systems. *Phys. Rev. B* **1988**, *37*, 6991.
51. Chrobak, D.; Kim, K.H.; Kurzydlowski, K.J.; Nowak, R. Nanoindentation experiments with different loading rate distinguish the mechanism of incipient plasticity. *Appl. Phys. Lett.* **2013**, *103*, 072101.
52. Chrobak, D.; Tymiak, N.; Beaber, A.; Ugurlu, O.; Gerberich, W.W.; Nowak, R. Deconfinement leads to changes in the nanoscale plasticity of silicon. *Nat. Nanotechnol.* **2011**, *6*, 480–484.
53. Goel, S.; Luo, X.; Agrawal, A.; Reuben, R.L. Diamond machining of silicon: A review of advances in molecular dynamics simulation. *Int. J. Mach. Tools Manuf.* **2015**, *88*, 131–164.
54. Erhart, P.; Albe, K. Analytical potential for atomistic simulations of silicon, carbon, and silicon carbide. *Phys. Rev. B* **2005**, *71*, 035211.
55. Richter, A.; Ries, R.; Smith, R.; Henkel, M.; Wolf, B. Nanoindentation of diamond, graphite and fullerene films. *Diam. Relat. Mater.* **2000**, *9*, 170–184.
56. Szlufarska, I.; Nakano, A.; Vashishta, P. A crossover in the mechanical response of nanocrystalline ceramics. *Science* **2005**, *309*, 911–914.

57. Walsh, P.; Kalia, R.K.; Nakano, A.; Vashishta, P.; Saini, S. Amorphization and anisotropic fracture dynamics during nanoindentation of silicon nitride: A multimillion atom molecular dynamics study. *Appl. Phys. Lett.* **2000**, *77*, 4332–4334.

58. Chen, Y.C.; Nomura, K.I.; Kalia, R.K.; Nakano, A.; Vashishta, P. Molecular dynamics nanoindentation simulation of an energetic material. *Appl. Phys. Lett.* **2008**, *93*, 171908.

59. Imran, M.; Hussain, F.; Rashid, M.; Cai, Y.; Ahmad, S.A. Mechanical behavior of Cu-Zr bulk metallic glasses (BMGs): A molecular dynamics approach. *Chin. Phys. B* **2013**, *22*, 096101.

60. Qiu, C.; Zhu, P.; Fang, F.; Yuan, D.; Shen, X. Study of nanoindentation behavior of amorphous alloy using molecular dynamics. *Appl. Surf. Sci.* **2014**, *305*, 101–110.

61. Zhao, D.; Zhao, H.; Zhu, B.; Wang, S. Investigation on hardening behavior of metallic glass under cyclic indentation loading via molecular dynamics simulation. *Appl. Surf. Sci.* **2017**, *416*, 14–23.

62. Feng, Q.; Song, X.; Xie, H.; Wang, H.; Liu, X.; Yin, F. Deformation and plastic coordination in WC-Co composite—Molecular dynamics simulation of nanoindentation. *Mater. Des.* **2017**, *120*, 193–203.

63. Zhang, Z.; Urbassek, H.M. Indentation into an Al/Si composite: Enhancèd dislocation mobility at interface. *J. Mater. Sci.* **2017**, 1–15.

64. Li, J.; Fang, Q.; Liu, B.; Liu, Y.; Liu, Y. Atomic-scale analysis of nanoindentation behavior of high-entropy alloy. *J. Micromech. Mol. Phys.* **2016**, *1*, 1650001.

65. Zhao, Y.; Peng, X.; Fu, T.; Huang, C.; Feng, C.; Yin, D.; Wang, Z. Molecular dynamics simulation of nano-indentation of (111) cubic boron nitride with optimized Tersoff potential. *Appl. Surf. Sci.* **2016**, *382*, 309–315.

66. Xu, S.; Wan, Q.; Sha, Z.; Liu, Z. Molecular dynamics simulations of nano-indentation and wear of the gamma-Ti-Al alloy. *Comput. Mater. Sci.* **2015**, *110*, 247–253.

67. Javaid, F.; Stukowski, A.; Durst, K. 3D Dislocation structure evolution in strontium titanate: Spherical indentation experiments and MD simulations. *J. Am. Ceram. Soc.* **2017**, *100*, 1134–1145.

68. Van Swygenhoven, H.; Derlet, P.M.; Hasnaoui, A. Atomic mechanism for dislocation emission from nanosized grain boundaries. *Phys. Rev. B* **2002**, *66*, 024101.

69. Van Vliet, K.J.; Tsikata, S.; Suresh, S. Model experiments for direct visualization of grain boundary deformation in nanocrystalline metals. *Appl. Phys. Lett.* **2003**, *83*, 1441–1443.

70. Feichtinger, D.; Derlet, P.M.; Van Swygenhoven, H. Atomistic simulations of spherical indentations in nanocrystalline gold. *Phys. Rev. B* **2003**, *67*, 024113.

71. Ma, X.L.; Yang, W. Molecular dynamics simulation on burst and arrest of stacking faults in nanocrystalline Cu under nanoindentation. *Nanotechnology* **2003**, *14*, 1208.

72. Hasnaoui, A.; Derlet, P.M.; Van Swygenhoven, H. Interaction between dislocations and grain boundaries under an indenter–a molecular dynamics simulation. *Acta Mater.* **2004**, *52*, 2251–2258.

73. Jang, H.; Farkas, D. Interaction of lattice dislocations with a grain boundary during nanoindentation simulation. *Mater. Lett.* **2007**, *61*, 868–871.

74. Liu, X.; Yuan, F.; Wei, Y. Grain size effect on the hardness of nanocrystal measured by the nanosize indenter. *Appl. Surf. Sci.* **2013**, *279*, 159–166.

75. Huang, C.C.; Chiang, T.C.; Fang, T.H. Grain size effect on indentation of nanocrystalline copper. *Appl. Surf. Sci.* **2015**, *353*, 494–498.

76. Li, J.; Guo, J.; Luo, H.; Fang, Q.; Wu, H.; Zhang, L.; Liu, Y. Study of nanoindentation mechanical response of nanocrystalline structures using molecular dynamics simulations. *Appl. Surf. Sci.* **2016**, *364*, 190–200.

77. Voyiadjis, G.Z.; Yaghoobi, M. Role of grain boundary on the sources of size effects. *Comput. Mater. Sci.* **2016**, *117*, 315–329.

78. Guleryuz, E.; Mesarovic, S.D. Dislocation nucleation on grain boundaries: Low angle twist and asymmetric tilt boundaries. *Crystals* **2016**, *6*, 77.

79. Talaei, M.S.; Nouri, N.; Ziaei-Rad, S. Grain boundary effects on nanoindentation of Fe bicrystal using molecular dynamic. *Mech. Mater.* **2016**, *102*, 97–107.

80. Dupont, V.; Sansoz, F. Grain Boundary Structure Evolution in Nanocrystalline Al by Nanoindentation Simulations. *MRS Online Proc. Libr. Arch.* **2005**, *903*, 0903–Z06–05.

81. Faghihi, D.; Voyiadjis, G.Z. Determination of nanoindentation size effects and variable material intrinsic length scale for body-centered cubic metals. *Mech. Mater.* **2012**, *44*, 189–211.

82. Esqué-de los Ojos, D.; Pellicer, E.; Sort, J. The Influence of Pore Size on the Indentation Behavior of Metallic Nanoporous Materials: A Molecular Dynamics Study. *Materials* **2016**, *9*, 355.

83. Ukwatta, A.; Achuthan, A. A molecular dynamics (MD) simulation study to investigate the role of existing dislocations on the incipient plasticity under nanoindentation. *Comput. Mater. Sci.* **2014**, *91*, 329–338.

84. Ziegenhain, G.; Urbassek, H.M. Nanostructured surfaces yield earlier: Molecular dynamics study of nanoindentation into adatom islands. *Phys. Rev. B* **2010**, *81*, 155456.

85. Van Vliet, K.J.; Suresh, S. Simulations of cyclic normal indentation of crystal surfaces using the bubble-raft model. *Philos. Mag. A* **2002**, *82*, 1993–2001.

86. Gouldstone, A.; Van Vliet, K.J.; Suresh, S. Nanoindentation: Simulation of defect nucleation in a crystal. *Nature* **2001**, *411*, 656.

87. Zarudi, I.; Zhang, L.C.; Swain, M.V. Microstructure evolution in monocrystalline silicon in cyclic microindentations. *J. Mater. Res.* **2003**, *18*, 758–761.

88. Komvopoulos, K.; Yan, W. Molecular dynamics simulation of single and repeated indentation. *J. Appl. Phys.* **1997**, *82*, 4823–4830.

89. Cheong, W.C.D.; Zhang, L. Effect of repeated nano-indentations on the deformation in monocrystalline silicon. *J. Mater. Sci. Lett.* **2000**, *19*, 439–442.

90. Shiari, B.; Miller, R.E. Multiscale modeling of ductile crystals at the nanoscale subjected to cyclic indentation. *Acta Mater.* **2008**, *56*, 2799–2809.

91. Cordill, M.J.; Lund, M.S.; Parker, J.; Leighton, C.; Nair, A.K.; Farkas, D.; Moody, N.R.; Gerberich, W.W. The Nano-Jackhammer effect in probing near-surface mechanical properties. *Int. J. Plast.* **2009**, *25*, 2045–2058.

92. Deng, C.; Schuh, C.A. Atomistic mechanisms of cyclic hardening in metallic glass. *Appl. Phys. Lett.* **2012**, *100*, 251909.

93. Imran, M.; Hussain, F.; Rashid, M.; Ahmad, S.A. Dynamic characteristics of nanoindentation in Ni: A molecular dynamics simulation study. *Chin. Phys. B* **2012**, *21*, 116201.

94. Salehinia, I.; Wang, J.; Bahr, D.F.; Zbib, H.M. Molecular dynamics simulations of plastic deformation in Nb/NbC multilayers. *Int. J. Plast.* **2014**, *59*, 119–132.

95. Wang, N.; Yan, F.; Li, L. Mesoscopic examination of cyclic hardening in metallic glass. *J. Non-Cryst. Solids* **2015**, *428*, 146–150.

96. Göring, G.; Dietrich, P.I.; Blaicher, M.; Sharma, S.; Korvink, J.G.; Schimmel, T.; Koos, C.; Hölscher, H. Tailored probes for atomic force microscopy fabricated by two-photon polymerization. *Appl. Phys. Lett.* **2016**, *109*, 063101.

97. Commercial Diamond Tips from SCD Probe (D300 Series). Available online: www.scdprobes.com/D300.pdf (accessed on 5 September 2017).

98. Alabd Alhafez, I.; Brodyanski, A.; Kopnarski, M.; Urbassek, H.M. Influence of Tip Geometry on Nanoscratching. *Tribol. Lett.* **2017**, *65*, 26.

99. Fischer-Cripps, A.C. Critical review of analysis and interpretation of nanoindentation test data. *Surf. Coat. Technol.* **2006**, *200*, 4153–4165.

100. Oliver, D.J.; Paul, W.; El Ouali, M.; Hagedorn, T.; Miyahara, Y.; Qi, Y.; Grütter, P.H. One-to-one spatially matched experiment and atomistic simulations of nanometre-scale indentation. *Nanotechnology* **2013**, *25*, 025701.

101. Paul, W.; Oliver, D.; Grütter, P. Indentation-formed nanocontacts: An atomic-scale perspective. *Phys. Chem. Chem. Phys.* **2014**, *16*, 8201–8222.

102. Tavazza, F.; Senftle, T.P.; Zou, C.; Becker, C.A.; van Duin, A.C.T. Molecular Dynamics Investigation of the Effects of Tip-Substrate Interactions during Nanoindentation. *J. Phys. Chem. C* **2015**, *119*, 13580–13589.

103. Tavazza, F.; Levine, L.E. DFT Investigation of Early Stages of Nanoindentation in Ni. *J. Phys. Chem. C* **2016**, *120*, 13249–13255.

104. Tavazza, F.; Kuhr, B.; Farkas, D.; Levine, L.E. Ni Nanoindentation at the Nanoscale: Atomic Rearrangements at the Ni-C Interface. *J. Phys. Chem. C* **2017**, *121*, 2643–2651.

105. Gao, Y.F.; Yang, Y.; Sun, D.Y. Wetting of Liquid Iron in Carbon Nanotubes and on Graphene Sheets: A Molecular Dynamics Study. *Chem. Phys. Lett.* **2011**, *28*, 036102.

106. Spearot, D.E.; Sangid, M.D. Insights on slip transmission at grain boundaries from atomistic simulations. *Curr. Opin. Solid State Mater. Sci.* **2014**, *18*, 188–195.

Review

A Novel Approach to Modelling Nanoindentation Instabilities

Garani Ananthakrishna * and Srikanth Krishnamoorthy

Materials Research Centre, Indian Institute of Science, Bangalore 560012, India; srikanthk@iisc.ac.in
* Correspondence: garani@iisc.ac.in; Tel.: +91-80-2293-2780

Received: 17 January 2018; Accepted: 27 April 2018; Published: 3 May 2018

Abstract: We review the recently developed models for load fluctuations in the displacement controlled mode and displacement jumps in the load controlled mode of indentation. To do this, we devise a method for calculating plastic contribution to load drops and displacement jumps by setting-up a system of coupled nonlinear time evolution equations for the mobile and forest dislocation densities by including relevant dislocation mechanisms. These equations are then coupled to the equation defining constant displacement rate or load rate. The model for the displacement controlled mode using a spherical indenter predicts all the generic features of nanoindentation such as the elastic branch followed by several force drops of decreasing magnitudes and residual indentation depth after unloading. The stress corresponding to the elastic force maximum is close to the yield stress of an ideal solid. The predicted numbers for all the quantities match experiments on single crystals of Au using a spherical indenter. We extend the approach to model the load controlled nanoindentation experiments that employ a Berkovich indenter. We first identify the dislocation mechanisms contributing to different regions of the $F - z$ curve as a first step for obtaining a good fit to a given experimental $F - z$ curve. This is done by studying the influence of the parameters associated with various dislocation mechanisms on the model $F - z$ curves. The study also demonstrates that the model predicts all the generic features of nanoindentation such as the existence of an initial elastic branch followed by several displacement jumps of decreasing magnitudes and residual plasticity after unloading for a range of model parameter values. Furthermore, an optimized set of parameter values can be easily determined that give a good fit to the experimental load–displacement curves for Al single crystals of (110) and (133) orientations. Our model also predicts the indentation size effect in a region where the displacement jumps disappear. The good agreement of the results of the models with experiments supports our view that the present approach can be used as an alternate method to simulations. The approach also provides insights into several open questions.

Keywords: plastic deformation; dislocation mechanisms; nonlinear dynamical approach; intermittent plastic flow; indentation hardness

---

1. Introduction

The fact that mechanical properties of small volume systems are different from the bulk has been evident in a number of early studies. For instance, the high strength of whiskers is due to the absence of dislocations [1,2]. Similarly, the grain boundary strengthening mechanism is due to the fact that dislocation motion is limited by the grain size, and the dynamic friction (wear) is controlled by the plastic deformation of micrometer or sub-micrometer asperities [3]. There has been a spurt of activity in size dependent studies on plastic deformation of small volume systems in the last three decades due to technological importance as well as the scientific challenges it offers. For instance, intermittent flow is observed when the diameter of micrometer rods are below a certain value while it is smooth when it is large implying the instability manifests when the aspect ratio is reduced [4–6]. Similar intermittent plastic flow in the form of load fluctuations or displacement jumps is reported when the indentation depth is less than 100 nm both in thin and bulk samples [7–12]. Thus, nanoindentation experiments fall into the class of experiments where plastic instability manifests due to small deformed volume [7–12].

Traditionally, two modes of nanoindentation experiments are employed, namely, load/force controlled (LC) mode or displacement controlled (DC) mode [7,9,11,13–15]. In both modes, experiments measure the force response of the sample (to the applied force) as a function of indentation depth. While several load drops of decreasing magnitudes are seen in the DC mode experiments [7,9,11,13–15], several displacement jumps of decreasing magnitudes are seen beyond the elastic limit in the LC mode experiments [10,12,14,16,17]. The maximum load on the elastic branch is close to the theoretical yield stress. In both modes, residual plasticity is seen after unloading the indenter. While the presence of surface defects, oxide coatings, precipitates, etc. complicate the interpretation of the results, careful and controlled experiments on well-prepared single crystals have demonstrated that the above generic features are reproducible [11,14,15,18]. Since the deformed volume prior to the onset of intermittent plastic flow is a few nm^3, the standard explanation for the high yield strength is the low probability of finding dislocations in such small deformed volumes. The sequence of load drops in the DC mode and displacement jumps in the LC mode beyond the elastic branch have been considered as signatures of instabilities triggered by bursts of plasticity.

Three distinct types of studies have provided much insight into the nanoindentation process, namely, bubble raft indentation[10], colloidal crystals, which are soft matter equivalent of the crystalline phase [19], and in situ transmission electron microscope studies. The latter, in particular, has been useful in visualizing the dislocation nucleation mechanism in real materials [20]. Finally, considerable theoretical understanding has come from various types of simulation studies such as molecular dynamics (MD) simulations [21–26], dislocation dynamics simulations and multiscale modeling simulations (using MD together with dislocation dynamics simulations) [26]. While there are a number of MD simulations for the DC mode, *there are no simulations of any kind that predict displacement jumps reported in the LC mode experiments.*

However, a serious limitation of these simulations is the limited size of the simulated volumes and short time scales used. More importantly, the indentation rates are often several orders of magnitude higher than those in experiments, raising questions of relevance of these simulations to real materials [23,27]. Furthermore, simulation approaches cannot employ experimental parameters such as the indentation rates or the geometrical parameters defining the indenter, thickness of the sample, etc. As a result, the predicted values of load, indentation depth, etc. differ considerably from those reported by experiments.

The above limitations of simulation methods prompted us to develop *an alternate framework that has the ability to adopt laboratory time and length scales* to predict numbers that agree with experiments. The purpose of this paper is to review the recently introduced dynamical approach and the models that predict the results for the DC and LC mode nanoindentation experiments [28,29]. The approach is based on our previous experience in modeling a significantly more complex spatio-temporal instability, namely, the Portevin–Le Chatelier (PLC) effect seen in bulk samples of dilute alloys under constant strain rate conditions [30–34]. The approach is applicable to ideal single crystals without any kind of defects.

Our approach to nanoindentation is based on two observations. First, the sequence of load drops (in the DC mode) and the sequence of displacement jumps (in the LC mode) seen beyond the elastic branch are triggered by collective dislocation activity. Second, the intermittent deformation is a signature of an instability. The most suitable mathematical platform for describing instabilities and collective behavior of constituent entities is nonlinear dynamics. Furthermore, since the basic defects contributing to plastic flow are dislocations, a proper description of collective dislocation behavior demands that we include all the relevant dislocation mechanisms contributing to the bursts of plasticity. On the basis of these observations, we recently developed a novel approach [28,29] by combining the power of nonlinear dynamics with a method for calculating the plastic contribution to the indentation depth. The latter is calculated by developing time evolution equations for the mobile and forest dislocation densities based on the knowledge of the dislocation mechanisms contributing to plastic flow [28,29]. In addition, the approach allows us to employ experimental indentation rates, the geometrical parameters defining the indenter, thickness of the film, etc. The efficacy of the approach is illustrated by developing a model for the DC mode nanoindentation experiments (carried out using a spherical indenter) [28]. The model predicts all the generic features of the DC mode with numbers matching experimental results on single crystals of gold [11,15]. We have also developed a model for predicting the displacement jumps in the LC mode indentation even though the exercise is even more challenging due to the conceptual difficulty in enforcing a constant load rate during displacement jumps. Indeed, this difficulty appears to be the reason for a complete absence of any kind of simulations to explain the displacement jumps in the LC mode. Since we plan to get a good fit to the experimental $F - z$ plots [12], we first study the influence of the parameters associated with dislocation mechanisms on the model $F - z$ curves with a view to identify the dislocation mechanisms that control different regions of the $F - z$ curve. The study further demonstrates that the model predicts all the generic features of nanoindentation such as the existence of an initial elastic branch followed by several displacement jumps of decreasing magnitudes and residual plasticity after unloading *for a range of values of model parameters associated with the dislocation mechanisms.* This also helps us to obtain optimized parameter values that give a good fit to the experimental load–displacement curves for Al single crystals of (110) and (133) orientations [12]. While hardness is not well defined at nanometer scales where displacement jumps dominate, for larger indentation depths where intermittency disappears, the hardness decreases with depth. *Thus, indentation size effect is also predicted.* These results also justify our expectation that the present approach can be used as an alternate method to simulations in modeling nanoindentation instabilities. These studies also provide insights into several open questions related to the collective behavior of dislocations: (a) Which dislocation mechanisms determine the first and subsequent load drops or displacement jumps? (b) Can all load drops and displacement jumps be explained on the basis of a common physical and mathematical mechanism? (c) What physical and mathematical mechanisms contribute to the decreasing magnitudes of the load drops or displacement jumps?

2. Insights from Simulations on Nanoindentation

At the outset, it must be mentioned that, apart from bubble raft simulation [10], only MD simulations predict load drops in DC mode of nanoindentation. There are no MD simulations (or any other for that matter) for the displacement jumps in the LC mode. In general, simulation methods have always served the purpose of understanding dislocation mechanisms underlying the evolution of complex dislocation microstructures in various dislocation mediated plasticity. The purpose of this section is to briefly summarize the results of MD simulations and multiscale modeling method, and discuss their advantages and limitations.

In the context of simulating the DC mode nanoindentation instability, MD simulations have played an important role in providing a good understanding of initial stages of plastic deformation. However, because of the limited volume that can be simulated, subsequent evolution of the plastic zone becomes prohibitive as indentation proceeds. This has motivated researchers to develop a multi-scale modeling method that uses MD results as a starting point for understanding further evolution of dislocation microstructure using discrete dislocation dynamical (DDD) simulations in conjunction with finite element methods (FEM).

The suitability of these simulations depends on the stage of evolution of dislocation microstructure that is being targeted. Here, we note that the load drops manifest at a few tens of nm scale corresponding to the initial stages of dislocation microstructure. Furthermore, the initial indented volume prior to the onset of plasticity is of the order of 10 nm^3. On the other hand, typical dislocation density in a well grown single crystals is $\sim 10^{10}/\text{m}^2$. Thus, the probability of finding a dislocation is close to zero. The nucleated dislocation loops act as the trigger for the pop-in events and therefore represent elastic to plastic transition. Then, MD simulations are best suited for this initial stages of deformation. The typical size of the simulated volume is a few million atoms. However, they are clearly not suitable for simulations when describing further development of plastic zone (beyond say 50 nm). A multi-scale method has been developed by Chang et al. [26] that incorporates MD results into discrete dislocation dynamic(DDD) simulations in conjunction with finite element methods to take into account the boundary conditions relevant for the loading condition.

There are a number of MD simulations that capture the evolution of dislocation microstructure to various degrees [21,23–26]. Features that emerge from these simulations include the nucleation of dislocation loops, their expansion and interaction, detachment from the source, and their propagation [26].

Almost all studies use a spherical indenter, which is modeled using a repulsive potential. Different kinds of potentials have been used (to simulate the desired crystal) in MD simulations. Typical large MD simulations use a million atoms and the radius of the indenter can range from about 0.8 nm to 15 nm [21,23–26]. Typical rates of indentation are ~0.1 Å/step with each step typically a few ps duration required to attain minimum energy configuration. Typical indentation depths are about 10 Å and the force is ~0.1 μN to a few μN.

For illustration, we use the results of Van Vliet et al. [23]. These authors use Ercolessi–Adamssi potential to simulate Aluminum. The load–displacement curve with several load drops is shown in Figure 1a. Due to the inherent advantage of MD simulations, dislocation activity (see Figure 9 of Ref. [23]) can be correlated with the load drops in the load–displacement curve. As can be seen, nucleation occurs at 20 nN, which corresponds to a nucleation stress of $\sigma_m \sim 3.4$ GPa. The glide loops expand along {111} planes. In addition to nucleation, other dislocation mechanisms the authors realize are the formation of sessile locks, cross-slip, and propagation after detachment from the source. The maximum depth of indentation is 20 Å. One can also notice small scale fluctuations in the load–displacement curve, an effect well known in MD simulations, even in the context of MD simulations of equilibrium properties. Such fluctuations tend to smooth out as the system size is increased.

In the following, we briefly summarize the results of the multi-scale modeling approach due to Change et al. [26]. To begin with, we briefly recall their MD results. They use an Embedded-atom method type of interatomic potential typical to several other MD simulations. The authors use a repulsive sphere that is pressed into the sample at a rate of 0.1 Å at each step of a few ps duration. The top surface is traction-free and the atoms located on the bottom surface are fixed. Periodic boundary conditions are applied in the direction perpendicular to the indentation axis. Figure 1b shows the load–displacement curve with several pop-in events. The first load drop has been identified with the nucleation of three prismatic interstitial dislocation loops. The nucleation occurs at 1 nm and subsequent evolution of dislocation microstructure with indentation depth consists of expansion of the loops, their multiplication followed by detachment from the surface (see Figure 2a–f of Ref. [26]).

The DDD simulations employed by the authors use an Edge-screw code with an FEM solver. The FEM is used to compute the highly heterogeneous stresses during the indentation process. The information about nucleation of dislocation loops seen in MD simulation are incorporated by a set of rules that specify the shape of loops and position of the loops to be introduced in the simulation cell. The imposed displacement rate determines the number of loops introduced. The load–displacement curve using a spherical indenter of radius $R = 150$ nm is shown in Figure 2. The figure shows the total load, load contribution from dislocations and the MD result extrapolated to the maximum depth of 100 nm. The evolution of dislocation microstructure becomes increasingly complex as illustrated in Figure 3a–d. For details, see Chang et al. [26].

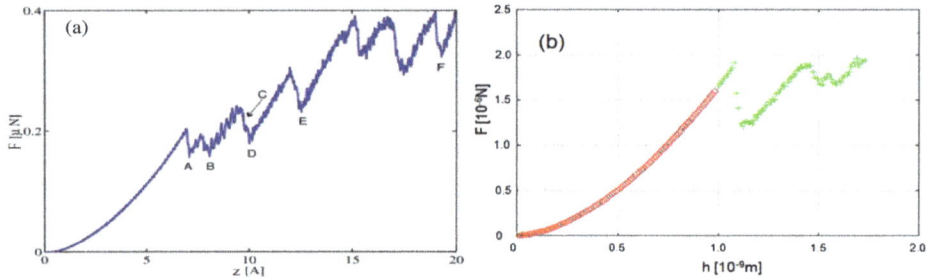

Figure 1. (**a**) Load–displacement curve obtained using MD simulations. Reproduced with permission from Van Vliet et al. [23]; (**b**) load–displacement curve obtained using MD simulations. Reproduced with permission from H-J. Chang et al. [26].

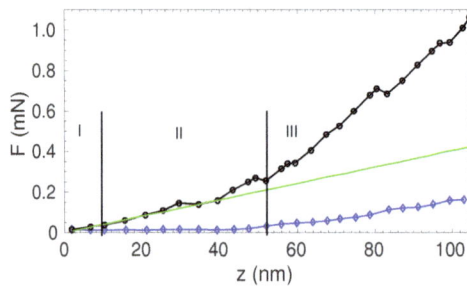

Figure 2. Load–displacement curve from DDD simulations using a spherical indenter of radius $R = 150$ nm. Region I is elastic, region II obtained from the DDD simulations is seen to follow the MD results and region III corresponds to the evolution beyond the MD region. Reproduced with permission from Chang et al. [26].

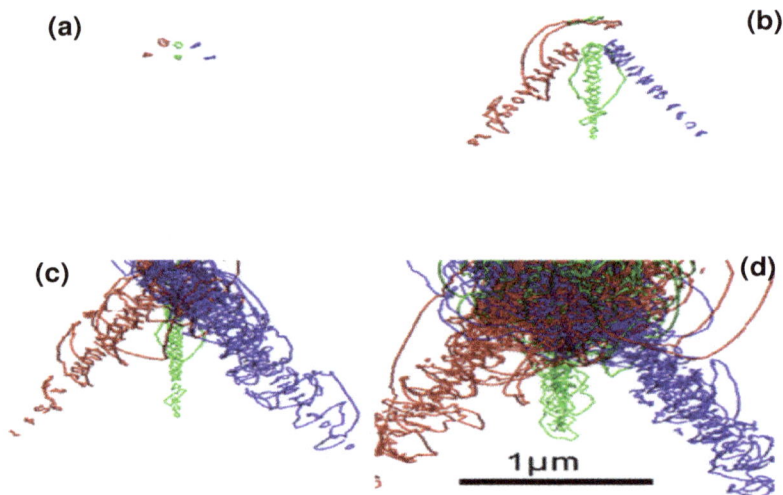

Figure 3. Dislocation microstructure evolution obtained from DDD simulations of Chang et al. (**a**) Nucleation of prismatic dislocation loops at early stages of indentation process. (**b**) Propagation of nucleated dislocation loops followed by initial stages of multiplication. (**c**), (**d**) Further multiplication of dislocations and formation of junctions. Reproduced with permission from Chang et al. [26].

As can be seen, a major advantage of the MD simulations is their ability to include a range of dislocation mechanisms starting from the nucleation of dislocations, their multiplication, formation of locks, junctions, propagation of loops, etc. [21,23–26]. *Note that these dislocation mechanisms are all used in our model.* The most obvious limitation of MD simulations is the limited volume they can simulate. This also means that the maximum radius of a spherical indenter is also limited. This also means that the predicted depths at which nucleation occurs is typically 6–10 Å significantly smaller than 65 Å in experiments [11,15]. One other important limitation is the length and time scales that are inherent to the small scale simulations. The relaxation time scale to attain the minimum energy configuration is typically a few ps. This coupled with the fact that the loading rates are a fraction of an Å means that the imposed displacement $\sim$m/s. This is at least 10^8 orders higher than that used in experiments by Kiely et al., who use 0.5 Å/s [11,15]. On the other hand, while DDD simulations are useful in tracking further development of dislocation microstructure, these simulations do not model the intermittent region. These limitations prompted us to devise an alternate framework that can use experimental conditions such as the imposed rates, geometrical parameters defining the indenter and other experimental parameters.

3. Background Material

This review specifically targets two most popular quasi-static modes of indentation, namely, the displacement controlled indentation where imposed displacement is increased at a constant rate and load controlled mode where the load is increased at a constant rate. These two modes of indentation are primarily chosen because the equations governing the DC and LC modes take a simple form compared to the dynamic loading mode, not considered here. The simplicity of these equations helps us establish a good correlation between imposed rates and plastic response of the material. In addition, for each of these modes of indentation, we can use different types of indenters. Here, we consider (a) spherical indenter for the DC mode with a view to compare the model results with those of Kiely et al. [11,15]), and (b) Berkovich indenter in the LC mode so as to compare our model results with Gouldstone et al. [12]. Here, we consider a single crystal without surface defects, inclusions, precipitates, etc. (Other complications in nanoindentation such as pile-up and sink-in effects are beyond the scope of our approach.). We begin by collecting relevant background material.

In our approach, the indentation depth z measured from the undeformed surface $z = 0$ is a dynamical variable, i.e., a variable that evolves with time. The total indentation depth z is the sum of elastic indentation depth z_e and plastic depth z_p. Then,

$$z = z_e + z_p. \tag{1}$$

The elastic depth is essentially known since it is governed by the indenter specific load–depth expression. However, calculating the plastic displacement z_p has been the obstacle for developing any nanoindentation model. The strength of our approach is that it provides a way of calculating z_p by first calculating the plastic strain rate $\dot{\epsilon}_p(t)$ using the Orowan equation [35]. This also means that we use strain as a fundamental variable that can be calculated. Noting that plastic deformation occurs within the sample of thickness T, we define the strain variable $\epsilon = \frac{z}{T}$. Then, we have the total strain $\epsilon = \epsilon_e + \epsilon_p$, where ϵ_e and ϵ_p refer, respectively, to the elastic and plastic components of the strain. Taking the time derivative, we have

$$\dot{\epsilon} = \frac{\dot{z}}{T} = \frac{\dot{z}_e}{T} + \frac{\dot{z}_p}{T} = \dot{\epsilon}_e + \dot{\epsilon}_p. \tag{2}$$

Then, Equation (1) takes the form

$$z = z_e + T \int \dot{\epsilon}_P(t)dt. \tag{3}$$

Note that z is well defined for any T. The plastic strain rate itself is given by the Orowan expression [35]

$$\dot{\epsilon}_p(t) = bV(\sigma)\rho_m(t). \tag{4}$$

Here, b is the magnitude of the Burgers vector, $V(\sigma)$ is the mean velocity of dislocations, ρ_m the mobile dislocation density and σ the stress. It is clear that this requires calculating ρ_m as a function of time. This is obtained by developing the time evolution equations for the mobile ρ_m and forest ρ_f densities and coupling them to equations defining the DC or the LC modes of indentation.

Note that Equation (4) is a function of stress. Therefore, σ must be expressed as a function of the load F and area A supporting the load since both evolve with time during indentation. This requires expressions for the load $F(z)$ and area $A(z)$ as a function of z. Both are indenter specific.

3.1. Displacement Controlled Indentation Using a Spherical Indenter

Since we plan to address nanoindentation experiments on single crystals of Au carried out in the DC mode using a spherical indenter [11,15], we first recall a few equations. Consider a spherical indenter of radius R. Then, force response of a sample is given by the Hertzian expression

$$F = \frac{4}{3}E^*R^{1/2}z_e^{3/2}. \tag{5}$$

Here, E^* is the effective modulus of the indenter and the sample given by

$$\frac{1}{E^*} = \frac{1 - \nu_s^2}{E_s} + \frac{1 - \nu_i^2}{E_i}, \tag{6}$$

where ν and E refer to the Poisson's ratio and Young's modulus of the sample s and indenter i.

Pile-up and sink-in effects are routinely seen in indentation experiments. In terms of material property, pile-up or sink-in effects depend on the ratio of the elastic modulus to the yield tress and strain hardening property of the material [13]. Pile-up is characterized by larger contact area compared to the ideal case while a sink-in effect is characterized by a smaller contact area. Thus, $\frac{z_e}{z_{e_c}}$, the ratio of the indentation depth z_e measured from $z = 0$ to the contact depth z_{e_c}, is taken as a measure of pile-up or sink-in effects. Thus, area–depth relations are usually expressed in terms of contact depth z_{e_c} measured from the contact point.

The area expression for a spherical indenter is given by

$$A_H(z_e) = \pi R z_{e_c}. \tag{7}$$

Note that Equations (5) and (7) are equations obtained in mechanical equilibrium and therefore do not have any time dependence at this point. However, once these equations are used in equations governing the loading condition and evolution equations for dislocation densities, they acquire time dependence.

Now, consider a spherical indenter of radius R driven into a sample of thickness T at a rate $\dot{r}$ (m/s). Rearranging Equation (2) and substituting, $\dot{z} = \dot{r}$ and $\dot{z}_p = T\dot{\epsilon}_p$, the machine equation corresponding to the DC mode nanoindentation reads

$$\dot{z}_e = \dot{r} - T\dot{\epsilon}_p. \tag{8}$$

Note that this equation is applicable to any indenter. For a spherical indenter of radius R used by Kiely et al. [11,15], we can express z_e in terms of F. Then, Equation (8) takes the form

$$\frac{d}{dt}\left[\frac{3F}{4E^*R^{1/2}}\right]^{2/3} = \dot{r} - T\dot{\epsilon}_p. \tag{9}$$

This equation is the force response of a sample subjected to constant displacement rate.

3.2. Load Controlled Indentation Using a Berkovich Indenter

Now, consider the LC mode nanoindentation using a Berkovich indenter with a view to compare our model results with the LC mode experiments on single crystals of Al that employs a Berkovich indenter [12]. Since our equations are functions of stress, the first step is to adopt phenomenological expressions for the load the area supporting the load as functions of indentation depth for a Berkovich indenter. Here, we use the phenomenological expression for the load [12] given by

$$F = 2.189E^*[1 - 0.21\nu_s - 0.01\nu_s^2 - 0.41\nu_s^3]z_e^2 = CE^*z_e^2. \tag{10}$$

Unlike Equation (7), the area function for a non-ideal Berkovich indenter is complicated due to the complicated geometry of the blunted the tip of the indenter [36–39]. Several expressions have been suggested in the literature for the area of a non-ideal Berkovich indenter. The most general form is the power law expression $A = \sum_{n=0} c_n z^{2/2^n}$ [37]. The calibrated shape of the tip of the indenter can be approximated by a spherical shape with a nominal radius R determined by SEM measurements [23,38,39]. Here, we follow Bei et al. The area of a blunted Berkovich indenter is taken to be sum of the area of an ideal Berkovich indenter and the area of a spherical indenter of radius R. Using the area of a sharp Berkovich indenter given by $A_B = 24.54z_{e_c}^2$, the area function of a real Berkovich intender is given by

$$A(z_{e_c}) = 24.54z_{e_c}^2 + 2\pi R z_{e_c}. \tag{11}$$

Bei et al. [38] show that Equation (11) fits the measured area closely for Cr_3Si single crystals. The geometry of the indenter also provides a relationship between z_e and the depth from the contact point z_{e_c}, i.e., $z_e = z_e(z_{e_c})$ [38]. A linear relation between z_e and z_{e_c}, i.e., $z_e = sz_{e_c}$ holds for the range of indentation depths used in nanoindentation studies.

While Equations (7) and (11) strictly hold in the elastic region only, we assume that they hold over the entire duration of indentation that includes *the elasto-plastic region*. i.e., $A(z_e) \rightarrow A(z_e + z_p)$. Indeed, in MD simulations, the measured area increases in steps at each burst of plasticity (see [23]). This assumption is further supported by the fact that the load increases after every load drop or displacement jump. This can only happen if the area supporting the load increases abruptly to make contact with the indenter surface after every load drop or displacement jump.

For the LC mode of indentation, the equation for a constant load rate is given by

$$\frac{dF(z)}{dt} = \dot{F}_0 = \text{constant}, \tag{12}$$

where $\dot{F}_0$ is the constant load rate. (In experiments, the maximum desired load is achieved in a predetermined number of steps.) This equation is valid for any indenter. For the Berkovich indenter, the expression for the load given by Equation (10) is used.

It is worthwhile to comment on the possible influence of the pile-up or sink-in effects on the properties addressed in present context of intermittent indentation process. It is clear that, since the area supporting the load is larger (pile-up) or smaller (sink-in) than the ideal case and the area expression is a function of z_{e_c}, the pile-up or sink-in effects are incorporated in the scale factor, s in $z_e = sz_{e_c}$. However, the measured area (or equivalently the scale factor s) is seldom given in experiments on nanoindentation instabilities. Thus, one needs to use a judicious choice of s. Since the focus of the work is in formulating a theoretical basis for describing the two (DC and LC mode) nanoindentation instabilities, any judicious choice of the scale factor, s that accounts for all aspects relating to intermittent indentation process is adequate. We shall later comment on hardness calculation in the two modes of nanoindentation instabilities. However, since pile-up or sink-in effects require spatial degrees of freedom for their characterization, calculating pile-up or sink-in effects independently is beyond the scope of our approach since our approach does not include spatial degrees of freedom.

4. Dynamical Approach to Modeling Nanoindentation Instabilities

The basic premise of our approach is that specimen or volume averaged dislocation densities are adequate to describe the measured load-indentation depth curves. This is based on the observation that experimental load–displacement curves are *also specimen averages* of the dislocation activity in the sample. For the same reason, experimental $F - z$ curves do not have any information about the inhomogeneous nature of the plastic deformation occurring within the sample. Thus, as long as we include all relevant dislocation mechanisms contributing to bursts of plasticity, it should be possible to model the time development of the load and indentation depth.

Further support for this premise comes from the fact that similar time evolution equations for the volume averaged dislocation densities have been *effectively* used in the literature to explain the *temporal aspects* of several *spatio-temporal plastic deformation instabilities and patterns* [30,32,34,40–43]. For example, the 'storage recovery' model for the evolution of the forest dislocation density (with respect to shear strain or time) has been used to obtain good insight into several characteristic features of dislocation cell pattern seen in stage III deformation of F.C.C materials [40]. The equation successfully predicts a number of experimental results such as the saturation stress, the hardening rate in the stage II, etc. Similarly, the temporal aspects of the PLC effect manifesting in the form of three types of serrations found with increasing strain rate has been well captured by the Ananthakrishna (AK) model that uses three types of volume averaged dislocation densities (see, for instance, Refs. [30,32,41,42]). Similar volume averaged dislocation densities have also been used to explain the stress–strain curve for a smooth yield phenomenon in the context of acoustic emission [34]. The same type of approach has been used to explain the inverse power law dependence of the yield stress on the diameter of nano-pillars [43]. *All these phenomena are spatio-temporal in nature, yet using time dependent equations have proved quite effective in predicting the salient features.* While this kind of mean field approach is not common in the plasticity area, it is routinely used in several areas of physics, chemistry and engineering [44,45].

General Form of Time Evolution Equations for Dislocation Densities

Following our earlier publications [28–30,32–34,41,46], we develop the time evolution equations for the mobile ρ_m and forest ρ_f densities based on relevant dislocation mechanisms contributing to the plastic flow. Dislocation mechanisms can be broadly categorized into dislocation production and transformation mechanisms. The former includes dislocation nucleation and multiplication. (In our model, all mechanisms leading to the line length increase are regarded as dislocation multiplication processes). Furthermore, since the sample is expected to be dislocation free within the initial indented nanometer volume, nucleation of dislocation loops has to occur before their multiplication. A few simulations [16,21,23,26] also show detachment of the loops from the source followed by their propagation to the boundary (see Figure 9 of Ref. [23] and Figure 5 of Ref. [26].). Each of these dislocation mechanisms can only be activated beyond a certain threshold stress. We denote the threshold stresses for nucleation, multiplication and propagation, respectively, by σ_n, σ_m and σ_p. The rate of nucleation of dislocations per unit area is given by $\frac{v_n}{\pi b^2} exp\,(\sigma/\sigma_n)$. For the sake of simplicity, we assume that a single loop is nucleated, i.e., $\frac{v_n}{\pi b^2} = 1$.

The rate of multiplication of dislocations is traditionally written as $\theta V_m(\sigma)\rho_m$, where $V_m(\sigma)$ represents the average velocity of dislocations [47] and θ the inverse of an appropriate length scale [28,30]. Commonly used phenomenological expressions [28,29,48] are

$$V_m(\sigma) = V_0 exp[\frac{\sigma - h\rho_f^{1/2}}{\sigma_m}], \tag{13}$$

and

$$V_m(\sigma) = V_0 [\frac{\sigma - h\rho_f^{1/2}}{\sigma_m}]^m. \tag{14}$$

In Equations (13) and (14), $h\rho_f^{1/2}$ is the back stress with $h = \alpha Gb$, $\alpha \sim 0.3$ and G is the shear modulus. In Equation (14), m is the velocity exponent. We shall use Equation (13), for the DC mode experiments using a Spherical indenter and Equation (14) for the LC mode where a Berkovich indenter is used.

In time sequence, detachment of dislocations from their source and their subsequent propagation to the boundary follows dislocation multiplication. The rate of propagation of dislocation loops is proportional to the time required for dislocations of velocity $v_p(\sigma)$ to reach the boundary taken to be located at $z = T$. Then, $v_p = \frac{v_p}{T}$. We shall also assume that $v_p(\sigma)$ obeys a power law of the form $v_p = v_{p0}[\frac{\sigma}{\sigma_p}]^p$, where p is an exponent. Then, $v_p = \frac{v_{p0}}{T}[\frac{\sigma}{\sigma_p}]^p = v_{p0}[\frac{\sigma}{\sigma_p}]^p$. Note that $v_p(\sigma)$ is the propagation velocity of dislocations through an undeformed part of the specimen. Therefore, p is necessarily different from the exponent m in $V_m(\sigma)$.

As dislocations multiply, they begin to interact with each other to form junctions. As discussed in a number of earlier publications, two distinct types of short range interactions can be identified that act as the source for the growth of the forest density ρ_f. These act as loss terms to the mobile density ρ_m [28,30,33,34,42,49]. When two dislocations moving in nearby glide planes approach a minimum distance (typically a few nanometers), they can form dipoles. The corresponding loss term to ρ_m is $\beta\rho_m^2$. Then, $\beta\rho_m^2$ is a source (storage) term for the growth of ρ_f. Here, β is a 'rate constant' that has the dimension of the area covered per unit time. A mobile dislocation can annihilate a forest dislocation. This is represented by $f\beta\rho_m\rho_f$. The parameter f is a dimensionless constant typically $\sim 10^{-2}$–10. This is a common loss term for both ρ_m and ρ_f. Finally, unlike the dipole source term that has a fixed short range interaction of a few nanometers, there is another source term for ρ_f whose interaction range evolves with deformation. This happens when dislocations moving in different glide planes intersect each other to form junctions. This can be written as $\Lambda\rho_m\rho_f$. Here, Λ refers to the mean separation between the junctions that also evolves with deformation (or time) since ρ_f itself evolves with time. Noting that the mean distance between the forest dislocations is proportional to $1/\rho_f^{1/2}$, we may write $\Lambda = \delta\rho_f^{-1/2}$. Here, δ is taken to be a constant. Then, the loss term for ρ_m is $\delta\rho_m\rho_f^{1/2}$. Clearly, this is a gain term to ρ_f. (δ has the dimension of velocity.) This storage term represents the forest mechanism. The storage and recovery terms ($\delta\rho_m\rho_f^{1/2}$ and $f\beta\rho_m\rho_f$, respectively) are the only mechanisms used in the storage-recovery model (except for the difference in the notation) [40]. Collecting the loss and gain terms for ρ_m and ρ_f, the time evolution equations can be written as

$$\dot{\rho}_m = \frac{\nu_n}{\pi b^2}exp\frac{\sigma}{\sigma_n} + \theta V_m(\sigma)\rho_m - v_{p0}\rho_m\left[\frac{\sigma}{\sigma_p}\right]^p - \beta\rho_m^2 - \delta\rho_m\rho_f^{1/2} - f\beta\rho_m\rho_f, \quad (15)$$

$$\dot{\rho}_f = \beta\rho_m^2 - f\beta\rho_m\rho_f + \delta\rho_m\rho_f^{1/2}. \quad (16)$$

The first two terms in Equation (15) refer respectively to the nucleation and multiplication of dislocations. The third term represents loss to the boundary due to propagation of dislocations. The next three terms are the dislocation transformation terms. Note also that, unlike the common loss term $f\beta\rho_m\rho_f$ in Equation (15) and (16), $\beta\rho_m^2$ and $\delta\rho_m\rho_f^{1/2}$ are the storage terms for the forest density that contribute to the growth of ρ_f. Thus, a competition between the parameters $f\beta$ corresponding to the common loss term, and the parameters corresponding to storage terms, β and δ, control the relative magnitudes of ρ_m and ρ_f.

5. A Dislocation Dynamical Model for Load Fluctuations Using a Spherical Indenter

We now recast the general form of Equations (15) and (16) suitable for the DC mode indentation using a spherical indenter [28] and compare the predicted results with the results for single crystals of gold [11,15]. This means that we use Equation (13) for $V_m(\sigma)$ and express $\sigma = F/A_H$ in Equations (15) and (16) using Equations (5) and (7). A few comments are in order at this point. Indentation is carried out on Au samples that have high E^*/σ_y ratio. Therefore, one expects pile-up effects are seen. However, micrograph Figure 1b of Ref. [11,15] shows a mild pile-up effect. The pile-up effect is reflected in the increased area (with respect to an ideal spherical indenter) given by $A_H = \pi R z_{e_c}$ in terms of contact depth z_{e_c}. However, no information is available about the measured area or the scale factor s in $z_e = s z_{e_c}$. Moreover, intermittent indentation process is seen until the maximum depth of 25 nm where hardness is ill defined as we shall see. For these reasons, we have used $s = 1$ in the following calculations. However, it must be stated that the basic results and conclusion remain valid for any s.

Then, the evolution equations for ρ_m and ρ_f for the Hertzian indenter of radius R take the form

$$\frac{d\rho_m}{dt} = \frac{v_n}{\pi b^2} exp \frac{\frac{F}{\pi R[z_e+z_p]}}{\sigma_n} + \theta V_0 \rho_m exp \frac{\frac{F}{\pi R[z_e+z_p]} - h\rho_f^{1/2}}{\sigma_m} - \beta\rho_m^2 - f\beta\rho_m\rho_f - \delta\rho_m\rho_f^{1/2}$$
$$- v_{p0}\rho_m \left[\frac{\frac{F}{\pi R[z_e+z_p]}}{\sigma_p} \right]^p, \tag{17}$$

$$\frac{d\rho_f}{dt} = \beta\rho_m^2 - f\beta\rho_m\rho_f + \delta\rho_m\rho_f^{1/2}. \tag{18}$$

These equations are coupled to the machine equation (Equation (9)) expressed in terms of load and area. Then, we have

$$\frac{d}{dt}\left[\frac{3F}{4E^* R^{1/2} T^{3/2}} \right]^{2/3} = \frac{\dot{r}}{T} - b\rho_m V_0 exp\left\{ \frac{F}{\pi R[z_e+z_p]\sigma_m} \right\}. \tag{19}$$

Equations (17) and (18) and Equation (19) constitute a coupled set of nonlinear equations for the nanoindentation problem.

5.1. Estimation of Parameters

Since there are several parameters, we begin by summarizing the estimated values of all the parameters detailed in Refs. [28,29]. Note that the model parameters can be classified as experimental and theoretical. As mentioned in the Introduction, since we work at laboratory (length and time) scales, we can directly adopt experimental parameters such as indentation rate $\dot{r}, R, T, E^*, b$ and h. For the present case, we use the experimental parameters used in Ref. [11] corresponding to the gold samples. For this case, the imposed experimental displacement rate is 0.5 Å/s while the imposed rate in MD simulations are several orders higher [21–26].

The model has several theoretical parameters. A comment is desirable in this context. In dislocation based plasticity models, the number of parameters is determined by the number of dislocation mechanisms required to model the phenomenon. Thus, the more complex the phenomenon, the more the number of dislocation mechanisms and the more the number of parameters. For instance, the AK model [30,32,33,46,50] for the complex spatio-temporal PLC instability has ten parameters. Even in simple models such as the storage recovery model [40] and the model for plastic deformation of micro-pillars [43], there are three and six respectively. In the present case, we have ten parameters corresponding to six dislocation mechanisms [28,29].

Now, consider estimating the theoretical parameters $\theta V_0, f\beta, \delta$ and v_{p_0}. Estimation of the parameter values for the two nanoindentation models has been carried out in detail in Refs. [28,29]. Very briefly, θV_0 is the basic time scale for the growth of the mobile dislocation density that determines the plastic flow. We set $\theta V_0 = 1$ to ensure that it matches the experimental time scale. (Here we use $V_0 \sim 10^{-6}$ m/s and $\theta \sim 10^6$/m). The three threshold stresses σ_n, σ_m and σ_p are fixed based on the basis of other studies or on physical grounds. The nucleation threshold is known to be $\sigma_n \sim E^*/10$. Similarly, the threshold for multiplication of dislocations is typically $\sigma_m \sim E^*/300$. As for the propagation threshold σ_p, since the propagation can only occur after dislocation multiplication, we assume $\sigma_m < \sigma_p$. The upper limit of the velocity prefactor v_{p_0} in the expression for propagation velocity is limited by V_0. While the exponent value p is not known, we fix it to be $p = 1$.

As for the parameters $f\beta$ and δ, we have demonstrated that the orders of magnitudes of these two parameters are determined once the asymptotic (long time) values of the two densities are given or vice versa [28,29]. See Ref. [28,29] for details.

A further requirement on the allowed ranges of parameter values is that Equations (17)–(19) should exhibit an instability. Since these equations form a coupled set of nonlinear equations, such equations often exhibit instabilities for a domain of values of the model parameters. As demonstrated in Ref. [28], the stability matrix corresponding to the steady state of Equations (17)–(19) exhibits a pair of complex conjugate roots with a negative real part. Thus, the DC mode nanoindentation is a transient instability rather than a true instability. Indeed, this is the underlying mathematical mechanism for the decreasing magnitudes of the load drops. (See also Appendix of Ref. [28]). The instability domain of parameter values has been determined numerically. The analysis shows that the instability occurs over a wide range of values of the parameters [28]. In general, the physically relevant values of the parameters form only a small subset of the instability domain. The precise values of the parameters used in the our calculation are given in the Table 1.

Table 1. Parameter values used for the model. The experimental parameters used are drawn from experiments on single crystals of Au [11].

E^* (GPa)	σ_n (GPa)	σ_p (GPa)	σ_m (GPa)	v_0 (m/s)
75	7.5	1.91	0.955	10^{-6}
$\dot{r}$ (Å/s)	$R = T$ (Å)	δ (m/s)	β (m^2/s)	f
0.5	1075	3.92×10^{-6}	10^{-13}	2.5

5.2. Results

Equations (17)–(19) have been solved using an adaptive time step Runge–Kutta solver (ODE15S MATLAB, (MathWorks, Natick, MA, USA)) with initial conditions $\rho_m = 0$ and $\rho_f = 0$ consistent with the absence of dislocations in the indentation volume. Using the parameters given in Table 1, we calculate the load as a function of indentation depth. A plot of $F - z$ is shown in Figure 4a. The loading run is shown by the blue curve and the unloading run by the red curve. It is clear that a large load drop is seen after the initial Hertzian response. The load maximum on the Hertzian branch is $F = 15.1$ μN and the first load drop is ∼10.6 μN. These values are close to the experimental values of 15 μN and 10 μN, respectively [11]. Subsequent load drops show a decreasing trend as in experiments. The corresponding stress as a function of z is shown in Figure 4b. The maximum stress on the Hertzian branch is ≈7.5 GPa (see Figure 4b), the same as used in experiment [11].

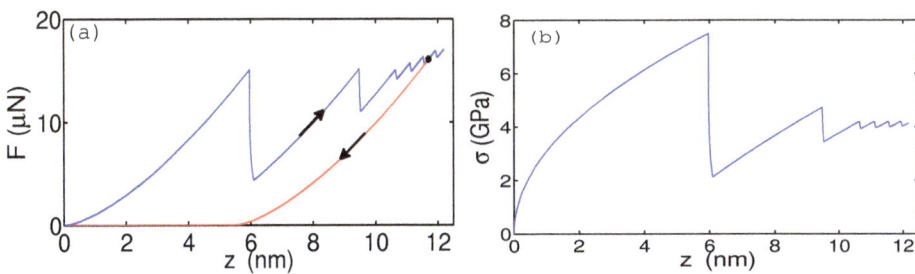

Figure 4. (a) Plots of load as a function of z for the loading (blue) and unloading (red) runs; (b) corresponding plot of stress. Parameters used are given in the Table 1, after [28].

Our approach automatically allows us to calculate ρ_m and ρ_f as a function of time or as a function of depth. Two features are evident from the plot of $\rho_m - z$ shown in Figure 5a. First, the nature of ρ_m is of burst type with the bursts occurring at each load drop with successive bursts decreasing rapidly from an initial high value of $\sim$7 $\times$ 10^{13}/m^2. Second, the duration of the bursts is very short with the durations of the successive bursts increasing concomitantly. *The large magnitudes of ρ_m bursts together with their short durations are the characteristic features of collective behavior of dislocations.* Another conclusion that can be drawn is that collective effects become weaker with successive load drops and are eventually lost as is clear from the decreasing magnitudes and increasing durations of the bursts of ρ_m. In contrast, the corresponding forest density ρ_f increases in steps (not shown, see Figure 2b of [28]). The magnitude of the steps decrease with z, eventually reaching a near saturation level. The positions of these steps are found to be well correlated with the positions of the ρ_m bursts. The stepped increasing nature of ρ_f and their decreasing magnitudes is due to the fact that a large part of ρ_m bursts are transformed to ρ_f. This also implies depletion of the peak heights of ρ_m and consequent decrease of the stepped growth of ρ_f.

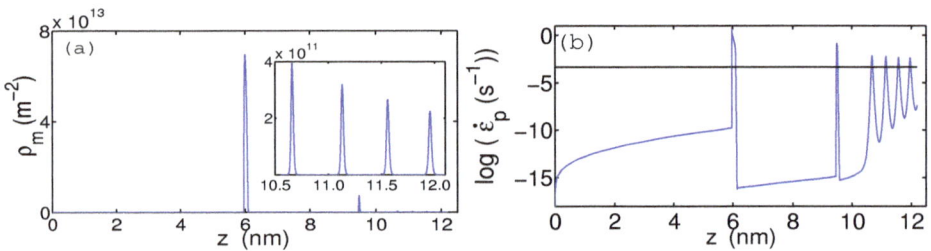

Figure 5. (a) Plot of ρ_m verses z. Inset in (a) shows bursts of ρ_m beyond the second; (b) plot of log($\dot{\epsilon}_p$) as a function of z. Continuous line corresponds to $\dot{r}/T = 4.65 \times 10^{-4} \text{s}^{-1}$. Parameters used are given in the Table 1, after [28].

Now, consider the physical mechanism underlying the load drops. To understand this, we have calculated the plastic strain rate $\dot{\epsilon}_p$ as a function depth. This is shown in Figure 5b. As can be seen, these bursts of $\dot{\epsilon}$ occurs whenever $\dot{\epsilon}_p$ exceeds $\frac{\dot{r}}{T}$, the imposed rate. It may also be noted that these bursts are well correlated with the bursts of ρ_m as should be expected. The decreasing peak heights of ρ_m or the load drops or $\dot{\epsilon}_p$ is not due to the increasing back stress, but due to the depletion of ρ_m as the indentation proceeds (see for details Ref. [28]).

Since our approach is based on dislocation mechanisms contributing to plastic flow, we can calculate the residual indentation depth by unloading the indenter from any point on the $F - z$ curve. When the indenter is unloaded beyond the first force drop, we find that the residual indentation depth, and consequently the residual imprint area is finite, another important feature of nanoindentation experiments. The red curve in Figure 4a shows the unloading curve of the indenter from the point marked ● (after the fifth force drop). The residual imprint area is $\pi a^2 \sim 18.78 \times 10^{-16}$ m^2, a value comparable to the estimated residual contact area (from Figure 2a of Ref. [11]). Note however that the values of residual indentation depth z_{pr} at the bottom of a load drop and the top of the following loading branch are nearly the same, but their load values are different. In contrast, z_{pr} is different for points on the top and bottom of a load drop. Thus, assuming hardness is defined as the ratio of the load to the residual imprint area, hardness assumes two values and therefore is not well defined in the intermittent region. As a consequence, pile-up or sink-in effects are not relevant in the region of intermittent nanoindentation.

5.3. Summary and Discussion

In summary, we have demonstrated that the model not only *predicts all the generic features of nanoindentation but also predicts numbers that match* the experimentally measured load, stress, depth of indentation, the maximum stress on the Hertzian branch and residual plasticity under unloading. The good match between the numbers predicted by model and experiments can be attributed partly to the fact that our approach allows for a direct adoption of experimental parameters such as $\dot{r}, R, T, b, E^*$ and $h = \alpha G b$ and partly to the fact that we have included all the relevant dislocation mechanisms contributing to the intermittent indentation process. Indeed, the strength of our approach lies in providing an innovative method for calculating z_p from the evolution equations for ρ_m (and ρ_f) through the Orowan expression for the plastic strain rate.

Our model also provides insight into the physical and mathematical origin of the decreasing magnitudes of the load drops, a feature seen in experiments and in our model. This feature is not due to the back stress $\sigma_b = \alpha G b \rho_f^{1/2}$ since the peak value of ρ_f ranges from ~ 5 to $\sim 7 \times 10^{13}$ m^{-2} giving $\sigma_b \sim$40–47 MPa. At a physical level, this is purely due to the fact that every time there is a burst of ρ_m, much of it is transformed to ρ_f depleting ρ_m as indentation proceeds. Note also that there is no reverse transformation of ρ_f to ρ_m. Such a reverse transformation is improbable due to the high stress required to break the junction. At a mathematical level, *nanoindentation is a transient instability*. The physical consequence of this is reflected in the decreasing magnitudes of bursts of ρ_m and consequent load drops.

6. A Dislocation Dynamical Model for Displacement Jumps Using a Berkovich Indenter

For the LC mode indentation, we plan to address experimental results on Al single crystals using a Berkovich indenter. For this, we use appropriate expressions for the velocity of dislocation, the load, and the area specific to Berkovich indenter in Equations (15) and (16). For this case, we use the velocity expression given by Equation (14). Then, using this and Equations (10) and (11) in Equations (15) and (16), we obtain the evolution equations:

$$
\begin{aligned}
\dot{\rho}_m &= \frac{v_n}{\pi b^2} exp\left[-\frac{z_{e_c}^2}{\frac{\sigma_n}{C'E^*}\left[\alpha_1(z_{e_c}+(z_p/s))^2+\alpha_2(z_{e_c}+(z_p/s))\right]} \right] + \theta V_0 \rho_m \left[\frac{\frac{C'E^* z_{e_c}^2}{[\alpha_1(z_{e_c}+z_p/s)^2+\alpha_2(z_{e_c}+z_p/s)]}-h\rho_f^{1/2}}{\sigma_m} \right]^m \\
&\quad - v_{p_0}\rho_m\left[\frac{z_{e_c}^2}{\frac{\sigma_p}{C'E^*}\left[\alpha_1\left(z_{e_c}+z_p/s\right)^2+\alpha_2(z_{e_c}+z_p/s)\right]} \right]^p - \beta\rho_m^2 - \delta\rho_m\rho_f^{1/2} - f\beta\rho_m\rho_f,
\end{aligned}
\tag{20}
$$

$$
\dot{\rho}_f = \beta\rho_m^2 - f\beta\rho_m\rho_f + \delta\rho_m\rho_f^{1/2}.
\tag{21}
$$

Using the scale factor s between z_e and z_{e_c}, Equation (12) takes the form

$$
\frac{d(C'E^* z_{e_c}{}^2)}{dt} = \dot{F}_0.
\tag{22}
$$

Note that we have assumed that the area-depth relation holds for the elasto-plastic region, i.e., $A(z) = A(z_{e_c} + z_{p_c})$, where z_{p_c} is the contact depth contribution from plastic deformation.

The contribution arising from the plastic displacement z_p is calculated by integrating

$$
\dot{z}_p = TbV_0\rho_m\left[\frac{\frac{C'E^* z_{e_c}^2}{[\alpha_1(z_{e_c}+z_p/s)^2+\alpha_2(z_{e_c}+z_p/s)]}-h\rho_f^{1/2}}{\sigma_m} \right]^m.
\tag{23}
$$

Note that we have used $z_{p_c} = z_p/s$. Using the value of $\nu_s = 0.33$ in Equation (10) and using the value of $s = 1.6$ obtained by using Equation (8) of Ref. [38], we get $C' = 5.12704$.

Equations (20)–(23) constitute a coupled set of nonlinear ordinary differential equations for the LC mode nanoindentation problem. Such equations often exhibit instabilities for a domain of values of the model parameters. (See also Appendix of Ref. [28] where details of the stability analysis are given).

In the present case, the instability domain of parameter values has been determined numerically. This shows that the instability occurs over a wide range of values of the parameters [29]. In general, the physically relevant values of the parameters form only a small subset of the instability domain. In the following, we briefly summarize the range of physically allowed values determined in Ref. [29].

6.1. Estimation of Parameters

As in the case of the DC mode indentation model [28], we adopt experimental parameters such as $E^*, \dot{F}_0, R, b, T$ and $h = \alpha G b$ and other shape parameters defining the Berkovich indenter geometry for single crystals of Aluminum [12]. The ranges of the parameters have been determined in our earlier publications [28,29] and is listed in Table 2. Very briefly, the nucleation stress $\sigma_n \sim E^*/10$, the multiplication stress $\sigma_m \sim E^*/300 - E^*/100$, the propagation stress $\sigma_p < \sigma_m$. Parameter ranges of β, δ and f have been determined along the lines mentioned for the DC mode in Section 5.1.

Recall that, for the LC mode, we have used the power law expression for the dislocation velocities given by Equation (14). This contains another parameter, namely, the velocity exponent m. This expression has been suggested on the basis of fits of dislocation velocities for bulk specimens subject to various loading conditions such as tensile and pulse loading using transmission electron microscopy, etch-pit technique, and slip line cinematography. (See Figure 19 of Ref. [48]). Under these conditions, stress is nearly uniform within the sample. Equation (14) is valid for single and groups of dislocations. Since the value of m depends on the material, the allowed range of values of m is difficult to estimate on physical grounds, particularly when conditions of deformation are very different (as for nanoindentation) from the conditions where it is measured. On the other hand, often, an inverse correspondence of the velocity exponent m with strain rate exponent n in $n = \left[\frac{d\ln\sigma}{d\ln\dot{\epsilon}}\right]$ is suggested. For single crystals of Al, the value of n is ~ 0.02 or less. This implies that the range of values of the velocity exponent m can be as high as 50. However, this correspondence breaks down in nanoindentation experiments at indentation depths ~ 50 nm where displacement jumps dominate. Moreover, at such small length scales, it is well known that the yield stress increases rapidly, which is also suggestive of the fact that m values for bulk may not be valid for small indentation depth. In view of this, we have used m to range from 2–15.

Similarly, the exponent p (in the expression for velocity of propagation) is taken to be in the range from 1–10. However, p turns out to be an insensitive parameter. The range of values of the parameters are listed in Table 2.

Interestingly, it is possible to calculate the depth at which nucleation occurs given σ_m and R the nominal radius of the blunted tip of the indenter or vice versa [29]. This is given by

$$z_{e_c}(t_n)\left[1 - \frac{\sigma_n}{E^*}\frac{24.54}{s^2 C}\right] = \frac{2\pi R}{s^2 C}\frac{\sigma_n}{E^*}. \tag{24}$$

(Here, $s = 1.6$.) *Thus, the maximum elastic indentation depth $z_e(t_n)$ with nucleation occurring at time t_n is completely determined by σ_n/E^* and R or vice versa.* Note that this statement is independent of the mode of deformation. Thus, Equation (24) allows determination of either σ_n or $z_e(t_n)$ given the other assuming R is given by experiments. For example, if we use $\sigma_n/E^* = 1/10$ and $R = 40$ nm (taken from Ref. [12]), we get $z_n = 14.9$ nm (see Table 2). Furthermore, given a value of σ_n, we can use $z_e(t_n)$ in Equation (10) to determine F. Then, Equation (12) determines the time $t = t_n$ at which the nucleation occurs. *Thus, both F and $z_e(t_n)$ are also uniquely determined by Equation (24).*

We can take the arguments further to show that the first displacement jump is controlled by $\left[(\sigma_n/\sigma_m)\right]^m$. To do this, we note that dislocations are nucleated only when the stress σ exceeds the nucleation stress σ_n. On the other hand, the magnitude of the displacement jump or load drop is directly controlled by $[\sigma_n/\sigma_m]^m \rho_m(t = t_n)$, where $\rho_m(t = t_n)$ is the dislocation density at the nucleation point. Note that we have made use of the fact at nucleation $\sigma = \sigma_n$ and that σ_n is much larger than σ_m. Therefore, $\left[(\sigma_n/\sigma_m)\right]^m$ determines the magnitude of the first displacement jump or vice versa.

However, since σ_n is already determined, σ_m and m are completely determined by the magnitude of the first displacement jump. (See Ref. [29] for details.)

One interesting consequence of this is that the first displacement jump (or load drop) can only exist provided the elastic branch exists or nucleation should occur prior to the jump. This can be seen by noting that the factor $[\sigma_n/\sigma_m]^m \rho_m(t = t_n)$ that controls the magnitude of the first displacement jump decreases as we decrease σ_n and eventually vanishes as $\sigma_n \to 0$. (Note that the elastic branch gets shorter and shorter is equivalent to $\sigma_n \to 0$.)

Table 2. Range of parameter values in the model equations using Berkovich indenter keeping the nominal tip radius $R = 40$ nm and $\sigma_n = 7.0$ GPa.

E^* (GPa)	σ_m (GPa)	σ_p (GPa)	$\dot{F}_0$ (µN/s)
73.7	$0.7 - 1.0$	$\sigma_m - 6.5$	0.3
v_{p_0} (m/s)	β (m²/s)	δ (m/s)	f
10^{-9}–5×10^{-6}	10^{-13}	$(0.01$–$4) \times 10^{-6}$	10^{-2}–10

6.2. Results

Equations (20)–(23) are solved using adaptive step Runge–Kutta solver (MATLAB 'ODE 15s') with initial conditions $\rho_m = 0$ and $\rho_f = 0$ to mimic the absence of dislocations in the indented volume.

6.3. Influence of Parameter Variation on the Model Force–Displacement Curves

The two main objectives, namely, to build a model for the LC mode indentation that would predict all the generic features of the LC mode experiments and to examine the ability of the model to provide a good fit to a given experimental load–displacement data require that we first identify the dislocation mechanisms (or equivalently the corresponding rate constants) controlling the different regions of the load–displacement curve. Towards this end, we begin by investigating the influence of each of these parameters on the model $F - z$ curves. Recognizing that it is a multi-parameter space, it is convenient to fix the parameters at specific values (for instance, those listed in Table 3) and study the influence of each of the parameters on the model $F - z$ curve. The study helps us to evaluate the relative importance of various dislocation mechanisms. In addition, the general trends of the influence of the parameters on the model $F - z$ curves would also be helpful to obtain an optimized set of parameters that provides a good fit to a given experimental $F - z$ curve.

Table 3. Optimized parameter values used for obtaining the best fit with the experimental load–displacement curve for single crystals of (110) orientation [12]. Other parameters are fixed at $\sigma_n = 7.0$ GPa, $\sigma_m = 0.91$ GPa and $m = 11$. Although the parameters σ_p, v_{p_0} and p have little influence on the model $F - z$ curves, the values cited are the values used for numerical fit with the experiment.

σ_p (GPa)	v_{p_0} (m/s)	p	β (m²/s)	δ (m/s)	f
1.5	4×10^{-7}	4	10^{-13}	1.58×10^{-6}	10

Recall that the analysis reported in the previous section demonstrates that σ_n can be taken to be fixed. Thus, we first consider the influence of σ_m on the model $F - z$ curve with all other parameters fixed as in Table 3. (Note that σ_m and m together specifies the dislocation multiplication mechanism. Therefore, while studying the influence of one, the other must be held fixed). Noting that, for small volume systems, σ_m is larger than the bulk value, we have varied σ_m from $\sim E^*/100$ to $E^*/70$. Plots of the $F - z$ curves for $E^*/\sigma_m = 105, 92, 82$ and 74, are shown in Figure 6a. It is clear that increasing σ_m decreases the magnitude of the first displacement jump and also the secondary jumps. More importantly, *the model $F - z$ curves for all values of σ_m exhibit displacement jumps of decreasing magnitudes,* a characteristic feature of the LC mode indentation. Note that the load remains practically constant during the first displacement jump.

Now, consider the influence of the velocity exponent m on the model $F - z$ curve. Plots of $F - z$ curves for $m = 4, 8, 10$ and 12 are shown in Figure 6b. The figure shows that the higher the value of m, the larger is the first displacement jump while the subsequent displacement jumps decrease marginally.

Note also that increasing σ_m and increasing m have the opposite effect on the magnitude of the first displacement jump. This feature can be easily understood by noting that the first displacement jump is determined by $\left[\frac{\sigma_n}{\sigma_m}\right]^m$. Clearly, as σ_m increases, the factor $\left[\frac{\sigma_n}{\sigma_m}\right]^m$ decreases while it increases with increasing m values. *The opposing influence of σ_m and m on the model $F - z$ curves turns out to be very helpful in obtaining the optimal values of σ_m and m that match the magnitude of the first displacement jump for a given experimental data, in particular while fitting the experimental plots of single crystals of Al.*

We have also examined the influence of δ (in the range $\delta = 0.06$ to 3.16×10^{-6} m/s) for a range of f values in the interval 10^{-2} to 10. (Note that we can vary the product $f\beta$ or vary f keeping β fixed). For small f, say 0.01 (or 0.1), as we vary δ, the model $F - z$ curves exhibit several displacement jumps of decreasing magnitudes as can be seen in Figure 6c. Similarly, we have studied the influence of f on the model $F - z$ curve keeping δ fixed in the range $\delta = 0.06$ to 3.16×10^{-6} m/s . Plots of the $F - z$ curves for various value of f are shown in Figure 6d. Again, all model $F - z$ curves exhibit several displacement jumps of decreasing magnitudes. Furthermore, this study also shows that while larger values of δ lead to more rapid increase in the load after each displacement jump, larger f leads to larger displacement jumps with smaller load increase. Note the opposing trends of the influence of f and δ on the $F - z$ curves.

On the other hand, several other parameters σ_p, p and v_{p_0} have very little influence on the model $F - z$ curves. This is fortuitous considering the fact that these parameters were difficult to estimate.

In conclusion, the study of the influence of the parameters on the model $F - z$ curve demonstrates that *for a wide range of values of the parameters, the model $F - z$ curves capture the characteristic features of the LC mode nanoindentation.* In addition, the study determines the relative importance of different dislocation mechanisms, or equivalently, the relative importance of the corresponding parameters on the nanoindentation process.

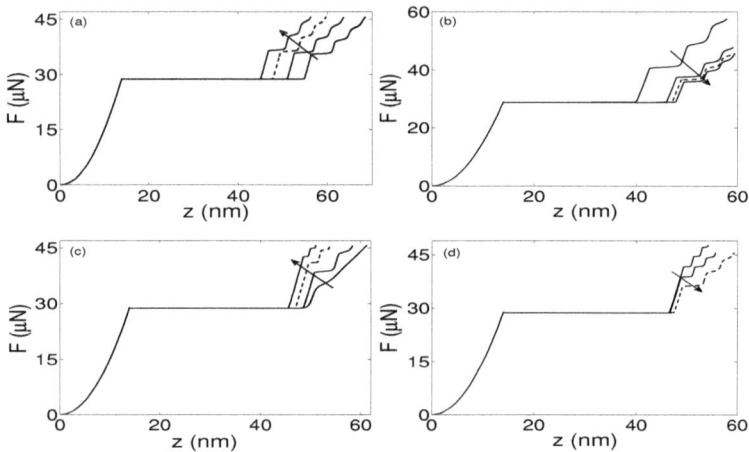

Figure 6. Variation of one parameter keeping other parameters fixed as in Table 3. The direction of the arrow in each of these plots represents increasing values of the parameters. (**a**) plots of load as a function of z for $\sigma_m / E^* = 1/105, 1/92, 1/82$ and $1/74$; (**b**) plots of $F - z$ curves for velocity exponent $m = 4, 8, 10, 12$; (**c**) plots of $F - z$ curves for four values of $\delta = (0.06, 0.158, 0.32, 3.16) \times 10^{-6}$ m/s keeping $f = 0.01$; (**d**) plots of $F - z$ curves for $f = 0.01, 5, 10$ for $\delta = 1.58 \times 10^{-6}$ m/s, after [29].

6.4. Determination of Optimal Parameter Values for Best Fit to Experimental Load–Displacement Curve

The results of the previous section will now be used to obtain an optimized set of parameters that give best fit to the desired experimental load–displacement curve. The following four results from the previous section will be used for this purpose: (a) There is a range of values of the parameters in the instability domain for which the model $F − z$ curves exhibit all the characteristic features of experimental $F − z$ curves. This also implies that one should expect to find an optimized set of parameters within this range that fit the experimental data; (b) The study identifies three dislocation mechanisms contributing to the load–displacement curve in a sequential way. The maximum elastic depth $z_e(t_n)$ at which the nucleation occurs is determined by Equation (24). The nucleation stress σ_n is determined by the maximum elastic depth z_n or vice versa; (c) The magnitude of the first displacement jump is determined by $[\sigma_n/\sigma_m]^m$. However, since σ_n is already determined, σ_m and m are completely determined by the magnitude of the first displacement jump; and (d) The rest of the $F − z$ curve with multiple displacement jumps is determined by f and δ given the maximum experimental load and indentation depth.

We shall use this information to fit the load–displacement curve for single crystals of Al for (110) orientation [12]. In experiments, the thickness of the samples used are $400, 600$ and 1000 nm, and the radius of the indenters used are $50, 100, 150$ nm. For the (110) case, we have used $T = 400$ nm and $R = 40$ nm. To do this, we shall use the following data extracted from the experimental $F − z$ curve (Figure 7 of Ref. [12]) to determine the optimal parameters for the best fit. The maximum depth on the elastic branch is $z_e(t_n) \approx 14$ nm and the first displacement jump of $\Delta z_1 \approx 33$ nm. The maximum load and maximum displacement are respectively $F_{max} \approx 44$ μN and $z_{max} \approx 60$ nm. Using $z_e(t_n) \approx 14$ nm (and $R = 40$ nm) in Equation (24) gives the nucleation stress $\sigma_n = 7.0$ GPa. Now, consider finding the optimal values of σ_m and m. Noting that increasing σ_m decreases the first displacement jump and increasing m increases the magnitude of the first displacement jump, and using $\Delta z_1 \sim 33$ nm gives $\sigma_m = 0.91$ GPa and $m = 11$. (Indeed, a careful perusal of the plots shown in Figure 6a,b also suggest value close to these values). These values give a good fit to the experimental $F − z$ curve until the end of the first displacement jump as is clear from Figure 7a. (Model $F − z$ curve is shown by the continuous line. Points marked • in the figure are experimental points extracted from Figure 7 of Ref. [12].)

We now consider getting a good fit to the rest of the experimental curve until $F_{max} \approx 44$ μN and $z_{max} \approx 60$ nm. Recall that the portion of the $F − z$ curve beyond the first displacement jump is controlled by f and δ, and therefore we need to find the optimized values of f and δ subject to the condition that the model F_{max} and z_{max} match closely the experimental values. To do this, we use the fact that δ and f have opposing influence on the model $F − z$ curve. Then, the values of f and δ that satisfy this condition are listed in Table 3.

Figure 7a shows the model $F − z$ curve (continuous curve) obtained by using the optimized parameter values along with the experimental points •. It is clear from the figure that the experimental points fall on the predicted curve except for the last few nanometers. The magnitudes of the successive displacement jumps decrease and so do the magnitudes of the load steps. Note also that the predicted load remains practically constant during the first displacement jump unlike in experiments where it decreases from 29.5 to 28.5 μN [12]. The latter can be attributed to the inability to enforce constant load rate in experiments during the displacement jump.

Following the above procedure, we have also attempted to get the best fit to the experimental $F − z$ curve for Al single crystals of (133) orientation. Both the model $F − z$ curve (continuous curve) and experimental points • are shown in Figure 7b. As can be seen, the fit is very good. Thus, *the model captures not just the generic features of the experimental $F − z$ curve for a range of parameter values, the model $F − z$ curve corresponding to the optimized parameter values provide very good fit to the experimental data.*

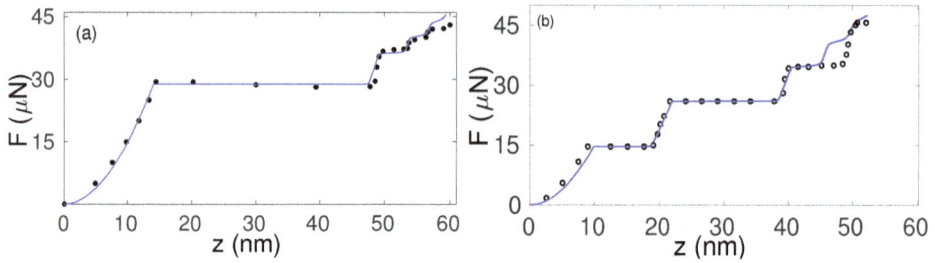

Figure 7. (**a**) plot of model $F - z$ curve (continuous curve) for Al single crystal of (110) orientation along with experimental points (•) extracted from Ref. [12], after [29]; (**b**) plot of $F - z$ curve (continuous curve) t for Al single crystal of (133) orientation along with experimental points (•) [12]. Original contribution of the authors.

Our approach allows us to compute the stress as a function of time or indentation depth. This is shown in Figure 8a corresponding to the (110) orientation. As can be seen, the maximum stress on the elastic branch ~ 7.1 GPa is close to the theoretical strength of the material. Thereafter, the stress relaxes largely during the first displacement jump. However, it must be noted that this relaxation occurs in a short duration of time, which also corresponds to the duration of the first displacement jump. Subsequently, the stress relaxation slows down eventually reaching the asymptotic value of $\sim$1 GPa. Note that this value is higher than the multiplication threshold stress $\sigma_m = 0.91$ GPa.

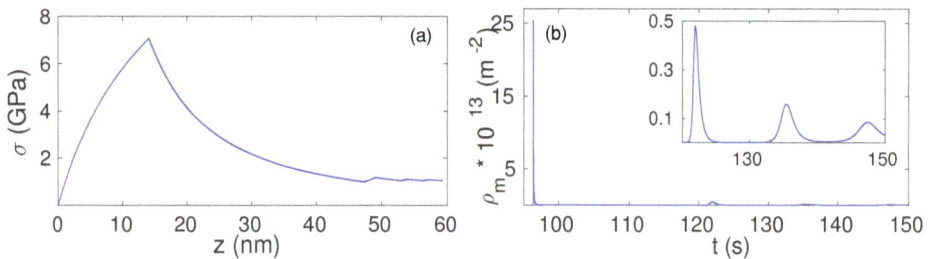

Figure 8. (**a**) Plot of stress as a function of z corresponding to the (110) orientation corresponding to Figure 4a; (**b**) plot of ρ_m as a function of time corresponding to Figure 4a. The inset in (**b**) shows secondary bursts on an expanded scale. Note that the first burst of ρ_m is nearly 50 times the second, after [29].

As emphasized earlier, the first displacement jump beyond the elastic branch is triggered by the abrupt multiplication of the nucleated dislocations. Figure 8b shows the plot of the mobile density as a function of time. As can be seen, the first burst is large ($\rho_m \sim 2.5 \times 10^{14}/\text{m}^2$) and its duration is short. The magnitudes of successive bursts decrease rapidly to a value 2.3×10^{11} m^{-2} with their durations increasing as is clear from the inset of Figure 8b. These features are characteristic features of collective dislocation activity. The decreasing magnitudes of the ρ_m bursts and increasing duration of these bursts clearly suggests that collective effects become weak and eventually disappear in the region where displacement jumps disappear. From a physical point of view, the decreasing magnitudes of the burst of ρ_m is caused by the common loss term, $f\beta\rho_m\rho_f$ that depletes both ρ_m and ρ_f while the storage terms $\beta\rho_m^2$ and $\delta\rho_f^{1/2}\rho_m$ depletes ρ_m. The net effect is that both ρ_m and ρ_f decrease to their asymptotic values.

Since our approach is based on dislocation mechanisms, we can calculate the residual indentation depth by unloading the indenter. This also allows us to calculate the hardness as the ratio of the load to residual imprint area. However, experimental load-depth plots for Al are limited to depths where displacement jumps dominate and therefore hardness is not well defined. On the other hand, model calculation can be performed for indentation depths where displacement jumps disappear. Hardness in this regime is shown in Figure 9a. It is clear that hardness decreases as a function of depth. In experiments, small pile-up is visible in the micrograph shown in Figure 9 of Ref. [12]. However, the authors do not report hardness and therefore no comparison is possible. We emphasize that the decreasing trend of H is not altered for other values of s. Thus, *our model predicts the indentation-size effect as well*. The asymptotic values of $H = 1.98$ GPa is almost twice the stress at large z (see Figure 8a).

Here, a few comments are in order on the existing models for hardness and our own radically different approach for calculating hardness. All traditional models of hardness are based on extending the Taylor relation for hardness to include the additional resistance arising from geometrically necessary dislocations (GNDs). The Nix–Gao model and its variants are based on Taylor relation, and therefore these models *do not calculate hardness as a function of indentation depth*, because, this amounts to calculating the density of geometrically necessary dislocations (GNDs) and statistically stored dislocations (SSDs) as a function of indentation depth. Indeed, these models use GNDs and SSDs (or equivalently asymptotic hardness and the slope of hardness as a function of indentation depth) as fitting parameters to experimental data.

We have recently extended our approach to calculate hardness as a function of indentation depth from small depths ∼nm to few μm [51]. While the traditional models are based on Taylor relation for flow stress, a property that reflects the resistance to dislocation motion, our approach relies on calculating residual indentation depth, a property that is a measure of ease of dislocation motion. Our model includes GNDs in addition to mobile and forest dislocations (or equivalently SSDs). The model [51] *predicts all the characteristic features of hardness such as the decreasing nature of hardness with increasing depths, square of the hardness scaling inversely with indentation depth larger than 150 nm and its deviation for smaller depths*. In addition, we have obtained good fits to the two widely cited hardness data for strain hardened polycrystalline Cu [52] and single crystals of Silver [53]. To the best of our knowledge, this is the first indentation size effect model that calculates hardness as a function of indentation depth independently.

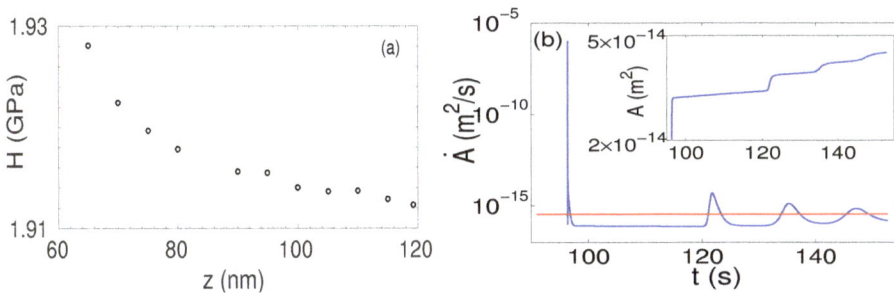

Figure 9. (a) Hardness calculated by unloading the indenter from points that have nearly the same load values but two distinct values of z_{pr} and points that have two distinct values of the load for nearly the same value of z_{pr}; (b) plot of $\frac{dA(z)}{dt}$ as a function of time. The red line corresponds to the rate of area increase due to applied force rate $\frac{\dot{F}_0}{\sigma_m} = 3.3 \times 10^{-16}/\text{m}^2$. The inset shows that the A increases in steps, after [29].

6.5. Physical and Mathematical Mechanisms for the Pop-in Events

One important feature of the LC mode nanoindentation is the existence of several displacement jumps of decreasing magnitudes. While the origin of displacement jumps is attributed to bursts of plasticity, explaining the decreasing magnitudes of the displacement jumps quantitatively has remained a puzzle. Clearly, this feature can be addressed within the scope of our model since the displacement jumps are controlled by bursts of ρ_m. Recall that studies on the influence of the parameters on the model $F - z$ curves (Section 6.3) has demonstrated that three different dislocation mechanisms operate in a time sequence. The initial elastic branch is controlled by nucleation of dislocations while the first displacement jump is controlled by the multiplication of the nucleated dislocations. On the other hand, the subsequent displacement jumps of decreasing magnitudes are controlled by the storage and recovery dislocation mechanisms. While the magnitudes of the displacement jumps are determined by the magnitudes of ρ_m bursts, these themselves decrease. The latter is due to the fact that a large part of every ρ_m burst is transformed to ρ_f. Furthermore, noting that the stress relaxes largely during the first displacement jump, the magnitudes of subsequent bursts of ρ_m are determined by the excess stress above σ_m. Then, the decreasing over stress leads to decreasing magnitudes of the secondary ρ_m bursts. This in turn implies decreasing magnitudes of the displacement jumps. This explanation should be contrasted with the one suggested in the literature, namely, increasing back stress. This is not the case since the back stress $\sigma_b = \alpha G b \rho_f^{1/2}$ is ~25 MPa even when ρ_f assumes a peak value of ~$3 \times 10^{13}/\text{m}^2$.

6.6. Common Mechanism Underlying All Displacement Jumps

Recall that, in the DC mode of indentation, the load drops occur whenever the plastic displacement rate $\dot{z}_p$ exceeds the applied displacement rate $\dot{r}$ as is clear from the Equation (19). The equation explains all load drops on the same footing. However, an equivalent mechanism is absent for the case of the LC mode nanoindentation. In view of this, we attempt to explain all displacement jumps on the basis of a common dynamical mechanism. Some insight can be obtained by examining the stress plot in Figure 8a together with the $F - z$ plot in Figure 7a. Figure 8a shows that the stress relaxes mostly during the first pop-in event itself, eventually reaching its asymptotic value of ~1 GPa. However, this value is larger than the dislocation multiplication stress $\sigma_m = 0.91$ GPa. Since the stress remains higher than σ_m even after the first displacement jump, one should expect a continuous increase in ρ_m or equivalently a continuous increase in z_p. In contrast, Figure 7a shows that the load increases quite sharply beyond the first displacement jump suggesting a dominant elastic contribution. This raises a question as to why the load should increase in a near elastic manner when the stress remains higher than σ_m? As we shall see, this question is also relevant for subsequent displacement jumps.

To see this, recall that both load and area evolve with time. The area increase can occur either due to imposed loading rate or due to plastic displacement rate. However, the time scales of these two processes are very different. While the load increases linearly, the rate of area increase $\dot{A}$ arising from plastic deformation can be very short as is clear from the short time scale of ρ_m bursts. From a dynamical point of view, this observation suggests that, while the underlying dislocation mechanisms for the first and secondary displacement jumps are different, we can describe all displacement jumps as resulting from a competition between the two well separated time scales, namely, the slow time scale corresponding to $\dot{F}_0$ and the fast time scale corresponding to ρ_m bursts. The rate of area change arising from imposed load rate $\dot{F}_0$ is given $\frac{\dot{F}_0}{\sigma_m}$, where σ_m is a natural choice for converting the applied load rate to the rate of area increase. Figure 9b shows a plot of $\dot{A}(z)$ as a function of time, which increases in bursts due to ρ_m bursts. Also shown is the rate of area increase due to applied load rate $\frac{\dot{F}_0}{\sigma_m} = $ constant (red line). It is clear that, whenever $\dot{A}(z)$ overshoots $\frac{\dot{F}_0}{\sigma_m}$, we see a displacement jump (triggered by bursts of ρ_m) that is almost fully plastic. Otherwise, the elastic component dominates over the plastic contribution. *This feature is very similar to the mechanism underlying the force drops in the DC mode of indentation where a force drop occurs whenever the plastic displacement rate exceeds the applied displacement rate.* Note that the duration of the first burst of $\dot{A}(z)$ (due to the first burst of ρ_m) is very short while the durations of the subsequent bursts get longer and longer until the magnitudes of $\dot{A}(z)$ and $\frac{\dot{F}_0}{\sigma_m}$ are comparable. Figure 9b can also be used to obtain quantitative estimates of the magnitudes of the load steps seen in Figure 7a. (See Ref. [29] for details.)

The above analysis demonstrates that *the mathematical mechanism causing the instability is a result of a competition between the slow applied time scale corresponding to load rate (strictly $\frac{\dot{F}_0}{\sigma_m}$) and another fast time scale corresponding to $\dot{A}(z)$(controlled by bursts of ρ_m).* The existence of one slow loading time scale and one fast time scale of response are typical of all relaxational oscillations (see Refs. [30,44,45]). However, in the present case, the instability is a transient one as is clear from the decreasing magnitudes of the displacement jumps and load steps. The analysis also demonstrates that imposing a constant load rate would be sufficient to trigger displacement jumps provided it is much larger than the short duration of ρ_m bursts. However, in experiments, enforcing constant force rate condition during displacement jumps may not be easy due to finite response time of the machine.

6.7. Conclusions

In summary, the model for the LC mode nanoindentation predicts all the generic features such as the existence of an elastic branch followed by several displacement jumps of decreasing magnitudes, and residual plasticity under unloading *for a range of values of the parameters*. Furthermore, the predicted values of the load, the magnitudes of displacement jumps, etc. are also similar to those seen in experiments. The study of the influence of the parameters on the model $F - z$ curves show that the elastic depth is determined by σ_n / E^* and the magnitude of the first displacement jump is controlled by the ratio of the nucleation stress to the multiplication stress, more precisely $(\sigma_m / \sigma_n)^m$. Subsequent displacement jumps are controlled by the storage and recovery mechanisms. This identification also allows us to determine the optimized parameters that fit closely the experimental $F - z$ curves corresponding to single crystals of Al for (110) and (133) orientations.

Even though the underlying dislocation mechanisms for the first and subsequent jumps are different, we have demonstrated explicitly that all displacement jumps result whenever the short duration bursts of area increase due to plastic deformation exceed the slow rate of area increase due to applied load rate. Otherwise, elastic response dominates. Note that the short duration of bursts of ρ_m is also a signature of collective behavior of dislocations. The mathematical mechanism for the instability is attributed to the competition between the slow time scale corresponding to the load rate and the fast time scale corresponding to bursts of mobile density. More specifically, the decreasing magnitudes of the displacement jumps or the load steps are signatures of transient nature of the instability. The underlying physical mechanism for decreasing magnitudes of the displacement jumps is the depletion of the mobile density that transforms to the forest density during the short duration of ρ_m bursts.

7. General Conclusions on the Dynamical Approach to Nanoindentation

The present dynamical approach has been developed as an alternate approach to simulations with a view to overcome their inherent limitations. While these methods have been routinely used due to their ability to include a range of dislocation mechanisms [21–26], this advantage is offset by the serious limitations due to the short time scales inherent to several simulations and their inability to adopt experimental parameters, particularly the imposed rate of deformation. Consequently, the predicted numbers differ considerably from those reported by experiments. Our idea is to develop an alternate method to overcome the above limitations by devising a method that employs experimental parameters such as the rates, the geometrical parameters defining the geometry of the indenter, thickness of the film, etc., to predict numbers that match the experiments. The fact that the load drops or displacement jumps are signatures of an instability arising from collective dislocation activity motivated us to use nonlinear dynamics to model the nanoindentation instabilities. This also means that a natural interpretation of the dislocation densities used in the model is that they represent collective modes of the plastic deformation (as in the case of the AK model for the PLC effect).

The fact that the two models predict all the generic features with the predicted numbers matching those from experiments is clearly a support for both the dynamical approach and the dislocation mechanisms used. In this context, the success of the models in predicting all the generic features along with good fits to the two experimental data sets must be viewed against the background of the absence of simulations of any kind for the LC mode nanoindentation experiments. The fact that the model $F - z$ curves exhibit the characteristic features of the experiments for a range of parameter values can only be attributed to the fact that our approach allows *for a direct adoption of the experimental parameters*. More importantly, the strength of our approach lies in providing an innovative method for calculating the contribution from plastic deformation to the total indentation depth using the time evolution equations for ρ_m and ρ_f. Note that obtaining a good fit to experiments requires that we match the model time and length scales with those in experiments. Noting that the total indentation depth is the sum of the elastic and plastic contributions, this match has been accomplished by demanding that the basic time scale for the growth of mobile density (that determines the extent of plastic deformation) matches the experimental time scale, i.e., we demand θV_0 is set to unity. (Note that imposed deformation such as the displacement rate or load rate or strain rate, are measured per second). This allows us to match the model z_{max} with that in experiment. Note that matching model F_{max} with that in experiment is straightforward since it is determined by load-elastic depth expression. Thus, our approach provides a consistent scheme to match the model time scale and length scale with those of the experiments. Note that the maximum depth is essentially determined by the plastic contribution to the depth.

One important consequence of our ability to calculate the plastic contribution to the indentation depth is that for larger depths where the load drops or displacement jumps disappear, the model predicts that hardness decreases with depth. Indeed, since our model uses dislocation density for the calculation of plastic contribution, our method provides a way to describe indentation size effect and constitutes an alternate approach to the strain gradient theories [54–57].

This kind of dislocation density based approach opens-up the possibility of studying nanoindentation on experimental length and time scales that cannot be achieved in finer scales simulation techniques. Furthermore, unlike simulations that require heavy computational resources, the model equations can be solved on a desktop computer.

The results predicted by the models for the DC mode and LC mode nanoindentation, in particular, the good fit to the two experimental load–displacement curves for the LC mode indentation where no simulations exist, clearly supports our view that the proposed method can be used as an alternate approach to simulation methods. This view is further supported by the success of the recent dislocation mechanism based dynamical model for calculating hardness as a function of indentation depth. The Indentation size effect model predicts all characteristic features of hardness in addition to providing good fits to the two widely used hardness data on cold worked polycrystalline Cu [52] and single crystals of Ag [53].

The approach may be criticized for ignoring the spatially inhomogeneous nature of the deformation. (Note that, although the indentation depth z is used as a dynamical variable, the model does not explicitly include spatial degrees of freedom.) Therefore, it might come as a surprise that, despite the absence of spatial degrees of freedom in the model, the model $F - z$ curve matches the experimental data very well. To appreciate this, it is important to recognize that dislocation densities used here represent the spatial/sample averages over the indented volume. They also represent collective dislocation modes that are responsible for load drops or displacement jumps. Much the same way, all experimentally measured quantities such as the force, stress, depth of indentation, residual plasticity under unloading, etc. are *also* the volume averages of the dislocation activity in the sample. Since both theoretically computed and experimentally measured quantities represent sample averages, *the good match is not all that surprising.* Moreover, it is well known that averages are quite insensitive to the details of the distribution. This statement is applicable to spatial averages as well. However, if spatial degrees of freedom are introduced, displacement jumps will exhibit some stochasticity, i.e., displacement jumps as well as the steps on force will be different in different runs. This is a general result in dynamical systems that has been well illustrated in the case of the PLC effect (see Ref. [30,33] and in particular Figure 2.1a of Ref. [42]). Inclusion of spatial degrees of freedom for the present problem is nontrivial since this involves a moving boundary that goes by the name Stefan's problem in mathematics. Finally, it would be interesting to explore the possibility of using the current approach along with finite element methods that calculate the stress distribution under the indenter.

This then would allow the possibility of obtaining spatial distribution of the dislocation activity in the sample.

Author Contributions: G.A. identified, conceptualised and built the models. Numerical solutions was carried out by S.K.; Ananlysis was done by G.A. and S.K.; G.A. and S.K. wrote the paper.

Acknowledgments: G.A. acknowledges the Board of Research in Nuclear Sciences Grant No. 2012/36/18—*BRNS* and the support from the Indian National Science Academy through the Honorary Scientist position.

Conflicts of Interest: The authors declare no conflict of interest.

References and Note

1. Brenner, S.S. Tensile strength of Whiskers. *J. Appl. Phys.* **1956**, *27*, 1484–1491. [CrossRef]
2. Brenner, S.S. Plastic deformation of Copper ans Silver Whiskers. *J. Appl. Phys.* **1957**, *28*, 1023–1026. [CrossRef]
3. Persson, B.N.J. *Sliding Friction: Physcal Prinicples and Applications*, 2nd ed.; Springer: Berlin, Germany, 2000; ISBN 978-3-642-08652-6.
4. Dimiduk, D.M.; Woodward, C.; LeSar R.; Uchic, M.D. Scale-Free Intermittent Flow in Crystal Plasticity. *Science* **2006**, *312*, 1188–1190. [CrossRef] [PubMed]
5. Zaiser, M.; Schwerdtfeger, J.; Schnieder, A.S.; Frick, C.P.; Clark, B.G.; Gruber, P.A.; Arzt, A. Strain bursts in plastically deforming molybdenum micro- and nanopillars. *Philos. Mag.* **2008**, *88*, 3861–3874. [CrossRef]
6. Kiener, D.; Grosinger, W.; Dehm, G.; Pippan, R. A further step towards an understanding of size-dependent crystal plasticity: In situ tension experiments of miniaturized single-crystal copper samples. *Acta Mater.* **2008**, *56*, 580–592. [CrossRef]
7. Gouldstone, A.; Chollacoop, N.; Doa, M.; Li, J.; Minor, A.M.; Shen, Y. Indentation across size scales and disciplines: Recent developments in experimentation and modeling. *Acta Mater.* **2007**, *55*, 4015–4039. [CrossRef]
8. Gerberich, W.W.; Venkataraman, S.K.; Huang, H.; Harvey, S.E.; Kohlstedt, D.L. The injection of plasticity through millinewton contacts. *Acta Metall. Mater.* **1995**, *43*, 1569–1576. [CrossRef]
9. Michalske, T.A.; Houston, J.E. Dislocation nucleation at nanoscale mechanical contacts. *Acta Mater.* **1998**, *46*, 391–396. [CrossRef]
10. Gouldstone, A.; Van Vliet, K.J.; Suresh, S. Simulation of defect nucleation in a crystal. *Nature* **2001**, *411*, 656. [CrossRef] [PubMed]
11. Kiely, J.D.; Hwang, R.Q.; Houston, J.E. Effect of Surface Steps on the Plastic Threshold in Nanoindentation. *Phys. Rev. Lett.* **1998**, *81*, 4424–4427. [CrossRef]
12. Gouldstone, A.; Koh, H.J.; Zeng, K.Y.; Giannakopoulos, A.E.; Suresh, S. Discrete and continuous deformation during nanoindentation of thin films. *Acta Mater.* **2000**, *48*, 2277–2295. [CrossRef]
13. Fischer-Cripps, C. *Nano-Indentation*; Springer: New York, NY, USA, 2011; ISBN 978-1-4419-9872-9
14. Corcoran, S.G.; Colton, R.J.; Lilleodden, E.T.; Gerberich, W.W. Anomalous plastic deformation at surfaces: Nanoindentation of gold single crystals. *Phys. Rev. B* **1997**, *55*, R16057–R16060. [CrossRef]
15. Kiely, J.D.; Houston, J.E. Nanomechanical properties of Au (111), (001), *and* (110) surfaces. *Phys. Rev. B* **1998**, *57*, 12588–12594. [CrossRef]
16. Li, J.; Van Vliet, K.J.; Zhu, T.; Yip, S.; Suresh, S. Atomistic mechanisms governing elastic limit and incipient plasticity in crystals. *Nature* **2002**, *418*, 307–310. [CrossRef] [PubMed]
17. Shibutani, Y.; Tsuru, T.; Koyama, A. Nanoplastic deformation of nanoindentation: Crystallographic dependence of displacement bursts. *Acta Mater.* **2007**, *55*, 1813–1822. [CrossRef]
18. Dietiker, M.; Nyilas, R.D.; Solenthaler, C.; Spolenak, R. Nanoindentation of single-crystalline gold thin films: Correlating hardness and the onset of plasticity. *Acta Mater.* **2008**, *56*, 3887–3899. [CrossRef]
19. Schall, P.; Cohen, I.; Weitz, D.A.; Spaepen, F. Visualizing dislocation nucleation by indenting colloidal crystals. *Nature* **2006**, *440*, 319–323. [CrossRef] [PubMed]
20. Minor, A.M.; Syed Asif, S.A.; Shan, Z.; Stach, K.A.; Cyrankowski, E.; Wyrobek, T.J.; Warren, O.L. A new view of the onset of plasticity during the nanoindentation of aluminium. *Nat. Mater.* **2006**, *5*, 697–702. [CrossRef] [PubMed]
21. Kelchner, C.L.; Plimpton, S.J.; Hamilton, J.C. Dislocation nucleation and defect structure during surface indentation *Phys. Rev. B* **1998**, *58*, 11085–11088. [CrossRef]

22. Zimmerman, J.A.; Kelchner, C.L.; Klein, P.A.; Hamilton, J.C.; Foiles, S.M. Surface Step Effects on Nanoindentation. *Phys. Rev. Lett.* **2001**, *87*, 165507. [CrossRef] [PubMed]
23. Van Vliet, K.J.; Li, J.; Zhu, T.; Yip, S.; Suresh, S. Quantifying the early stages of plasticity through nanoscale experiments and simulations. *Phys. Rev. B* **2003**, *67*, 104105. [CrossRef]
24. Lilleodden, E.T.; Zimmerman, J.A.; Foiles, S.M.; Nix, W.D. Atomistic simulations of elastic deformation and dislocation nucleation during nanoindentation. *J. Mech. Phys. Solids* **2003**, *51*, 901–920. [CrossRef]
25. Tsuru, T.; Shibutani, Y. Atomistic simulations of elastic deformation and dislocation nucleation in Al under indentation-induced stress distribution. *Model. Simul. Mater. Sci. Eng.* **2006**, *14*, S55–S62. [CrossRef]
26. Chang, H.J.; Fivel, M.; Rodney, D.; Verdier, M. Multiscale modelling of indentation in FCC metals: From atomic to continuum. *Comptes Rendus Phys.* **2010**, *11*, 285–292. [CrossRef]
27. Zhu, T.; Li, J.; Samanta, A.; Leach, A.; Gall, K. Temperature and Strain-Rate Dependence of Surface Dislocation Nucleation. *Phys. Rev. Lett.* **2008**, *100*, 025502. [CrossRef] [PubMed]
28. Ananthakrishna, G.; Katti, R.; Srikanth, K. Dislocation dynamical approach to force fluctuations in nanoindentation experiments. *Phys. Rev. B* **2014**, *90*, 094104. [CrossRef]
29. Srikanth, K.; Ananthakrishna, G. Dynamical approach to displacement jumps in nanoindentation experiments. *Phys. Rev. B* **2017**, *95*, 014107. [CrossRef]
30. Ananthakrishna, G. Current theoretical approaches to collective behavior of dislocations. *Phys. Rep.* **2007**, *440*, 113–259, 10.1016/j.physrep.2006.10.003. [CrossRef]
31. Ananthakrishna, G.; Bharathi, M.S. Dynamical approach to the spatiotemporal aspects of the Portevin-Le Chatelier effect: Chaos, turbulence, and band propagation. *Phys. Rev. E.* **2004**, *70*, 026111. [CrossRef] [PubMed]
32. Ananthakrishna, G.; Valsakumar, M.C. Repeated yield drop phenomena: A temporal dissipative structure *J. Phys. D Appl. Phys.* **1982**, *15*, L171–L175. [CrossRef]
33. Sarmah, R.; Ananthakrishna, G. Correlation between band propagation property and the nature of serrations in the Portevin–Le Chatelier effect. *Acta Mater.* **2015**, *91*, 192–201. [CrossRef]
34. Kumar, J.; Sarmah, R.; Ananthakrishna, G. General framework for acoustic emission during plastic deformation. *Phys. Rev. B* **2015**, *92*, 144109. [CrossRef]
35. Kubin, L.P. *Dislocations, Mesoscale Simulations and Plastic Flow*; Clarendon Press: Oxford, UK, 2012; ISBN-13: 9780198525011.
36. Oliver, W.C.; Pharr, G.M. An improved technique for determining hardness and elastic modulus using load and displacement sensing indentation experiments. *J. Mater. Res.* **1992**, *7*, 1564–1583. [CrossRef]
37. VanLandingham, M.R.; Juliano, T.F.; Hagon, H.J. Measuring tip shape for instrumented indentation using atomic force microscopy. *Meas. Sci. Technol.* **2005**, *16*, 2173–2185. [CrossRef]
38. Bei, H.; George, E.P.; Hay, J.L.; Pharr, G.M. Influence of Indenter Tip Geometry on Elastic Deformation during Nanoindentation. *Phys. Rev. Lett.* **2005**, *95*, 045501. [CrossRef] [PubMed]
39. Sakharova, N.A.; Fernandes, J.V.; Antunes, J.M.; Oliveira, M.C. Comparison between Berkovich, Vickers and conical indentation tests: A three-dimensional numerical simulation study. *Int. J. Solids Struct.* **2009**, *46*, 1095–1104. [CrossRef]
40. Kocks, U.F.; Mecking, H. Physics and phenomenology of strain hardening: The FCC case. *Prog. Mater. Sci.* **2003**, *48*, 171–273. [CrossRef]
41. Rajesh, S.; Ananthakrishna, G. Relaxation oscillations and negative strain rate sensitivity in the Portevin–Le Chatelier effect. *Phys. Rev. E* **2000**, *61*, 3664–3674. [CrossRef]
42. Srikanth, K. A Dynamical Appraoch to Plastic Deformation of Nano-Scale Materials: Nano- and Micro-Indentation. Ph.D. Thesis, Indian Institute of Science, Bangalore, India, July 2016.
43. Nix, W.D.; Lee, S.W. Micro-pillar plasticity controlled by dislocation nucleation at surfaces. *Philos. Mag.* **2011**, *91*, 1084–1096. [CrossRef]
44. Strogatz, S.H. *Nonlinear Dynamics and Chaos: With Applications to Physics, Biology, Chemistry, and Engineering (Studies in Nonlinearity)*; Westview Press: Boulder, CO, USA, 2001; ISBN 9780813349114.
45. Haken, H. *Advanced Synergetics, Instability Hierarchies of Self-organizing Systems and Devices*; Springer: Berlin/Heidelberg, Germany, 1987; ISBN 978-3-642-45553-7.
46. Ananthakrishna, G.; Sahoo, D. A model based on nonlinear oscillations to explain jumps on creep curves. *J. Phys. D* **1981**, *14*, 2081–2090. [CrossRef]

47. In general, dislocations travel in the medium of other dislocations and form locks and junctions where they are arrested. They are also arrested by other pinning points. Therefore, their motion is intermittent with waiting periods at junctions followed by near free flight between them once they get unpinned beyond a certain stress. The mean travel time between any two points is dominated by the waiting periods. Thus, the mean velocity V_m is the average over the distance covered during the time interval, which will be a function of stress, i.e., $V_m = V_m(\sigma)$. See Ref. [48] for details. This velocity should not confused with the velocity of individual dislocations given by $VB = \sigma b$ with B defining the drag coefficient in a defect free crystal.)

48. Neuhäusser, H. Slip-Line Formation and Collective Dislocation Motion. In *Dislocations in Solids*; Nabarro, F.R.N., Ed.; Elsevier Science: Amsterdam, The Netherlands, 1983; Volume 6, p. 319.

49. Kubin, L.P.; Estrin, Y. Evolution of dislocation densities and the critical conditions for the Portevin-Le Châtelier effect. *Acta Metall. Mater.* **1990**, *38*, 697–708. [CrossRef]

50. Ananthakrishna, G. *Dislocations in Solids*; Nabarro, F.R., Hirth, J.P., Eds.; Elsevier Science: Amsterdam, The Netherlands, 2007; Volume 13, pp. 81–223; ISBN: 978-0-444-51888-0.

51. Ananthakrishna, G.; Srikanth, K. An alternate approach for hardness based on residual indentation depth: Comparison with experiments. *Phys. Rev. B* **2018**, *97*, 104103. [CrossRef]

52. McElhaney, K.W.; Vlassak, J.J.; Nix, W.D. Determination of indenter tip geometry and indentation contact area for depth-sensing indentation experiments. *J. Mat. Res.* **1997**, *13*, 1300–1306. [CrossRef]

53. Ma, Q.; Clarke, D.R. Size dependent hardness of silver single crystals. *J. Mater. Res.* **1995**, *10*, 853–863. [CrossRef]

54. Pharr, G.M.; Herbert, E.G.; Gao, Y. Indentation size effect: A critical examination of experimental observations and Mechanistic interpretations. *Ann. Rev. Mater. Res.* **2010**, *40*, 271–292. [CrossRef]

55. Nix, W.D.; Goa, H. Indentation size effects in crystalline materials: A Law for strain gradient plasticity. *J. Mech. Phys. Solids* **1998**, *465*, 411–425. [CrossRef]

56. Gao, H.; Huang, Y.; Nix, W.D.; Hutchinson, J.W. Mechanism based strain gradient plasticity—I. Theory. *J. Mech. Phys. Solids* **1999**, *47*, 1239–1263. [CrossRef]

57. Hutchinson, J.W. Plasticity at the micron scale. *Int. J. Solids Struct.* **2000**, *37*, 225–238. [CrossRef]

crystals

MDPI

Review

Review of Nanoindentation Size Effect: Experiments and Atomistic Simulation

George Z. Voyiadjis * and Mohammadreza Yaghoobi

Computational Solid Mechanics Laboratory, Department of Civil and Environmental Engineering,
Louisiana State University, Baton Rouge, LA 70803, USA; myagho1@lsu.edu
* Correspondence: voyiadjis@eng.lsu.edu

Academic Editors: Helmut Cölfen and Ronald W. Armstrong
Received: 26 September 2017; Accepted: 20 October 2017; Published: 23 October 2017

Abstract: Nanoindentation is a well-stablished experiment to study the mechanical properties of materials at the small length scales of micro and nano. Unlike the conventional indentation experiments, the nanoindentation response of the material depends on the corresponding length scales, such as indentation depth, which is commonly termed the size effect. In the current work, first, the conventional experimental observations and theoretical models of the size effect during nanoindentation are reviewed in the case of crystalline metals, which are the focus of the current work. Next, the recent advancements in the visualization of the dislocation structure during the nanoindentation experiment is discussed, and the observed underlying mechanisms of the size effect are addressed. Finally, the recent computer simulations using molecular dynamics are reviewed as a powerful tool to investigate the nanoindentation experiment and its governing mechanisms of the size effect.

Keywords: nanoindentation; size effect; atomistic simulation; dislocation; grain boundary

1. Introduction

The size effect in material science is the variation of material properties as the sample characteristic length changes. In the current work, the focus is on the size effect on the nanoindentation response of crystalline metals. Nanoindentation is a well-stablished experiment to investigate the mechanical properties of material samples of small volumes. During the nanoindentation, a very hard tip is pressed into the sample, and the variation of load versus the penetration depth is recorded. In the case of nanoindentation, the dependency of material hardness on the corresponding characteristic length is termed as the size effect. The underlying mechanism of the size effect during nanoindentation depends on the material nature. In the case of crystalline metals, the size effect is governed by the dislocation-based mechanisms. The nanoindentation size effects in other materials such as ceramics, amorphous metals, polymers and semiconductor materials are controlled by cracking, non-dislocation-based mechanisms, shear transformation zones and phase transformations [1–4].

A very common size effect during the nanoindentation occurs in the case of the geometrically self-similar indenter tips. The conventional plasticity predicts that the hardness should be independent of the penetration depth. Experimental results, however, have shown that the hardness is a function of indentation depth, which is commonly termed as the indentation size effect [4–22]. The common size effect is the increase in hardness by decreasing the indentation depth. Most of the size effect studies support this trend. However, few studies have shown the inverse size effects in which as the indentation depth decreases, the hardness also decreases, which is commonly termed as inverse size effects [4–7,9,10,23]. However, the inverse size effect is commonly attributed to the artifacts of the experiment such as instrument vibration and the error in contact area measurements [4,8–10]. In addition to the size effect during the nanoindentation using a geometrically self-similar tip, another indentation size effect occurs in the case of a spherical indenter tip. However, the characteristic length of this size effect is the indenter radius itself in a way that the hardness increases as the tip radius decreases [14–18].

Besides the experimental observations, computer simulations have greatly contributed to the investigation of the response of materials during nanoindentation. The common modeling methods are finite element [24–41], crystal plasticity [42–50], discrete dislocation dynamics [51–62], the quasicontinuum method [63–76] and molecular dynamics (MD) [77–88]. Going down from simulations with larger length scales, i.e., finite element, to that of the smallest length scales, i.e., MD, the simulation accuracy increases while the required resources hugely increases. The most accurate method to model the nanoindentation experiment and investigate the underlying physics of the indentation size effect is to model the sample as a cluster of atoms using MD simulation. Accordingly, the MD simulation of nanoindentation can be envisaged as a pseudo-nanoindentation experiment in which the sample is modeled with the accuracy down to the atomic scale.

In the current review, the focus is on the nanoindentation size effect in crystalline metals, in which the deformation mechanisms are governed by the nucleation and evolution of the dislocation network. The size effect trend in which the hardness increases as the indentation depth decreases is considered in the current review. The aim of this study is to address recent advancements in experiments and atomistic simulation to capture the underlying mechanisms of the size effect during nanoindentation. To do so, first, the classical experimental observations and theoretical models of size effects during nanoindentation are reviewed. Next, the interaction of size effects during nanoindentation with the effects of grain size is summarized. The recent advancements for nanoindentation experiments with the focus on various methods of dislocation density measurement are then reviewed. Advanced size effect models, which have been proposed based on the recent experimental observations, are addressed here. Finally, the atomistic simulation of nanoindentation is introduced as a powerful tool to investigate the underlying mechanisms of the size effect during nanoindentation. The dislocation nucleation and evolution pattern observed during the MD simulation of nanoindentation are also summarized. The details and methodology of atomistic simulations, however, are beyond the scope of the current work.

2. Classical Experimental Observations and Theoretical Models

The older generation of size effect observations during indentation have been reported usually as the variation of hardness versus the indentation load (see, e.g., Mott, [5]) during the Vickers microhardness experiment, which has a square-based diamond pyramid. In the early stages of the size effect observations, however, the observed trends have been attributed to the experimental artifacts such as sample surface preparation or indenter tip imperfections. Advancing the nanoindentation techniques and load and depth sensing instruments, however, the obtained results show that the indentation size effects are not artifacts of the experiment, and it has underlying physical mechanisms. Figure 1 shows the common size effect trend for several nanoindentation experiments available in the literature in which the hardness increases as the indentation depth decreases. In Figure 1, for each set of experiments, the hardness is normalized using the hardness value at large indentation depths H_0, at which the hardness becomes independent of the indentation depth. To unravel the governing mechanisms of the size effect during nanoindentation, the concept of geometrically necessary dislocations (GNDs), which was introduced by Ashby [89], has been commonly adopted [11–14,90]. Ashby [89] stated that dislocations could be categorized into two groups. The first one, which is called geometrically necessary dislocations (GNDs), is formed to sustain the imposed displacement for the sake of compatibility. The second type, which is called statistically stored dislocations (SSDs), is a group of dislocations trapping each other in a random way. The model proposed by Nix and Gao [13] is a milestone in this family of models in which they predicted the size effect during nanoindentation using a conical tip (Figure 2). In Figure 2, the SSDs are excluded; however, they contribute to the indentation hardness. During the nanoindentation, the compatibility of deformation between the sample and indenter is guaranteed by the nucleation and movement of GNDs. Accordingly, the total length of GNDs λ_G can be obtained based on the indentation depth h, contact radius a_p, Burgers vector of GNDs b and indenter geometry as follows (Nix and Gao [13]):

$$\lambda_G = \int_0^{a_p} \frac{2\pi rh}{ba_p}\mathrm{d}r = \frac{\pi a_p h}{b} = \frac{\pi a_p^2 \tan(\theta)}{b} \tag{1}$$

where $\tan(\theta) = h/a_p$. In the next step, Nix and Gao [13] assumed that the plastic zone is a hemisphere with the radius of a_p. Accordingly, the density of GNDs ρ_G can be described as follows:

$$\rho_G = \frac{\lambda_G}{V} = \frac{3h}{2ba_p^2} = \frac{3}{2bh}\tan^2(\theta) \tag{2}$$

where V is the volume of the plastic zone.

○ De Guzman et al. (1993) - Cu	✗ De Guzman et al. (1993) - Ni
+ Ma and Clarke (1995) - Ag [1 0 0]	☐ Ma and Clarke (1995) - Ag [110]
▲ Poole et al. (1996) - Cu (Work-hardened)	△ Poole et al. (1996) - Cu (Annealed)
-------- McElhaney et al. (1998) - Cu (Strain-hardened)	•••• McElhaney et al. (1998) - Cu (Annealed)
◆ Nix and Gao (1998) - Cu [1 1 1]	◇ Lim and Chaudhri (1999) - Cu (Annealed)
✦ Liu and Ngan (2001) - Cu (111) (Mechanical polish)	◉ Liu and Ngan (2001) - Cu (111) (Electropolished)
◘ Swadener et al. (2002) - Ir (Annealed)	■ Bull (2003) - Fe (Mechanical Polish)
✗ Bull (2003) - Fe (Electropolished)	▲ Bull (2003) - TiN (Blunt Berkovich tip)
● McLaughlin and Clegg (2008) - Cu (0 0 1)	– – – Rester et al. (2008) - Cu {1 1 1}
– · – Rester et al. (2008) - Ni {1 1 1}	– ·· – Rester et al. (2008) - Ag {1 1 1}

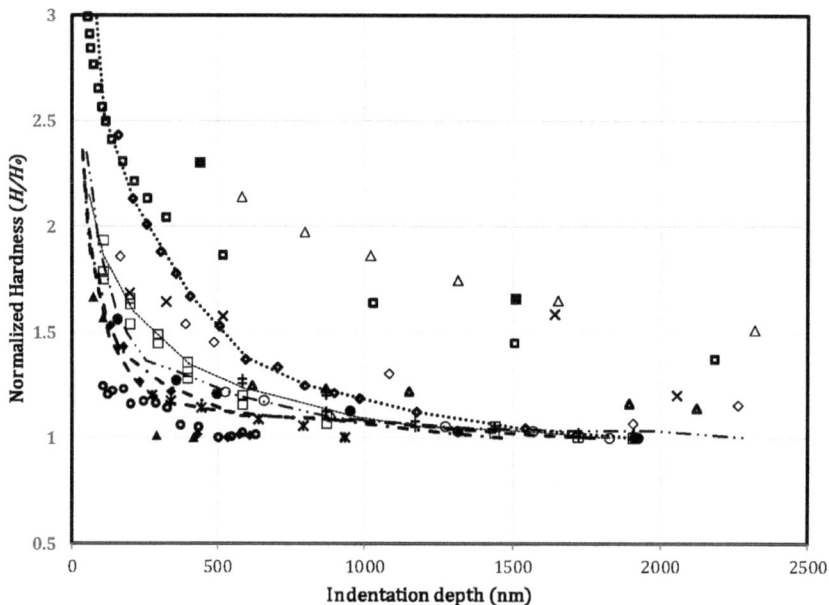

Figure 1. Size effect during nanoindentation. The original experimental data have been reported by De Guzman [12], Ma and Clarke [91], Poole et al. [92], McElhaney et al. [93], Nix and Gao [13], Lim and Chaudhri [15], Liu and Ngan [94], Swadener [14], Bull [20], McLaughlin and Clegg [95] and Rester et al. [96].

Nix and Gao [13] further neglected the interaction between SSDs and GNDs and assumed that the total dislocation density ρ is simply the summation of SSDs and GNDs densities, i.e., $\rho = \rho_S + \rho_G$, where ρ_S and ρ_G are the SSD and GND densities, respectively. Finally, the Taylor hardening model relates the dislocation density ρ to the indentation hardness H as follows [13]:

$$H = 3\sqrt{3}\alpha Gb\sqrt{\rho} = 3\sqrt{3}\alpha Gb\sqrt{\rho_G + \rho_S} \tag{3}$$

where G is the shear modulus and α is a material constant. Accordingly, the variation of hardness versus the indentation depth can be defined as follows:

$$\frac{H}{H_0} = \sqrt{1 + \frac{h^*}{h}} \tag{4}$$

where H_0 is the hardness due to SSDs and h^* is the characteristic length that governs the dependency of the hardness on the penetration depth. H_0 and h^* can be described as follows:

$$H_0 = 3\sqrt{3}\alpha Gb\sqrt{\rho_S} \tag{5}$$

$$h^* = \frac{81}{2}b\alpha^2 \tan^2(\theta)\left(\frac{G}{H_0}\right)^2 \tag{6}$$

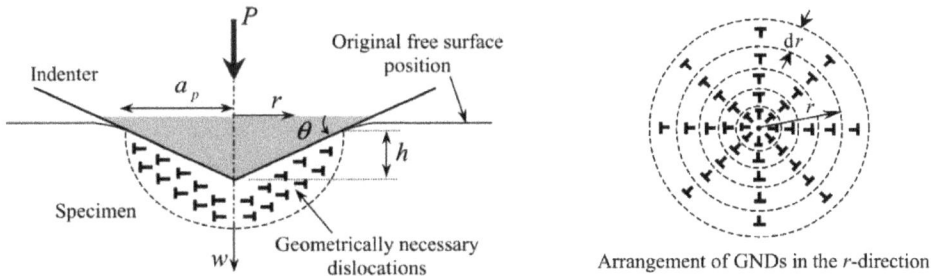

Figure 2. Axisymmetric rigid conical indenter. Geometrically necessary dislocations (GNDs) created during the indentation process. The dislocation structure is idealized as circular dislocation loops (after Abu Al-Rub and Voyiadjis [97]).

In the next step, Nix and Gao [13] connected the physical mechanism of indentation size effect to the strain gradient plasticity model by defining the strain gradient during the nanoindentation as follows:

$$\chi \equiv \frac{\tan(\theta)}{a_p} \tag{7}$$

Accordingly, the length scale for a strain gradient model can be obtained as follows:

$$l \equiv b\left(\frac{G}{\sigma_0}\right)^2 \tag{8}$$

where $\sigma_0 = H_0/3$.

Swadener et al. [14] extended the proposed model of Nix and Gao [13] to include the general indenter geometry of $h = A_r{}^n$, where $n > 1$ and A is a constant. Accordingly, the total dislocation length can be described as follows:

$$\lambda_G = \int_0^{a_p} \frac{2\pi r}{b}\left(\frac{dh}{dr}\right)dr = \frac{2\pi nA}{b(n+1)}a_p^{n+1} \tag{9}$$

Accordingly, the GND density can be described as follows:

$$\rho_G = \frac{\lambda_G}{V} = \frac{3nA}{b(n+1)}a_p^{n-2} = \frac{3nA^{(2/n)}}{b(n+1)}h^{(1-2/n)} \tag{10}$$

Pugno [90] extended the framework proposed by Swadener et al. [14] for the indenter with the general geometry by approximating the indentation surface as a summation of discrete steps due to the formation of dislocation loops (Figure 3). Accordingly, the total GND length can be calculated as follows:

$$\lambda_G = \frac{\Omega - A}{b} = \frac{S}{b} \tag{11}$$

where A and S are the summation of horizontal and vertical surfaces, respectively, and $\Omega = S + A$ (Figure 3). The GND density can be described as follows:

$$\rho_G = \frac{\lambda_G}{V} = \frac{S}{bV} \tag{12}$$

Accordingly, the size effect law can be described based on the surface to volume ratio of the indentation domain.

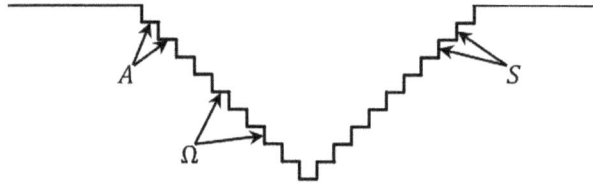

Figure 3. Approximating the indentation surface as a summation of discrete steps: A is the projected contact area; Ω is the contact surface; and $S = \Omega - A$.

Swadener et al. [14] also addressed the size effect during indentation using a spherical tip. The geometry of the spherical tip was approximated by parabola, i.e., $h = r^2/2R_p$. According to Equation (10), $\rho_G = 1/bR_p$, i.e., the density of GNDs is independent of penetration depth for a spherical indenter. However, the size effect for a spherical indenter is governed by another characteristic length, which is the indenter radius R_p. Using the similar methodology as Equation (4), the size effect for spherical indenters can be described as follows:

$$\frac{H}{H_0} = \sqrt{1 + \frac{R^*}{R_p}} \tag{13}$$

where $R^* = \bar{r}/b\rho_S$. Figure 4 compares the theoretical predictions presented by Nix and Gao [13] and Swadener et al. [14]. The results show that the theoretical model can capture the size effects for large indentation depths and indenter radii. In the case of small depths and radii, however, the theoretical models noticeably deviate from the experimental results.

Feng and Nix [98] also showed that the Nix and Gao model [13] cannot capture the size effect in MgO during nanoindentation at lower indentation depths (Figure 5). They investigated three reasons that may induce the Nix and Gao model [13] to break down at small indentation depths: indenter tip bluntness, effect of Peierls stress and dislocation pattern underneath the indenter. They showed that the first two reasons are not responsible for the observed discrepancies. In order to consider the dislocation distribution beneath the indenter, they showed that the modified radius of the plastic zone should be incorporated as follows:

$$R_{pz} = fa_p \tag{14}$$

where f varies as the indentation depth increases. For a large indentation depths, $f \rightarrow 1$, the model reduces to the Nix and Gao model [13]. Accordingly, the volume of plastic volume can be obtained as below:

$$V = \frac{2}{3}\pi R_{pz}^3 = \frac{2}{3}\pi f^3 a_p^3 \tag{15}$$

The size effect model based on the modified definition of plastic volume is depicted in Figure 6, which captures the size effect both at large and small indentation depths. Durst et al. [36] incorporated the same method to modify the Nix and Gao model [13]. Instead of variable f, which is fitted using the experiment (Feng and Nix [98]), they proposed that f is a constant. They incorporated $f = 1.9$ in their work and showed that it can capture the size effects in single crystal and ultrafine-grained copper [36].

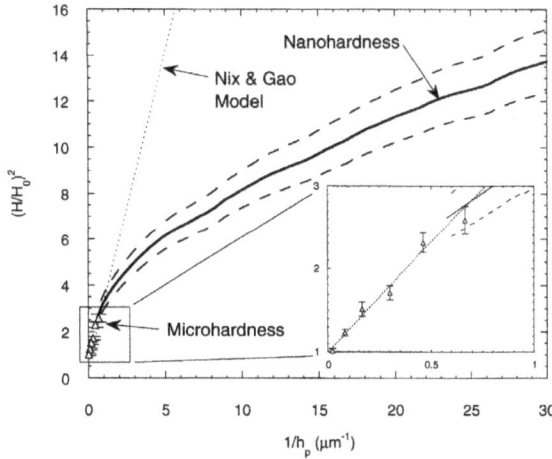

Figure 4. Indentation size effect in annealed iridium measured with a Berkovich indenter (Δ and solid line) and comparison of experiments with the Nix and Gao [13] model ($H_0 = 2.5$ GPa; $h^* = 2.6$ μm). The dashed lines represent ±1 standard deviation of the nanohardness data (after Swadener et al. [14]).

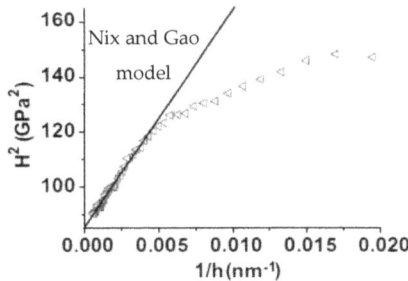

Figure 5. Indentation size effect in polished MgO: experimental results versus the Nix and Gao [13] model ($H_0 = 9.24$ GPa; $h^* = 92.5$ nm) (after Feng and Nix [98]).

In order to modify the Nix and Gao model [13], Huang et al. [99] proposed an analytical model by defining a cap for GND density ρ_G^{max}, which is a material constant. Accordingly, in the case of the conical indenter, the local GND density can be defined as follows:

$$(\rho_G)_{local} = \rho_G^{max} \quad \text{if } h < h_{nano} \tag{16a}$$

$$(\rho_G)_{\text{local}} = \begin{cases} \rho_G^{\max} & \text{for} & r < \frac{h_{\text{nano}}}{\tan(\theta)} \\ \frac{\tan(\theta)}{br} & \text{for} & \frac{h_{\text{nano}}}{\tan(\theta)} \leq r \leq a_p = \frac{h}{\tan(\theta)} \end{cases} \tag{16b}$$

where $h_{\text{nano}} = \tan^2(\theta)/(b\rho_G^{\max})$ is a nanoindentation characteristic length. The average GND density can be obtained by averaging $(\rho_G)_{\text{local}}$ over the plastic zone volume, which is a hemisphere with the radius of a_p, as follows [99]:

$$\rho_G = \rho_G^{\max} \begin{cases} 1 & \text{if } h < h_{\text{nano}} \\ \frac{3h_{\text{nano}}}{2h} - \frac{h_{\text{nano}}^3}{2h^3} & \text{if } h < h_{\text{nano}} \end{cases} \tag{17}$$

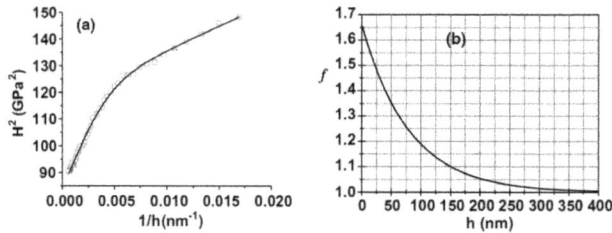

Figure 6. Indentation size effect in polished MgO: (**a**) experimental results versus the modified Nix and Gao model presented by Feng and Nix [98] ($H_0 = 9.19$ GPa; $h^* = 102$ nm); (**b**) variation of f versus h (after Feng and Nix [98]).

Based on the modified GND density, Huang et al. [99] developed a strain gradient plasticity model and captured the indentation size effect in MgO and Ir (Figure 7). The values of $\rho_G^{\max}$ for MgO and Ir are 1.28×10^{16} m^{-2} and 9.68×10^{14} m^{-2}, respectively.

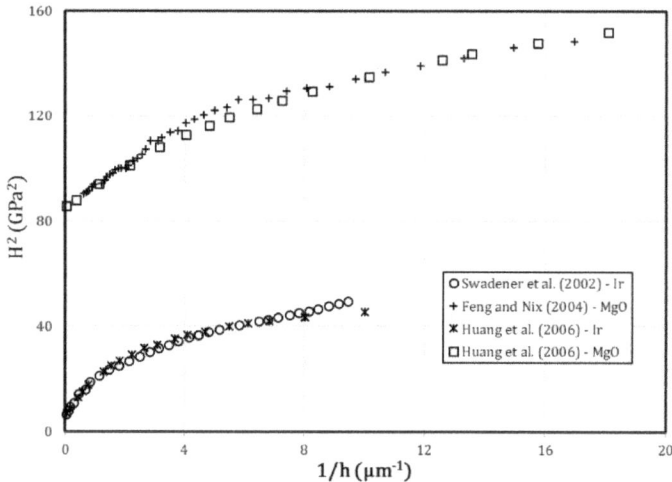

Figure 7. Comparison between the indentation size effect captured using the strain gradient plasticity model developed by Huang et al. [99] using a cap for GND density versus the experimental results in Ir and MgO ($\rho_G^{\max}{}_{\text{Ir}} = 9.68 \times 10^{14}$ m^{-2}, $\rho_G^{\max}{}_{\text{MgO}} = 1.28 \times 10^{16}$ m^{-2}). The original simulation and experimental data have been reported by Swadener [14], Feng and Nix [98] and Huang et al. [99].

3. Interaction of Size Effects during Nanoindentation and Grain Size Effects

The interaction between the effects of grain size with indentation size effects have been addressed both theoretically and experimentally [100–108]. Here, the experimental and theoretical studies conducted by Voyiadjis and his coworkers [103–108] are discussed in more detail. Figure 8 compares the response of single crystalline and polycrystalline Al during nanoindentation. In the case of single crystalline Al, the conventional size effects trend can be observed. In the case of polycrystalline Al, however, a local hardening can be observed, which is due to the effect of the grain boundary (GB). In order to capture the effect of GB, the theoretical model has been proposed by Voyiadjis and his coworkers [103–108] based on the concept of GNDs and the methodology developed by Nix and Gao [13]. Figure 9 depicts the effect of GB on the dislocation movement pattern during nanoindentation. Unlike the single crystalline metals, the GNDs movements are blocked by grain boundaries, which induce a dislocation pile-up against the GB. Accordingly, a local hardening is induced in the nanoindentation response due to the GB resistance. The dislocation can move to the next grain when the stress concentration induced by dislocation pile-up reaches a critical value, and the nanoindentation response follows the conventional size effect pattern. Accordingly, the schematic of effects of GB on the nanoindentation response can be divided into three regions (Figure 10). In Region I, the indenter does not feel the GB, and the initial nucleation and evolution of GNDs govern the size effect, which is similar to the response of a single crystal sample. Eventually, the dislocations reach the GB at which their movements are blocked, and they pile-up against the GB, which leads to a local hardening (Region II). The pile-up stress eventually reaches a critical value, and dislocations will dissociate to the next grain; additionally, the nanoindentation response follows that of a single crystal sample (Region III).

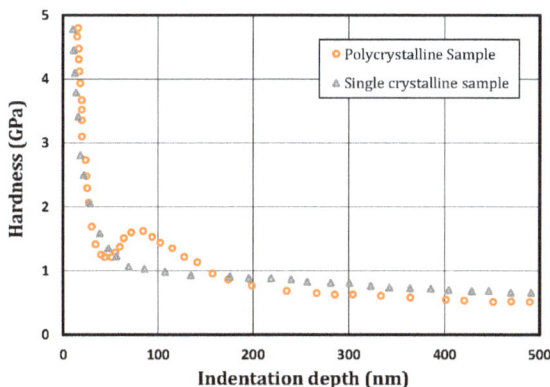

Figure 8. Comparison of the nanoindentation response of a single crystalline Al sample with that of a polycrystalline one at strain rates of $0.1\ s^{-1}$. The original data were reported by Voyiadjis et al. [106].

Voyiadjis and Zhang [107] and Zhang and Voyiadjis [108] investigated the effects of GB on the nanoindentation response of bicrystalline Al and Cu samples, respectively. They varied the distance of indentation from the GB and compared the observed nanoindentation responses (Figure 11). The results showed that the distance from the GB controls the depths at which different regions of hardening occur in a way that the local hardening occurs at lower indentation depths for indentations closer to the GB. As the indentation distance becomes large enough, the response becomes similar to that of a single crystalline sample. Accordingly, Zhang and Voyiadjis [108] proposed a material length scale that incorporates the effect of distance between the GB and indenter tip *r* based on the framework developed by Nix and Gao [13], Abu Al-Rub and Voyiadjis [97], Voyiadjis and Abu Al-Rub [109] and Voyiadjis et al. [106] as follows:

$$l = \left(\frac{\alpha_G}{\alpha_S}\right)^2 \left(\frac{b_G}{b_S}\right) \overline{M}r \left\{ \frac{\delta_1 r e^{-(E_r/RT)}}{\left[1 + \delta_2 r p^{(1/\tilde{m})}\right]\left[1 + \delta_3 (\dot{p})^q\right]} \right\} \tag{18}$$

where d_g is the average grain size, $\overline{M}$ is the Schmid factor, $1/\tilde{m}$ and q are the hardening exponents, δ_1, δ_2 and δ_3 are the material constants calibrated using the nanoindentation experiment, E_r is the activation energy, R is the gas constant and T is the absolute temperature.

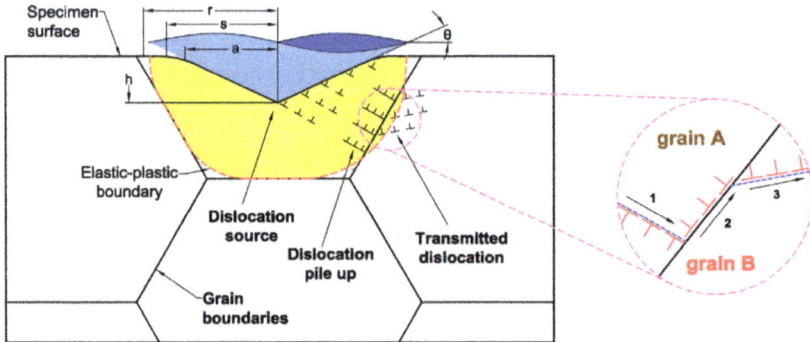

Figure 9. The schematic of the interaction between the GNDs and grain boundaries for polycrystalline metals during nanoindentation (after Voyiadjis and Zhang [107]).

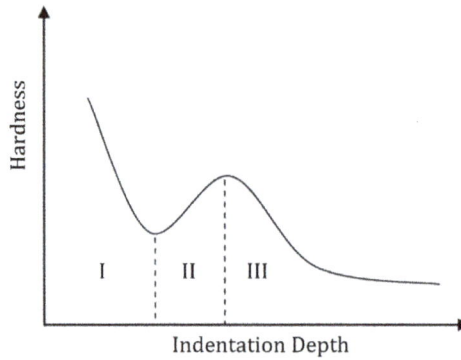

Figure 10. The typical nanoindentation response of polycrystalline samples.

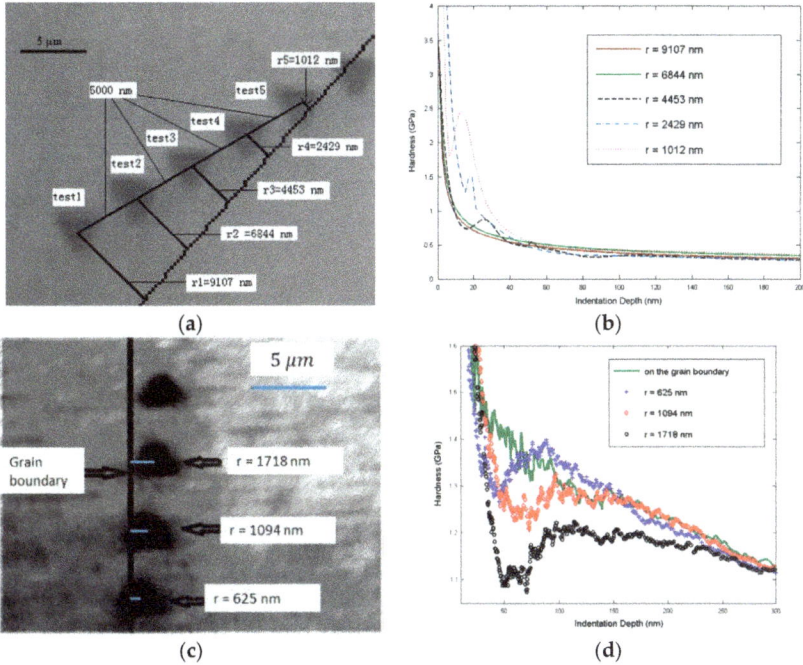

Figure 11. Effect of grain boundary on the nanoindentation response of crystalline metals: (**a**) distances between the grain boundary and indentations for Al bicrystal sample; (**b**) the effect of the distance from indentation to the grain boundary on the nanoindentation response of the Al bicrystal sample; (**c**) distances between the grain boundary and indentations for the Cu bicrystal sample; (**d**) the effect of the distance from indentation to the grain boundary on the nanoindentation response of the Cu bicrystal sample (after Voyiadjis and Zhang [107] and Zhang and Voyiadjis [108]).

4. Recent Experimental Observations and Theoretical Models

The fundamental assumption in the size effects model based on the density of GNDs following the Nix and Gao model [13] is that the forest hardening is the governing mechanism. Accordingly, they relate the size effect to the GNDs induced by the strain gradient, which enhances the nanoindentation response. As described by Swadener et al. [14], Feng and Nix [98] and Huang et al. [99], the Nix and Gao model [13] cannot capture the size effects at shallow indentation depths. Although different methods of modifying the plastic zone volume [98] and assigning a GND saturation cap [99] solved the model shortcomings to some extent, both methods are phenomenological in nature, and they are using fitting procedures to solve the problem. Furthermore, these models still incorporate the forest hardening mechanism to capture the relation between the material strength and dislocation density.

Uchic and his coworkers [110,111] introduced the micropillar compression experiment and showed that the crystalline metallic samples show strong size effects even in the absence of any strain gradients. They have tested pure Ni pillars at room temperature and showed that the sample strength increases by decreasing the pillar diameter (Figure 12). In order to capture the size effect in pillars, however, other hardening mechanisms, including source truncation, weakest link theory and source exhaustion, have been introduced rather than the forest hardening mechanism (see, e.g., Uchic et al. [112] and Kraft et al. [113]) In micropillars, the double-ended dislocation sources become single-ended ones due to the interaction of dislocations with the pillar-free surfaces. Accordingly, the pillar size governs the single-ended sources leading to shorter dislocation sources for smaller pillars. Consequently, decreasing the pillar size increases the sample strength, which is commonly called the source truncation mechanism (Parthasarathy et al. [114]; Rao et al. [115]). Another hardening mechanism is the weakest link theory (Norfleet et al. [116]), which describes that decreasing the pillar size increases the strength of the weakest deformation mechanism and leads to the increase in the sample strength. The third hardening mechanism is source exhaustion (see, e.g., Rao et al. [117]), which states that the material strength increases when not enough dislocation sources are available to sustain the applied plastic flow. In the case of pillars, dislocation starvation (Greer et al. [118]) is one of the mechanism that can lead to the source exhaustion in a way that all the dislocations escape from the pillar-free surfaces, leaving the sample without any dislocations. Accordingly, the sample strength is controlled by source-limited activations.

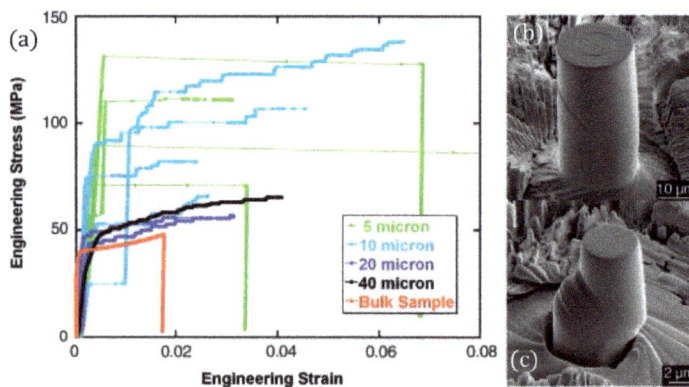

Figure 12. Response of pure Ni pillars during the uniaxial compression experiment at room temperature. (**a**) Stress-strain curves for pillars with diameters that vary from 40 to 5 µm and a bulk single crystal having approximate dimensions of 2.6 × 2.6 × 7.4 mm; (**b**) A scanning electron micrograph (SEM) image of a pillar with the diameter of 20 µm at 4% strain; (**c**) A SEM image of 5 µm diameter pillar at 19% strain during a rapid burst of deformation (after Uchic et al. [111]).

In order to unravel the hardening mechanism at small indentation depths, the dislocation microstructure should be monitored. Three experimental techniques of backscattered electron diffraction (EBSD) [22,43,45,49,50,96,119–121], convergent beam electron diffraction (CBED) [95] and X-ray microdiffraction (μXRD) [122–125] have been used so far to observe the dislocation microstructure in metallic samples. These techniques can map the local lattice orientations with high spatial resolutions. Accordingly, the Nye dislocation density tensor and the associated GND density can be calculated. Due to the complex nature of these experimental schemes, the number of works addressing the size effects during nanoindentation is very limited. Kiener et al. [119], McLaughlin and Clegg [95] and Demir et al. [121] have reported anomalies regarding the description based on the forest hardening mechanism and strain gradient theory to capture the size effects during nanoindentation. Kiener et al. [119] indented the tungsten and copper samples using a Vickers indenter and used EBSD to quantify the lattice misorientation. They incorporated the misorientation angle as a represented parameter of GND density. They showed that the maximum misorientation angle ω_f increases as the indentation depth increases, while the nanoindentation hardness decreases (Figure 13), which cannot be described based on the strain gradient theory and forest hardening mechanism. McLaughlin and Clegg [95] investigated the size effects in a copper single crystal sample indented by a Berkovich tip using CBED. They measured the total misorientation for two indentation loads of 5 mN and 15 mN. It was observed that both the magnitude of the overall misorientation and the neighbor-to-neighbor misorientations underneath the 15 mN indentation are significantly larger than underneath the 5 mN indentation. However, the hardness for indentation load of 5 mN is larger than that of 15 mN. Again, McLaughlin and Clegg [95] showed that the models based on the strain gradient theory and forest hardening mechanisms, such as Nix and Gao [13], are not applicable to the shallow indentations. They attributed the observed size effects to the lack of dislocation sources at lower indentation loads, which is similar to the source exhaustion mechanism.

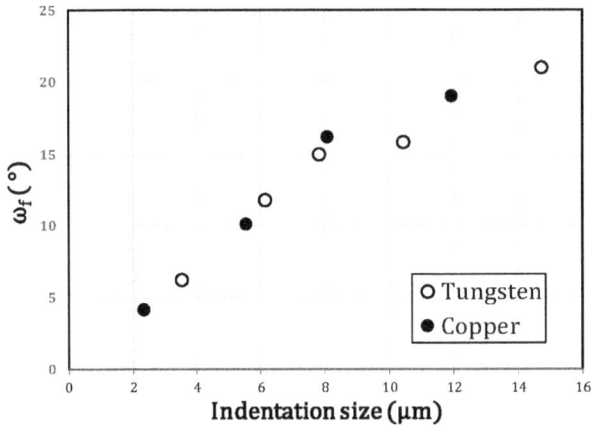

Figure 13. Variation of the maximum misorientation angle ω_f versus the indentation size in tungsten and copper. The original data were reported by Kiener et al. [119].

Demir et al. [121] investigated the size effects at low indentation depths incorporating the GND density obtained using EBSD tomography. They conducted the nanoindentation test on a single crystal Cu sample using a conical indenter with a spherical tip. Figure 14 illustrates the GND density pattern for five equally-spaced cross-sections at four different indentation depths. The results show that the assumption that the dislocations are homogenously distributed beneath the indenter, which has been commonly incorporated in the literature, is not accurate. Furthermore, Demir et al. [121] obtained the total dislocation density using the information obtained from 50 individual 2D EBSD sections at different indentation depths. Figure 15 shows the variation of both dislocation density and hardness as the indentation depth varies. As the results show, the density of GNDs increases as the indentation depth increases. According to the forest hardening mechanisms and consequently the Nix and Gao model [13], the hardness should also increase. However, as the results show, the hardness decreases as the indentation depth increases. This means that the forest hardening mechanism breaks down at shallow indentation depths, and the models developed based on the premise, i.e., Nix and Gao model [13] and its relating models, are not valid in this region. Demir et al. [121] attributed the observed hardening to the decreasing dislocation segment lengths as the indentation depth decreases, which has a similar nature to the source truncation and weakest link theory mechanisms. One should note that the conducted experiments solely consider the GNDs and not SSDs. Accordingly, the result may not be dependable. However, Demir et al. [121] stated that the amount of imposed deformation decreases by decreasing the indentation depth, and consequently, SSDs are not responsible for the enhanced hardening at lower indentation depths.

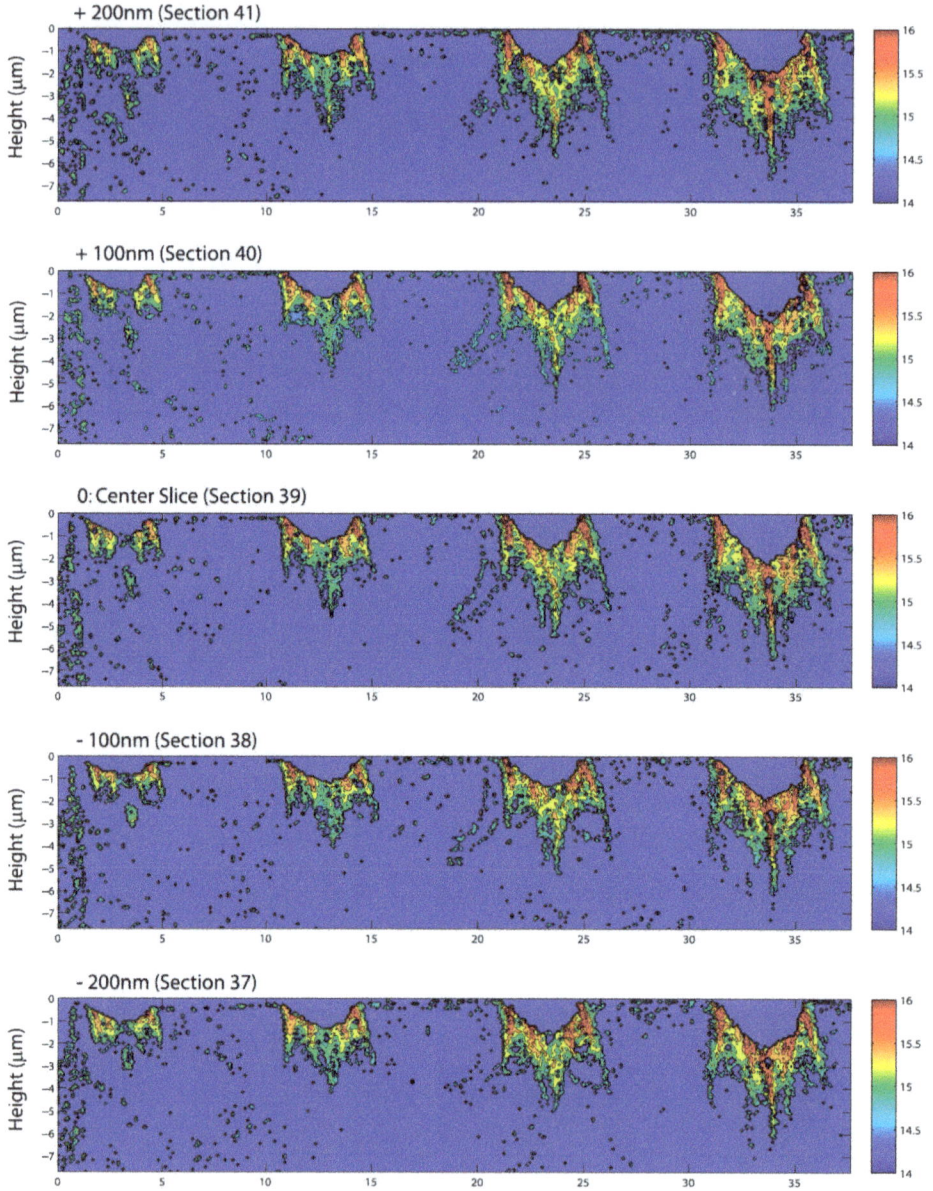

Figure 14. Five equally-spaced cross-sections (center slice, ±100 nm, ±200 nm) through the four indents. Color code: GND density in decadic logarithmic scale (m^{-2}) (after Demir et al. [121]).

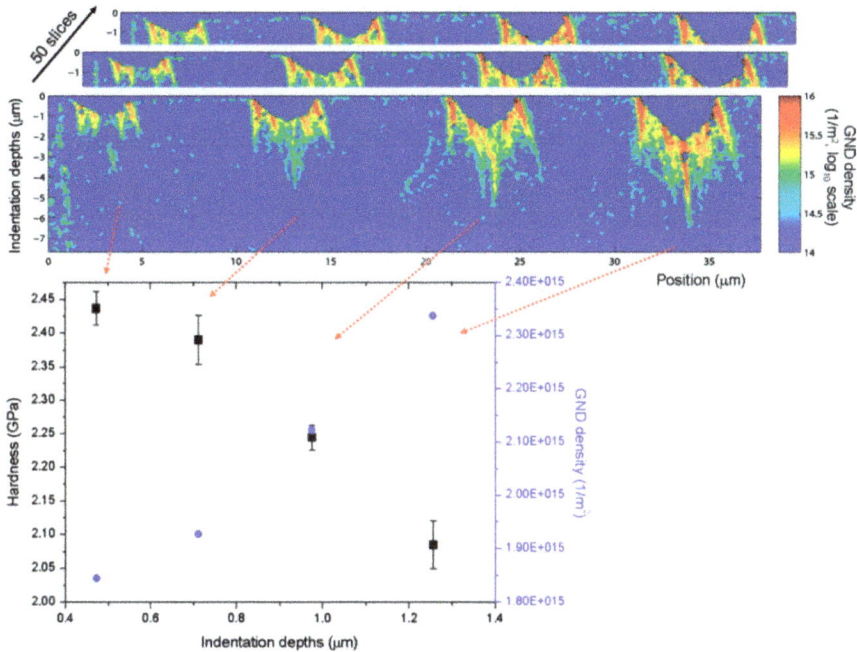

Figure 15. The variations of hardness and GND density versus the indentation depth (after Demir et al. [121]).

Feng et al. [125] incorporated the μXRD scheme and studied the evolution of defects in Cu single crystal during nanoindentation indented using a Berkovich indenter. Feng et al. [125] stated that the size effects are in line with the strain gradient theory and forest hardening mechanism even at shallow indentation depths, which is in contradiction with the results reported by Kiener et al. [119], McLaughlin and Clegg [95] and Demir et al. [121]. They observed that as the indentation depth decreases, both GND density and hardness increases. However, unlike Demir et al. [121], which obtained the GND density pattern below the indenter, Feng et al. [125] approximated the strain gradient. Accordingly, the equation of total GND density depends on the radius of the plastic zone, which was assumed to be the contact radius multiplied by a constant. Accordingly, the effect of heterogeneity reported by Demir et al. [121] was not included in the methodology presented by Feng et al. [125] to obtain the total GND density. In order to investigate whether the trend reported by Kiener et al. [119], McLaughlin and Clegg [95] and Demir et al. [121] is accurate or the one presented by Feng et al. [125], the meso-scale simulations, such as the atomistic simulation, should be incorporated.

5. Atomistic Simulation of Nanoindentation

The microstructural observations during the nanoindentation indicate some contradictions in a way that both the increase and decrease in GND density have been reported as the indentation depths decrease [95,119,121,125]. In order to shed light on these discrepancies and unravel the governing mechanism of size effects during nanoindentation, one way is to model the sample as a cluster of atoms. Molecular dynamics (MD) is a powerful tool that can be incorporated as a pseudo-experimental tool to address the size effect during the nanoindentation. In the case of crystalline metals, many different aspects of nanoindentation experiment, such as initial dislocation nucleation and evolution pattern [79,81,83], the effect of surface step [80], effects of GB [84,88,126,127], the effect of film thickness [128], the effect of substrate [129], the effect of residual stress [130], the effect of boundary conditions [85] and size effects during nanoindentation [86,87] have been investigated using MD. Szlufarska [131] summarized the advances in atomistic modelling of the nanoindentation experiment. She addressed the challenges of atomistic simulation of nanoindentation and how it can increase the current understanding of nanoindentation [131].

During the MD simulation, Newton's equations of motion are solved for all the atoms. Even for nano-sized samples, the number of degrees of freedom is enormous, and massive parallel code should be incorporated to integrate Newton's equations of motion. The interaction of atoms with each other is described using predefined potentials. In the case of crystalline metals, two potentials of the embedded-atom method (EAM) (Daw and Baskes [132]) and modified embedded-atom method (MEAM) (Baskes [133]) are commonly incorporated.

The indenter itself can be modeled as a cluster of atoms. However, the indenter is commonly modeled as a repulsive potential to reduce the computational cost. Selecting an appropriate set of boundary conditions to precisely mimic the considered phenomenon is an essential part of the MD simulation. Up to now, four different sets of boundary conditions have been used to model the nanoindentation experiment, which can be described as follows:

- Fixing some atomic layers at the sample bottom to act as a substrate, using the free surface for the top and periodic boundary conditions for the remaining surfaces (see, e.g., Kelchner et al. [79]; Zimmerman et al. [80]; Nair et al. [128]).

- Fixing some atomic layer at the surrounding surfaces and using free surfaces for the sample top and bottom (see, e.g., Medyanik and Shao [134]; Shao and Medyanik [135]).

- Using the free surface for the sample top and bottom, incorporating the periodic boundary conditions for the remaining surfaces and putting a substrate under the thin film (see, e.g., Peng et al. [129]).

- Incorporating the free surfaces for the sample top and bottom, using periodic boundary conditions for the remaining surfaces and equilibrating the sample by adding some forces (see, e.g., Li et al. [81]; Lee et al. [83]).

Yaghoobi and Voyiadjis [85] studied the effects of selected boundary conditions on the nanoindentation response of FCC crystalline metals. They showed that the boundary conditions may alter the plasticity initiation and defect nucleation pattern depending on the film thickness and indenter radius.

In order to study the size effects using MD, indentation depth, indentation force, contact area and dislocation length and density should be precisely obtained. In the case of hardness, unlike the experiment, the precise contact can be obtained much more easily. To do so, the precise contact area (A) should be captured to calculate the hardness. Saraev and Miller [136] proposed a method using a projection of atoms that are in contact with the indenter to obtain the precise contact area. A 2D-mesh is produced from the projections of atoms in contact with the indenter. The total contact area is then calculated using the obtained 2D mesh. The indenter force can be derived from the repulsive potential used to model the indenter. Voyiadjis and his co-workers [85–87] showed that the true indentation depth h is different from the tip displacement d during nanoindentation. They developed the required geometrical equations to obtain the precise indentation depth.

MD provides the atomic trajectories and velocities. However, the dislocation structure is an essential part of the size effects' investigation. Accordingly, a post-processing scheme should be incorporated to capture the dislocation structure. In order to visualize the defects, several methods have been introduced such as energy filtering, bond order, centrosymmetry parameter, adaptive common neighbor analysis, Voronoi analysis, and neighbor distance analysis, which have been compared with each other by Stukowski [137]. Furthermore, the Crystal Analysis Tool (Stukowski and Albe [138]) can extract the dislocation structure from the atomistic data. The common-neighbor analysis method is the basic idea of this code. The code is able to calculate the dislocations information such as the Burgers vector and total dislocation length. To extract the required information, the Crystal

Analysis Tool constructs a Delaunay mesh, which connects all atoms. Next, using the constructed mesh, the elastic deformation gradient tensor is obtained. The code defines the dislocations using the fact that the elastic deformation gradient does not have a unique value when a tessellation element intersects a dislocation.

Voyiadjis and Yaghoobi [86] conducted large-scale MD simulation of Ni thin film during nanoindentation using different indenter geometries of right square prismatic, conical and cylindrical to investigate the proposed theoretical models for indentation size effects. First, Voyiadjis and Yaghoobi [86] investigated the dislocation nucleation and evolution pattern during nanoindentation. As an example, Figure 16 shows the dislocation nucleation and evolution for Ni thin film indented by a cylindrical indenter during nanoindentation. The dislocations and stacking faults are visualized while the perfect atoms are removed. The color of Shockley, Hirth and stair-rod partial dislocations and perfect dislocations are green, yellow, blue and red, respectively. Figure 17 illustrates the dislocation loop formation and movement along three directions of $[\bar{1}\,0\,\bar{1}]$, $[\bar{1}\,\bar{1}\,0]$ and $[0\,\bar{1}\,\bar{1}]$.

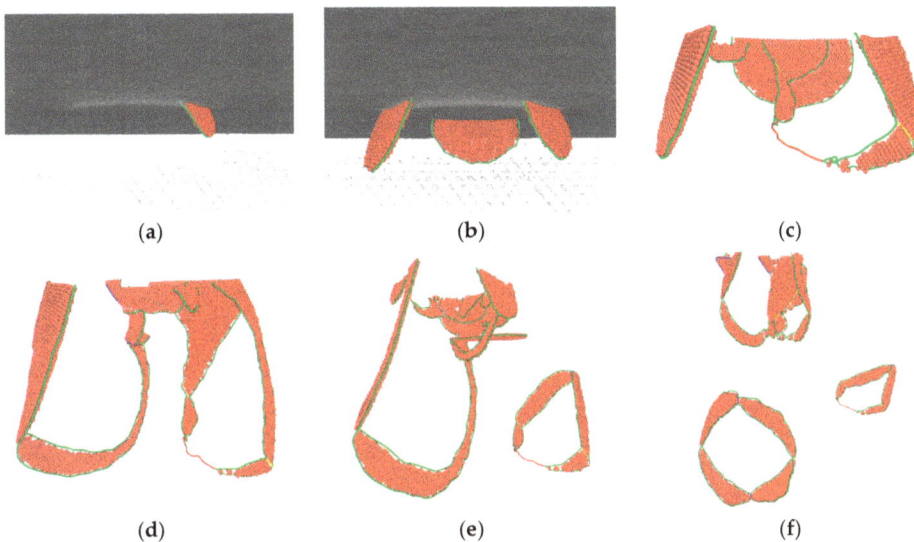

Figure 16. Defect nucleation and evolution of Ni thin film indented by a cylindrical indenter at different indenter tip displacements of (**a**) $d \approx 0.70$ nm; (**b**) $d \approx 0.86$ nm; (**c**) $d \approx 0.96$ nm; (**d**) $d \approx 1.02$ nm; (**e**) $d \approx 1.05$ nm; (**f**) $d \approx 1.12$ nm (after Voyiadjis and Yaghoobi [86]).

Voyiadjis and Yaghoobi [86] investigated the plastic zone size measured directly from MD. The dislocation prismatic loops glide toward the bottom of the sample. Accordingly, Voyiadjis and Yaghoobi [86] assumed that theses dislocation loops should not be considered as a measure of the plastic zone size. Otherwise, the size of the plastic zone becomes unreasonably large. In other words, the plastic zone size is determined based on the furthest dislocation that is attached to the main body of dislocations beneath the indenter. It was assumed that the plastic zone is a hemisphere with a radius of $R_{pz} = fa_c$. They showed that f is a function of indentation depth. In the case of the conical indenter, which is a self-similar indenter, the maximum value of f observed during the nanoindentation simulation is 5.2, which is larger than those that have been previously proposed (see Durst et al. [36]). However, in the case of cylindrical indenter, the maximum value of f is nine, which is much larger than that of the self-similar indenter.

Figure 18 compares the dislocation lengths obtained from atomistic simulation with those predicted by theoretical models of Nix and Gao [13], Swadener [14] and Pugno [90]. The results show that the theoretical predictions can accurately capture the dislocation lengths during nanoindentation. However, some discrepancies are observed, which can be related to the fact that atomistic simulation captures the total dislocation length including both geometrically necessary and statistically stored dislocations, while the theoretical models only calculate geometrically necessary dislocations. Furthermore, the dislocations that are detached from the main dislocations network as the prismatic loops and leave the plastic zone around the indenter are not considered in the total dislocation length calculations. Gao et al. [139] investigated the plastic zone volume size during the nanoindentation in FCC and BCC metals using a spherical indenter. They showed that the plastic zone size is not strongly influenced by the crystallographic orientation, crystal structure and the selected parameter for indentation. Furthermore, they observed that the plastic zone sizes after indent are larger than the values reported in the literature. These findings are in line with those reported by Voyiadjis and Yaghoobi [86]. However, they showed that after the indenter retraction, the plastic zone shrinks, and its size becomes closer with those reported in the literature (see Durst et al. [36]). Figure 19 shows the plastic zone volume changes after indent and after retraction during nanoindentation of Ta. The plastic zone coefficient f changes from 2.7 to 2.3.

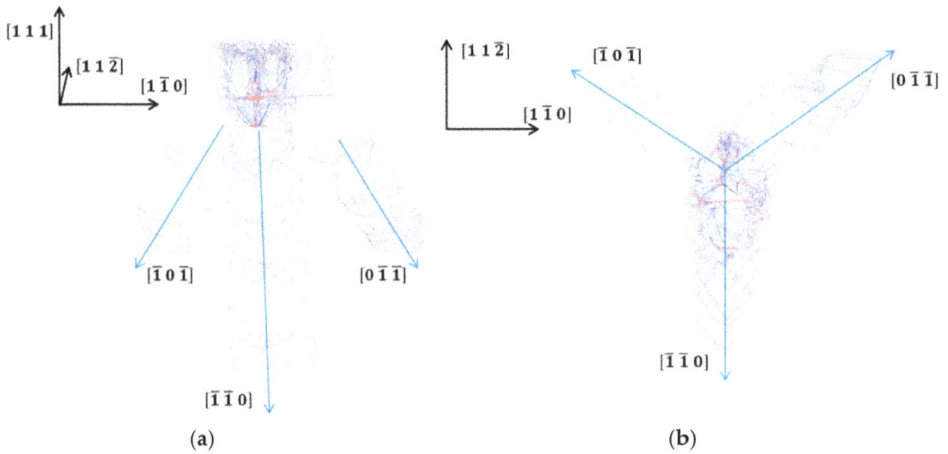

Figure 17. Prismatic loops forming and movement in Ni thin film indented by the cylindrical indenter during nanoindentation: (**a**) side view and (**b**) top view (after Voyiadjis and Yaghoobi [86]).

Yaghoobi and Voyiadjis [87] incorporated large-scale MD simulation of Ni thin films during nanoindentation to investigate which size effect trend occurs during atomistic simulation, i.e., the one reported by Kiener et al. [119], McLaughlin and Clegg [95] and Demir et al. [121] or the one presented by Feng et al. [125]. They selected a conical indenter with an angle of $\theta = 60°$ and a spherical tip, which was used by Demir et al. [121]. Figure 20a presents the variation of the mean contact pressure ($p_m = P/A$), which is equivalent to the hardness H in the plastic region, as a function of indentation depth h. In the elastic region, Figure 20a shows that p_m increases as the indentation depth increases. However, after the initiation of plasticity, Figure 20a shows that the mean contact pressure, i.e., hardness, decreases as the indentation depth increases. Yaghoobi and Voyiadjis [87] assumed that the plastic zone is a hemisphere with the radius of $R_{pz} = f a_c$ in which f is a constant. They incorporated different values of $f = 1.5$, 2.0, 2.5, 3.0 and 3.5, which are chosen to investigate the effect of plastic zone size on the dislocation density during nanoindentation. The dislocation density ρ versus the indentation depth h is plotted in Figure 20b. The results indicate that as the indentation depth increases, the dislocation density also increases for different values of f. Based on the Taylor hardening model, since the dislocation density increases as the indentation depth increases, the hardness should also increase. However, Figure 20a shows that as the indentation depth increases, the hardness decreases, while the dislocation density increases. The observed trend is in line with the experimental observation of Kiener et al. [119], McLaughlin and Clegg [95] and Demir et al. [121]. In other words, the results show that the forest hardening model cannot capture the size effect in the case of the simulated sample.

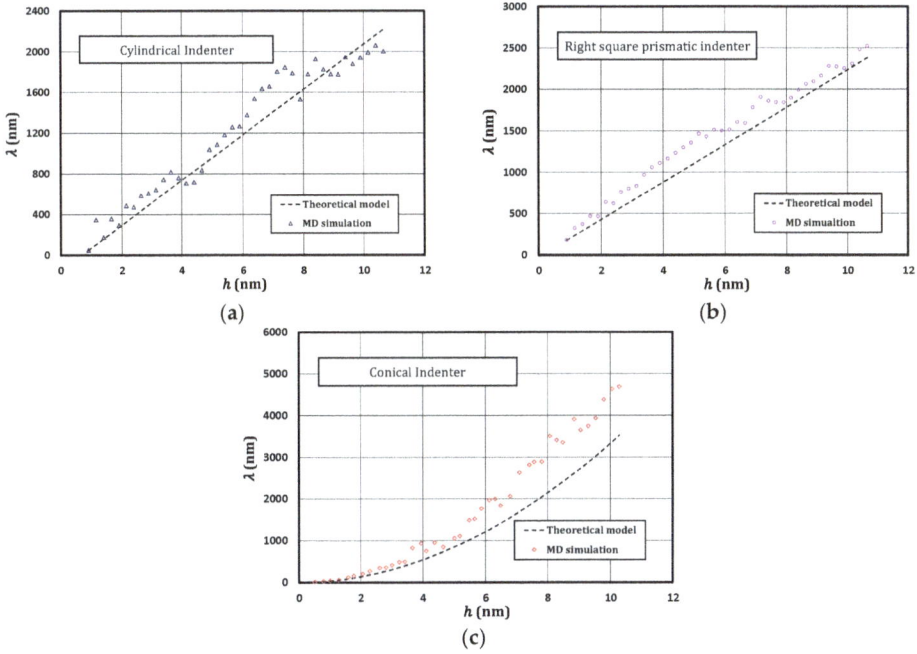

(a)

(b)

(c)

Figure 18. Total dislocation length obtained from the simulation and theoretical models of Nix and Gao [13], Swadener [14] and Pugno [90] in samples indented by the (**a**) cylindrical indenter; (**b**) right square prismatic indenter and (**c**) conical indenter (after Voyiadjis and Yaghoobi [86]).

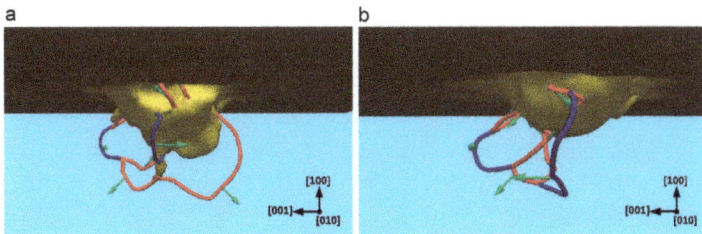

Figure 19. Dislocation structure beneath the indenter during the indentation of Ta (100) surface at 0 K: (**a**) after indent $f = 2.7$; (**b**) after retraction of the indenter $f = 2.3$ (after Gao et al. [139]).

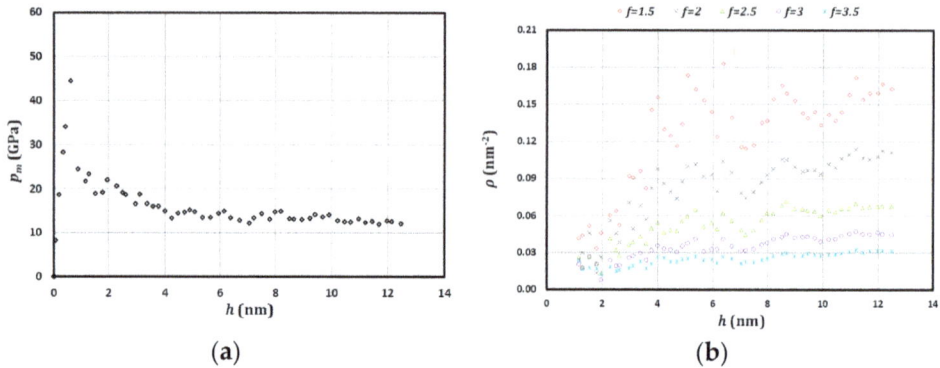

Figure 20. Large-scale atomistic simulation of a Ni thin film during nanoindentation: (**a**) variation of the mean contact pressure p_m versus the indentation depth h; (**b**) variation of the dislocation density ρ versus the indentation depth h for different values of f (after Yaghoobi and Voyiadjis [87]).

Similar to the compression and tension experiments on micropillars of small length scales, the observed results show that the forest hardening mechanism does not govern size effects in the nanoscale samples during nanoindentation. To unravel the sources of size effects, the initial phases of indentation should be studied. Yaghoobi and Voyiadjis [87] showed that the source exhaustion hardening is the governing mechanism of size effects at shallow indentation depths. By increasing the indentation depth, the dislocation length and density increase, as well, which provides more dislocation sources. Furthermore, the source lengths increase, which reduces the critical resolved shear stress. Consequently, the applied stress required to sustain flow during nanoindentation decreases, i.e., hardness decreases as indentation depth increases.

The interaction between the effects of grain size with indentation size effects have been addressed using MD [84,88,126,127]. However, the observed trends do not show a unified trend. Jang and Farkas [84] incorporated MD simulation and studied a bicrystal nickel thin film with $\sum 5$ (2 1 0) [0 0 1] GB during the nanoindentation. It was shown that the GB induced some resistance to the indentation due to the stacking fault expansion [84]. Kulkarni et al. [127] compared the response of the coherent twin boundary (CTB) with that of $\sum 9$ (2 2 1) tilt GB during nanoindentation using atomistic simulation. They stated that unlike the $\sum 9$ (2 2 1) GB, the CTB does not considerably reduce the strength of the sample. However, unlike Jang and Farkas [84], they did not observe noticeable enhancement in strength. In order to address the discrepancies observed related to the effect of grain boundaries on the indentation response of FCC metals during nanoindentation, Voyiadjis and Yaghoobi [88] investigated the nanoindentation response of bicrystal Ni samples using large-scale MD simulation. They incorporated different symmetric and unsymmetric CSL GBs and two large and small samples with the dimensions of 24 nm × 24 nm × 12 nm and 120 nm × 120 nm × 60 nm. The GBs are located at a third of the thickness from the top surface. Voyiadjis and Yaghoobi [88] compared the nanoindentation response of the bicrystal samples with the corresponding single crystal ones and observed that the effect of GB on the governing mechanism of hardening depends on the grain size. Figure 21 compares the effects of $\sum 11$ (3 3 2) $[1\,\bar{1}\,0]$ on the nanoindentation responses of the small and large samples. GB does not change the general pattern of size effects; however, it can contribute to some specific mechanisms depending on the sample size. The size effects are initially controlled by the source exhaustion mechanism during the nanoindentation. In the case of a small sample, Voyiadjis and Yaghoobi [88] showed that the GB contributes to the dislocation nucleation and reduces the material strength. In the case of a large sample, GB does not have any noticeable effect on the source exhaustion mechanism. This is due to the fact that the total dislocation length of the bicrystal sample is close to that of the single crystal sample. The forest hardening mechanism becomes dominant at larger indentation depths. In the cases of small samples, however, the responses of bicrystal and single-crystal samples are similar in this region. It is worth mentioning that the dislocation movements are blocked by GBs during nanoindentation in small bicrystal samples. However, the density of piled-up dislocation is not enough to change the nanoindentation response. In the cases of large samples, the total dislocation content is much greater than that of the small samples. Accordingly, the interaction of dislocations and GBs plays a key role. Hence, the increase in bicrystal samples can be justified by blockage of the dislocation movements according to the forest hardening mechanism. Moreover, the results show that the total dislocation length in the upper grain is a better representative factor to investigate the strength size effects in the case of the forest hardening mechanism. In other words, according to the forest hardening mechanism, the dislocations in the plastic zone that are closer to the indenter are the effective ones. Hence, in the cases of large samples, the most important role of GBs is to change the pattern of the dislocation structure by blocking their movement, which increases the resulting hardness at higher indentation depths. Voyiadjis and Yaghoobi [88] also investigated the effect of GB on the dislocation nucleation and evolution during nanoindentation. For example, Figure 22 depicts the dislocation nucleation and evolution for the $\Sigma 3$ (1 1 1) $[1 \bar{1} 0]$ bicrystal and its related single crystal large samples. Figure 22 supports the underlying mechanism suggested by Voyiadjis and Yaghoobi [88] to capture the effect of GB on the nanoindentation response of large samples. One should note that the strain rates incorporated in the atomistic simulation are much higher than those selected for experiments. Accordingly, the interpretation of the obtained results should be carefully handled. The applied strain rate can influence both hardening mechanisms and dislocation network properties (see, e.g., [140–144]). In other words, one should note that the observed mechanisms are not an artifact of the high strain rates used in the atomistic simulation.

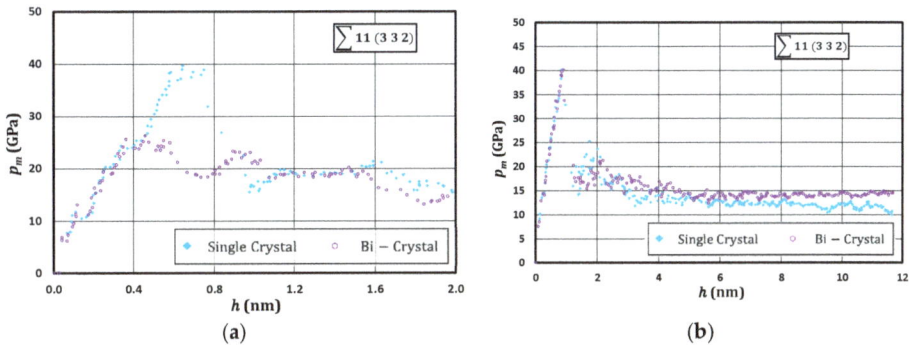

Figure 21. Effect of $\Sigma 11$ (3 3 2) grain boundary on the nanoindentation response of Ni thin film: (**a**) small sample and (**b**) large sample (after Voyiadjis and Yaghoobi [88]).

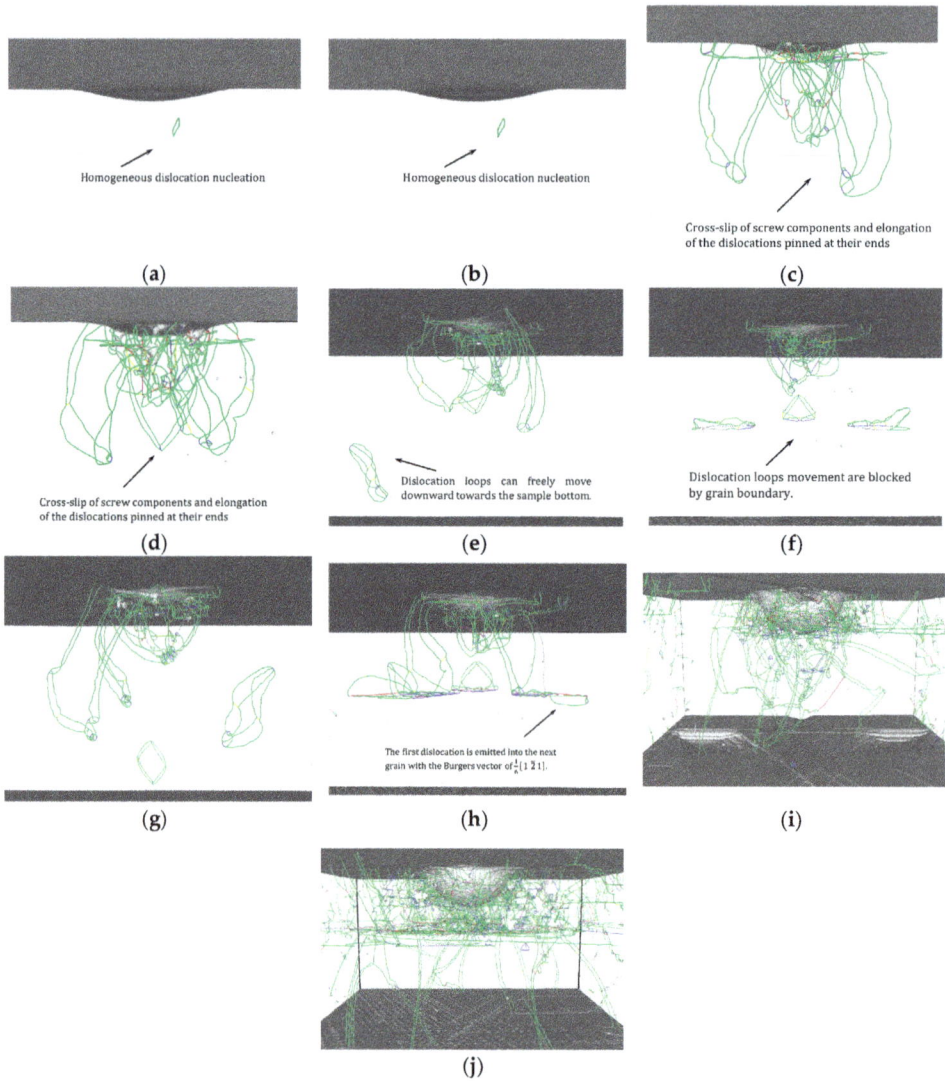

Figure 22. Effect of $\sum 3$ (1 1 1) $\left[1\,\bar{1}\,0\right]$ GB on the dislocation nucleation and evolution of large sample at different indentation depths: (**a**) single crystal sample, $h \approx 0.88$ nm; (**b**) bicrystal sample, $h \approx 0.88$ nm; (**c**) single crystal sample, $h \approx 1.15$ nm; (**d**) bicrystal sample, $h \approx 1.15$ nm; (**e**) single crystal sample, $h \approx 1.44$ nm; (**f**) bicrystal sample, $h \approx 1.44$ nm; (**g**) single crystal sample, $h \approx 2.03$ nm; (**h**) bicrystal sample, $h \approx 2.03$ nm; (**i**) single crystal sample, $h \approx 11.5$ nm; (**j**) bicrystal sample, $h \approx 11.5$ nm (after Voyiadjis and Yaghoobi [88]).

6. Summary and Conclusions

The current work reviews the size effect in crystalline metals during nanoindentation, with emphasis on the underlying mechanisms governing this phenomenon. The indentation size effects in crystalline metals including the variation of hardness versus the indentation depth for geometrically self-similar indenter tips and variation of hardness versus the indenter radius for spherical indenters are well documented and have been observed by many researchers. The size effects have been successfully captured using the concept of geometrically necessary dislocations (GNDs). Accordingly, the increase in hardness by decreasing the indentation depth is attributed to the increase of GNDs density, which leads to the enhanced strength according to the forest hardening mechanism. A similar methodology has been incorporated to capture the size effects in the case of spherical indenters as the indenter radius varies. Furthermore, a material length scale can be obtained for gradient plasticity models using the same methodology as $l \equiv b(G/\sigma_0)^2$ [13]. The presented mechanism defines the variation of hardness H versus the indentation depth h in the form of $(H/H_0)^2 = 1 + h^*/h$, where H_0 and h^* are the material constants. However, the conducted studies have shown that the developed relation cannot capture the reported size effect for small indentation depths. Furthermore, the coupling effects of indentation depth and grain size effects can be successfully captured using the same methodology. The experimental results show that the grain boundary (GB) may enhance the hardness by blocking the movement of nucleated GNDs. Accordingly, a new material length scale can be obtained for nonlocal plasticity, which incorporates the effect of material grain size [108].

When it comes to the small indentation depths, some discrepancies and contradictory explanations have been presented. The understanding of the underlying mechanisms of size effects during the nanoindentation for shallow indentation depths is becoming more and more complete with the help of extensive experimental and computational investigations. Up to now, three experimental techniques of backscattered electron diffraction (EBSD), convergent beam electron diffraction (CBED) and X-ray microdiffraction (μXRD) have been introduced to measure the dislocation density in metallic samples. These methods have been incorporated to investigate the governing mechanisms of size effects during nanoindentation for small indentation depths. However, two types of trends for the variation of dislocation density have been reported as the indentation depth decreases. Some experimental studies (see, e.g., [95,119,121]) have shown that the dislocation density decreases as the indentation depth decreases, while the hardness increases, which is in contradiction with the conventional size effect theory and its underlying hardening mechanism, i.e., the forest hardening mechanism. The observed hardening as the indentation depth decreases has been attributed to the decrease in the length of dislocation segments. On the other hand, some experimental observations validate the conventional theory of size effects by reporting that the dislocation density increases by decreasing the indentation depth, which increases the hardness according to the forest hardening mechanism (see, e.g., [125]). In the case of small dislocation depths, more experiments need to be conducted to study the nanoindentation size effects by measuring the variation in dislocation density during the indentation and to test which trend is valid, i.e., the dislocation density increases or decreases as the indentation depth decreases. Finally, the atomistic simulation can contribute to the better understanding of the nature of size effects during nanoindentation at smaller indentation depths. The atomistic simulations have investigated many aspects of the nanoindentation response of metallic samples. The dislocation nucleation and evolution can be visualized during the nanoindentation, which can be of great help to investigate the indentation size effect. The conducted simulations have shown that the dislocation density decreases as the indentation depth decreases, while the hardness increases (see, e.g., [87]). The source exhaustion is introduced as the governing mechanism of size effects for shallow indentations where not enough dislocation content is available to sustain the applied plastic flow. Accordingly, as the indentation depth increases, the dislocation density increases, and less stress is required to sustain the plastic flow, which leads to lower hardness. Furthermore, the effect of GB on the indentation size effect can be studied using atomistic simulations. The results have shown that the GB may increase or decrease the indentation hardness depending on the grain size. This is due to the fact that in the case of metallic samples with very fine grains, the GB contributes to the dislocation nucleation, which provides more dislocations at shallow depths and decreases the indentation hardness according to the source exhaustion mechanism. In the case of larger grain sizes, on the other hand, the blockage of dislocation movements by GB enhances the hardness according to the forest hardening mechanism.

Acknowledgments: The current work is partially funded by the NSF EPSCoR CIMM project under Award #OIA-1541079.

Conflicts of Interest: The authors declare no conflict of interest.

References

1. Bull, S.J.; Page, T.F.; Yoffe, E.H. An explanation of the indentation size effect in ceramics. *Philos. Mag. Lett.* **1989**, *59*, 281–288. [CrossRef]
2. Li, H.; Bradt, R.C. The microhardness indentation load/size effect in rutile and cassiterite single crystals. *J. Mater. Sci.* **1993**, *28*, 917–926. [CrossRef]
3. Zhu, T.T.; Bushby, A.J.; Dunstan, D.J. Size effect in the initiation of plasticity for ceramics in nanoindentation. *J. Mech. Phys. Solids* **2008**, *56*, 1170–1185. [CrossRef]
4. Pharr, G.M.; Herbert, E.G.; Gao, Y. The Indentation Size Effect: A Critical Examination of Experimental Observations and Mechanistic Interpretations. *Annu. Rev. Mater. Res.* **2010**, *40*, 271–292. [CrossRef]
5. Mott, B.W. *Microindentation Hardness Testing*; Butterworths: London, UK, 1957.
6. Bückle, H. Progress in microindentation hardness testing. *Metall. Rev.* **1959**, *4*, 49–100. [CrossRef]
7. Gane, N. The direct measurement of the strength of metals on a submicrometer scale. *Proc. R. Soc. Lond. Ser. A* **1970**, *317*, 367–391. [CrossRef]
8. Upit, G.P.; Varchenya, S.A. The size effect in the hardness of single crystals. In *The Science of Hardness Testing and Its Research Applications*; Westbrook, J.H., Conrad, H., Eds.; American Society for Metals: Metals Park, OH, USA, 1973; Volume 10, pp. 135–146.
9. Chen, C.C.; Hendrickson, A.A. Microhardness phenomena in silver. In *The Science of Hardness Testing and Its Research Applications*; Westbrook, J.H., Conrad, H., Eds.; American Society for Metals: Metals Park, OH, USA, 1973; Volume 21, pp. 274–290.
10. Tabor, D. Indentation hardness and its measurement: Some cautionary comments. In *Microindentation Techniques in Materials Science and Engineering. ASTM STP 889*; Blau, P.J., Lawn, B.R., Eds.; ASTM: Philadelphia, PA, USA, 1986; pp. 129–159.
11. Stelmashenko, N.A.; Walls, M.G.; Brown, L.M.; Milman, Y.V. Microindentation on W and Mo oriented single crystals: An STM study. *Acta Metall. Mater.* **1993**, *41*, 2855–2865. [CrossRef]
12. De Guzman, M.S.; Neubauer, G.; Flinn, P.; Nix, W.D. The role of indentation depth on the measured hardness of materials. *Mater. Res. Soc. Symp. Proc.* **1993**, *308*, 613–618. [CrossRef]
13. Nix, W.D.; Gao, H. Indentation size effects in crystalline materials: A law for strain gradient plasticity. *J. Mech. Phys. Solids* **1998**, *46*, 411–425. [CrossRef]
14. Swadener, J.G.; George, E.P.; Pharr, G.M. The correlation of the indentation size effect measured with indenters of various shapes. *J. Mech. Phys. Solids* **2002**, *50*, 681–694. [CrossRef]
15. Lim, Y.Y.; Chaudhri, M.M. The effect of the indenter load on the nanohardness of ductile metals: An experimental study on polycrystalline work-hardened and annealed oxygen-free copper. *Philos. Mag. A* **1999**, *79*, 2979–3000. [CrossRef]
16. Bushby, A.J.; Dunstan, D.J. Plasticity size effects in nanoindentation. *J. Mater. Res.* **2004**, *19*, 137–142. [CrossRef]
17. Spary, I.J.; Bushby, A.J.; Jennett, N.M. On the indentation size effect in spherical indentation. *Phil. Mag.* **2006**, *86*, 5581–5593. [CrossRef]
18. Durst, K.; Göken, M.; Pharr, G.M. Indentation size effect in spherical and pyramidal indentations. *J. Phys. D Appl. Phys.* **2008**, *41*, 074005. [CrossRef]
19. Gerberich, W.W.; Tymiak, N.I.; Grunlan, J.C.; Horstemeyer, M.F.; Baskes, M.I. Interpretations of indentation size effect. *J. Appl. Mech.* **2002**, *69*, 433–442. [CrossRef]
20. Bull, S.J. On the origins of the indentation size effect. *Z. Metallkd.* **2003**, *94*, 787–792. [CrossRef]
21. Zhu, T.T.; Bushby, A.J.; Dunstan, D.J. Materials mechanical size effects: A review. *Mater. Tech.* **2008**, *23*, 193–209. [CrossRef]
22. Kiener, D.; Durst, K.; Rester, M.; Minor, A.M. Revealing deformation mechanisms with nanoindentation. *JOM* **2009**, *61*, 14–23. [CrossRef]
23. Sangwal, K. On the reverse indentation size effect and microhardness measurement of solids. *Mater. Chem. Phys.* **2000**, *63*, 145–152. [CrossRef]

24. King, R.B. Elastic analysis of some punch problems for a layered medium. *Int. J. Solids Struct.* **1987**, *23*, 1657–1664. [CrossRef]

25. Giannakopoulos, A.E.; Larsson, P.L.; Vestergaard, R. Analysis of Vickers indentation. *Int. J. Solids Struct.* **1994**, *31*, 2679–2708. [CrossRef]

26. Larsson, P.L.; Giannakopoulos, A.E.; SÖderlund, E.; Rowcliffe, D.J.; Vestergaard, R. Analysis of Berkovich indentation. *Int. J. Solids Struct.* **1996**, *33*, 221–248. [CrossRef]

27. Bolshakov, A.; Oliver, W.C.; Pharr, G.M. Influences of stress on the measurement of mechanical properties using nanoindentation: Part II. Finite element simulations. *J. Mater. Res.* **1996**, *11*, 760–768. [CrossRef]

28. Bolshakov, A.; Oliver, W.C.; Pharr, G.M. Finite element studies of the influence of pile-up on the analysis of nanoindentation data. *Mater. Res. Soc. Symp. Proc.* **1996**, *436*, 141–146. [CrossRef]

29. Bolshakov, A.; Pharr, G.M. Influences of pileup on the measurement of mechanical properties by load and depth sensing indentation techniques. *J. Mater. Res.* **1998**, *13*, 1049–1058. [CrossRef]

30. Lichinchi, M.; Lenardi, C.; Haupt, J.; Vitali, R. Simulation of Berkovich nanoindentation experiments on thin films using finite element method. *Thin Solid Films* **1998**, *312*, 240–248. [CrossRef]

31. Huber, N.; Tsakmakis, C. Experimental and theoretical investigation of the effect of kinematic hardening on spherical indentation. *Mech. Mater.* **1998**, *27*, 241–248. [CrossRef]

32. Cheng, Y.-T.; Cheng, C.-M.J. Scaling approach to conical indentation in elastic-plastic solids with work hardening. *Appl. Phys.* **1998**, *84*, 1284–1291. [CrossRef]

33. Cheng, Y.-T.; Cheng, C.-M. Scaling relationships in conical indentation of elastic-perfectly plastic solids. *Int. J. Solids Struct.* **1999**, *36*, 1231–1243. [CrossRef]

34. Hay, J.C.; Bolshakov, A.; Pharr, G.M. A critical examination of the fundamental relations used in the analysis of nanoindentation data. *J. Mater. Res.* **1999**, *14*, 2296–2305. [CrossRef]

35. Dao, M.; Chollacoop, N.; Van Vliet, K.J.; Venkatesh, T.A.; Suresh, S. Computational modeling of the forward and reverse problems in instrumented sharp indentation. *Acta Mater.* **2001**, *49*, 3899–3918. [CrossRef]

36. Durst, K.; Backes, B.; Göken, M. Indentation size effect in metallic materials: Correcting for the size of the plastic zone. *Scr. Mater.* **2005**, *52*, 1093–1097. [CrossRef]

37. Warren, A.W.; Guo, Y.B. Machined surface properties determined by nanoindentation: Experimental and FEA studies on the effects of surface integrity and tip geometry. *Surf. Coat. Technol.* **2006**, *201*, 423–433. [CrossRef]

38. Beghini, M.; Bertini, L.; Fontanari, V. Evaluation of the stress–strain curve of metallic materials by spherical indentation. *Int. J. Solids Struct.* **2006**, *43*, 2441–2459. [CrossRef]

39. Walter, C.; Antretter, T.; Daniel, R.; Mitterer, C. Finite element simulation of the effect of surface roughness on nanoindentation of thin films with spherical indenters. *Surf. Coat. Technol.* **2007**, *202*, 1103–1107. [CrossRef]

40. Kalidindi, S.R.; Pathak, S. Determination of the effective zero-point and the extraction of spherical nanoindentation stress–strain curves. *Acta Mater.* **2008**, *56*, 3523–3532. [CrossRef]

41. Sakharova, N.A.; Fernandes, J.V.; Antunes, J.M.; Oliveira, M.C. Comparison between Berkovich, Vickers and conical indentation tests: A three-dimensional numerical simulation study. *Int. J. Solids Struct.* **2009**, *46*, 1095–1104. [CrossRef]

42. Wang, Y.; Raabe, D.; Klüber, C.; Roters, F. Orientation dependence of nanoindentation pile-up patterns and of nanoindentation microtextures in copper single crystals. *Acta Mater.* **2004**, *52*, 2229–2238. [CrossRef]

43. Zaafarani, N.; Raabe, D.; Singh, R.N.; Roters, F.; Zaefferer, S. Three dimensional investigation of the texture and microstructure below a nanoindent in a Cu single crystal using 3D EBSD and crystal plasticity finite element simulations. *Acta Mater.* **2006**, *54*, 1863–1876. [CrossRef]

44. Casals, O.; Očenášek, J.; Alcalá, J. Crystal plasticity finite element simulations of pyramidal indentation in copper single crystals. *Acta Mater.* **2007**, *55*, 55–68. [CrossRef]

45. Zaafarani, N.; Raabe, D.; Roters, F.; Zaefferer, S. On the origin of deformation-induced rotation patterns below nanoindents. *Acta Mater.* **2008**, *56*, 31–42. [CrossRef]

46. Alcalá, J.; Casals, O.; Ocenasek, J. Micromechanics of pyramidal indentation in fcc metals: Single crystal plasticity finite element analysis. *J. Mech. Phys. Solids* **2008**, *56*, 3277–3303. [CrossRef]

47. Britton, T.B.; Liang, H.; Dunne, F.P.E.; Wilkinson, A.J. The effect of crystal orientation on the indentation response of commercially pure titanium: Experiments and simulations. *Proc. R. Soc. Lond. A Math. Phys. Eng. Sci.* **2010**, *466*, 695–719. [CrossRef]

48. Eidel, B. Crystal plasticity finite-element analysis versus experimental results of pyramidal indentation into (001) fcc single crystal. *Acta Mater.* **2011**, *59*, 1761–1771. [CrossRef]

49. Dahlberg, C.F.O.; Saito, Y.; Öztop, M.S.; Kysar, J.W. Geometrically necessary dislocation density measurements associated with different angles of indentations. *Int. J. Plast.* **2014**, *54*, 81–95. [CrossRef]

50. Dahlberg, C.F.O.; Saito, Y.; Öztop, M.S.; Kysar, J.W. Geometrically necessary dislocation density measurements at a grain boundary due to wedge indentation into an aluminum bicrystal. *J. Mech. Phys. Solids* **2017**, *105*, 131–149. [CrossRef]

51. Fivel, M.; Verdier, M.; Canova, G. 3D simulation of a nanoindentation test at a mesoscopic scale. *Mater. Sci. Eng. A* **1997**, *234–236*, 923–926. [CrossRef]

52. Fivel, M.C.; Robertson, C.F.; Canova, G.R.; Boulanger, L. Three-dimensional modeling of indent-induced plastic zone at a mesoscale. *Acta Mater.* **1998**, *46*, 6183–6194. [CrossRef]

53. Kreuzer, H.G.M.; Pippan, R. Discrete dislocation simulation of nanoindentation: The effect of moving conditions and indenter shape. *Mater. Sci. Eng. A* **2004**, *387–389*, 254–256. [CrossRef]

54. Kreuzer, H.G.M.; Pippan, R. Discrete dislocation simulation of nanoindentation. *Comput. Mech.* **2004**, *33*, 292–298. [CrossRef]

55. Kreuzer, H.G.M.; Pippan, R. Discrete dislocation simulation of nanoindentation: Indentation size effect and the influence of slip band orientation. *Acta Mater.* **2004**, *55*, 3229–3235. [CrossRef]

56. Kreuzer, H.G.M.; Pippan, R. Discrete dislocation simulation of nanoindentation: The effect of statistically distributed dislocations. *Mater. Sci. Eng. A* **2005**, *400*, 460–462. [CrossRef]

57. Widjaja, A.; Van Der Giessen, E.; Needleman, A. Discrete dislocation modeling of submicron indentation. *Mater. Sci. Eng. A* **2005**, *400*, 456–459. [CrossRef]

58. Balint, D.S.; Deshpande, V.; Needleman, A.; Van Der Giessen, E. Discrete dislocation plasticity analysis of the wedge indentation of films. *J. Mech. Phys. Solids* **2006**, *54*, 2281–2303. [CrossRef]

59. Widjaja, A.; Van Der Giessen, E.; Deshpande, V.; Needleman, A. Contact area and size effects in discrete dislocation modeling of wedge indentation. *J. Mater. Res.* **2007**, *22*, 655–666. [CrossRef]

60. Rathinam, M.; Thillaigovindan, R.; Paramasivam, P. Nanoindentation of aluminum (100) at various temperatures. *J. Mech. Sci. Technol.* **2009**, *23*, 2652–2657. [CrossRef]

61. Tsuru, T.; Shibutani, Y.; Kaji, Y. Nanoscale contact plasticity of crystalline metal: Experiment and analytical investigation via atomistic and discrete dislocation models. *Acta Mater.* **2010**, *58*, 3096–3102. [CrossRef]

62. Po, G.; Mohamed, M.S.; Crosby, T.; Erel, C.; El-Azab, A.; Ghoniem, N. Recent progress in discrete dislocation dynamics and its applications to micro plasticity. *J. Mater.* **2014**, *66*, 2108–2120. [CrossRef]

63. Tadmor, E.B.; Miller, R.; Philipps, R.; Ortiz, M. Nanoindentation and incipient plasticity. *J. Mater. Res.* **1999**, *14*, 2233–2250. [CrossRef]

64. Smith, G.; Tadmor, E.; Bernstein, N.; Kaxiras, E. Multiscale simulations of silicon nanoindentation. *Acta Mater.* **2001**, *49*, 4089–4101. [CrossRef]

65. Iglesias, R.A.; Leiva, E.P. Two-grain nanoindentation using the quasicontinuum method: Two-dimensional model approach. *Acta Mater.* **2006**, *54*, 2655–2664. [CrossRef]

66. Fanlin, Z.; Yi, S. Quasicontinuum simulation of nanoindentation of nickel film. *Acta Mech. Solida Sin.* **2006**, *19*, 283–288. [CrossRef]

67. Jin, J.; Shevlin, S.; Guo, Z. Multiscale simulation of onset plasticity during nanoindentation of Al (001) surface. *Acta Mater.* **2008**, *56*, 4358–4368. [CrossRef]

68. Jiang, W.G.; Su, J.J.; Feng, X.Q. Effect of surface roughness on nanoindentation test of thin films. *Eng. Fract. Mech.* **2008**, *75*, 4965–4972. [CrossRef]

69. Li, J.; Ni, Y.; Wang, H.; Mei, J. Effects of crystalline anisotropy and indenter size on nanoindentation by multiscale simulation. *Nanoscale Res. Lett.* **2009**, *5*, 420–432. [CrossRef] [PubMed]

70. Yu, W.; Shen, S. Effects of small indenter size and its position on incipient yield loading during nanoindentation. *Mater. Sci. Eng. A* **2009**, *526*, 211–218. [CrossRef]

71. Yu, W.; Shen, S. Multiscale analysis of the effects of nanocavity on nanoindentation. *Comput. Mater. Sci.* **2009**, *46*, 425–430. [CrossRef]

72. Yu, W.; Shen, S. Initial dislocation topologies of nanoindentation into copper (001) film with a nanocavity. *Eng. Fract. Mech.* **2010**, *77*, 3329–3340. [CrossRef]

73. Li, J.; Lu, H.; Ni, Y.; Mei, J. Quasicontinuum study the influence of misfit dislocation interactions on nanoindentation. *Comput. Mater. Sci.* **2011**, *50*, 3162–3170. [CrossRef]

74. Lu, H.; Li, J.; Ni, Y. Position effect of cylindrical indenter on nanoindentation into Cu thin film by multiscale analysis. *Comput. Mater. Sci.* **2011**, *50*, 2987–2992. [CrossRef]

75. Lu, H.; Ni, Y. Effect of surface step on nanoindentation of thin films by multiscale analysis. *Thin Solid Films* **2012**, *520*, 4934–4940. [CrossRef]

76. Lu, H.; Ni, Y.; Mei, J.; Li, J.; Wang, H. Anisotropic plastic deformation beneath surface step during nanoindentation of fcc al by multiscale analysis. *Comput. Mater. Sci.* **2012**, *58*, 192–200. [CrossRef]

77. Landman, U.; Luedtke, W.; Burnham, N.A.; Colton, R.J. Atomistic mechanisms and dynamics of adhesion, nanoindentation, and fracture. *Science* **1990**, *248*, 454–461. [CrossRef] [PubMed]

78. Hoover, W.G.; De Groot, A.J.; Hoover, C.G.; Stowers, I.F.; Kawai, T.; Holian, B.L.; Boku, T.; Ihara, S.; Belak, J. Large-scale elastic-plastic indentation simulations via nonequilibrium molecular dynamics. *Phys. Rev. A* **1990**, *42*, 5844–5853. [CrossRef] [PubMed]

79. Kelchner, C.L.; Plimpton, S.J.; Hamilton, J.C. Dislocation nucleation and defect structure during surface indentation. *Phys. Rev. B* **1998**, *58*, 11085–11088. [CrossRef]

80. Zimmerman, J.A.; Kelchner, C.L.; Klein, P.A.; Hamilton, J.C.; Foiles, S.M. Surface step effects on nanoindentation. *Phys. Rev. Lett.* **2001**, *87*, 165507. [CrossRef] [PubMed]

81. Li, J.; Van Vliet, K.J.; Zhu, T.; Yip, S.; Suresh, S. Atomistic mechanisms governing elastic limit and incipient plasticity in crystals. *Nature* **2002**, *418*, 307–310. [CrossRef] [PubMed]

82. Zhu, T.; Li, J.; Van Vliet, K.J.; Ogata, S.; Yip, S.; Suresh, S.J. Predictive modeling of nanoindentation-induced homogeneous dislocation nucleation in copper. *J. Mech. Phys. Solids* **2004**, *52*, 691–724. [CrossRef]

83. Lee, Y.; Park, J.Y.; Kim, S.Y.; Jun, S. Atomistic simulations of incipient plasticity under Al (111) nanoindentation. *Mech. Mater.* **2005**, *37*, 1035–1048. [CrossRef]

84. Jang, H.; Farkas, D. Interaction of lattice dislocations with a grain boundary during nanoindentation simulation. *Mater. Lett.* **2007**, *61*, 868–871. [CrossRef]

85. Yaghoobi, M.; Voyiadjis, G.Z. Effect of boundary conditions on the MD simulation of nanoindentation. *Comput. Mater. Sci.* **2014**, *95*, 626–636. [CrossRef]

86. Voyiadjis, G.Z.; Yaghoobi, M. Large scale atomistic simulation of size effects during nanoindentation: Dislocation length and hardness. *Mater. Sci. Eng. A* **2015**, *634*, 20–31. [CrossRef]

87. Yaghoobi, M.; Voyiadjis, G.Z. Atomistic simulation of size effects in single-crystalline metals of confined volumes during nanoindentation. *Comput. Mater. Sci.* **2016**, *111*, 64–73. [CrossRef]

88. Voyiadjis, G.Z.; Yaghoobi, M. Role of grain boundary on the sources of size effects. *Comput. Mater. Sci.* **2016**, *117*, 315–329. [CrossRef]

89. Ashby, M.F. The deformation of plastically non-homogeneous materials. *Philos. Mag.* **1970**, *21*, 399–424. [CrossRef]

90. Pugno, N.M. A general shape/size-effect law for nanoindentation. *Acta Mater.* **2007**, *55*, 1947–1953. [CrossRef]

91. Ma, Q.; Clarke, D.R. Size dependent hardness of silver single crystals. *J. Mater. Res.* **1995**, *10*, 853–863. [CrossRef]

92. Poole, W.J.; Ashby, M.F.; Fleck, N.A. Micro-hardness of annealed and work-hardened copper polycrystals. *Scr. Mater.* **1996**, *34*, 559–564. [CrossRef]

93. McElhaney, K.W.; Valssak, J.J.; Nix, W.D. Determination of indenter tip geometry and indentation contact area for depth sensing indentation experiments. *J. Mater. Res.* **1998**, *13*, 1300–1306. [CrossRef]

94. Liu, Y.; Ngan, A.H.W. Depth dependence of hardness in copper single crystals measured by nanoindentation. *Scr. Mater.* **2001**, *44*, 237–241. [CrossRef]

95. McLaughlin, K.K.; Clegg, W.J. Deformation underneath low-load indentations in copper. *J. Phys. D* **2008**, *41*, 074007. [CrossRef]

96. Rester, M.; Motz, C.; Pippan, R. Indentation across size scales: A survey of indentation-induced plastic zones in copper {111} single crystals. *Scr. Mater.* **2008**, *59*, 742–745. [CrossRef]

97. Abu Al-Rub, R.K.; Voyiadjis, G.Z. Analytical and experimental determination of the material intrinsic length scale of strain gradient plasticity theory from micro- and nano-indentation experiments. *Int. J. Plast.* **2004**, *20*, 1139–1182. [CrossRef]

98. Feng, G.; Nix, W.D. Indentation size effect in MgO. *Scr. Mater.* **2004**, *51*, 599–603. [CrossRef]

99. Huang, Y.; Zhang, F.; Hwang, K.C.; Nix, W.D.; Pharr, G.M.; Feng, G. A model for size effects in nanoindentation. *J. Mech. Phys. Solids* **2006**, *54*, 1668–1686. [CrossRef]

100. Soer, W.A.; De Hosson, J.T.M. Detection of grain-boundary resistance to slip transfer using nanoindentation. *Mater. Lett.* **2005**, *59*, 3192–3195. [CrossRef]

101. Cao, Y.; Allameh, S.; Nankivil, D.; Sethiaraj, S.; Otiti, T.; Soboyejo, W. Nanoindentation measurements of the mechanical properties of polycrystalline Au and Ag thin films on silicon substrates: Effects of grain size and film thickness. *Mater. Sci. Eng. A* **2006**, *427*, 232–240. [CrossRef]

102. Hou, X.D.; Bushby, A.J.; Jennett, N.M. Study of the interaction between the indentation size effect and Hall–Petch effect with spherical indenters on annealed polycrystalline copper. *J. Phys. D Appl. Phys.* **2008**, *41*, 074006. [CrossRef]

103. Voyiadjis, G.Z.; Peters, R. Size effects in nanoindentation: An experimental and analytical study. *Acta Mech.* **2010**, *211*, 131–153. [CrossRef]

104. Voyiadjis, G.Z.; Almasri, A.H.; Park, T. Experimental nanoindentation of BCC metals. *Mech. Res. Commun.* **2010**, *37*, 307–314. [CrossRef]

105. Almasri, A.H.; Voyiadjis, G.Z. Nano-indentation in FCC metals: Experimental study. *Acta Mech.* **2010**, *209*, 1–9. [CrossRef]

106. Voyiadjis, G.Z.; Faghihi, D.; Zhang, C. Analytical and experimental determination of rate-and temperature-dependent length scales using nanoindentation experiments. *J. Nanomech. Micromech.* **2011**, *1*, 24–40. [CrossRef]

107. Voyiadjis, G.Z.; Zhang, C. The mechanical behavior during nanoindentation near the grain boundary in a bicrystal FCC metal. *Mater. Sci. Eng. A* **2015**, *621*, 218–228. [CrossRef]

108. Zhang, C.; Voyiadjis, G.Z. Rate-dependent size effects and material length scales in nanoindentation near the grain boundary for a bicrystal FCC metal. *Mater. Sci. Eng. A* **2016**, *659*, 55–62. [CrossRef]

109. Voyiadjis, G.Z.; Abu Al-Rub, R.K. Gradient Plasticity Theory with a Variable Length Scale Parameter. *Int. J. Solids Struct.* **2005**, *42*, 3998–4029. [CrossRef]

110. Uchic, M.D.; Dimiduk, D.M.; Florando, J.N.; Nix, W.D. Exploring specimen size effects in plastic deformation of Ni3(Al,Ta). *Mater. Res. Soc. Symp. Proc.* **2003**, *753*, 27–32. [CrossRef]

111. Uchic, M.D.; Dimiduk, D.M.; Florando, J.N.; Nix, W.D. Sample dimensions influence strength and crystal plasticity. *Science* **2004**, *305*, 986–989. [CrossRef] [PubMed]

112. Uchic, M.D.; Shade, P.A.; Dimiduk, D.M. Plasticity of micrometer-scale single crystals in compression. *Annu. Rev. Mater. Res.* **2009**, *39*, 361–386. [CrossRef]

113. Kraft, O.; Gruber, P.; Mönig, R.; Weygand, D. Plasticity in confined dimensions. *Annu. Rev. Mater. Res.* **2010**, *40*, 293–317. [CrossRef]

114. Parthasarathy, T.A.; Rao, S.I.; Dimiduk, D.M.; Uchic, M.D.; Trinkle, D.R. Contribution to size effect of yield strength from the stochastics of dislocation source lengths in finite samples. *Scr. Mater.* **2007**, *56*, 313–316. [CrossRef]

115. Rao, S.I.; Dimiduk, D.M.; Tang, M.; Parthasarathy, T.A.; Uchic, M.D.; Woodward, C. Estimating the strength of single-ended dislocation sources in micron-sized single crystals. *Philos. Mag.* **2007**, *87*, 4777–4794. [CrossRef]

116. Norfleet, D.M.; Dimiduk, D.M.; Polasik, S.J.; Uchic, M.D.; Mills, M.J. Dislocation structures and their relationship to strength in deformed nickel microcrystals. *Acta Mater.* **2008**, *56*, 2988–3001. [CrossRef]

117. Rao, S.I.; Dimiduk, D.M.; Parthasarathy, T.A.; Uchic, M.D.; Tang, M.; Woodward, C. Athermal mechanisms of size-dependent crystal flow gleaned from three-dimensional discrete dislocation simulations. *Acta Mater.* **2008**, *56*, 3245–3259. [CrossRef]

118. Greer, J.R.; Oliver, W.C.; Nix, W.D. Size dependence of mechanical properties of gold at the micron scale in the absence of strain gradients. *Acta Mater.* **2005**, *53*, 1821–1830. [CrossRef]

119. Kiener, D.; Pippan, R.; Motz, C.; Kreuzer, H. Microstructural evolution of the deformed volume beneath microindents in tungsten and copper. *Acta Mater.* **2006**, *54*, 2801–2811. [CrossRef]

120. Rester, M.; Motz, C.; Pippan, R. Microstructural investigation of the volume beneath nanoindentations in copper. *Acta Mater.* **2007**, *55*, 6427–6435. [CrossRef]

121. Demir, E.; Raabe, D.; Zaafarani, N.; Zaefferer, S. Investigation of the indentation size effect through the measurement of the geometrically necessary dislocations beneath small indents of different depths using EBSD tomography. *Acta Mater.* **2009**, *57*, 559–569. [CrossRef]

122. Yang, W.; Larson, B.C.; Pharr, G.M.; Ice, G.E.; Budai, J.D.; Tischler, J.Z.; Liu, W. Deformation microstructure under microindents in single-crystal Cu using three-dimensional X-ray structural microscopy. *J. Mater. Res.* **2004**, *19*, 66–72. [CrossRef]

123. Larson, B.C.; El-Azab, A.; Yang, W.; Tischler, J.Z.; Liu, W.; Ice, G.E. Experimental characterization of the mesoscale dislocation density tensor. *Philos. Mag.* **2007**, *87*, 1327–1347. [CrossRef]

124. Larson, B.C.; Tischler, J.Z.; El-Azab, A.; Liu, W. Dislocation density tensor characterization of deformation using 3D X-ray microscopy. *J. Eng. Mater. Technol.* **2008**, *130*, 021024. [CrossRef]

125. Feng, G.; Budiman, A.S.; Nix, W.D.; Tamura, N.; Patel, J.R. Indentation size effects in single crystal copper as revealed by synchrotron X-ray microdiffraction. *J. Appl. Phys.* **2008**, *104*, 043501. [CrossRef]

126. Hasnaoui, A.; Derlet, P.M.; Van Swygenhoven, H. Interaction between dislocations and grain boundaries under an indenter–a molecular dynamics simulation. *Acta Mater.* **2004**, *52*, 2251–2258. [CrossRef]

127. Kulkarni, Y.; Asaroo, R.J.; Farkas, D. Are nanotwinned structures in fcc metals optimal for strength, ductility and grain stability? *Scr. Mater.* **2009**, *60*, 532–535. [CrossRef]

128. Nair, A.K.; Parker, E.; Gaudreau, P.; Farkas, D.; Kriz, R.D. Size effects in indentation response of thin films at the nanoscale: A molecular dynamics study. *Int. J. Plast.* **2008**, *24*, 2016–2031. [CrossRef]

129. Peng, P.; Liao, G.; Shi, T.; Tang, Z.; Gao, Y. Molecular dynamic simulations of nanoindentation in aluminum thin film on silicon substrate. *Appl. Surf. Sci.* **2010**, *256*, 6284–6290. [CrossRef]

130. Sun, K.; Shen, W.; Ma, L. The influence of residual stress on incipient plasticity in single-crystal copper thin film under nanoindentation. *Comput. Mater. Sci.* **2014**, *81*, 226–232. [CrossRef]

131. Szlufarska, I. Atomistic simulations of nanoindentation. *Mater. Today* **2006**, *9*, 42–50. [CrossRef]

132. Daw, M.S.; Baskes, M.I. Embedded-atom method: Derivation and application to impurities, surfaces, and other defects in metals. *Phys. Rev. B* **1984**, *29*, 6443–6453. [CrossRef]

133. Baskes, M.I. Modified embedded-atom potentials for cubic materials and impurities. *Phys. Rev. B* **1992**, *46*, 2727–2742. [CrossRef]

134. Medyanik, S.N.; Shao, S. Strengthening effects of coherent interfaces in nanoscale metallic bilayers. *Comput. Mater. Sci.* **2009**, *45*, 1129–1133. [CrossRef]

135. Shao, S.; Medyanik, S.N. Dislocation–interface interaction in nanoscale fcc metallic bilayers. *Mech. Res. Commun.* **2010**, *37*, 315–319. [CrossRef]

136. Saraev, D.; Miller, R.E. Atomic-scale simulations of nanoindentation-induced plasticity in copper crystals with nanometer-sized nickel coatings. *Acta Mater.* **2006**, *54*, 33–45. [CrossRef]

137. Stukowski, A. Structure identification methods for atomistic simulations of crystalline materials. *Model. Simul. Mater. Sci. Eng.* **2012**, *20*, 045021. [CrossRef]

138. Stukowski, A.; Albe, K. Extracting dislocations and non-dislocation crystal defects from atomistic simulation data. *Model. Simul. Mater. Sci. Eng.* **2010**, *18*, 085001. [CrossRef]

139. Gao, Y.; Ruestes, C.J.; Tramontina, D.R.; Urbassek, H.M. Comparative simulation study of the structure of the plastic zone produced by nanoindentation. *J. Mech. Phys. Solids* **2015**, *75*, 58–75. [CrossRef]

140. Armstrong, R.W.; Walley, S.M. High strain rate properties of metals and alloys. *Int. Mater. Rev.* **2008**, *53*, 105–128. [CrossRef]

141. Armstrong, R.W.; Li, Q. Dislocation mechanics of high-rate deformations. *Metall. Mater. Trans. A* **2015**, *46*, 4438–4453. [CrossRef]

142. Yaghoobi, M.; Voyiadjis, G.Z. Size Effects in FCC Crystals During the High Rate Compression Test. *Acta Mater.* **2016**, *121*, 190–201. [CrossRef]

143. Voyiadjis, G.Z.; Yaghoobi, M. Size and Strain Rate Effects in Metallic Samples of Confined Volumes: Dislocation Length Distribution. *Scr. Mater.* **2017**, *130*, 182–186. [CrossRef]

144. Yaghoobi, M.; Voyiadjis, G.Z. Microstructural investigation of the hardening mechanism in FCC crystals during high rate deformations. *Comput. Mater. Sci.* **2017**, *138*, 10–15. [CrossRef]

crystals

MDPI

Article

The Effect of the Vertex Angles of Wedged Indenters on Deformation during Nanoindentation

Xiaowen Hu and Yushan Ni *

Department of Aeronautics and Astronautics, Fudan University, Shanghai 200433, China;
xwhu16@fudan.edu.cn
* Correspondence: niyushan@fudan.edu.cn

Academic Editors: Ronald W. Armstrong, Stephen M. Walley and Wayne L. Elban
Received: 30 September 2017; Accepted: 12 December 2017; Published: 14 December 2017

Abstract: In order to study the effect of the angle of wedged indenters during nanoindentation, indenters with half vertex angles of 60°, 70° and 80° are used for the simulations of nanoindentation on FCC aluminum (Al) bulk material by the multiscale quasicontinuum method (QC). The load-displacement responses, the strain energy-displacement responses, and hardness of Al bulk material are obtained. Besides, atomic configurations for each loading situation are presented. We analyze the drop points in the load-displacement responses, which correspond to the changes of microstructure in the bulk material. From the atom images, the generation of partial dislocations as well as the nucleation and the emission of perfect dislocations have been observed with wedged indenters of half vertex angles of 60° and 70°, but not 80°. The stacking faults move beneath the indenter along the direction $[1\bar{1}0]$. The microstructures of residual displacements are also discussed. In addition, hardness of the Al bulk material is different in simulations with wedged indenters of half vertex angles of 60° and 70°, and critical hardness in the simulation with the 70° indenter is bigger than that with the 60° indenter. The size effect of hardness in plastic wedged nanoindentation is observed. There are fewer abrupt drops in the strain energy-displacement response than in the load-displacement response, and the abrupt drops in strain energy-displacement response reflect the nucleation of perfect dislocations or extended dislocations rather than partial dislocations. The wedged indenter with half vertex angle of 70° is recommended for investigating dislocations during nanoindentation.

Keywords: nanoindentation; multiscale simulation; wedged indenter; dislocation nucleation and emission; size effect of hardness during indentation

1. Introduction

With the increase of application of micro-electronic devices, nanoindentation has been widely used as a standard method to obtain mechanical properties such as nanohardness and elastic modulus of materials [1–4]. Both experimental and numerical research have been implemented to observe the microstructures near the indenter tip in recent years. In terms of experimental studies, for instance, C. A. Schuh et al. [5,6] used a Triboindenter to investigate the dislocation nucleation and incipient plasticity during high-temperature nanoindentation on Pt single crystal, and obtained mechanical properties of standard fused silica at temperature up to 405 °C. D. Ge et al. [7] employed a three-sided pyramidal Berkovich tip to investigate the size effect in the nanoindentation of silicon. Sandra Korte et al. [8] implemented a high-temperature nanoindentation of Au in vacuum, and obtained values of properties like Young moduli, yield and flow stresses. However, experimental nanoindentation tests are sensitive to the environment and instrument [5–8]. Numerical simulation becomes an effective method to study the basic mechanical behaviors in nanoindentation.

When it comes to numerical methods, the finite element (FE) method is often employed to simulate nanoindentation. For example, J. A. Knapp et al. [9] developed procedures of nanoindentation based on finite-element modeling, accurately calculated some mechanical properties and compared them with experiments. Kaushal K Jha et al. [10] studied the physical meaning of the total and elastic energy constants in elasto-plastic indentations by Finite-element method, and discussed their applications in the evaluation of nanomechanical quantities such as the indenter tip radius and the nominal hardness. However, it is hard to validly simulate the changes of microstructures near the indenter tip, such as dislocations and twinnings. Moreover, some drop points of load-displacement curve are hard to obtain. Molecular dynamics (MD) is another widely employed method to study the microscopic mechanisms. Many researchers studied nanoindentation with this method. Landman et al. [11], for the first time, simulated nanocontact problems using Ni tips with MD method. Yen-Hung Lin et al. [12] simulated the nanoindentation on monocrystalline silicon with both spherical and Berkovich indenters, and discussed the nanoindentation-induced deformation and the phase transformation during the process. Jiang et al. [13] simulated the nanoindentations on a binary metallic glass under various strain rates, and validated that the serration is not directly dependent on the resultant shear-banding spatiality. Although the MD method can provide a lot of details of micro-deformation, such like dislocation nucleation and emission, it requires lots of calculations that will limit the simulation both on the time scale and the model size. The quasi-continuum (QC) method is one of the multiscale methods which allows relatively large model to simulate the mechanical behaviors in nanoscale. Many researchers have studied micro behaviors of nanomaterials with this method. Huaibao Lu et al. [14,15] investigated the effect of surface step on nanoindentation in various orientations by the QC method. Aibin Zhu et al. [16] simulated the nanoindentation on single crystal cooper by QC method, and discussed the effect of radius of sphere indenters on nanohardness. Mei and Ni [17] studied the anisotropic behavior during nano-adhesive contact by the QC method, which proves the importance of crystal orientation of micro-devices. Fanlin Zeng et al. [18] investigated the titled flat-ended nanoindentation with different titled angles by the QC method, and simulation results agreed well with analytical ones.

Wedge indentation is an important process for determining the mechanical properties of materials, and angles of conical and wedged indenters affect the measurement of mechanical properties greatly. D. S. Dugdale [19] conducted the wedge indentation experiments with three different cold-worked metals, and found that the measured hardnesses was independent of size when the metal was larger than a certain size. K. Eswar Prasad et al. [20] investigated the role of the angle of conical indenters on the plastic deformation during nanoindentation by using the FE method, and compared the results with experimental data. Fan-lin Zeng et al. [21] simulated the wedged nanoindentation on nickel by using indenters with different vertex angles, and studied the dislocation emission beneath the indenter tip. However, the unloading nanoindentation processes with different angles of wedged indenters have not been studied, and the influence of the angle of wedged indenters on the hardness of materials has not been discussed yet. In this paper, we simulate the loading and unloading nanoindentation processes on Al bulk material by using wedged indenters with different vertex angles by QC method. Load-displacement responses and energy-displacement responses were obtained. Hardness of materials in different wedged nanoindentations were calculated too. In particular, the mechanism of dislocation emission will be discussed, and the residual dislocation during the unloading process will also be investigated.

2. Methodology and Simulation Model of Nanoindentation

The QC method is an effective multiscale approach to simulate the mechanical behaviors of crystalline materials, which couples the continuum and atomic simulation. It is established that discrete atomic descriptions are only necessary at highly deformed regions and vicinity of defects or interfaces, while the linear elastic continuum method is employed in other regions. In the QC method, there are two kinds of representative atoms called local atoms and non-local atoms. The local atoms gather the deformation behavior of atoms that go through a nearly homogeneous deformation whose energies are computed from the local deformation gradients based on Cauchy-Born continuum rule. On the other hand, the non-local atoms are treated by Ercolessi-Adams Al potential [22] to describe the atomic movement in the area where heterogeneous deformations occur. The QC method runs through molecular static energy minimization of all atoms over the atomic (non-local) domain and the finite element (local) domain. Meanwhile, the QC method automatically reduces the degrees of freedom by implementing an automatic adaption scheme with no atomistic detail loss in the core region. The sizes of atomistic region and finite element region are constantly updated and expands during the simulation, and the selection of representative atoms and their local versus nonlocal status is automatically carried out by a formulation using a certain criteria [23] In this case, the atomistic zone of the model is big enough to capture plastic deformation and nucleation and emission of dislocations. The dislocations in the material will always be located in the atomistic region. In addition, the models used in the QC method are pseudo-2D models [24]. This means that although the analysis is performed in a two-dimestional coordinate system, the displacements in the z-direction are allowed and all atomistic calculations are three-dimensional. Within this setting, only dislocations with line directions perpendicular to the plane of analysis can be nucleated and the displacement fields were constrained to have no variation in the out-of-plane z-direction. These constraints appear to be compatible with the two-dimensional nature of the indenter. This is a form of generalized plane strain, which could effectively simplify our simulation. More details of QC method are discussed in [23,24].

The nanoindentation model of this work is illustrated in Figure 1. Wedged indenters with different semi-angles (α = 60°, 70° and 80°) were pressed into aluminum (Al) single crystal bulk material. The model is 100 nm thick, 200 nm wide, and infinite in the out-of-plane direction with periodic boundary conditions. In addition, the size of the initial atomistic region is 10 nm wide and 2 nm thick. The model is much larger than most of models used in MD nanoindentation simulations. The crystal orientations of the model along x-, y-, and z-axis are chosen to be [111], [$\bar{1}$10] and [$\bar{1}\,\bar{1}$2], respectively, which could promote the emissions of Shockley partial dislocations and deformation twinning [25]. A fixed boundary condition is applied to the bottom surface, and free boundary conditions are applied to the top, left and right surfaces. The indenter was settled on the middle of the top surface, and the indenter was gradually pushed into the film along y-direction with an increment of 0.01 nm per step. The interactions between the atoms are described by Ercolessi-Adams Al potential [22], and the elastic moduli predicted by this potential are C_{11} = 117.74 GPa, C_{12} = 62.02 GPa and C_{44} = 36.76 GPa. The effective values of shear modulus G = 33.14 GPa and Poisson's ratio ν = 0.319 are calculated by a Voigt average [24]:

$$G = \tfrac{1}{5}(C_{11} - C_{12} + 3C_{44})$$
$$\nu = \tfrac{1}{2}\left[\frac{C_{11}+4C_{12}-2C_{44}}{2C_{11}+3C_{12}+C_{44}}\right] \tag{1}$$

Thus lattice constant a = 0.4032 nm. Finally, indenters penetrated 1.2 nm deep into the material.

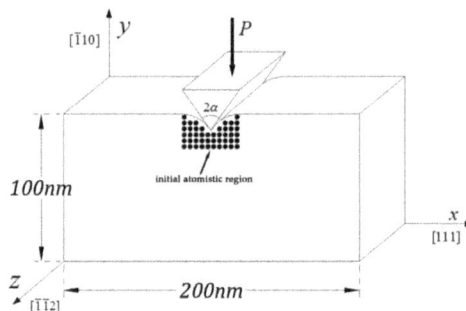

Figure 1. Schematic representation of wedged nanoindentation.

3. Results and Discussion

3.1. Load-Displacement Responses

3.1.1. Load-Displacement Responses during Different Loading Processes

We present the load-displacement responses of nanoindentation loading processes using wedged indenters with half vertex angles (α in Figure 1) of $60°$, $70°$ and $80°$ in Figure 2. Loads are calculated in Newton per meter length in the z direction of the y directional resultant force on the indenters. At the very beginning of indentations, the curves of three different loading processes are almost the same. This is because the contact parts of three indenters are almost the same at the beginning, which is called the tip radius effect [26]. When the contact width exceeds the tip width, the curves become different. It is clear that load goes up with the increase of indentation depth at the beginning of three different indentations. These initial deformation stages are regarded as the elastic stages. As indenters are pressed deeply, the pop-in effect is investigated in nanoindentation with both half vertex angles of $60°$ and $70°$. As it is shown, the load drops suddenly at a depth of 4.7 Å (Point A_1) with a critical load $P_{cr60} = 3.22$ N/m during nanoindentation with the indenter of $60°$. As the indenter proceeds more deeply, the load continues to increase with several drops. Three different drop points that happen at 4.7 Å, 7.7 Å and 10.7 Å (A_1, B_1 and C_1) are marked, while three abrupt drops occur at 5.6 Å, 8.1 Å and 10.4 Å (A_2, B_2, and C_2) during nanoindentation using the wedged indenter of $70°$, as shown in Figure 2. The critical load during nanoindentation with the $70°$ indenter is $P_{cr70} = 3.99$ N/m. However, in nanoindentation with the $80°$ indenter, the load smoothly increases with the increase of indentation depth and no abrupt drop occurs. In addition, the load increases with the increase of angle of wedged indenter when the indentation depth is the same. The elastic stage of nanoindentation with the indenter of $70°$ continues longer than that with the indenter of $60°$. It is known that the drop points in the load-displacement response indicate the dislocation nucleation and emission, which will be discussed below.

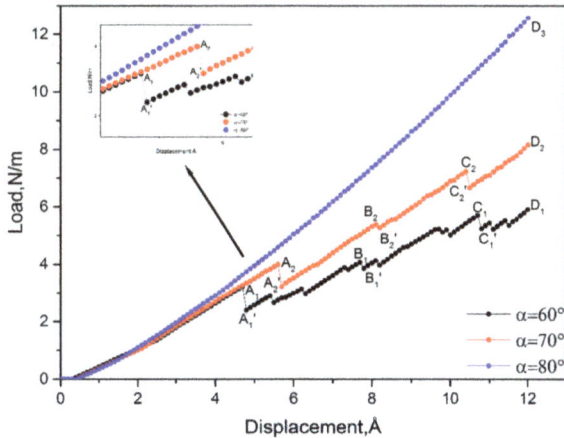

Figure 2. Load-displacement responses for wedged nanoindentation on Al single crystal by using indenters with different vertex angles.

3.1.2. Load-Displacement Responses during Different Unloading Processes

As we discussed above, several drop points are observed in loading processes. It is essential to confirm whether the plastic deformation happens at these points. The indenter with half vertex angle of $60°$ is retracted from A_1' and D_1, as shown in Figure 3, while the indenter of $70°$ is retracted from A_2', B_2' and C_2', as shown in Figure 4.

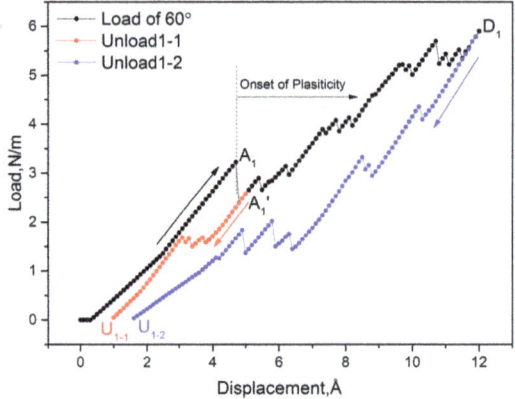

Figure 3. Load-displacement responses during unloading processes using the wedged indenter with half vertex angle of 60°.

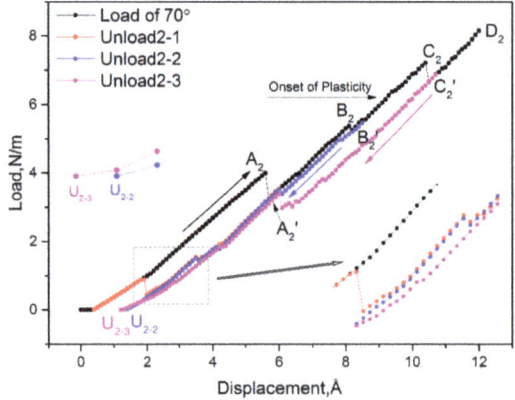

Figure 4. Load-displacement responses during unloading processes using the wedged indenter with half vertex angle of 70°.

In Figure 3, we perform unloading processes with the indenter of half vertex angle of 60° at A_1' and D_1 as "Unload1-1" and "Unload1-2", respectively. During "Unload1-1", the load decreases as the indentation depth decreases. The value of load fluctuates when indentation depth declines from 3.8 Å to 3.1 Å, and then the load linearly decreases to zero with a residual displacement $U_{1-1} = 0.9$Å. The residual displacement of unloading process is the indentation depth when the load on the indenter decreases to zero. During "Unload1-2" process, two abrupt jumps of load occur at 10.2 Å and 8.5 Å. After the displacement fluctuates from 6.4 Å to 5.0 Å, the load smoothly drops to zero with a residual displacement $U_{1-2} = 1.5$ Å. These residual displacements prove that irreversible plastic deformation occurs in indentation loading process after point A_1'.

In Figure 4, retractions from A_2', B_2' and C_2' were exhibited as "Unload2-1", "Unload2-2", and "Unload2-3", respectively. As is shown, the load from A_2' decreases with the decrease of indentation depth. When indentation depth falls to 1.9 Å, the load abruptly jumps up to the curve of loading process and then decreases back to the origin with no residual displacement. This means the deformation happened at point A_2 is recoverable. During "Unload2-2" process, the load decreases from B_2' and no obviously abrupt jump occurs, which is different from "Unload2-1" process. Finally, the load decreases to zero with a residual displacement U_{2-2} = 1.3 Å. As for "Unload2-3", the load decreases gradually where an abrupt jump happens at 6.0 Å. Then, the load decreases to zero with a residual displacement U_{2-3} = 1.2 Å, which is almost the same as U_{2-2}. The residual displacements are attributed to the changes of microscopic structures of the material during unloading processes and will be discussed below. The pop-out effect that happens during the unloading process also agrees with many experiments, and it is considered to be relevant to the atomic rearrangement of the bulk material [7,27].

The load-displacement response of unloading process with the indenter of 80° is also obtained. The curve of unloading process completely coincides with the curve of loading process. This suggests that the loading process with the indenter of 80° is an elastic process. On the other hand, the load-displacement responses of three different nanoindentations show good linearity at the initial stage of unloading processes, which also agrees with many experimental results [28,29].

3.2. Atom Configurations of Dislocation Nucleation and Emission

3.2.1. Atomic Configurations during Nanoindentation Using the Indenter with Half Vertex Angle of 60°

In order to understand the mechanism of the drop points in the load-displacement responses, the atomic details beneath the indenter tip need to be observed. The Von Mises effective strain distribution and atomic configurations of abrupt drops during nanoindentation using the wedged indenter with half vertex angle of 60° are shown in Figure 5a–d correspond to points A_1', B_1', C_1' and D_1 in Figure 2, respectively). As we mentioned in Section 2, the line directions of dislocations in QC method are all perpendicular to the plane of analysis. By standard, we use "⊥" symbol to label the dislocations in atom configurations [30]. The projection of Burgers vector is parallel to the bottom line of "⊥", and the direction of this projection vector is from left to right. The Burgers vectors of the dislocations are determined by the directions of their projection vectors and the Thompson tetrahedron in FCC metals [31]. We also plot the slip direction of dislocations in the configurations. The discussion about dislocation nucleation and emission goes as follow.

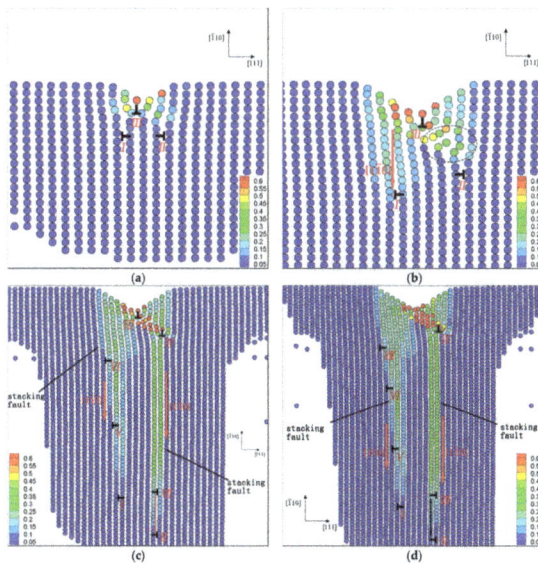

Figure 5. Strain distribution and dislocations beneath the wedged indenter with vertex angle of 60° during loading nanoindentation process: (**a**) h = 4.8 Å, (**b**) h = 7.8 Å, (**c**) h = 10.8 Å, (**d**) h = 12.0 Å.

During loading process of 60°, when the indenter is pressed into material 4.6 Å, a dislocation dipole (a pair of Shockley partials, $1/6[2\bar{1}\bar{1}]$ and $1/6[\bar{2}11]$) nucleates beneath the indenter shoulder at point **I** and **II** (Figure 5a). This is the first emergence of geometrically necessary dislocations [32]. In addition, a perfect dislocation $1/2[110]$ emerges at point **III**. This dislocation nucleation corresponds to the drop point A_1' (in Figure 2), which means that the nucleation of dislocations will reduce the force on the indenter and release the stress concentration on the zone around the indenter tip. When the indentation depth comes to 7.7 Å (point B_1' in Figure 2), the dislocation dipole is dissociated, as shown in Figure 5b. Shockley partial **II** moves below, and a lot of disordered atoms stack above the dislocation **II**. Shockley partial **I** is emitted on slip plane $[\bar{1}\bar{1}\bar{1}]$ along $[1\bar{1}0]$ direction because the direction $[1\bar{1}0]$ is a favorable slip direction for FCC metals [24]. When indentation depth continues to increase to 10.8 Å (point C_1'), a $1/2[110]$ perfect dislocation emerges at point **IV**, as shown in Figure 5c. The formation of this dislocation is due to the stacking of disordered atoms in Figure 5b. The Shockley partials **I** and **II** continue to slide on $[\bar{1}\bar{1}\bar{1}]$ plane along $[\bar{1}10]$ direction. Three new Shockley partials appear, among which two ($1/6[\bar{2}11]$, marked as **V** and **VI**) appear beside the slip planes, which are generated by the emission of Shockley partial **I**, and the third Shockley partial **VII** ($1/6[1\bar{2}1]$) appears in the stacking fault 18.9 Å above the Shockley partial **II**. In addition, the Shockley partial **II** and **VII** could be considered as dissociated from a perfect dislocation $1/2[1\bar{1}0]$. These two dislocations and stacking fault between them (as the line connected in Figure 5c) constitute an extended dislocation. In Figure 5d, the perfect dislocation **III** disappears, and other dislocations continue moving down. Then a new Shockley partial **VIII** emerges. As it can be seen from Figure 5c,d, the distance between the stacking faults is four atoms' width. Thus, the width of the zone where partial dislocations move is eight atoms' width in the loading process with the indenter of 60°.

To confirm the residual displacements after retractions, the atom configurations of material after unloading processes by using the indenter with half vertex angle of 60° at A_1' and D_1 are plotted in Figure 6 ((a) and (b) correspond to U_{1-1} and U_{1-2} in Figure 3, respectively).

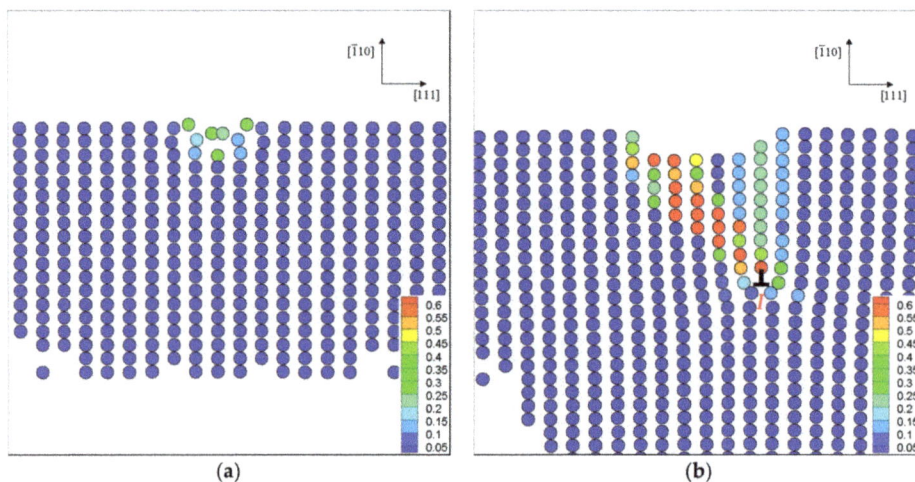

Figure 6. Strain distribution and atom configurations of Al bulk material under the wedged indenter with half vertex angle of 60° after retractions at A_1' and D_1': (a) U_{1-1}; (b) U_{1-2}.

After retraction process at A_1', several disordered atoms stack at the surface as shown in Figure 6a. The dislocations have disappeared, which could be explained by the following model [24]. The forces on the dislocations include the Peach–Koehler force (F_{PK}) due to the indenter stress field pushing the dislocation into the bulk material, and the image force F_I pulling the dislocation back to the surface. These two forces and Peierls stress σ_p maintain a balance when dislocation is stationary. It is described as

$$F_{PK} + F_I = b\sigma_p \tag{2}$$

where b is module of Burgers vector. When the indenter is pulled back, the Peach-Koehler force applied on the dislocation decreases, the image force would dominate and the dislocation would return to the surface and disappear.

In Figure 6b, the atomic configurations of bulk material after retraction from D_1' are presented. Compared to the atom configuration at D_1 in loading process, only one perfect dislocation **I** is trapped in the bulk material, and other dislocations have moved back to the surface and disappeared. Because the module of Burgers vector of perfect dislocation is bigger than that in partial dislocation, the Peierls stress σ_p applied on this perfect dislocation dominates. Thus, the image force from the surface cannot pull it back, and the dislocation remains in the material. In this sense, perfect dislocation is a stable microstructure in residual displacement.

3.2.2. Atomic Configurations during Nanoindentation Using the Indenter with Half Vertex Angle of 70°

In Figure 2, three different abrupt drops of load are investigated during nanoindentation by using the indenter with half vertex angle of 70°. The atom configurations of these points are extracted to observe the microstructure changes at these moments, as shown in Figure 7 ((a), (b), (c) correspond to points A_2', B_2', C_2' in Figure 2, respectively). In Figure 7a, the atomic configuration of the zone beneath the wedged indenter with half vertex angle of 70° at point A_2' is presented. A dislocation dipole (two Shockley partials **I** and **II**) emerges beneath the indenter shoulder, and a $1/2[110]$ perfect dislocation nucleates at point **III**. Both are also observed in the loading process with the indenter of 60°. As shown in Figure 8b, when displacement comes to 8.2 Å (point B_2'), dislocation **I** slides on slip plane $[\bar{1}\,\bar{1}\,\bar{1}]$ along $[1\bar{1}0]$ direction, while former dislocation **II** in Figure 7a disappears. When the indenter is pressed into the bulk material 10.5 Å (point C_2'), as shown in Figure 7c, a new Shockley partial **II** $(1/6[2\,\bar{1}\,\bar{1}])$ is emitted, and two stacking fault zones emerge beneath the indenter shoulder. The dislocation **I** continues moving down, and two new Shockley partials nucleate on new slip planes $[\bar{1}\,\bar{1}\,\bar{1}]$ beside the slip plane where dislocation **I** locates. This also happens in the loading process with the indenter of 60°. Another Shockley partial **VI** $(1/6[1\bar{2}1])$ nucleates 17.5 Å above dislocation **II**. Dislocations **II** and **VI** and the stacking fault between them constitute an extended dislocation. In addition, the distance between two stacking faults in loading process with the indenter of 70° is seven atoms' width. The width of zone that partial dislocations move is 11-atoms width, which is wider than that in nanoindentation with the indenter of 60°.

The atomic configurations of the bulk material after unloading processes with the indenter of 70° retracted from B_2' and C_2' are plotted in Figure 8 ((a) and (b) correspond to U_{2-2} and U_{2-3} in Figure 4, respectively). As shown in Figure 8a, when retraction from B_2' ends, one perfect dislocation remains at surface. This means that unrecoverable plastic deformation has happened at B_2' during loading process. In Figure 8b, three perfect dislocations remain in the film when retraction from C_2' ends, which are arrayed as a line parallel to the surface. The pattern they are arranged leads to the fact that the residual displacement of these three dislocations is almost the same as that of one dislocation (U_{2-2} and U_{2-3}), as we mentioned in Section 3.1.2. From the discussion of atomic configurations of residual displacements, it is clear that perfect dislocations rather than partial dislocations will remain in the material when retraction process ends. Additionally, more dislocations in bulk material with the indenter of 70° will remain than those with the indenter of 60°.

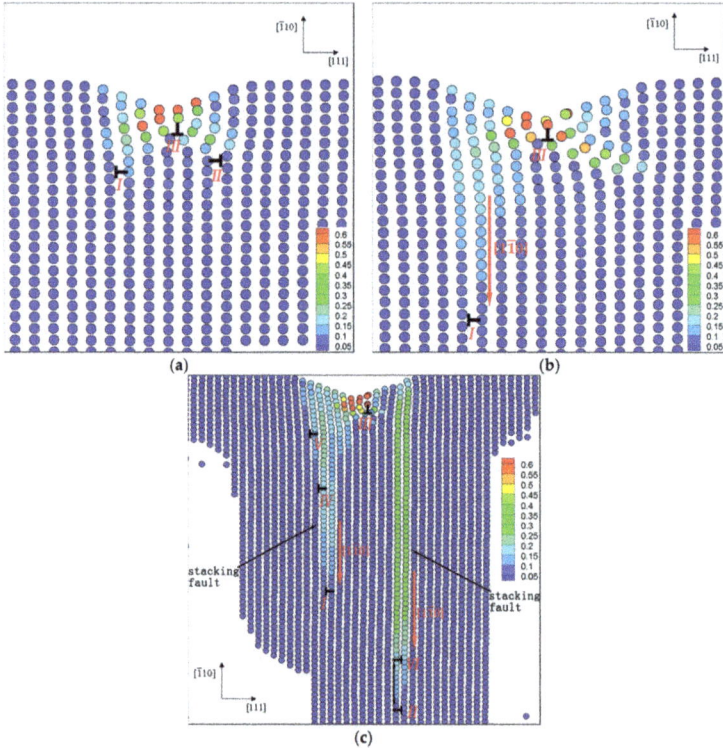

Figure 7. Strain distribution and dislocations beneath the wedged indenter with half vertex angle of 70° during loading nanoindentation process: (**a**) h = 5.7 Å, (**b**) h = 8.2 Å, (**c**) h = 10.5 Å.

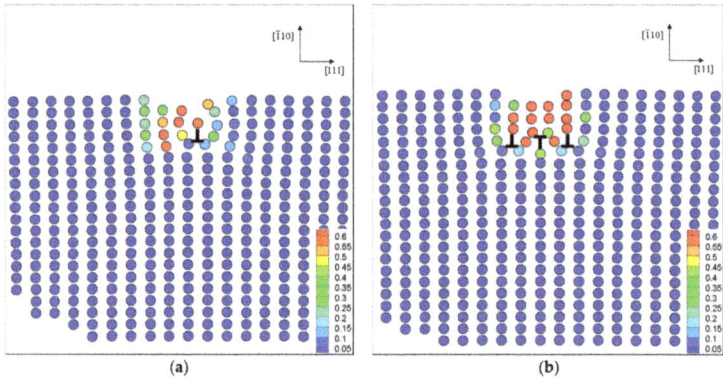

Figure 8. Strain distribution and atom configurations of Al bulk material under the wedged indenter with half vertex angle of 70° after retractions at B_2' and C_2': (**a**) $U_{2\text{-}2}$; (**b**) $U_{2\text{-}3}$.

As the analysis of atomic configurations is shown above, the mechanisms of dislocation nucleation and emission in wedged nanoindentations with indenters of 60° and 70° are similar. The dislocation zones in nanoindentation with both indenters of 60° and 70° extend deeply inside the bulk material, but the width would not exceed the contact width, which is defined as the width of the segment where indenter is in contact with the material. This phenomenon also agrees with many indentation experiments [33,34]. The width of the zone where partial dislocations move in the loading process with the indenter of 70° is wider than that in loading process of 60°, and the amount of dislocations that remain in the material after retraction with the indenter of 70° is larger than that with the indenter of 60°. Thus, the indenter with half vertex angle of 70° is a better choice for investigating dislocation nucleation and emission and for understanding residual displacement during wedged nanoindentation.

3.3. Contact Hardness Responses

Hardness of solids is a measure of resistance that a solid matter shows to various kinds of local permanent deformation [35], and hardness during indentation is defined as [1,36]:

$$H = \frac{P}{A_p} \tag{3}$$

where P is the load, and A_p is the projected area. In our simulation, the projected area is contact width multiplied by the length of the indenter in z-direction. The hardness can be calculated by dividing load P by contact width. The projected area is a discrete parameter that increases periodically with the increase of indentation depth, because it is related to the number of atoms which are in contact with the indenter. Thus, there is a critical hardness during nanoindentation. The critical hardness of wedged nanoindentations with different half vertex angles of 60°, 70° are listed in Table 1. Results show that critical hardness of wedge nanoindentation with half vertex angle of 70° is larger than that with half vertex angle of 60°. This also agrees with the result of conical indenter studied by Alhafez et al [37].

Table 1. Critical hardness of wedged nanoindentations

Half Vertex Angle of Wedged Indenter	Critical Hardness (GPa)
$\alpha = 60°$	7.018
$\alpha = 70°$	7.295

The size effect of contact hardness with contact depth is a typical characteristic of nanoindentation. As discuss above, the contact width and depth in our simulations are discrete parameters. More than one value of hardness is obtained at the same contact depth. Thus, the calculated hardness H is a range for each , and this range is shown as an error bar. The calculated results of hardness are plotted in Figure 9. As shown, calculating results of hardness in loading processes with both indenters of 60° and 70° show dependence on the contact depth. As the contact depth increases, the indentation hardness has a tendency to decrease. This phenomenon also agrees with many other researchers' work [38–42].

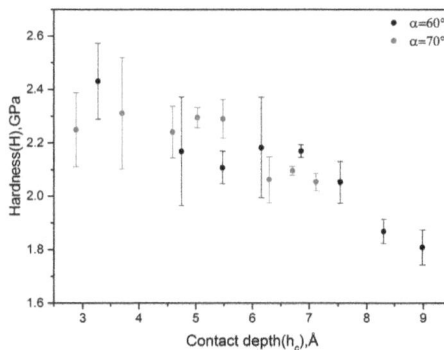

Figure 9. Correlation between hardness and contact depth in wedged nanoindentation with different half vertex angles of 60° and 70°.

3.4. Strain Energy-Displacement Responses

As we mentioned before, the QC method runs through the minimization of energy of the system. The strain energy in the QC method is the total energy of all atoms over the atomic (non-local) domain and the finite element (local) domain [23]. The strain energy versus displacement responses for wedged nanoindentations with half vertex angles of 60°, 70° and 80° were obtained, as shown in Figure 10. The strain energy increases with the increase of indentation depth, and two obvious drops occur at point A_1' and C_1' in response to indentation with the indenter of 60°, where strain energy drops down 0.593 eV and 0.924 eV, respectively. The drop points of strain energy in Figure 10 correspond to the drop points of load in Figure 2, so we use same letters. Two drops occur in strain energy-displacement response when half vertex angle of the indenter is 70°, in which strain energy drops down 0.850 eV and 2.016 eV at point A_2' and C_2'. It is obvious that there are fewer drop points in strain energy-displacement response than in the load-displacement response. According to the atomic configurations shown in Section 3.2, these drop points in strain-energy responses all correspond to the nucleation of perfect dislocations or emergence of extended dislocations rather than partial dislocations. However, in load-displacement response, the drop points correspond to the nucleation and emission of all kinds of dislocations, including partial dislocations. Thus, in wedged nanoindentation, the abrupt drops in strain energy-displacement response only reflect the nucleation of perfect dislocations or extended dislocations.

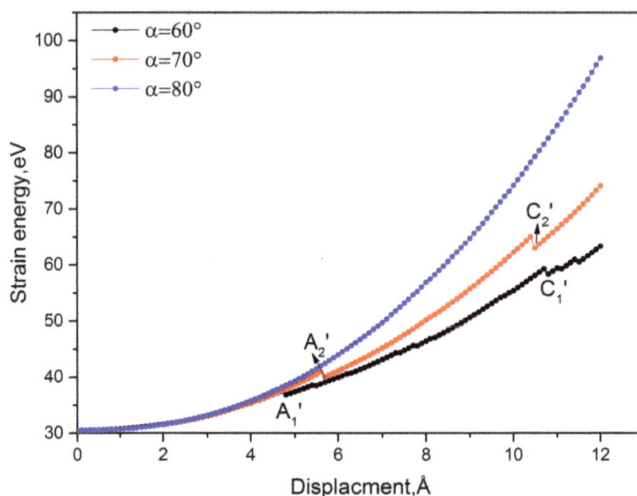

Figure 10. Strain energy-displacement responses of wedged nanoindentations with half vertex angles of 60°, 70° and 80°.

4. Conclusions

The effect of the vertex angle of wedged indenters on deformation of Al single crystal bulk material during nanoindentation was investigated by the QC method. Wedged indenters with half vertex angles of 60°, 70° and 80° are used to simulate the loading and unloading nanoindentation processes. Load-displacement responses of loading and unloading processes, strain energy-displacement responses, hardness, and dislocation nucleation and emission for three kinds of indenters are analyzed. Results show that some abrupt drops occur in the load-displacement responses of nanoindentation with indenters of 60° and 70°, and these points correspond to the dislocation nucleation and emission. No abrupt drop occurs during the loading process of the indenter of 80°, and the unloading curve overlaps the loading curve, suggesting that in this simulation with that included angle the indentation is purely an elastic process. From the analysis of atomic configurations, the mechanisms of dislocation nucleation and emission in wedged nanoindentation with indenters of 60° and 70° are discussed, and the movements of partial dislocations are investigated. Results show that only perfect dislocations will remain when the retraction of indenter ends. Although residual displacements are almost the same, the microstructures of them can be different. It is clear that the dislocation zone beneath the indenter extends deeply, but the width of plastic zone does not exceed the contact width. The width of zone where partial dislocations move during indentation with the indenter of 70° is wider than that in indentation with the indenter of 60°. In addition, the size effect of hardness in plastic wedged nanoindentations with indenters of 60° and 70° is observed. There are fewer abrupt drops in the strain energy-displacement response than in the load-displacement response, and the abrupt drops in strain energy-displacement response in wedged nanoindentation only reflect the nucleation of perfect dislocations or extended dislocations rather than partial dislocations. In summary, the wedged indenter with half vertex angle of 70° is recommended for investigating the dislocations and for understanding residual displacements during elastic-plastic nanoindentation.

Acknowledgments: This work is supported by the National Natural Science Foundation of China (Grant No. 11572090).

Author Contributions: Xiaowen Hu conducted the simulation, analyzed the data and wrote the paper. Yushan Ni conceived and supervised the whole work, and also revised the paper.

Conflicts of Interest: The authors declare no conflict of interest.

References

1. Oliver, W.C.; Pharr, G.M. An improved technique for determining hardness and elastic modulus using load and displacement sensing indentation experiments. *J. Mater. Res.* **1992**, *7*, 1564–1583. [CrossRef]
2. Pethicai, J.B.; Hutchings, R.; Oliver, W.C. Hardness measurement at penetration depths as small as 20 nm. *Philos. Mag. A* **2006**, *48*, 593–606. [CrossRef]
3. Tao, S.; Li, D.Y. Tribological, mechanical and electrochemical properties of nanocrystalline copper deposits produced by pulse electrodeposition. *Nanotechnology* **2006**, *17*, 65–78. [CrossRef]
4. Krauss, A.R.; Auciello, O.; Gruen, D.M.; Jayatissa, A.; Sumant, A.; Tucek, J.; Mancini, D.C.; Moldovan, N.; Erdemir, A.; Ersoy, D.; et al. Ultrananocrystalline diamond thin films for MEMS and moving mechanical assembly devices. *Diam. Relat. Mater.* **2001**, *10*, 1952–1961. [CrossRef]
5. Schuh, C.A.; Mason, J.K.; Lund, A.C. Quantitative insight into dislocation nucleation from high-temperature nanoindentation experiments. *Nat. Mater.* **2005**, *4*, 617–621. [CrossRef] [PubMed]
6. Schuh, C.A.; Packard, C.E.; Lund, A.C. Nanoindentation and contact-mode imaging at high temperatures. *J. Mater. Res.* **2006**, *21*, 725–736. [CrossRef]
7. Ge, D.; Minor, A.M.; Stach, E.A.; Morris, J.W. Size effects in the nanoindentation of silicon at ambient temperature. *Philos. Mag.* **2006**, *86*, 4069–4080. [CrossRef]
8. Korte, S.; Stearn, R.J.; Wheeler, J.M.; Clegg, W.J. High temperature microcompression and nanoindentation in vacuum. *J. Mater. Res.* **2012**, *27*, 167–176. [CrossRef]
9. Knapp, J.A.; Follstaedt, D.M.; Myers, S.M.; Barbour, J.C.; Friedmann, T.A. Finite-element modeling of nanoindentation. *J. Appl. Phys.* **1999**, *85*, 1460–1474. [CrossRef]
10. Jha, K.K.; Zhang, S.; Suksawang, N.; Wang, T.L.; Agarwal, A. Work-of-indentation as a means to characterize indenter geometry and load-displacement response of a material. *J. Phys. D Appl. Phys.* **2013**, *46*, 415501. [CrossRef]
11. Landman, U.; Luedtke, W.D.; Burnham, N.A.; Colton, R.J. Atomistic mechanisms and dynamics of adhesion, nanoindentation, and fracture. *Science* **1990**, *248*, 454–461. [CrossRef] [PubMed]
12. Lin, Y.H.; Jian, S.R.; Lai, Y.S.; Yang, P.F. Molecular dynamics simulation of nanoindentation-induced mechanical deformation and phase transformation in monocrystalline silicon. *Nanoscale Res. Lett.* **2008**, *3*, 71–75. [CrossRef]
13. Jiang, S.Y.; Jiang, M.Q.; Dai, L.H.; Yao, Y.G. Atomistic origin of rate-dependent serrated plastic flow in metallic glasses. *Nanoscale Res. Lett.* **2008**, *3*, 524–529. [CrossRef] [PubMed]

14. Lu, H.B.; Ni, Y.S. Effect of surface step on nanoindentation of thin films by multiscale analysis. *Thin Solid Films* **2012**, *520*, 4934–4940. [CrossRef]

15. Lu, H.B.; Ni, Y.S.; Mei, J.F.; Li, J.W.; Wang, H.S. Anisotropic plastic deformation beneath surface step during nanoindentation of FCC Al by multiscale analysis. *Comput. Mater. Sci.* **2012**, *58*, 192–200. [CrossRef]

16. Zhu, A.B.; He, D.Y.; He, R.J.; Zou, C. Nanoindentation simulation on single crystal copper by quasi-continuum method. *Mater. Sci. Eng. A-Struct.* **2016**, *674*, 76–81. [CrossRef]

17. Mei, J.F.; Ni, Y.S. The study of anisotropic behavior of nano-adhesive contact by multiscale simulation. *Thin Solid Films* **2014**, *566*, 45–53. [CrossRef]

18. Zeng, F.L.; Zhao, B.; Sun, Y. Multiscale simulation of incipient plasticity and dislocation nucleation on nickel film during tilted flat-ended nanoindentation. *Acta Mech. Solida Sin.* **2015**, *28*, 484–496. [CrossRef]

19. Dugdale, D.S. Wedge indentation experiments with cold-worked metals. *J. Mech. Phys. Solids* **1953**, *2*, 14–26. [CrossRef]

20. Prasad, K.E.; Chollacoop, N.; Ramamurty, U. Role of indenter angle on the plastic deformation underneath a sharp indenter and on representative strains: An experimental and numerical study. *Acta Mater.* **2011**, *59*, 4343–4355. [CrossRef]

21. Zeng, F.L.; Sun, Y.; Liu, Y.Z.; Zhou, Y. Multiscale simulations of wedged nanoindentation on nickel. *Comput. Mater. Sci.* **2012**, *62*, 47–54. [CrossRef]

22. Ercolessi, F.; Adams, J.B. Interatomic potentials from first-principles calculations: The force-matching method. *EPL-Europhys. Lett.* **1994**, *26*, 583–588. [CrossRef]

23. Shenoy, V.B.; Miller, R.; Tadmor, E.B.; Rodney, D.; Phillips, R.; Ortiz, M. An adaptive finite element approach to atomic-scale mechanics—The quasicontinuum method. *J. Mech. Phys. Solids* **1999**, *47*, 611–642. [CrossRef]

24. Tadmor, E.B.; Miller, R.; Phillips, R.; Ortiz, M. Nanoindentation and incipient plasticity. *J. Mater. Res.* **1999**, *14*, 2233–2250. [CrossRef]

25. Li, J.W.; Ni, Y.S.; Wang, H.S.; Mei, J.F. Effects of crystalline anisotropy and indenter size on nanoindentation by multiscale simulation. *Nanoscale Res. Lett.* **2010**, *5*, 420–432. [CrossRef] [PubMed]

26. Qu, S.; Huang, Y.; Nix, W.D.; Jiang, H.; Zhang, F.; Hwang, K.C. Indenter tip radius effect on the Nix-Gao relation in micro- and nanoindentation hardness experiments. *J. Mater. Res.* **2004**, *19*, 3423–3434. [CrossRef]

27. Cross, G.; Schirmeisen, A.; Stalder, A.; Grutter, P.; Tschudy, M.; Durig, U. Adhesion interaction between atomically defined tip and sample. *Phys. Rev. Lett.* **1998**, *80*, 4685–4688. [CrossRef]

28. Rubio, G.; Agrait, N.; Vieira, S. Atomic-sized metallic contacts: Mechanical properties and electronic transport. *Phys. Rev. Lett.* **1996**, *76*, 2302–2305. [CrossRef] [PubMed]

29. Stalder, A.; Durig, U. Study of yielding mechanics in nanometer-sized Au contacts. *Appl. Phys. Lett.* **1996**, *68*, 637–639. [CrossRef]

30. Hull, D.; Bacon, D.J. Chapter 1—Defects in crystals. In *Introduction to Dislocations*, 5th ed.; Butterworth-Heinemann: Oxford, UK, 2011; pp. 1–20.

31. Hull, D.; Bacon, D.J. Chapter 5—Dislocations in face-centered cubic metals. In *Introduction to Dislocations*, 5th ed.; Butterworth-Heinemann: Oxford, UK, 2011; pp. 85–107.

32. Ashby, M.F. The deformation of plastically non-homogeneous materials. *Philos. Mag.* **1970**, *21*, 399–424. [CrossRef]

33. Chaudhri, M.M. Subsurface strain distribution around Vickers hardness indentations in annealed polycrystalline copper. *Acta Mater.* **1998**, *46*, 3047–3056. [CrossRef]

34. Srikant, G.; Chollacoop, N.; Ramamurty, U. Plastic strain distribution underneath a Vickers Indenter: Role of yield strength and work hardening exponent. *Acta Mater.* **2006**, *54*, 5171–5178. [CrossRef]

35. Tabor, D. The hardness of solids. *Rev. Phys. Technol.* **1970**, *1*, 145. [CrossRef]

36. Oliver, W.C.; Pharr, G.M. Measurement of hardness and elastic modulus by instrumented indentation: Advances in understanding and refinements to methodology. *J. Mater. Res.* **2004**, *19*, 3–20. [CrossRef]

37. Alhafez, I.A.; Brodyanski, A.; Kopnarski, M.; Urbassek, H.M. Influence of tip geometry on nanoscratching. *Tribol. Lett.* **2017**, *65*, 1–13. [CrossRef]

38. Nix, W.D.; Gao, H. Indentation size effects in crystalline materials: A law for strain gradient plasticity. *J. Mech. Phys. Solids* **1997**, *46*, 411–425. [CrossRef]

39. Manika, I.; Maniks, J. Size effects in micro- and nanoscale indentation. *Acta Mater.* **2006**, *54*, 2049–2056. [CrossRef]

40. Durst, K.; Goken, M.; Pharr, G.M. Indentation size effect in spherical and pyramidal indentations. *J. Phys. D Appl. Phys.* **2008**, *41*, 074005. [CrossRef]

41. Pharr, G.M.; Herbert, E.G.; Gao, Y.F. The indentation size effect: A critical examination of experimental observations and mechanistic interpretations. In *Annual Review Of Materials Research*; Clarke, D.R., Ruhle, M., Zok, F., Eds.; Annual Reviews: Palo Alto, CA, USA, 2010; Volume 40, pp. 271–292.

42. Ma, L.; Morris, D.J.; Jennerjohn, S.L.; Bahr, D.F.; Levine, L.E. The role of probe shape on the initiation of metal plasticity in nanoindentation. *Acta Mater.* **2012**, *60*, 4729–4739. [CrossRef]

crystals

MDPI

Article

Nanoindentation-Induced Pile-Up in the Residual Impression of Crystalline Cu with Different Grain Size

Jiangjiang Hu [1], Yusheng Zhang [2,*], Weiming Sun [3] and Taihua Zhang [1]

1. College of Mechanical Engineering, Zhejiang University of Technology, Hangzhou 310014, China; jiangjiangwho@gmail.com (J.H.); zhangth@zjut.edu.cn (T.Z.)
2. Advanced Materials Research Center, Northwest Institute for Non-Ferrous Metal Research, Xi'an 710016, China
3. Key Laboratory of Automobile Materials, College of Materials Science and Engineering Jilin University, Nanling Campus, Changchun 130025, China; sunwm15@mails.jlu.edu.cn
* Correspondence: y.sh.zhang@163.com; Tel.: + 86-029-8622-1498

Received: 25 October 2017; Accepted: 19 December 2017; Published: 26 December 2017

Abstract: Nanoindentation morphologies of crystalline copper have been probed at the grain scale. Experimental tests have been conducted on nanocrystalline (NC), ultrafine-grained (UFG), and coarse-grained (CG) copper samples with a new Berkvoich indenter at the strain rate of 0.04/s without holding time at an indentation depth of 2000 nm at room temperature. As the grain size increases, the height of the pile-up around the residual indentation increases and then exhibits a slightly decrease in the CG Cu. The maximum of the pile-up in the CG Cu obviously deviates from the center of the indenter sides. Our analysis has revealed that the dislocation motion and GB activities in the NC Cu, some cross- and multiple-slip dislocations inside the larger grain in the UFG Cu, and forest dislocations from the intragranular Frank-Read sources in the CG Cu would directly induce this distinct pile-up effect.

Keywords: nanoindentation; pile-up effect; grain size; dislocation motion; grain boundary activities

1. Introduction

The nanoindentation technique has been widely used to characterize the mechanical properties of bulk- and thin-film materials on nano- and microscopic scales, such as elastic modulus (E) and hardness (H). In order to minimize the impact of blunting from the indenter, the involved mechanical parameters in this contact area calibration are acquired from the indentation of a standard fused silica sample. However, in some cases there exists an apparent upward extrusion at the edge of the contact with the indenter in some metals, known as pile-up, which means that the actual contact area is larger than the value calculated by the Oliver-Pharr method [1]. Some studies have reported that the true contact area is 60% larger than the measured value as serious pile-up occurs, leading to the overestimation of the E and H [2–7]. Therefore, an understanding of the formation of the indentation pile-up in various materials and loading conditions under nanoindentation testing is invaluable.

To date, there is no convince evidence to suggest that any new modified method considering pile-up at a larger indentation depth could bring more accurate measurement results, and the Oliver-Pharr method still is considered to provide relatively reliable E and H values. Even so, the morphology of residual impression could also present other useful information or parameters in deformed materials. For example, many studies have shown that the formation of pile-up is closely related to the ratio of the yield stress (σ_y) to E, the ratio of the final indentation depth h_f to the maximum indentation depth h_{max}, and the strain hardening exponent (n). Metallic materials with smaller values of σ_y / E and n usually exhibit larger pile-up height, but an accurate determination of these two parameters in the indentation process becomes impossible. Therefore, the easily measured h_f/h_{max} from the load-displacement curve can be used to determine the prevalence of the pile-up. It has been suggested that for materials with low work hardening ability, the amount of pile-up would become obvious as the ratio of h_f/h_{max} becomes larger than 0.7 [2,3]. Interestingly, most of the theoretical development reports were based on numerical modeling [5,8–12], with very limited experimental data being used to explain this phenomenon. As for crystalline materials, it has been widely accepted that the grain boundary (GB) mediated deformation process can operate effectively in materials with a grain size smaller than 500 nm. The concurrently occurred dislocation activity grouping with GB activities could promote the formation of micro (narrow) shear bands under tensile or compression testing. Although some reports have studied the effect of work-hardening grain orientation on pile-up formation in coarse-grained (CG) metals [3,13], other factors such as grain size and deformation mechanisms have rarely been mentioned.

The aim of the present work is to establish the relationship between the pile-up effect and grain size in crystalline materials. Nanocrystalline (NC), ultrafine-grained (UFG), and CG copper samples were detected by nanoindentation at room temperature at an indentation depth of 2000 nm, and their pile-up effect around the residual impressions were systemically studied. Deformation mechanism transformation was used to explain why the pile-up effect exhibits different morphologies in crystalline materials with different grain sizes.

2. Experiments Details

The bulk NC/UFG Cu specimens with different grain sizes used in this article were synthesized by electric brush-plating on a substrate of copper sheets with a bath only containing $CuSO_4 \cdot 5H_2O$ (180–220 g/L); the detailed processes are presented in References [14–16]. A commercial coarse-grained (CG) copper sheet with thickness of 2 mm and purity of 99.99 wt % was annealed at a temperature of 800 °C for 24 h, and prepared for a contrastive experiment. Foil samples for transmission electron microscope (TEM, JEM-2100F, JEOL, Tokyo, Japan) observation under an accelerating voltage of 200 kV were prepared by cutting, polishing, and dimpling by argon-ion milling (EMRES101) at 5 kV. Square-shaped nanoindentation specimens of NC/UFG/CG Cu with a gauge size of $30 \times 30 \times 2$ mm^3 were cut, and then mechanically grinded with SiC papers, and finally polished with a microcloth using a slurry of 0.5 μm alumina. To acquire reliable nanoindentation data, the surfaces of the test specimens were mechanically polished to mirror finishing, and then their mechanical properties were characterized at room temperature by a nanoindenter (Agilent-G200, Keysight, Santa Rosa, CA, USA) with a Berkovich diamond indenter. All of the samples were loaded at strain rates of 0.4/s, 0.04/s, and 0.004/s to the indentation depth of 2000 nm without holding time under the procedure of the continuous stiffness measurement (CSM). The thermal drift rate prior to testing was limited below 0.025 nm/s and the indentation under same condition was repeated at least 10 times. The surface morphology of the residual impression obtained at a strain rate of 0.04/s was observed by laser confocal microscopy (LSM, OLS4500, OLYMPUS, Tokyo, Japan).

3. Results

3.1. TEM Observation

Figure 1 gives the TEM bright images of the typical microstructures of the as-brush-plated NC/UFG Cu. It reveals that these materials consist of uniformly equiaxed grains with random crystalline orientations and predominant high-angle grain boundaries (GBs). Grain size was measured from the statistical analysis of 500 grains taken from several TEM images, and the average values of these materials are 30 nm, 150 nm, and 300 nm, respectively.

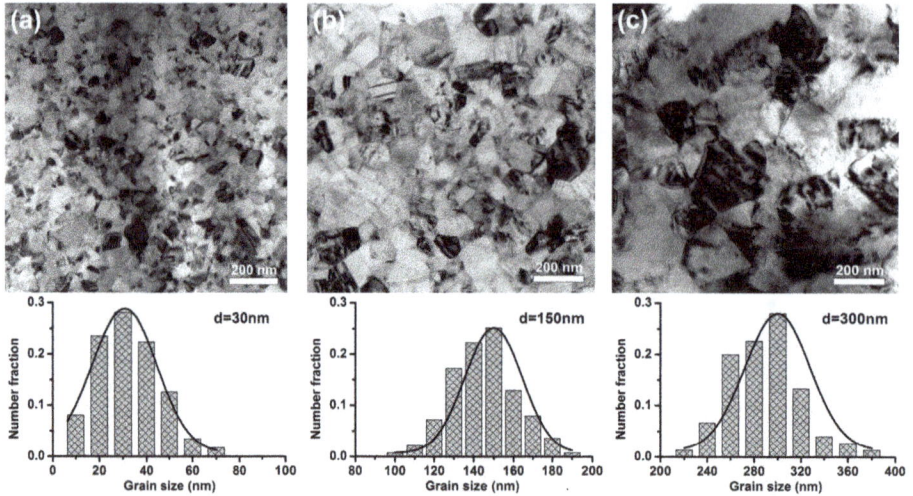

Figure 1. TEM bright-field images of the as-brush-plated NC/UFG Cu: (**a**) 30 nm Cu; (**b**) 150 nm Cu; (**c**) 300 nm Cu.

3.2. Residual Morphology of the Impression

Figure 2 shows the load-displacement (*P-h*) curves obtained at a strain rate of 0.04/s to the indentation depth of 2000 nm without holding time on these four materials. It can be seen that as the grain size decreases, the acquired load at a given displacement in the loading stage increases rapidly and the slope value of the *P-h* curves during the unloading regime decreases obviously. To assess the effect of the pile-up in the indenters, Table 1 summarizes the ratio of h_f/h_{max} in these four materials. Clearly, this ratio decreases steadily as the grain size decreases.

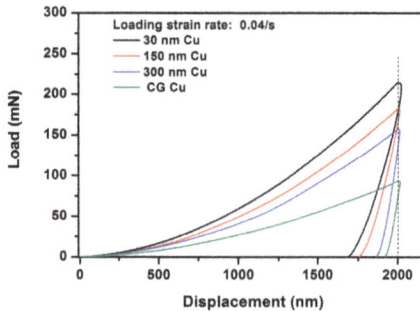

Figure 2. Several load-displacement (*P-h*) curves obtained at an indentation depth of 2000 nm at $\dot{\varepsilon}_L$ of 0.04/s in the NC/UFG/CG Cu.

Table 1. h_f/h_{max} of NC/UFG/CG Cu obtained at a strain rate of 0.04/s and indentation depth of 2000 nm.

Materials	h_{max} at P_{max}	h_f at 0 mN	h_f/h_{max}
30 nm Cu	2000 nm	1694 nm	0.847
150 nm Cu	2000 nm	1756 nm	0.878
300 nm Cu	2000 nm	1867 nm	0.934
CG Cu	2000 nm	1919 nm	0.960

LCM images can be used to determine the extent and nature of the pile-up in the deformed surface morphology of the impression. Figure 3 gives of the residual morphology of the Berkovich indenter after loading at a strain rate of 0.04/s and the same indentation depth of 2000 nm in these four NC/UFG/CG Cu samples. Clearly, except the indents of 30 nm Cu exhibiting very little protrusion, the other three materials exhibit an obvious pile-up effect in the vicinities of the impression. In order to quantify the pile-up height along the z-axis in the impression, 20 cross-sectional lines perpendicular to each side of the indentation edges were measured (the left graphs in the Figure 3). The right image in Figure 3 gives the corresponding three-dimensional graphs of these four materials at x, y, and z axis angles of 59°, 274°, and 4°, which could better display the pile-up morphology of the indenter edge. Figure 4 gives the detailed maximum height of the pile-up ($h_{pile\text{-}up}$) along the three different directions. It can be seen that the $h_{pile\text{-}up}$ would reach the maximum value as the grain size increases from 30 nm to 300 nm, followed by a slight decrease in the CG Cu. Besides that, there is another interesting result that the $h_{pile\text{-}up}$ in the CG Cu obviously deviates from the center of each of the indenter sides.

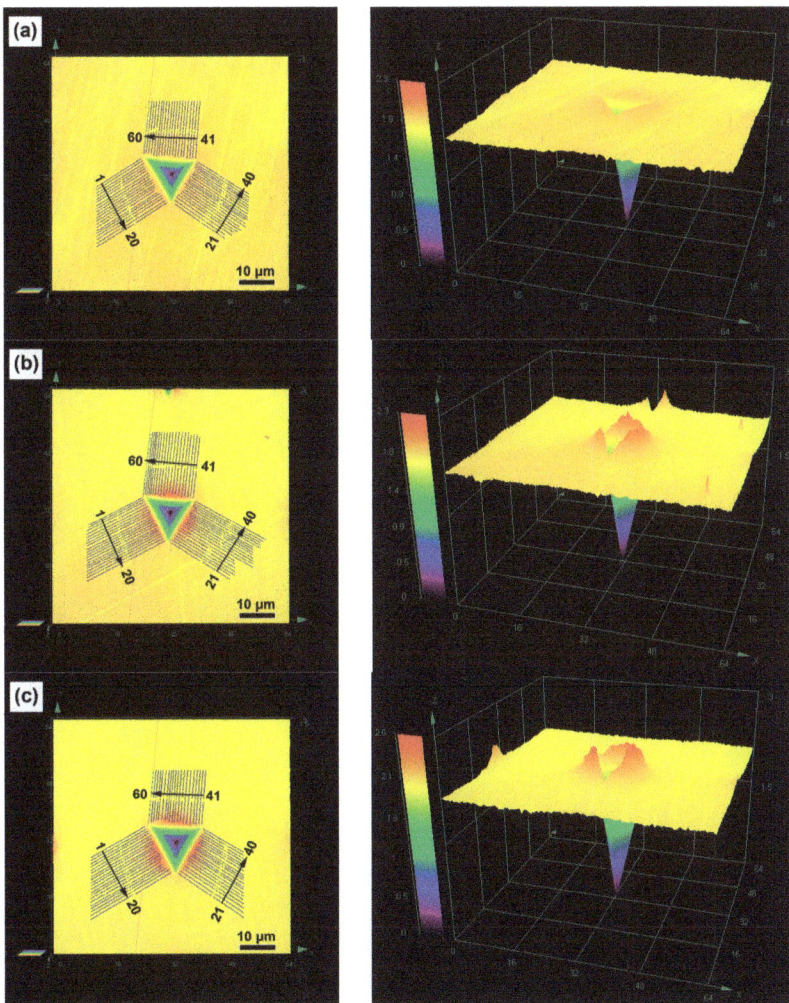

Figure 3. *Cont.*

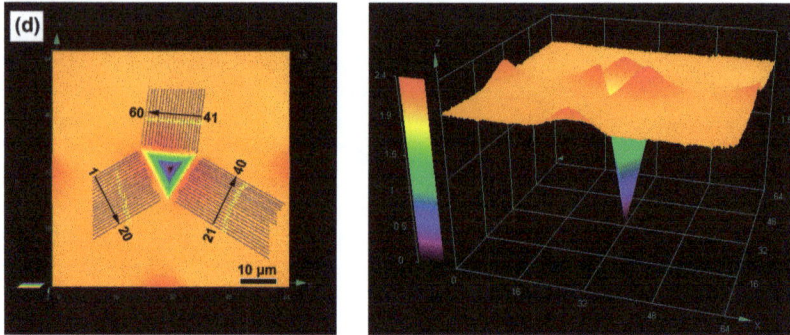

Figure 3. LCM images of 30 nm Cu (**a**); 150 nm Cu (**b**); 300 nm Cu (**c**); and CG Cu (**d**) obtained at a strain rate of 0.04/s and the indentation depth of 2000 nm. The left side gives the places where $h_{pile-up}$ is measured and the right gives the corresponding three-dimensional graphs.

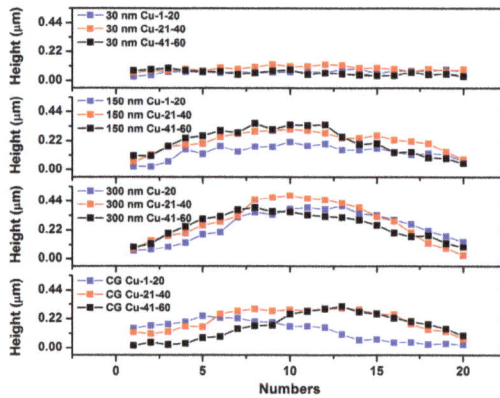

Figure 4. The values of the $h_{pile-up}$ around the three indenter edges in the 30 nm Cu, 150 nm Cu, 300 nm Cu, and CG Cu.

3.3. Correction to the Hardness and Modulus from Pile-up Measurements

Taking the pile-ups of the residual impression into account, the projected area should be calibrated in order to yield relatively accurate values of *E* and *H*. A method exploiting the topography measurements to correct the values of these two mechanical parameters was proposed by Renner as the following procedures [11]:

$$H = P_{max}/A_c \tag{1}$$

where A_c is the project contact area between the indenter and the material defined by the Oliver-Pharr method [1].

$$A_c = 24.56h_c^2 + \sum_{i=1}^{8} C_i h_c^{1/i^2} \tag{2}$$

where C_i is the constant dependent on the sharp of the indenter, and h_c is the contact depth at P_{max}.

$$h_c = h_{max} - \varepsilon P_{max}/S \tag{3}$$

where ε is the geometrical constant of 0.75 depending on the indenter shape and S is the stiffness measured at P_{max}.

$$E = \left(1 - v^2\right) \Big/ \left(1/E_r - \left(1 - v_i^2\right)\big/E_i\right) \tag{4}$$

E_r is the reduced elastic modulus, having a equation of $E_r = \sqrt{\pi}S/2\sqrt{A_c}$. E_i and v_i are the Young's modulus and the Poisson's ratio of the indenter, respectively.

It can be seen that the Oliver-Pharr method only considers that the sink-in event would occur at maximum displacement in the tested materials, while the pile-up issues are usually neglected. As shown in Figure 5, a better approximation of the real projected contact area can be defined by the following equations:

$$A_{real} = A_c + A_{pile-up} \tag{5}$$

$$A_{pile-up} = \frac{1}{2\tan\alpha}\sum_{i=1}^{N} h_{pile-upi}w_i \tag{6}$$

where N is the number of pile-ups formed around the indentation imprint and α is the projection angle of the pile-up contact area on the indented plane. $h_{pile-upi}$ and w_i are the maximum height and width of the ith pile-up, respectively. The contact surface between the pile-ups and the indenter can be treated as a triangle, and its dimensions are used to determine the values of $h_{pile-upi}$ and w_i. Figure 5 gives an example of the 300 nm Cu sample mentioned in the text. The pile-up corrections of the E and H in the NC/UFG/CG Cu by Equations (5) and (6) are drawn in Figure 6. It can be seen that a larger $h_{pile-up}$ would lower the values of the E and H.

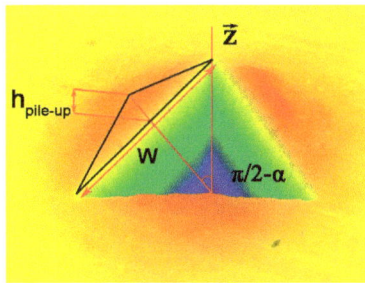

Figure 5. Approximation of the real projected contact area resulting from pile-up formation: the maximum height and width of each pile-up around the imprint are considered to determine the projected area using triangles, and the imprints depict one typical example from the 300 nm Cu mentioned in the text.

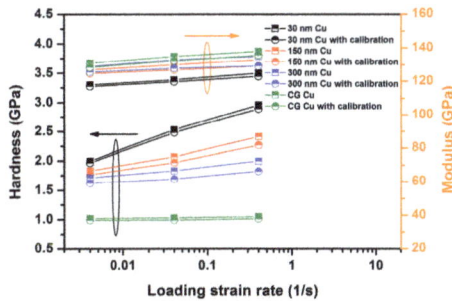

Figure 6. E and H values without and with pile-up corrections as a function of the loading strain rate for 30 nm, 150 nm, 300 nm, and CG Cu.

3.4. Strain Rate Sensitivity and Activation Volume

To give a quantitative measurement of the strain rate dependence of hardness in the different materials, the strain rate sensitivity (m) and activation volume (v) obtained by the equations of $m = \partial \log H / \partial \log \dot{\varepsilon}_L$ and $v = \sqrt{3} KT \partial \ln \dot{\varepsilon}_L / \partial H$. Figure 7a,b give the m and v of the NC/UFG/CG Cu obtained at three typically different loading strain rates of 0.4/s, 0.04/s, and 0.004/s. The m decreases with the grain size while v presents an opposite tendency, and their values in these four materials are 0.08414, 0.05482, 0.02564, and 0.00549, and $6.06b^3$, $11.47b^3$, $27.99b^3$, and $194.17b^3$, respectively, where $b = 0.256$ nm is the Burgers vector for the $1/2\{1\,1\,0\}$ dislocation in the Cu.

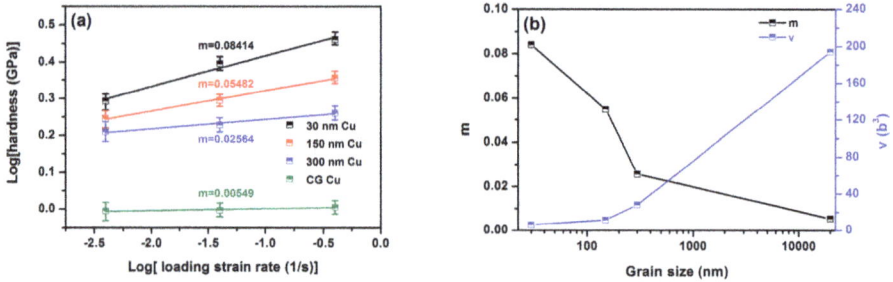

Figure 7. The hardness versus loading strain rate ($\log H - \log \dot{\varepsilon}_L$) curves for measuring the m values of NC/UFG/CG Cu measured at three different $\dot{\varepsilon}_L$ of 0.4/s, 0.04/s, and 0.004/s at room temperature (**a**); The corresponding m and v versus grain size curves (**b**).

4. Discussion

Finite element modeling (FEM) results have pointed out that [17], as the h_f/h_{max} is larger than 0.7, and/or in the materials without obvious work harden, a larger amount of the pile-up would directly induce an underestimation of the contact area, which further produces correspondingly large errors in the measurement (overestimation) of the E and H at smaller indentation depths. Our experimental results show that although all of the values of h_f/h_{max} in these four materials are larger than 0.7, the height of the pile-up does not exhibit a continuously increasing tendency as the values become larger. The pile-up evolution during the indentation is closely related to the relative amounts of elastic and plastic deformation as characterized by the ratio of the elastic modulus to yield stress, i.e., E/σ_y. As $E/\sigma_y = 0$, the contact is strictly elastic and dominated by sink-in in the way prescribed by Hertzian contact mechanics. As $E/\sigma_y = \infty$, the indentation process follows the rigid-plastic deformation and extensive pile-up would occur around the residual impression [7]. Our previous reports have already systematically studied the tensile properties of NC/UFG Cu with different grain sizes [14,16,18], and the corresponding values of E/σ_y in these four materials were found to be 175 (30 nm Cu), 264 (150 nm Cu), 386 (300 nm Cu), and 1930 (CG Cu). As expected, the elastic behavior is larger for the NC Cu because of smaller h_f/h_{max} value, but there is little recovery for materials with larger grain size due to the fact that sufficiently larger plasticity occurs in these materials. These different elastic-plastic behaviors are closely related to the distinct deformation mechanism transformation as the grain size decreases below 500 nm.

Many studies have reported that introducing a larger volume of the GB structure could enhance m values in NC metals, such as NC Ni [16,19], NC Cu [14–16], nanotwinned Cu [20], NC Ta [21], and NC Mg [22]. The rate sensitivity of the mechanical properties is related to the thermally activated process of overcoming the obstacles of the glissile dislocation movement, which are normally expressed by the activation volume (v) of the thermally activated event. If the estimated v is less than $2b^3$ or even on the order of one atomic volume (b^3), this may suggest that the deformation mechanism totally stems from the GB activities, such as the GB sliding/diffusion or even Coble creep, which can control the plasticity of NC metals. For the v in our 30 nm Cu, it is certainly much larger than the $2b^3$. This arises from the interactions between dislocations and GBs, including full/partial dislocation emission and absorption, de-pinning of dislocations impeded by GB ledges of impurities, and dynamic recovery of dislocations that reside along GBs. It should be noted that the activation volume of the face center cubic (FCC) metals can be also expressed as the equation of $v = b\chi l$ [23], where χ is the distance swept out by the glide dislocation during one activation event that can be treated as an approximate constant of b, and l is the length of dislocation segment involved in the thermal activation. As the grain size continuously increases, the forest dislocation density inside the grains is expected to increase, whereas the obstacle density associated with GBs decreases. This means that both the proportion of plastic deformation controlled by the dislocation motion and l become larger, which directly enhances the values of v. When the grain size increases to several micrometers, i.e., CG metals, conventional forest dislocation movement generated in the intragranular sources would dominate the plastic deformation and l is much larger, leading to the largest v obtained in our CG Cu. This deformation mechanism transformation varying with the grain size can well explain why the pile-up effect would gradually weaken or even disappear as the grain size decreases to a nano-regime.

For the NC Cu, more GBs are involved in the indentation deformation and both dislocation activity and GB sliding can be activated in the strain rate range studied. In this case, a large amount of slip planes are activated and dislocations emitted from GBs would propagate in the grain interior and be absorbed by the opposite GBs. This indicates that the highly mobile or unstable generated dislocations in the loading regime can only stay temporarily in the grain interior, because GB serves as a sink for dislocation absorption, and the materials exhibit very limited strain hardening capabilities (as shown in Figure 8a). Therefore, the effect of the pile-up induced by the strain hardening becomes invalid. Besides that, because of the limited intragranular space and GB activities such as GB sliding and rotation in NC metals, dislocation gliding occurs only on fewer slip systems and high stress concentration created in some local regions such as triple junctions would relax, which could reduce the number of the dislocations and further weaken pile-up lateral extension around the imprints.

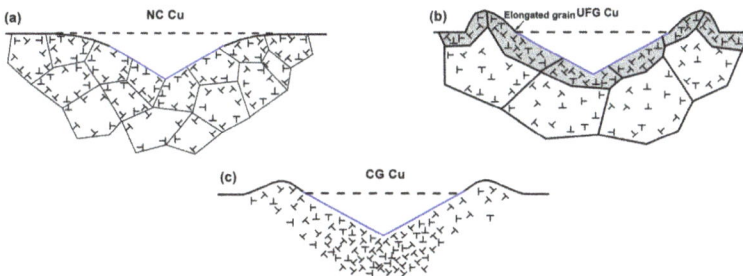

Figure 8. Schematic illustrations of local GB structures and dislocation distribution in the NC (**a**); UFG (**b**); and CG (**c**) Cu underneath the indenters, where "T" represents dislocation.

As the grain size enters the ranges of 100–1500 nm, although a number of GB mechanisms such as GB sliding and GB rotation would participate in their plastic deformation, the transition between intragranular and intergranular deformation mechanism is not supposed to be abrupt [14,16,19,21,22,24]. Dislocation activities such as the cross- and multiple-slips from the internal source can still be activated effectively in the large grain zones, and dislocations can be trapped inside the grains as a result of dislocation interactions between slip systems or with debris left by cross-slip. So, some parts of the pile-up can form in a way similar to that in the CG metals, where strain hardening induces plastic deformation around the surface (see below). However, previous studies have demonstrated that many grains in the UFG Cu would have severely elongated grains under larger plastic deformation in the tensile tests [25,26]. For the nanoindentation tests, the top surface around the indenter has the largest deformation during the total deformation process, and thus it can be deduced that the grains in these regions possibly have elongated grains (as shown in Figure 8b). This means that the pile-up lateral extension and propagation of dislocation slip are possibly confined in the regions around the plastic zone, and the severely deformed grains would produce additional plasticity around the indenter, leading to the highest $h_{pile-up}$ in the UFG Cu with a grain size of 300 nm.

For conventional CG metals, plastic deformation mainly involves the dislocation multiplication from the intragranular Frank-Read sources, which could produce the dislocation cells/walls/networks structure in the grain interior. These formed microstructures, representing a strong pinning point to the dislocation-bowing segment, would suppress dislocation cross slips, leading to strong strain hardening in the subsequent plasticity. FEM results have revealed that due to the higher stress applied underneath the tip of the indenter, the shear strain or dislocation density generated at the deepest areas is much higher compared with other areas, and materials around this region harden more severely [8]. The strain hardening generated by such dislocation activities is more effective in suppressing the strain localization. However, for the regions approaching the surface, the dislocation slips in the other <110> directions would become relatively easy, which contributes to the final topography of the pile-up. Unlike the above NC/UFG metals, the CG Cu has a larger grain size and thus is assumed that the local material behavior is within the single crystalline which cannot be influenced by the neighboring grains. So, the plastic zone around the imprints in the CG Cu cannot be confined by the GB structure in the UFG Cu and the pile-up lateral extension thus declines to som extent (as shown in Figure 8c). Additionally, Renner claimed that the pile-up sizes and distributions such as height and lateral extension depend on the crystallographic directions and indenter disorientation, which are also the main reasons for generating significant asymmetries around imprints in CG metals [11].

5. Conclusions

Nanoindentation tests were carried out on crystalline Cu with different grain sizes ranging from 30 nm to 300 nm as well as CG Cu, and the residual indenter morphologies of these four materials were systematically studied. It can be observed that the height of the pile-up reaches the maximum value as the grain size increases from 30 nm to 300 nm, and then decreases in the CG Cu. Our analysis argued that dislocation activity and GB activities in the NC metals, some cross- and multiple-slips dislocation insides the larger grain in the UFG Cu, and forest dislocations from the intragranular Frank-Read sources in the CG Cu are responsible for the distinct pile-up effect in these materials.

Acknowledgments: This work was supported by the National Natural Science Foundation of China (Grant Nos. 11727803, 11672356 and 51371089) and Innovation team in key areas of Shaanxi Province (2016KCT-30).

Author Contributions: Jiangjiang Hu conducted the experiments and wrote the paper. Yusheng Zhang supervised the revised the whole work. Weiming Sun processed the data. Taihua Zhang conceived the work and revised the paper.

Conflicts of Interest: The authors declare no conflict of interest.

References

1. Oliver, W.C.; Pharra, G.M. Measurement of hardness and elastic modulus by instrumented indentation: Advances in understanding and refinements to methodology. *J. Mater. Res.* **2004**, *19*, 3–20. [CrossRef]
2. McElhaney, K.W.; Vlassak, J.J.; Nix, W.D. Determination of indenter tip geometry and indentation contact area for depth-sensing indentation experiments. *J. Mater. Res.* **1998**, *13*, 1300–1306. [CrossRef]
3. Gale, J.D.; Achuthan, A. The effect of work-hardening and pile-up on nanoindentation measurements. *J. Mater. Sci.* **2014**, *49*, 5066–5075. [CrossRef]
4. Beegan, D.; Chowdhury, S.; Laugier, M.T. A nanoindentation study of copper films on oxidised silicon substrates. *Surf. Coat. Technol.* **2003**, *176*, 124–130. [CrossRef]
5. Rodríguez, J.; Maneiro, M.A.G. A procedure to prevent pile up effects on the analysis of spherical indentation data in elastic-plastic materials. *Mech. Mater.* **2007**, *39*, 987–997. [CrossRef]

6. Zhou, L.; Yao, Y. Single crystal bulk material micro/nano indentation hardness testing by nanoindentation instrument and AFM. *Mater. Sci. Eng.* **2007**, *460–461*, 95–100. [CrossRef]
7. Taljat, B.; Pharr, G.M. Development of pile-up during spherical indentation of elastic-plastic solids. *Int. J. Solids Struct.* **2004**, *41*, 3891–3904. [CrossRef]
8. Liu, Y.; Varghese, S.; Ma, J.; Yoshino, M.; Lu, H.; Komanduri, R. Orientation effects in nanoindentation of single crystal copper. *Int. J. Plast.* **2008**, *24*, 1990–2015. [CrossRef]
9. Demiral, M.; Roy, A.; El Sayed, T.; Silberschmidt, V.V. Influence of strain gradients on lattice rotation in nano-indentation experiments: A numerical study. *Mater. Sci. Eng.* **2014**, *608*, 73–81. [CrossRef]
10. Sánchez-Martín, R.; Pérez-Prado, M.T.; Segurado, J.; Molina-Aldareguia, J.M. Effect of indentation size on the nucleation and propagation of tensile twinning in pure magnesium. *Acta Mater.* **2015**, *93*, 114–128. [CrossRef]
11. Renner, E.; Gaillard, Y.; Richard, F.; Amiot, F.; Delobelle, P. Sensitivity of the residual topography to single crystal plasticity parameters in Berkovich nanoindentation on FCC nickel. *Int. J. Plast.* **2016**, *77*, 118–140. [CrossRef]
12. Petryk, H.; Stupkiewicz, S.; Kucharski, S. On direct estimation of hardening exponent in crystal plasticity from the spherical indentation test. *Int. J. Solids Struct.* **2017**, *112*, 209–221. [CrossRef]
13. Chen, T.; Tan, L.; Lu, Z.; Xu, H. The effect of grain orientation on nanoindentation behavior of model austenitic alloy Fe-20Cr-25Ni. *Acta Mater.* **2017**, *138*, 83–91. [CrossRef]
14. Hu, J.; Han, S.; Sun, G.; Sun, S.; Jiang, Z.; Wang, G.; Lian, J. Effect of strain rate on tensile properties of electric brush-plated nanocrystalline copper. *Mater. Sci. Eng.* **2014**, *618*, 621–628. [CrossRef]
15. Hu, J.; Sun, G.; Zhang, X.; Wang, G.; Jiang, Z.; Han, S.; Zhang, J.; Lian, J. Effects of loading strain rate and stacking fault energy on nanoindentation creep behaviors of nanocrystalline Cu, Ni-20 wt.% Fe and Ni. *J. Alloys Compd.* **2015**, *647*, 670–680. [CrossRef]
16. Hu, J.; Zhang, J.; Jiang, Z.; Ding, X.; Zhang, Y.; Han, S.; Sun, J.; Lian, J. Plastic deformation behavior during unloading in compressive cyclic test of nanocrystalline copper. *Mater. Sci. Eng.* **2016**, *651*, 999–1009. [CrossRef]
17. Bolshakova, A.; Pharr, G.M. Influences of pileup on the measurement of mechanical properties by load and depth sensing indentation techniques. *J. Mater. Res.* **1998**, *13*, 1049–1058. [CrossRef]
18. Zhang, H.; Jiang, Z.; Lian, S.; Jiang, Q. Strain rate dependence of tensile ductility in an electrodeposited Cu with ultrafine grain size. *Mater. Sci. Eng.* **2008**, *479*, 136–141. [CrossRef]
19. Li, H.; Liang, Y.; Zhao, L.; Hu, J.; Han, S.; Lian, J. Mapping the strain-rate and grain-size dependence of deformation behaviors in nanocrystalline face-centered-cubic Ni and Ni-based alloys. *J. Alloys Compd.* **2017**, *709*, 566–574. [CrossRef]
20. Wei, Y. The kinetics and energetics of dislocation mediated de-twinning in nano-twinned face-centered cubic metals. *Mater. Sci. Eng.* **2011**, *528*, 1558–1566. [CrossRef]
21. Hu, J.; Sun, W.; Jiang, Z.; Zhang, W.; Lu, J.; Huo, W.; Zhang, Y.; Zhang, P. Indentation size effect on hardness in the body-centered cubic coarse-grained and nanocrystalline tantalum. *Mater. Sci. Eng.* **2017**, *686*, 19–25. [CrossRef]
22. Hu, J.; Zhang, W.; Bi, G.; Lu, J.; Huo, W.; Zhang, Y. Nanoindentation creep behavior of coarse-grained and ultrafine-grained pure magnesium and AZ31 alloy. *Mater. Sci. Eng.* **2017**, *698*, 348–355. [CrossRef]
23. Wei, Q.; Cheng, S.; Ramesh, K.T.; Ma, E. Effect of nanocrystalline and ultrafine grain sizes on the strain rate sensitivity and activation volume: Fcc versus bcc metals. *Mater. Sci. Eng.* **2004**, *381*, 71–79. [CrossRef]
24. Jiang, Z.; Liu, X.; Li, G.; Jiang, Q.; Lian, J. Strain rate sensitivity of a nanocrystalline Cu synthesized by electric brush plating. *Appl. Phys. Lett.* **2006**, *88*, 143115. [CrossRef]
25. Wei, Q.; Jia, D.; Ramesh, K.T.; Ma, E. Evolution and microstructure of shear bands in nanostructured Fe. *Appl. Phys. Lett.* **2002**, *81*, 1240. [CrossRef]
26. Wei, Q.; Kecskes, L.; Jiao, T.; Hartwig, K.T.; Ramesh, K.T.; Ma, E. Adiabatic shear banding in ultrafine-grained Fe processed by severe plastic deformation. *Acta Mater.* **2004**, *52*, 1859–1869. [CrossRef]

Article

Atomistic Insights into the Effects of Residual Stress during Nanoindentation

Kun Sun [1], Junqin Shi [1] and Lifeng Ma [2],*

[1] State Key Laboratory for Mechanical Behavior of Materials, Xi'an Jiaotong University, Xi'an 710049, China; sunkun@mail.xjtu.edu.cn (K.S.); shijunqin2012@stu.xjtu.edu.cn (J.S.)
[2] S&V Laboratory, Department of Engineering Mechanics, Xi'an Jiaotong University, Xi'an 710049, China
* Correspondence: malf@mail.xjtu.edu.cn

Academic Editors: Ronald W. Armstrong
Received: 21 June 2017; Accepted: 30 July 2017; Published: 1 August 2017

Abstract: The influence of in-plane residual stress on Hertzian nanoindentation for single-crystal copper thin film is investigated using molecular dynamics simulations (MD). It is found that: (i) the yield strength of incipient plasticity increases with compressive residual stress, but decreases with tensile residual stress; (ii) the hardness decreases with tensile residual stress, and increases with compressive residual stress, but abruptly drops down at a higher compressive residual stress level, because of the deterioration of the surface; (iii) the indentation modulus reduces linearly with decreasing compressive residual stress (and with increasing tensile residual stress). It can be concluded from the MD simulations that the residual stress not only strongly influences the dislocation evolution of the plastic deformation process, but also significantly affects the size of the plastic zone.

Keywords: nanoindentation; residual stress; plastic deformation; molecular dynamics

1. Introduction

Nowadays, thin film materials at micro- and nano-scale play a significant role in a wide range of engineering applications, such as medical instruments, micro-/nano-electromechanical systems (MEMS/NEMS), and optical devices [1–4]. The mechanical properties of thin film materials, which strongly influence the reliability and service life of a nano-device, have been attracting both industrial and scientific interest [4]. Therefore, it is vital to characterize the mechanical properties and understand the deformation mechanism of a thin film material, in order to manufacture nano-devices with high reliability and service life. Nanoindentation is one of the most popular methods used to investigate the mechanical properties of materials at micro- and nano-scale [5].

During a nanoindentation test, residual stress is a particular non-ignorable factor, which can influence the thin film material's hardness and its plastic deformation behavior. Great efforts have been made in research [6–11]. Tsui et al. [7] found that hardness measured by a sharp indenter increases with compressive stress and decreases with tensile stress, but later on Yang et al. [8] claimed that the uniaxial tensile and compressive stress has a very small effect on the Vickers hardness. A subsequent finite element simulation also denied the change in hardness because of the pileup of the surface [9]. Interestingly, in comparison with the previously mentioned sharp indenter, Taljat and Pharr [11] reported that residual stress has an obvious influence on the load-depth curve by using spherical indentation. The controversy about residual stress function on the plastic deformation of nanoindentation in the literature suggests that more efforts need to be made, and new research approaches need to be employed, to reveal the deformation mechanism at micro-scale.

Molecular dynamics (MD) simulation is a very powerful approach to reveal micro-deformation in nano-materials [12–14]. For example, the plastic deformation behaviors of nanoindentation in

FCC (Cu and Al) and BCC (Fe and Ta) metals by a spherical indenter have been analyzed using MD [15]. Taking advantage of MD, size effects due to grain boundary, high rate compression, and other confining conditions have been systematically studied by Voyiadjis and Yaghoobi [16–19]. Research on thin film deformation under nanoindentation has also been carried out. The nanoindentation behavior of a metal film under a spherical indenter was first simulated using MD by Kelchner et al. [20]. Since then, by using MD simulations, many new results have been reported. For example, Liang et al. [21] investigated the crystal orientation influence on the plastic deformation of copper under nanoindentation. Yaghoobi et al. [22] found that the typically observed defect in Ni thin film is the Shockley partial dislocation at the onset of plasticity. Moreover, nanoindentation on complex Ti-V multilayered thin films has been initially explored [23]. No doubt, the above studies have made considerable progress in understanding the nanoindentation behavior of metal films. Recently, we [24] simulated the nanoindentation of a virtual sphere indenter for single-crystal copper thin film. We found that the residual stress has an effect on the local surface hardness and incipient plasticity. However, there are three problems remaining: (i) in our previous study, we mainly focused on the scenario of the incipient plasticity of indentation, and thus the full developing indentation process and deformation process is unknown; (ii) because the virtual sphere indenter employed in our previous study is an unrealistic one, which was oversimplified, a more realistic indentation model needs to be built; and (iii) in our previous study, localized and detailed dislocation nucleation and development was discussed, but the full deformation field is still not clear.

In this paper, we perform MD simulations of nanoindentation for single-crystal copper thin film with pre-existing equiaxial stress, to study the effect of residual stress on plasticity deformation. In particular, the mechanism of the elastic–plastic transition will be explored, and the dynamics of dislocation evolution and the indentation stress field will also be investigated.

2. Simulation Model and Methodology

In this work, the MD method was performed to simulate the nanoindentation process between a single-crystal copper thin film and a spherical diamond indenter. Displacement control by positioning the spherical indenter is used during loading along the crystal direction [001] of the copper. The software LAMMPS [25] is used to perform the simulation. An embedded atom method (EAM) potential [26] is adopted to define the Cu-Cu atoms' interaction. The Morse potential [27] is adopted to calculate the interaction between the copper atoms and carbon atoms of the diamond indenter, and the parameters are the same as in Ref. [4].

The schematic of the nanoindentation configuration with a spherical diamond indenter is shown in Figure 1. The lattice constant of the copper is 3.615 Å, and the models with pre-existing stress are of the same size, $32.5 \times 32.5 \times 21.3$ nm, consisting of about 1,955,200 atoms. The spherical diamond indenter of radius $R = 2.55$ nm is assumed to be perfectly rigid, and the deformable copper substrate includes three kinds of atoms: boundary atoms, thermostat atoms, and Newtonian atoms. Three layers of atoms at the bottom of the substrate (lower z plane) are boundary atoms fixed in their initial lattice positions to reduce the boundary effects. A thermostat is applied to the thermostatic zone. The motions of Newtonian atoms obey the classical Newton's second law, which are integrated with a velocity Verlet algorithm with a time step of 1 fs.

The initial stress-free model is relaxed through an energy minimization, followed by a zero-stress relaxation in the isothermal-isobaric (NPT) ensemble using a Nosé–Hoover thermostat at 0.1 K ignoring thermal vibrations for 20 ps. Then, to study the effect of residual stress on the nanoindentation behavior, pre-existing stress models are constructed by equiaxial compressing and/or tensioning the initial model in x and y directions. The pre-strain values are adopted from −2.5% to 2.0%. The pre-stress is listed in Table 1, and is normalized by the ideal shear strength ($G_{ideal} = 4.56$ GPa) for the Cu (001) single-crystal [28]. After the desired strain is reached, the simulations are performed at 0.1 K under the canonical (NVT) ensemble. Periodic boundary conditions are imposed in the x and y directions for all simulations. The centro-symmetry parameters are used to discern defects [20].

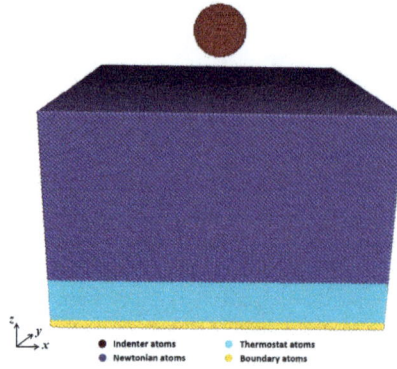

Figure 1. Schematic of nanoindentation atomic configuration with a spherical diamond indenter.

Table 1. Pre-strain vs. corresponding residual stress and normalized by the ideal shear strength of Cu single crystals.

Pre-strain (%)	−2.5	−2.0	−1.5	−1.0	−0.5	0	0.5	1.0	1.5	2.0
σ_{res} (GPa)	−2.42	−1.81	−1.26	−0.76	−0.32	0	0.46	0.79	1.10	1.36
λ	−0.53	−0.40	−0.28	−0.17	−0.07	0	0.10	0.17	0.24	0.30

σ_{res}, residual stress resulting from the molecular dynamics (MD) simulations; λ, relative residual stress, $\lambda = \sigma_{res} / G_{ideal}$, where G_{ideal} is the ideal shear strength (G_{ideal} = 4.56 GPa) of Cu single crystals [20].

At nano-scale, copper film shows strong anisotropy, i.e., the elasticity modulus is different along different crystal orientations [22]. From the material properties of copper suggested by Ref. [29], the related material elastic constants are C_{11} = 169.9 GPa, C_{12} = 122.6 GPa, and C_{44} = 76.2 GPa. Hence, the anisotropy factor, defined by $2C_{44}/(C_{11} - C_{12})$, is 3.22 for single Cu film, and the Young's modulus at [001] crystal orientation is calculated by following expression [29]:

$$E_{001} = (C_{11} - C_{12}) \times (C_{11} + 2C_{12})/(C_{11} + C_{12}) = 67.1 \text{ GPa} \qquad (1)$$

Consequently, the residual stress, shown in Table 1, is not symmetrical about the pre-strain because of anisotropy.

3. Results and Discussion

3.1. Load vs. Indention Depth Response at Different Residual Stresses

The load vs. indentation depth curves with residual stress are depicted in Figures 2 and 3. There are three regimes in the load vs. indentation depth curve during indentation [4], namely the quasi-elastic regime, the sudden load drop regime, and the strain burst regime. The three regimes are distinct as shown in Figures 2 and 3. In the quasi-elastic regime, the load raises linearly with indentation depth. Meanwhile, the residual stress on the load vs. indentation depth curve can be definitely neglected because of the exact overlapping for all curves, indicating that the residual stress has little effect on the elastic property of single-crystal copper thin film. Contrarily, in the sudden load drop regime, a sharp decrease of load occurs with increasing indentation depth, indicating the production of plastic yielding. In the strain burst regime, stress is continuously relaxed to a great extent, because of dislocation persistently gliding and propagating.

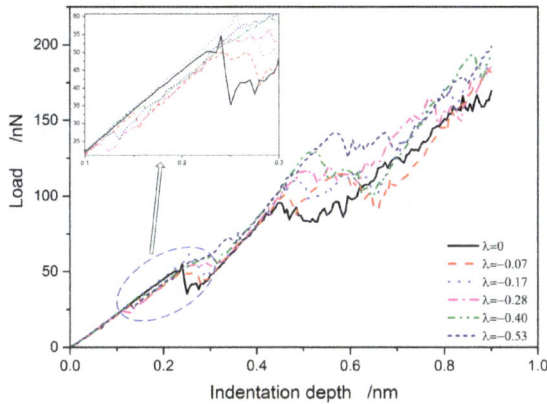

Figure 2. Load vs. indentation depth curves with different compressive residual stress.

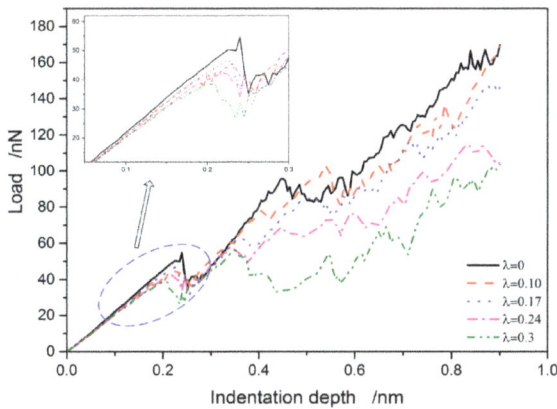

Figure 3. Load vs. indentation depth curves with different tensile residual stress.

It is observed from Figures 2 and 3 that an inflection point (a local maximum) corresponding to the elastic–plastic transition point [30] is observed before a large load drop occurs. By carefully examining the curves, the elastic–plastic transition point is increased with compressive residual stress, which moves towards the upper right relative to the case for no residual stress, as shown in Figure 2. On the contrary, the transition point is decreased with increasing tensile residual stress, which moves towards the bottom left as shown in Figure 3. Compared with the previous work [24], in which a virtual sphere indenter was used, the simulation also shows a similar trend. This suggests that a larger residual stress causes an increasing yield strength, and consequently postpones the incipient plasticity at the compressive state. For the tensile state the case is exactly to the contrary. Furthermore, in the strain burst regime, the load vs. indentation depth curve displays a less dispersive distribution in the compressive residual stress than that in the tensile residual stress. That difference between the compressive and tensile residual stresses is ascribed to a liable dislocation nucleation and propagation for tensile residual stress during indentation.

In Hertzian contact with residual stress, the load P by a spherical indenter is related to the indentation depth h, and can be defined as follows [2]:

$$P = \frac{4}{3}E^*R^{1/2}h^{3/2}, \tag{2}$$

where the indentation modulus E^* is the material's elastic response, and can be obtained from curve fitting in an MD simulation as shown in Figure 4. The indentation modulus E^* has, however, 142.6 GPa as calculated by the load vs. indentation depth curve, which is a bit larger than the 136.7 GPa reported in the literature [31]. The modulus E^*_{fit} shows a linear decrease with the reduction of compressive stress (and the increase of tensile stress). This indicates that the residual stresses have an important effect on the elastic property of single copper thin film, which is consistent with empirical estimates by experiments and finite element simulations [32].

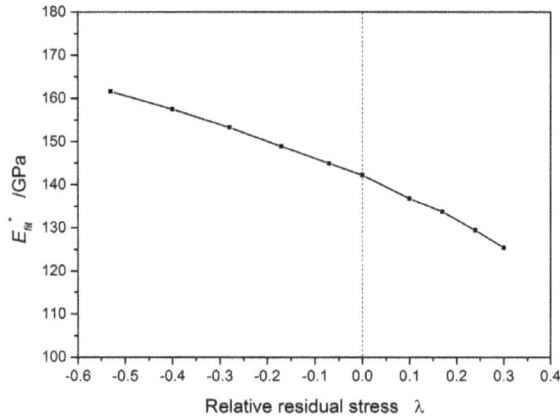

Figure 4. Indentation modulus E^*_{fit} as a function of relative residual stress.

The variation of the mean contact pressure versus the indentation depth under different relative residual stresses may reflect the contact surface properties. The mean contact pressure is defined as

$$p_m = \frac{P}{A},\tag{3}$$

where P is real contact load obtained from an MD simulation, and the corresponding contact area is A evaluated from the projected polygon.

The contact pressure against indentation depth is presented in Figure 5. Apparently, it can be observed that, under the same indentation depth, compressive residual stress can increase the mean contact pressure (Figure 5a), while tensile residual stress decreases the mean contact pressure (Figure 5b). With this result, we can define the maximum mean contact pressure as the indentation hardness corresponding to a given residual stress value in Figure 5, as

$$H = (p_m)_{max},\tag{4}$$

which is shown in Figure 6. It can be seen that the indentation hardness H is 14.32 GPa ($\lambda = 0$). Thus, from the well-known relationship between yield stress and hardness:

$$H = 3.23\sigma_{yield},\tag{5}$$

the yield stress can be derived as 4.43 GPa. This value agrees with the ideal shear strength of Cu [33]. When increasing tensile stress, the hardness rapidly decreases, while the indentation hardness increases with compressive residual stress in a moderate range. That is consistent with the experiments, finite element simulations, and MD simulations [34,35]. However, when the relative compressive residual stress λ is larger than 0.40, the hardness decreases rather than increases as shown in Figure 6, which will be discussed with dislocation evolution in the following subsection.

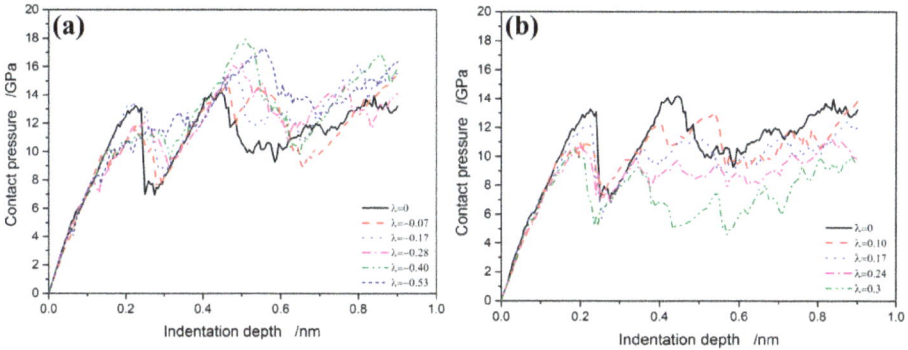

Figure 5. Mean contact pressure vs. indentation depth with different relative residual stress; (a) compressive residual stress $\lambda \leq 0$; (b) tensile residual stress $\lambda \geq 0$.

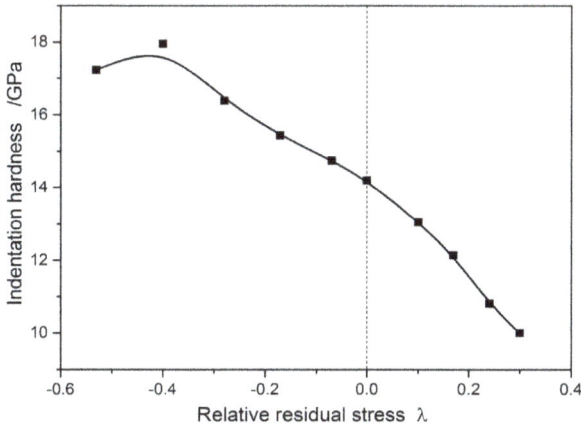

Figure 6. Hardness as a function of relative residual stress. (The curve is fitted by B-spline with the MD numerical result points).

As the indenter is pressed in the surface, as illustrated in Figure 7a, the material either is in a plastic pileup or in an elastic sink-in at the crater rim. Here, a relative contact area C^2 can be defined as

$$C^2 = A / A_{nom}, \tag{6}$$

where A is the real contact area from the MD simulation, and $A_{nom} = \pi(2Rh - h^2)$ is the nominal contact area. Substantial error can occur when measuring the contact radius after unloading due to the large elastic recovery [28]. The relative contact area C^2, as a function of relative residual stress λ, is plotted in Figure 7b. It can be seen that the relative contact area keeps a constant value greater than 1.0 in compressive stress ($\lambda < 0$), while the relative contact area also keeps unchanged value at 1.025 with increasing tensile residual stress at first, and then decreases continuously ($\lambda > 0.17$). This implies that the film around the indenter is in plastic pileup for compressive residual stress, while the sink-in effect appears when the residual stress exceeds a critical value. This is in good agreement with the MD simulations and the experimental results [28,36]. The in-plane residual stress has, therefore, a significant influence on the elastic-plastic property in single-crystal copper thin film.

(a)

(b)

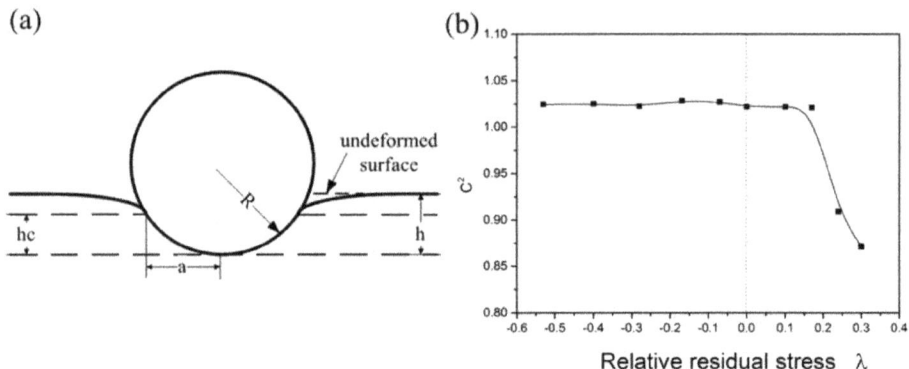

Figure 7. (**a**) A spherical indentation geometry, where a is contact radius, hc is contact depth, h is indentation depth, and (**b**) relative contact area curve versus residual stress.

3.2. Microstructure Evolution at Different Residual Stress

The atomistic visualization of defects is incorporated here in order to study the microstructure evolution of copper thin film with different residual stresses. The dislocation snapshots at relative residual stress λ = 0, 0.17, −0.17, 0.30, and −0.53 are presented in Figures 8–12, respectively. In those images, normal FCC atoms in the substrate are removed off.

It can be observed from Figures 8–12 that small atom clusters appear in the indented film with penetrating initiating, which can be regarded as precursors to dislocations. As the indentation depth moves beyond a critical indentation depth, dislocation embryos occur as shown in Figures 8–12, then, they evolve and transform to a tetrahedral structure, which corresponds to a sharp load drop as shown in Figures 2 and 3. In all cases, the Shockley partial dislocations bounding the stacking fault domains are nucleated and evolved in {111} planes (as shown in Figure 8d), which commonly occurs for FCC (111) nanoindentation. Similar results are also observed in the MD simulation of Cu in the (111) surface [24], although there exist three atom clusters in indented film. From b to c in Figures 8–12, a great quantity of dislocations initiate and propagate, and dislocation loops also appear beneath the indented film with the increasing indentation depth in the strain burst regime (Figures 2 and 3). However, compared with the dislocation distribution in compressive residual stress (in Figures 10d and 12d), the dislocations are extended to a deeper position beneath the indented surface in tensile residual stress, as shown in Figures 9c and 11d. In addition, it can be easily understood that a larger tensile residual stress accompanies a much smaller indentation depth for the incipient plasticity, which is contrary to the case of compressive residual stress, verifying the variation in the load drop regime in the load vs. indentation depth curves (Figures 2 and 3).

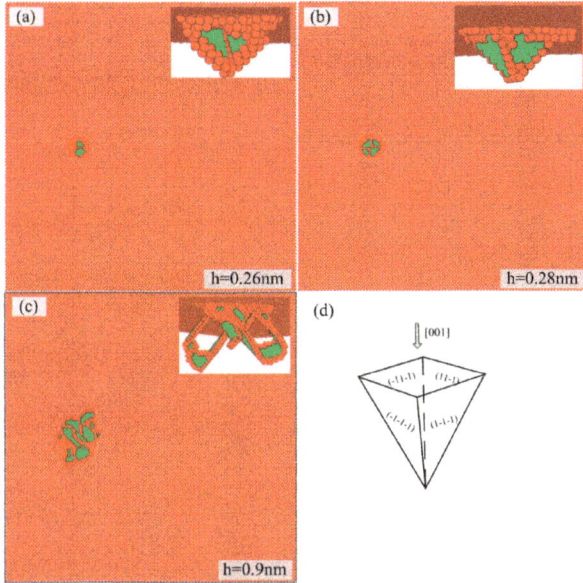

Figure 8. Atom configuration during indentation process without residual stress: (**a–d**) schematic of four inclined {111}-type slip planes for (100) indentation. The green color shows stacking faults.

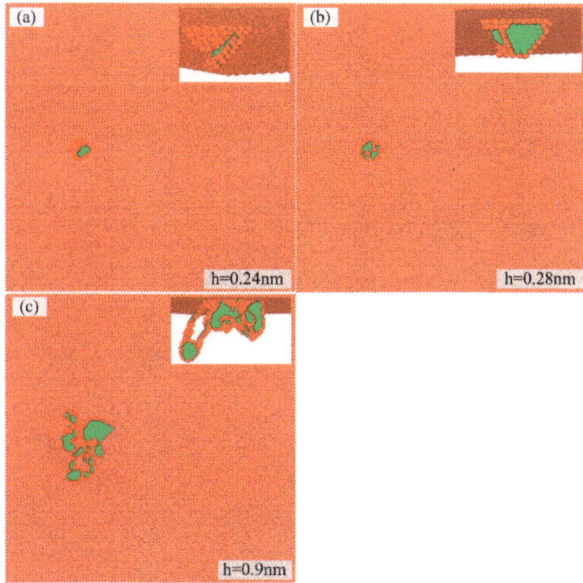

Figure 9. Atom configuration during indentation process with residual tensile stress $\lambda = 0.17$ at various indentation depths: (**a**) 0.24 nm; (**b**) 0.28 nm; (**c**) 0.9 nm. The green color shows stacking faults.

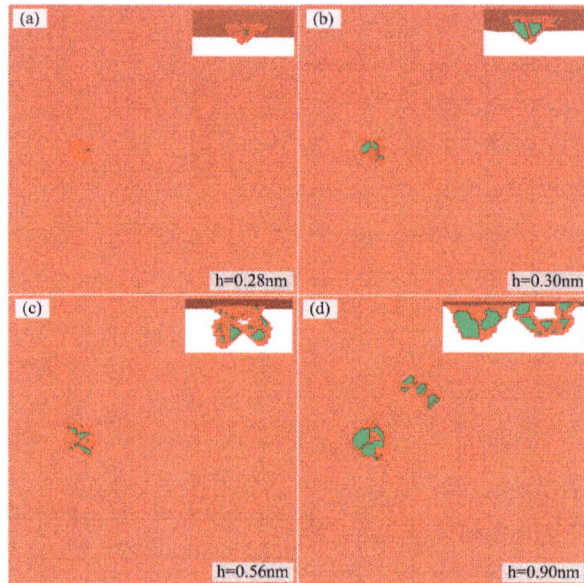

Figure 10. Atom configuration during indentation process with residual compress stress λ = −0.17 at various indentation depths: (**a**) 0.28 nm; (**b**) 0.30 nm; (**c**) 0.56 nm; (**d**) 0.90 nm.. The green color shows stacking faults.

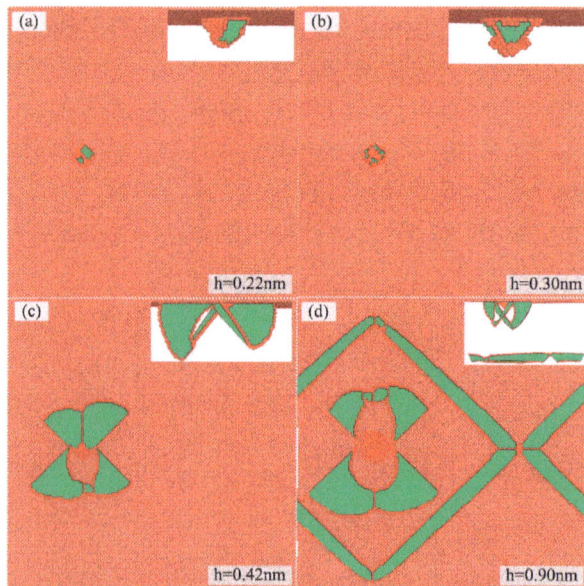

Figure 11. Atom configuration during indentation process with residual tensile stress λ = 0.30 at various indentation depths: (**a**) 0.22 nm; (**b**) 0.30 nm; (**c**) 0.42 nm; (**d**) 0.90 nm. The green color shows stacking faults.

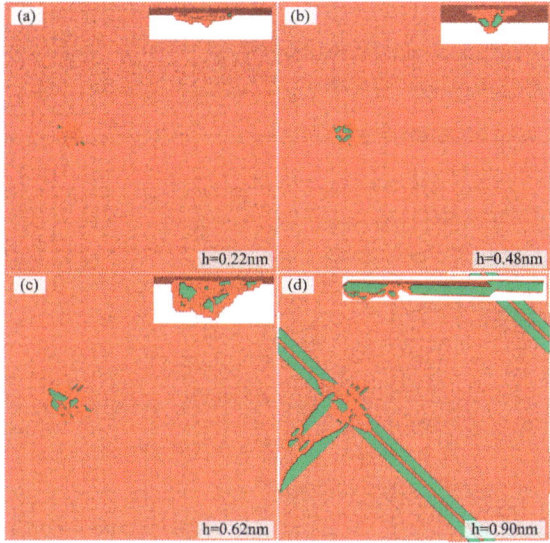

Figure 12. Atom configuration during indentation process with residual compress stress λ = −0.53 at various indentation depths: (**a**) 0.22 nm; (**b**) 0.48 nm; (**c**) 0.62 nm; (**d**) 0.90 nm. The green color shows stacking faults.

To further reveal the effect of residual stress on the dislocation distribution, a dislocation line length is introduced to describe the dislocation density in a certain volume. Dislocation lines are detected using the dislocation extraction algorithm (DXA) algorithm [37], and plotted as a function of relative residual stress in Figure 13. It can be seen that the dislocation line length increases linearly from λ = −0.28 to λ = 0.17, while it increases rapidly with the increasing tensile residual stress when the relative residual stress exceeds 0.17. This indicates that the larger tensile residual stress leads to a dramatic increase in dislocation density, and consequently deteriorates the indented surface film. When the compressive residual stress is smaller than −0.28, the dislocation density increases again as shown in Figure 13. By carefully examining the atomic configuration images, the increasing zone of dislocation density is mainly located at the subsurface, compared with Figure 12d, which definitely results in the dislocation vanishing in the surface during indenting. It also becomes a reason why a decrease of hardness occurs for a much larger compressive stress as shown in Figure 6.

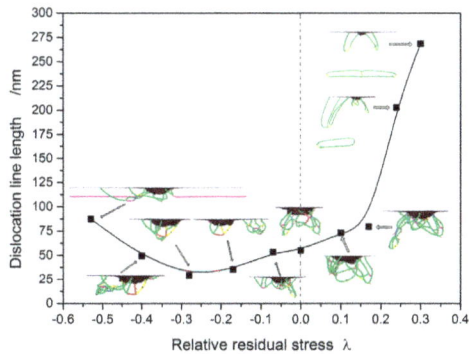

Figure 13. Dislocation line length in Cu samples with different residual stresses.

3.3. Indentation Stress Field with Different Residual Stresses

During indenting, dislocations are usually limited to a volume just beneath the indenter, which is here called the plastic zone. When dislocation prismatic loops glide away the indented film surface, they are not considered a part of the plastic zone in this paper. With this convention in place, the plastic zone is, therefore, calculated and plotted in Figure 14. In the literature, Samuels and Mulhearn [38,39] claimed that the plastic zone at macro-scale shows a spherical symmetry (usually hemispherical) shape if a conical indenter is used. Similarly, Chiang et al. [40] and Fischer-Cripps [41] developed a cavity model and demonstrated that the contacting surface of the indenter is encased by a hydrostatic "core", which is in turn surrounded by a hemispherical plastic zone. In contrast, the plastic zone at micro-scale in our simulations (Figure 14) exhibits apparent material anisotropy and a far different deformation shape from spherical symmetry beneath the indented surface. In addition, the plastic zone is obviously in a different shape with different residual stress levels. With the increase of tensile stress, the plastic zone is dispersed and extended from the subsurface to a deeper position in the film (Figure 14c), which is in good agreement with the dislocation distribution as shown in Figure 11. However, it can be found in Figure 14d,e that the indentation plastic zone shrinks under compressive stress.

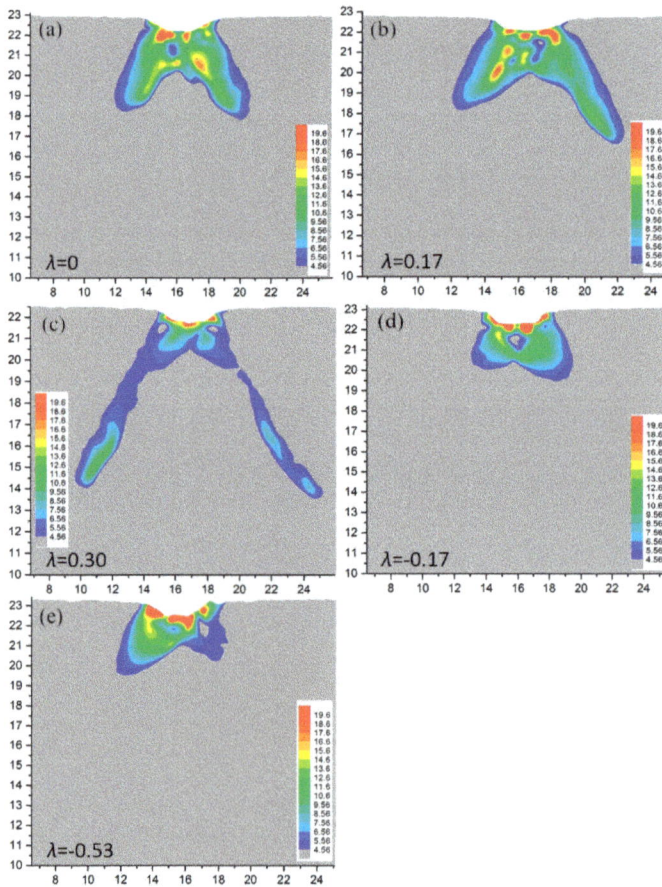

Figure 14. Indentation plastic zone with different residual stresses: (**a**) $\lambda = 0$; (**b**) $\lambda = 0.17$; (**c**) $\lambda = 0.30$; (**d**) $\lambda = -0.17$; (**e**) $\lambda = -0.53$.

3.4. Comparison of Simulation Results of Realistic and Virtual Indenters

In theory, it is believed that the simulation results obtained by a realistic indenter should be more practical than the ones by a virtual indenter. Here, it is helpful to conduct a comparison of the simulation results obtained by a realistic indenter and a virtual indenter, under the same indentation condition. The load versus indentation depth curves are presented in Figure 15. It can be seen from Figure 15a–c that the two load-depth curves are quite similar in general, but for the realistic indenter the incipient plasticity loads are higher than the ones for the virtual indenter, and a similar result can be found in the regime after plastic yielding. Similarly, it can be observed in Figure 15d that the hardness by the realistic indenter is higher than the one by the virtual indenter. These results are ascribed to the smoothness degree of the indenters. At the same time, the dislocation length vs. the relative residual stress under the same indentation depth (0.9 nm) is plotted in Figure 16. It can be found that the dislocation length by the realistic indenter is longer than the one by the virtual indenter for any residual stress level. Both Figures 15 and 16 provide a detailed description of the difference between the two indenter models.

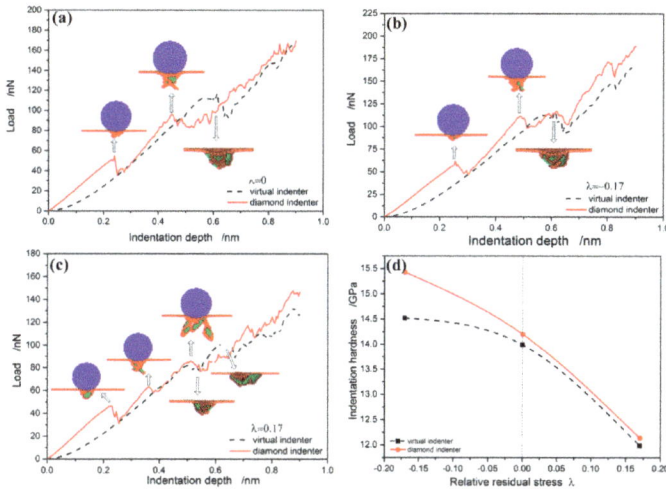

Figure 15. Comparison of indentation results of realistic and virtual sphere indenters. (**a**) without residual stress, (**b**) under compressive residual stress, (**c**) under tensile residual stress, (**d**) hardness vs. residual stress.

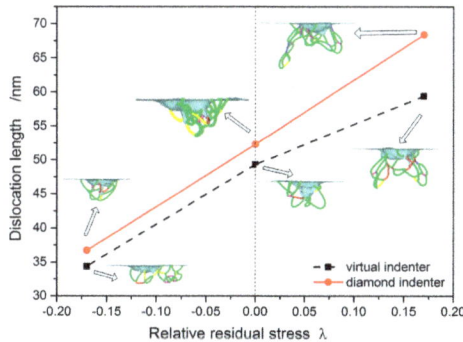

Figure 16. Dislocation length vs. relative residual stress under the same indentation depth ($h = 0.9$ nm).

4. Conclusions

The effects of in-plane residual stress on Hertzian nanoindentation behaviors for single-crystal copper thin film have been investigated through MD simulations. It has been found that the residual stresses give a significant influence on the plastic deformation, nominal indentation hardness, and indentation modulus of copper specimens:

(1) The threshold of incipient plasticity increases with compressive stress but decreases with tensile stress.
(2) The hardness decreases with tensile residual stress, while it increases with moderate compressive residual stress, but drops down with a higher compressive residual stress.
(3) The indentation modulus reduces linearly with decreasing compressive residual stress (and increasing tensile residual stress).
(4) The indentation plastic zone is extended from a concentrated shape into a deeper position beneath the indentation surface with increasing tensile stress, while the compressive residual stress is able to shrink the plastic zone.

Acknowledgments: The present authors appreciate the financial support from the National Natural Science Foundations of China (Grant No. 51475359 and 51375364) and the Natural Science Foundation of Shannxi Province of China (Grant No. 2014JM6219).

Author Contributions: Sun Kun conducted the MD simulation and wrote the paper. Junqin Shi calculated the data. Lifeng Ma conceived and supervised the whole work, and also revised the paper.

Conflicts of Interest: The authors declare no conflict of interest.

References

1. Zhang, L.; Zhao, H.; Dai, L.; Yang, Y.; Du, X.; Tang, P.; Zhang, L. Molecular dynamics simulation of deformation accumulation in repeated nanometric cutting on single-crystal copper. *RSC Adv.* **2015**, *17*, 12678–12685. [CrossRef]
2. Hansson, P. Influence of the crystallographic orientation of thin copper coatings during nano indentation. *Procedia Mater. Sci.* **2014**, *3*, 1093–1098. [CrossRef]
3. Voyiadjis, G.Z.; Yaghoobi, M. Large scale atomistic simulation of size effects during nanoindentation: Dislocation length and hardness. *Mater. Sci. Eng. A* **2015**, *634*, 20–31. [CrossRef]
4. Li, L.; Song, W.; Xu, M.; Ovcharenko, A.; Zhang, G. Atomistic insights into the loading—Unloading of an adhesive contact: A rigid sphere indenting a copper substrate. *Comp. Mater. Sci.* **2015**, *98*, 105–111. [CrossRef]
5. Lahr, G.; Shao, S.; Medyanik, S.N. Dynamic Effects on Nanoscale Indentation and Dislocation Ropagation. Available online: http://lfp.mme.wsu.edu/REU2009/files/07.pdf (accessed on 25 July 2009).
6. Kokubo, S. *On the Change in Hardness of a Plate Caused by Bending*; Science Report of Tohoku Imperial University; Tohoku Imperial University: Sendai, Japan, 1932.
7. Tsui, T.Y.; Oliver, W.C.; Pharr, G.M. Influences of stress on the measurement of mechanical properties using nanoindentation: Part I. Experimental studies in an aluminum alloy. *J. Mater. Res.* **1996**, *11*, 752–759. [CrossRef]
8. Yang, F.; Peng, L.; Okazaki, K. Microindentation of aluminum. *Metall. Mater Trans. A* **2004**, *35*, 3323–3328. [CrossRef]
9. Bolshakov, A.; Oliver, W.C.; Pharr, G.M. Influences of stress on the measurement of mechanical properties using nanoindentation: Part II. Finite element simulation. *J. Mater. Res.* **1996**, *11*, 760–768. [CrossRef]
10. Xu, Z.H.; Li, X. Influence of equi-biaxial residual stress on unloading behaviour of nanoindentation. *Acta Mater.* **2005**, *53*, 1913–1919. [CrossRef]
11. Taljat, B.; Pharr, G.M. Measurement of Residual Stresses by Load and Depth Sensing Spherical Indentation. *Mater. Res. Soc. Symp. Proc.* **2000**, *594*, 519–524. [CrossRef]
12. Peng, P.; Liao, G.L.; Shi, T.L.; Tang, Z.R.; Gao, Y. Molecular dynamic simulations of nanoindentation in aluminum thin film on silicon substrate. *Appl. Surf. Sci.* **2010**, *21*, 6284–6290. [CrossRef]
13. Weinberger, C.R.; Tucker, G.J. Atomistic simulations of dislocation pinning points in pure face- centered-cubic nanopillars. *Model. Simul. Mater. Sci. Eng.* **2012**, *20*, 075001. [CrossRef]

14. Tucker, G.J.; Aitken, Z.H.; Greer, J.R.; Weinberger, C.R. The mechanical behavior and deformation of bicrystalline nanowires. *Model. Simul. Mater. Sci. Eng.* **2013**, *53*, 015004. [CrossRef]

15. Gao, Y.; Ruestes, C.J.; Tramontina, D.R.; Urbassek, H.M. Comparative simulation study of the structure of the plastic zone produced by nanoindentation. *J. Mech. Phys. Solids* **2015**, *75*, 58–75. [CrossRef]

16. Voyiadjis, G.Z.; Yaghoobi, M. Role of grain boundary on the sources of size effects. *Comput. Mater. Sci.* **2016**, *117*, 315–329. [CrossRef]

17. Voyiadjis, G.Z.; Yaghoobi, M. Size and strain rate effects in metallic samples of confined volumes: Dislocation length distribution. *SCR Mater.* **2017**, *130*, 182–186. [CrossRef]

18. Yaghoobi, M.; Voyiadjis, G.Z. Size effects in FCC crystals during the high rate compression test. *Acta Mater.* **2016**, *121*, 190–201. [CrossRef]

19. Yaghoobi, M.; Voyiadjis, G.Z. Atomistic simulation of size effects in single-crystalline metals of confined volumes during nanoindentation. *Comput. Mater. Sci.* **2016**, *111*, 64–73. [CrossRef]

20. Kelchner, C.L.; Plimpton, S.J.; Hamilton, J.C. Dislocation nucleation and defect structure during surface indentation. *Phys. Rev. B* **1998**, *58*, 11085–11088. [CrossRef]

21. Liang, H.Y.; Woo, C.H.; Huang, H.C.; Ngan, A.H.W.; Yu, T.X. Crystalline plasticity on copper (001), (110), and (111) surfaces during nanoindentation. *CMES Comp. Model. Eng.* **2004**, *6*, 105–114.

22. Yaghoobi, M.; Voyiadjis, G.Z. Effect of boundary conditions on the MD simulation of nanoindentation. *Comp. Mater. Sci.* **2014**, *95*, 626–636. [CrossRef]

23. Feng, C.; Peng, X.; Fu, T.; Zhao, Y.; Huang, C.; Wang, Z. Molecular dynamics simulation of nano-indentation on Ti-V multilayered thin films. *Physica E* **2017**, *87*, 213–219. [CrossRef]

24. Sun, K.; Shen, W.; Ma, L. The influence of residual stress on incipient plasticity in single-crystal copper thin film under nanoindentation. *Comput. Mater. Sci.* **2014**, *81*, 226–232. [CrossRef]

25. Plimpton, S. Fast Parallel Algorithms for Short-Range Molecular Dynamics. *J. Comput. Phys.* **1995**, *117*, 1–9. [CrossRef]

26. Mishin, Y.; Mehl, M.J.; Papaconstantopoulos, D.A.; Voter, A.F.; Kress, J.D. Structural stability and lattice defects in copper: Ab initio, tight-binding, and embedded-atom calculations. *Phys. Rev. B* **2001**, *22*, 4106.

27. Morse, P. Diatomic molecules according to the wave mechanics. II. Vibrational levels. *Phys. Rev.* **1929**, *34*, 57–64. [CrossRef]

28. Dub, S.N.; Lim, Y.Y.; Chaudhri, M.M. Nanohardness of high purity Cu (111) single crystals: The effect of indenter load and prior plastic sample strain. *J. Appl. Phys.* **2010**, *107*, 043510. [CrossRef]

29. Simons, G.; Wang, H. *Single Crystal Elastic Constants and Calculated Aggregate Properties*; MIT Press: Cambridge, MA, USA, 1977.

30. Liang, H.Y.; Woo, C.H.; Huang, H.; Ngan, A.H.W.; Yu, T.X. Dislocation nucleation in the initial stage during nanoindentation. *Philos. Mag.* **2003**, *83*, 3609–3622. [CrossRef]

31. Zhang, L.; Huang, H.; Zhao, H.; Ma, Z.; Yang, Y.; Hu, X. The evolution of machining-induced surface of single-crystal FCC copper via nanoindentation. *Nanoscale Res. Lett.* **2013**, *8*, 21. [CrossRef] [PubMed]

32. Schall, J.D.; Brenner, D.W. Atomistic simulation of the influence of pre-existing stress on the interpretation of nanoindentation data. *J. Mater. Res.* **2004**, *19*, 3172–3180. [CrossRef]

33. Krenn, C.R.; Roundy, D.; Cohen, M.L.; Chrzan, D.C.; Morris, J.W. Connecting atomistic and experimental estimates of ideal strength. *Phys. Rev. B* **2002**, *65*, 134111. [CrossRef]

34. Huber, N.; Heerens, J. On the effect of a general residual stress state on indentation and hardness testing. *Acta Mater.* **2008**, *56*, 6205–6213. [CrossRef]

35. Suresh, S.; Giannakopoulos, A.E. A new method for estimating residual stresses by instrumented sharp indentation. *Acta Mater.* **1998**, *46*, 5755–5767. [CrossRef]

36. Timoshenko, S.P.; Goodier, J.N. *Theory of Elasticity*, 3rd ed.; McGraw-Hill: London, UK, 1970.

37. Stukowski, A.; Albe, K. Extracting dislocation and non-dislocation crystal defects from atomic simulation data. *Model. Simul. Mater. Sci. Eng.* **2010**, *18*, 085001. [CrossRef]

38. Samuels, L.E.; Mulhearn, T.O. An experimental investigation of the deformed zone associated with indentation hardness impressions. *J. Mech. Phys. Solids* **1957**, *5*, 125–134. [CrossRef]

39. Mulhearn, T.O. The deformation of metals by vickers-type pyramidal indenters. *J. Mech. Phys. Solids* **1959**, *7*, 85–88. [CrossRef]

40. Chiang, S.S.; Marshall, D.B.; Evans, A.G. The response of solids to elastic/plastic indentation. I. Stresses and residual stresses. *J. Appl. Phys.* **1982**, *53*, 298. [CrossRef]
41. Fischer-Cripp, A.C. *Introduction to Contact Mechanics*, 2nd ed.; Springer: New York, NY, USA, 2007.

crystals

MDPI

Review

Thickness-Dependent Strain Rate Sensitivity of Nanolayers via the Nanoindentation Technique

Jian Song [1], Yue Liu [1,*], Zhe Fan [2,*] and Xinghang Zhang [2]

[1] State Key Laboratory of Metal Matrix Composites, School of Materials Science and Engineering, Shanghai Jiao Tong University, Shanghai 200240, China; songjian2015@126.com
[2] School of Materials Engineering, Purdue University, West Lafayette, IN 47907, USA; xzhang98@purdue.edu
*   Correspondence: yliu23@sjtu.edu.cn (Y.L.); zfan.phd@gmail.com (Z.F.)

Received: 14 January 2018; Accepted: 21 February 2018; Published: 9 March 2018

Abstract: The strain rate sensitivity (SRS) and dislocation activation volume are two inter-related material properties for understanding thermally-activated plastic deformation, such as creep. For face-centered-cubic metals, SRS normally increases with decreasing grain size, whereas the opposite holds for body-center-cubic metals. However, these findings are applicable to metals with average grain sizes greater than tens of nanometers. Recent studies on mechanical behaviors presented distinct deformation mechanisms in multilayers with individual layer thickness of 20 nanometers or less. It is necessary to estimate the SRS and plastic deformation mechanisms in this regime. Here, we review a new nanoindentation test method that renders reliable hardness measurement insensitive to thermal drift, and its application on SRS of Cu/amorphous-CuNb nanolayers. The new technique is applied to Cu films and returns expected SRS values when compared to conventional tensile test results. The SRS of Cu/amorphous-CuNb nanolayers demonstrates two distinct deformation mechanisms depending on layer thickness: dislocation pileup-dominated and interface-mediated deformation mechanisms.

Keywords: thin film; nanoindentation; strain rate sensitivity; deformation mechanisms

1. Introduction

For most metallic materials, the plastic deformation can be regarded as a thermally-activated process which can be quantitatively characterized by the values of the strain rate sensitivity (SRS), m, and the dislocation activation volume, V^* [1,2]. SRS is an important indicator of the plasticity of metallic materials, which can be estimated by the amount of activation volume required for dislocation motions. These two inter-related material properties are important for understanding thermally-activated plastic deformation, such as creep.

In general, for crystalline metals, higher values of SRS are often accompanied by excellent ductility and deformability. The values of SRS are strongly size-dependent: for face-centered-cubic (fcc) metals, SRS normally increases with decreasing grain sizes, whereas the opposite holds for body-center-cubic (bcc) structures [3]. Metallic glasses (MGs) exhibit high strength, excellent abrasion and corrosion resistance [4], but they are generally brittle due to the shear band (SB) controlled-deformation mechanisms therefore the SRS is close to zero [5]. Prior studies have shown that adding a crystalline phase to the amorphous matrix can increase the toughness/plasticity of ZrTi-based MGs [6] and crystalline/amorphous (C/A) multilayers [7–14]. For the C/A multilayer composites, their thermally-activated plastic deformation mechanisms can also be revealed by quantifying their SRS. Thus, various methods have been developed for determining SRS in the past few decades [15–22].

The method for determining SRS can be roughly classified as macroscopic testing and "localized" testing, usually nanoindentation. Macroscopically, the SRS is often measured by uniaxial tension [15–20] and compression tests [21,22] on bulk specimens, whereas nanoindentation is widely adopted to measure SRS of small specimens, such as thin films. Figure 1 illustrates the differences between SRS measurement under tension/compression and nanoindentation. In addition to the different scope of applications, the key difference between uniaxial tension/compression and nanoindentation measurements lies in the determination of strain rate ($\dot{\varepsilon}$). In the tensile/compression testing method, $\dot{\varepsilon}$ is calculated based on the original length of sample (l_o), and its increment (Δl) under displacement-control mode. In comparison, the nanoindentation SRS measurements achieve a constant indentation strain rate $\dot{h}/h$ through maintaining $\dot{P}/P$ based on their relationship (h is the indentation depth, $\dot{h}$ is the displacement rate, P is the loading force, and $\dot{P}$ is the real-time loading rate) [23]. Although the nanoindentation method determines SRS on a local scale, many nanoindentation experiments are typically performed to ensure the results are representative of their overall mechanical properties.

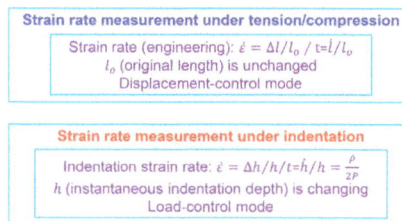

Figure 1. Differences between strain rate sensitivity measurement under tension/compression and nanoindentation. Different from measuring the dimension change of tension and compression specimens, nanoindentation quantifies the strain from a combination of load and indentation depth.

In 1999, the $\dot{P}/P$ technique for SRS measurement was first presented by Lucas and Oliver, and they applied this technique for measuring the hardness of high-purity indium under various strain rates [23]. This method shows good applicability during the high strain rate experiment, however, when it comes to low strain rates, the displacement of the indenter will be significantly affected by thermal drift. In 2010, the continuous stiffness measurement (CSM) that could decrease the influence of thermal drift was proposed by Hay et al. [24]. In 2011, an indentation strain-rate-jump test method for measuring SRS proposed by Maier et al. showed a significant mitigation of the thermal drift [25]. Unlike the previous methods, this special nanoindentation creep method alleviates thermal drift by first applying the highest strain rates and then the lowest strain rates. However, during the strain-rate-jump test, the indentation depths must be greater than 500 nm for reliable determination of SRS.

Recent studies on mechanical behaviors of multilayers focused on deformation mechanisms when the individual layer thickness is 20 nm or less where the majority of multilayer systems reach high strength [26–28]. However, there are limited studies on the size-dependent SRS in multilayers using the nanoindentation test, except some atomistic simulations on the deformation mechanisms of nano-sized thin films [29–32]. Here, we review a new nanoindentation test method that renders hardness nearly insensitive to thermal drift. Such a technique permits reliable determination of SRS for Cu films. The SRS measured using this technique reveals size-dependent variation of deformation mechanisms in Cu/a-CuNb (crystalline Cu/amorphous CuNb) multilayers.

2. Experimental Methods

The 1.5 μm Cu film, 1 μm thick a-CuNb single layer and the Cu/a-CuNb nanolayers with different individual layer thickness (h, ranges from 5 to 150 nm; and total thickness ranges from 1 to 2.4 μm) were deposited on HF-etched Si wafers by DC (direct current) magnetron sputtering. Pure Cu (99.995%) and CuNb alloy (Cu 50 at%-Nb 50 at%) targets were used to prepare these nanostructured materials. The microstructure was characterized by FEI Tecnai G2 F20 TEM (Thermo Fisher Scientific,

Hillsboro, OR, USA). All the nanoindentation tests were conducted on Agilent G200 (Agilent Technologies, Santa Clara, CA, USA) and Hysitron TI950 TriboIndenter (Bruker, Billerica, MA, USA) in the continuous stiffness measurement (CSM) mode. For a-CuNb and Cu/a-CuNb multilayers, at least 10 tests were conducted for each sample at each specific strain rates of 0.2, 0.05, and 0.01 s^{-1}, respectively. For the Cu single-layer, the specific testing strain rates are 0.05, 0.01, and 0.002 s^{-1}.

This new modified testing method developed by Liu et al. [33] involves directly measuring the contact stiffness and calculating the contact area from the measured stiffness and modulus. Figure 2 depicts the differences between the conventional and new modified method. It is easy to find that both the conventional and modified method are based on the relationship performed by Sneddon's stiffness equation [34]:

$$E_r = \frac{\sqrt{\pi}}{2} \frac{S}{\sqrt{A}}$$ (1)

where E_r is the reduced elastic modulus, A and S represent the contact area and contact stiffness, respectively [35,36]. The primary difference between these two methods is reflected in the different calculation process of A. As for the conventional method, h (indentation displacement), P (load), and S at different $\dot{\varepsilon}$ can be directly measured, the h_c (contact depth) can be calculated by using the equation:

$$h_c = h - \frac{0.75P}{S}$$ (2)

where $0.75P/S$ is a calculated value which represents h_s (the vertical distance from the contact point of the sample and the indenter to the sample surface) [35], then A could be derived by:

$$A = m_0 h_c^2 + m_1 h_c$$ (3)

where m_0 and m_1 are determined by the standard calibration process [35]. At last, E_r and H (hardness) could be derived by those parameters. At high strain rates, the measurement error of h is negligible. However, at lower strain rate, the measurement of h will be significantly affected by thermal drift which makes the later calculations become unreliable. Unlike the conventional method, the modified method calculates the contact area from the measured stiffness and modulus, and the process can be summarized as follows: Firstly, measured E_r at high $\dot{\varepsilon}$, then E_r as a known parameter can be used at lower $\dot{\varepsilon}$ later. Secondly, at low $\dot{\varepsilon}$ condition, A can be derived by Equation (1) (thus, H is calculated), then h_c is calculated from Equation (3), and at last, h can be calculated by using the Equation (2). The calculated h instead of directly measured can reduce the thermal-drift effects significantly and control the real-time indentation displacement [33].

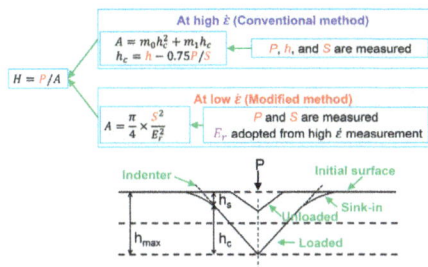

Figure 2. Flowcharts illustrating differences between conventional and modified methods. The conventional method measures h, P, and S at different strain rates ($\dot{\varepsilon}$). Values of h_c, A, E_r, H, and E can be derived. The modified method first determines E_r at high $\dot{\varepsilon}$, then applies E_r as a known parameter at lower $\dot{\varepsilon}$. The h and A value at low $\dot{\varepsilon}$ can be calculated instead of directly measured to reduce the thermal-drift effects.

3. Results and Discussion

The hardnesses of single-layer Cu and a-CuNb films were measured by using this modified method. Figure 3 shows the comparison of nanoindentation results obtained from conventional (blue) and modified (red) methods using the same sets of nanoindentation data at various strain rates. Both the methods show good stability during the high strain rate experiment, and lead to similar results. However, at the lower strain rate, for a-CuNb multilayers (0.01 s^{-1}), the scattering data presented by the conventional method is not acceptable, while the convergent and consistent hardness values were obtained from the modified method. For Cu films (0.002 s^{-1}), both methods returned similar results, except the hardness obtained by the conventional method is slightly larger than the modified method, and the convergence of the conventional method is slightly better than the modified method. This is because under the condition of a small load or small indentation depth, the determination of the contact area from only contact stiffness and load is less accurate than from contact stiffness, load, and indentation depth, despite the thermal drift.

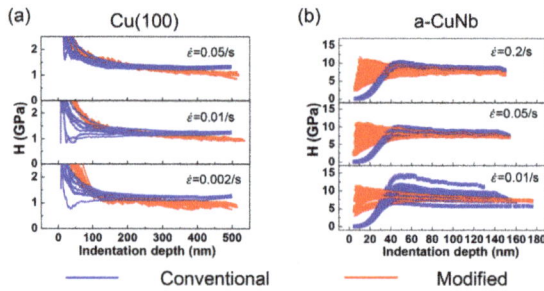

Figure 3. Comparison of nanoindentation results obtained from conventional (blue) and modified (red) methods using the same sets of indentation data at various strain rates for (**a**) Cu and (**b**) amorphous CuNb. At a high strain rate (0.05 s^{-1}), the indentation hardnesses calculated from both techniques are similar. However at low strain rate, the conventional analysis leads to more scattered results.

The SRS values of Cu film were calculated by using the hardness obtained at a depth of 300 nm from the two methods under different strain rates. As shown in Figure 4a, the SRS value ($m = 0.048 \pm 0.006$) produced from the modified method (half-filled red circles) are less scattered than he conventional method. In contrast, the conventional method (half-filled black squares) leads to relatively erroneous and unacceptable SRS values ($m = 0.112 \pm 0.020$), which are more than twice that of the modified method.

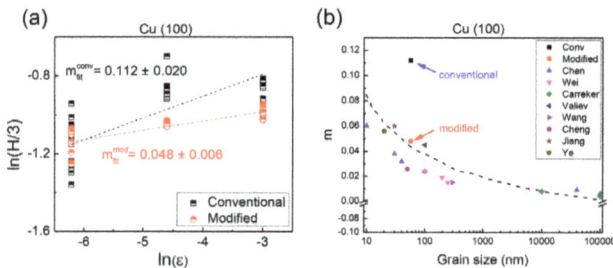

Figure 4. (**a**) The conventional method (half-filled black squares) yields erroneous SRS values for Cu. In contrast, the modified method (half-filled red circles) produces reliable SRS values. (**b**) Compiled plots of the SRS values (m) vs. grain size obtained from various techniques, including indentation jump, tensile and compression tests for Cu. The result obtained from the modified technique is consistent with the general trend reported in the literature, in contrast to the radically different result from the conventional method.

In order to further verify the reliability of the new method, we compare the measured SRS value of Cu film with the results that are obtained from bulk Cu samples or Cu films recorded in the previous literature, such as uniaxial tension, compression, and indentation jump tests [17–20,37–41], and plot the results in Figure 4b. For fcc metals, SRS normally increases with decreasing grain sizes. Here, the SRS value of Cu film with average grain size of ~70 nm (confirmed by TEM shown in Figure 5a) obtained from the modified technique is consistent with the general trend reported before, shown in Figure 4b. However, the conventional method shows very different results due to the influence of thermal drift at lower strain rates.

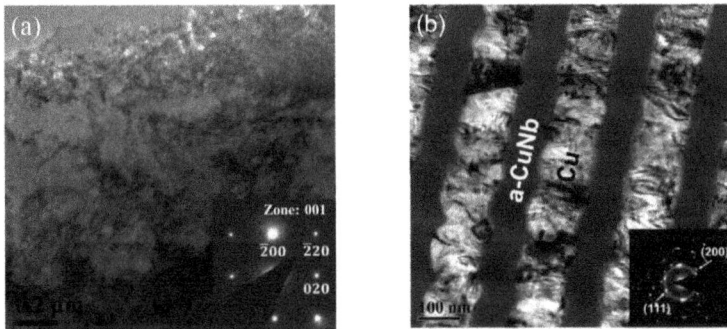

Figure 5. (a) Cross-section TEM micrograph of sputter-deposited nanocrystalline Cu film. (b) Cross-sectional TEM micrograph of sputter-deposited Cu 100 nm/a-CuNb 100 nm multilayer film.

After the new technique is validated through Cu films and returns expected SRS values, the SRS of C/A multilayers can be easily determined. Thus, in this section, the relationship between individual thickness (*h*), the plastic deformation mechanism, and the SRS of Cu/a-CuNb multilayers with different individual layer thickness (*h*, ranging from 5 to 150 nm) are systematically discussed. The image in Figure 5b is the cross-sectional microstructure of Cu 100 nm/a-CuNb 100 nm multilayers (referred as Cu/a-CuNb 100 nm) characterized by TEM. The layered structures can be clearly distinguished, and the selected area diffraction (SAD) pattern in the lower right corner shows diffuse diffraction halo and diffraction spots which represent the featureless amorphous layer and polycrystalline Cu layers, respectively.

Before conducting the nanoindentation experiments, we can briefly discuss the relationship between the SRS and individual layer thickness in crystalline (fcc metals)/amorphous multilayers. First, we assume that the C/A multilayers are under an isobaric stress condition during the nanoindentation experiment (i.e., the stress applied to crystalline layers and amorphous layers are equal). In this case, the total plastic deformation can be divided into two parts: the plastic deformation (strain) from the amorphous phase and from the crystalline phase. Moreover, the plastic deformation of these two phases is different partly due to the hardness of the amorphous phase being much higher than the crystalline phase. Thus, when the individual thickness of the amorphous phase and the crystalline phase are designed to the same value, the strain of C/A multilayers can be expressed as:

$$\varepsilon = \frac{1}{2}(\varepsilon_a + \varepsilon_c) \tag{4}$$

where ε is the total strain (indenter displacement divided by the film thickness: $\Delta L/L$), ε_a and ε_c are the strain of the amorphous layers and crystalline layers. Then, the strain rate can be shown as:

$$\dot{\varepsilon} = \frac{1}{2}(\dot{\varepsilon}_a + \dot{\varepsilon}_c) \tag{5}$$

Finally, according to [42], the SRS of C/A multilayers can be calculated as:

$$\frac{1}{m} = \frac{\partial \ln \dot{\varepsilon}}{\partial \ln \sigma} = \frac{\Delta L_a}{\Delta L_a + \Delta L_c} \times \frac{1}{m_a} + \frac{\Delta L_c}{\Delta L_a + \Delta L_c} \times \frac{1}{m_c} \quad (6)$$

where ΔL_a, m_a, ΔL_c, and m_c represent the displacement and SRS of the amorphous phase and the crystalline phase, respectively. Here, it is important to note that we should figure out the proportion of ε_a and ε_c in the total plastic deformation. That means the key in this study is to determine how the displacement is distributed. In Equation (6), both the SRS and the percentage of plastic deformation accommodated by the crystalline ($\frac{\Delta L_c}{\Delta L_a + \Delta L_c}$) and amorphous ($\frac{\Delta L_a}{\Delta L_a + \Delta L_c}$) layers should change with the change of h.

For the SRS of Cu/a-CuNb multilayers, we take the previously measured SRS of the 1.5 µm Cu film ($m = 0.048$, assuming that m does not change in this regime) into Equation (6) and treat m_a as a value close to zero, then we can qualitatively discuss the relationship between m and h. As shown in Figure 6a, the modeled curve (dashed line) indicates the evolution of co-deformation of both Cu and a-CuNb. In addition, the SRS of a-CuNb, Cu, their average value (dotted lines) and Cu/a-CuNb are also plotted on the graph.

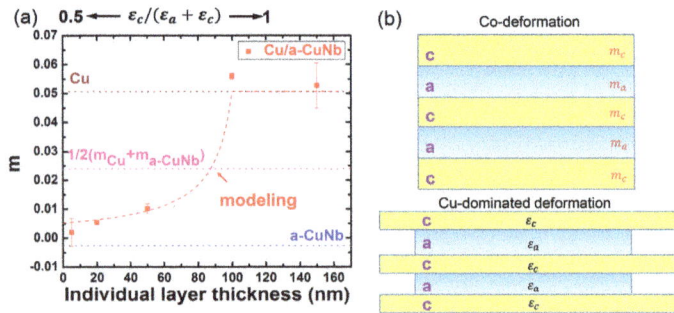

Figure 6. (a) SRS as a function of individual layer thickness (h) of Cu/a-CuNb multilayers. The SRS of a-CuNb, Cu, and their average values were added as dotted lines. When $h < 50$ nm, the SRS of Cu/a-CuNb multilayers decreases with decreasing h. When $h > 100$ nm, Cu/a-CuNb multilayers have an apparent SRS value similar to that of the single layer Cu film. The modeled curve (dashed line) indicates strain distribution from Cu and a-CuNb as a function of h. (b) A schematic showing the different deformation mechanisms at different h. When h is small (<50 nm), dislocation activities are limited and crystalline and amorphous layers can co-deform; when h is relatively large (>100 nm), deformation is dominated by dislocation activities, and crystalline layers accommodate more strain than amorphous layers.

The experimental data are consistent with the model discussed previously. At lager h, the plastic deformation is mainly generated by the crystalline phase, the value of $\frac{\Delta L_c}{\Delta L_a + \Delta L_c}$ will be close to 1 and $\frac{\Delta L_a}{\Delta L_a + \Delta L_c}$ close to zero. Therefore, the SRS should be similar to the single layer Cu. However, when h is smaller, the $\frac{\Delta L_c}{\Delta L_a + \Delta L_c}$ and $\frac{\Delta L_a}{\Delta L_a + \Delta L_c}$ would be close to 0.5 due to the co-deformation of crystalline phase and amorphous phase (the two phases share the deformation equally) [8,13], and the SRS would be closer to the single-layer amorphous film based on Equation (6).

Figure 6b is a schematic showing the different deformation mechanisms at different h. When $h > 100$ nm, Cu layers deform plastically due to its low strength and accommodate most of the strain, and thus Cu/a-CuNb multilayers have an apparent SRS value similar to that of the single layer Cu film. The mechanism of plastic deformation in this regime is dominated by dislocation activities inside the Cu layers.

When $h < 50$ nm, the SRS of Cu/a-CuNb multilayers decreases with decreasing h. The yield strength of Cu film increases with decreasing h due to the Hall-Petch effect (although there may occasionally be an inverse Hall-Petch effect when the grain size decreases to a few nanometers, here we leave this possibility aside). At the same time, the density of dislocations in the Cu grains decreases and the activation volume increases. The generation and propagation of dislocation is very difficult, and the stress concentration caused by the dislocations at the interface is not enough to transmit the dislocations across the interface. The limited dislocations will be confined within the Cu layers and slide along the interface [43]. On the other hand, for the a-CuNb layers, the deformation mechanism is the coalescence of shear transformation zones instead of major shear banding, and at this layer thickness, Cu and a-CuNb layers can co-deform.

4. Conclusions

A new nanoindentation test method that enables reliable determination of SRS is reviewed. The new method is applied in two model systems (single-layer Cu film and Cu/amorphous-CuNb multilayers) to reduce the thermal drift effect and yields reliable results. The method yields correct SRS of single-layer Cu film compared with the conventional method, and the layer thickness-dependent SRS of Cu/a-CuNb nanolayers under the iso-stress condition is systematically discussed: when $h > 100$ nm, the plastic deformation is mainly accommodated by the Cu layers through dislocation pile-up, and Cu/a-CuNb multilayers have an apparent SRS value similar to that of the single layer Cu film; when $h < 50$ nm, the crystalline and amorphous layers co-deform, and the SRS of Cu/a-CuNb multilayers decreases with decreasing h. The main deformation mechanism is interface-mediated. This new method is beneficial for the measurement of hardness and the SRS of nano-scale materials without the thermal drift error at low strain rate under nanoindentation.

Acknowledgments: X.Z. acknowledges the partial financial support by NSF-DMR 1642759.

Author Contributions: Y.L. and X.Z. conceived and designed the experiments; Y.L. and Z.F. performed the experiments; Z.F. and J.S. analyzed the data; J.S. wrote the paper.

Conflicts of Interest: The authors declare no conflict of interest.

References

1. Suo, T.; Ming, L.; Zhao, F.; Li, Y.L.; Fan, X. Temperature and strain rate sensitivity of ultrafine-grained copper under uniaxial compression. *Int. J. Appl. Mech.* **2013**, *5*, 1350016. [CrossRef]
2. Asaro, R.J.; Suresh, S. Mechanistic models for the activation volume and rate sensitivity in metals with nanocrystalline grains and nano-scale twins. *Acta Mater.* **2005**, *53*, 3369–3382. [CrossRef]
3. Wei, Q. Strain rate effects in the ultrafine grain and nanocrystalline regimes—Influence on some constitutive responses. *J. Mater. Sci.* **2007**, *42*, 1709–1727. [CrossRef]
4. Akihisa, I. Stabilization of metallic supercooled liquid and bulk amorphous alloys. *Acta Mater.* **2000**, *48*, 279–306.
5. Greer, A.L.; Cheng, Y.Q.; Ma, E. Shear bands in metallic glasses. *Mater. Sci. Eng. R Rep.* **2013**, *74*, 71–132. [CrossRef]
6. Hofmann, D.C.; Suh, J.Y.; Wiest, A.; Duan, G.; Lind, M.L.; Demetriou, M.D.; Johnson, W.L. Designing metallic glass matrix composites with high toughness and tensile ductility. *Nature* **2008**, *451*, 1085–1089. [CrossRef] [PubMed]
7. Knorr, I.; Cordero, N.M.; Lilleodden, E.T.; Volkert, C.A. Mechanical behavior of nanoscale Cu/PdSi multilayers. *Acta Mater.* **2013**, *61*, 4984–4995. [CrossRef]
8. Zhang, J.Y.; Liu, Y.; Chen, J.; Chen, Y.; Liu, G.; Zhang, X.; Sun, J. Mechanical properties of crystalline Cu/Zr and crystal–amorphous Cu/Cu–Zr multilayers. *Mater. Sci. Eng. A* **2012**, *552*, 392–398. [CrossRef]
9. Chu, J.P.; Jang, J.S.C.; Huang, J.C.; Chou, H.S.; Yang, Y.; Ye, J.C.; Wang, Y.C.; Lee, J.W.; Liu, F.X.; Liaw, P.K.; et al. Thin film metallic glasses: Unique properties and potential applications. *Thin Solid Films* **2012**, *520*, 5097–5122. [CrossRef]

10. Huang, L.; Zhou, J.; Zhang, S.; Wang, Y.; Liu, Y. Effects of interface and microstructure on the mechanical behaviors of crystalline Cu-amorphous Cu/Zr nanolaminates. *Mater. Des.* **2012**, *36*, 6–12. [CrossRef]

11. Fan, Z.; Xue, S.; Wang, J.; Yu, K.Y.; Wang, H.; Zhang, X.H. Unusual size dependent strengthening mechanisms of Cu/amorphous CuNb multilayers. *Acta Mater.* **2016**, *120*, 327–336. [CrossRef]

12. Wang, Y.M.; Li, J.; Hamza, A.V.; Barbee, T.W., Jr. Ductile crystalline-amorphous nanolaminates. *Proc. Natl. Acad. Sci. USA* **2007**, *104*, 11155–11160. [CrossRef] [PubMed]

13. Kim, J.Y.; Jang, D.C.; Greer, J.R. Nanolaminates utilizing size-dependent homogeneous plasticity of metallic glasses. *Adv. Funct. Mater.* **2011**, *21*, 4550–4554. [CrossRef]

14. Donohue, A.; Spaepen, F.; Hoagland, R.G.; Misra, A. Suppression of the shear band instability during plastic flow of nanometer-scale confined metallic glasses. *Appl. Phys. Lett.* **2007**, *91*, 241905. [CrossRef]

15. Torre, F.D.; Swygenhoven, H.V.; Victoria, M. Nanocrystalline electrodeposited Ni: Microstructure and tensile properties. *Acta Mater.* **2002**, *50*, 3957–3970. [CrossRef]

16. Lu, L.; Li, S.X.; Lu, K. An abnormal strain rate effect on tensile behavior in nanocrystalline copper. *Scr. Mater.* **2001**, *45*, 1163–1169. [CrossRef]

17. Valiev, R.Z.; Alexandrov, I.V.; Zhu, Y.T.; Lowe, T.C. Paradox of strength and ductility in metals processed by severe plastic deformation. *J. Mater. Res.* **2011**, *17*, 5–8. [CrossRef]

18. Wang, Y.M.; Ma, E. Temperature and strain rate effects on the strength and ductility of nanostructured copper. *Appl. Phys. Lett.* **2003**, *83*, 3165–3167. [CrossRef]

19. Wei, Q.; Cheng, S.; Ramesh, K.T.; Ma, E. Effect of nanocrystalline and ultrafine grain sizes on the strain rate sensitivity and activation volume: fcc versus bcc metals. *Mater. Sci. Eng. A* **2004**, *381*, 71–79. [CrossRef]

20. Gu, C.D.; Lian, J.S.; Jiang, Z.H.; Jiang, Q. Enhanced tensile ductility in an electrodeposited nanocrystalline Ni. *Scr. Mater.* **2006**, *54*, 579–584. [CrossRef]

21. May, J.; Höppel, H.W.; Göken, M. Strain rate sensitivity of ultrafine-grained aluminium processed by severe plastic deformation. *Scr. Mater.* **2005**, *53*, 189–194. [CrossRef]

22. Li, Y.J.; Mueller, J.; Höppel, H.W.; Göken, M.; Blum, W. Deformation kinetics of nanocrystalline nickel. *Acta Mater.* **2007**, *55*, 5708–5717. [CrossRef]

23. Lucas, B.N.; Oliver, W.C. Indentation power-law creep of high purity indium. *Metall. Mater. Trans. A* **1999**, *30A*, 601–610. [CrossRef]

24. Hay, J.; Agee, P.; Herbert, E. Continuous stiffness measurement during instrumented indentation testing. *Exp. Tech.* **2010**, *34*, 86–94. [CrossRef]

25. Maier, V.; Durst, K.; Mueller, J.; Backes, B.; Höppel, H.W.; Göken, M. Nanoindentation strain-rate jump tests for determining the local strain-rate sensitivity in nanocrystalline Ni and ultrafine-grained Al. *J. Mater. Res.* **2011**, *26*, 1421–1430. [CrossRef]

26. Liu, Y.; Bufford, D.; Wang, H.; Sun, C.; Zhang, X.H. Mechanical properties of highly textured Cu/Ni multilayers. *Acta Mater.* **2011**, *59*, 1924–1933. [CrossRef]

27. Liu, Y.; Bufford, D.; Rios, S.; Wang, H.; Chen, J.; Zhang, J.Y.; Zhang, X. A formation mechanism for ultra-thin nanotwins in highly textured Cu/Ni multilayers. *J. Appl. Phys.* **2012**, *111*, 073526. [CrossRef]

28. Liu, Y.; Chen, Y.; Yu, K.Y.; Wang, H.; Chen, J.; Zhang, X. Stacking fault and partial dislocation dominated strengthening mechanisms in highly textured Cu/Co multilayers. *Int. J. Plast.* **2013**, *49*, 152–163. [CrossRef]

29. Hasnaoui, A.; Derlet, P.M.; Van Swygenhoven, H. Interaction between dislocations and grain boundaries under an indenter—A molecular dynamics simulation. *Acta Mater.* **2004**, *52*, 2251–2258. [CrossRef]

30. Voyiadjis, G.Z.; Yaghoobi, M. Review of nanoindentation size effect: Experiments and atomistic simulation. *Crystals* **2017**, *7*, 321. [CrossRef]

31. Voyiadjis, G.Z.; Yaghoobi, M. Role of grain boundary on the sources of size effects. *Comput. Mater. Sci.* **2016**, *117*, 315–329. [CrossRef]

32. Nair, A.K.; Parker, E.; Gaudreau, P.; Farkas, D.; Kriz, R.D. Size effects in indentation response of thin films at the nanoscale: A molecular dynamics study. *Int. J. Plast.* **2008**, *24*, 2016–2031. [CrossRef]

33. Liu, Y.; Hay, J.; Wang, H.; Zhang, X.H. A new method for reliable determination of strain-rate sensitivity of low-dimensional metallic materials by using nanoindentation. *Scr. Mater.* **2014**, *77*, 5–8. [CrossRef]

34. Sneddon, I.N. The relation between load and penetration in the axisymmetric boussinesq problem for a punch of arbitrary profile. *Int. J. Eng. Sci.* **1965**, *3*, 47–57. [CrossRef]

35. Oliver, W.C.; Pharr, G.M. An improved technique for determining hardness and elastic modulus using load and displacement sensing indentation experiments. *J. Mater. Res.* **2011**, *7*, 1564–1583. [CrossRef]

36. Pharr, G.M.; Oliver, W.C.; Brotzen, F.R. On the generality of the relationship among contact stiffness, contact area, and elastic modulus during indentation. *J. Mater. Res.* **2011**, *7*, 613–617. [CrossRef]
37. Cheng, S.; Ma, E.; Wang, Y.M.; Kecskes, L.J.; Youssef, K.M.; Koch, C.C.; Trociewitz, U.P.; Han, K. Tensile properties of in situ consolidated nanocrystalline Cu. *Acta Mater.* **2005**, *53*, 1521–1533. [CrossRef]
38. Gray, G.T., III; Lowe, T.C.; Cady, C.M.; Vaile, R.Z.; Aleksandrov, I.V. Influence of strain rate & temperature on the mechanical response of ultrafine-grained Cu, Ni, and Al-4Cu-0.5Zr. *Nanostruct. Mater.* **1997**, *9*, 477–480.
39. Chen, J.; Lu, L.; Lu, K. Hardness and strain rate sensitivity of nanocrystalline Cu. *Scr. Mater.* **2006**, *54*, 1913–1918. [CrossRef]
40. Jiang, Z.H.; Liu, X.L.; Li, G.Y.; Jiang, Q.; Lian, J.S. Strain rate sensitivity of a nanocrystalline Cu synthesized by electric brush plating. *Appl. Phys. Lett.* **2006**, *88*, 143115. [CrossRef]
41. Ye, J.C.; Wang, Y.M.; Barbee, T.W.; Hamza, A.V. Orientation-dependent hardness and strain rate sensitivity in nanotwin copper. *Appl. Phys. Lett.* **2012**, *100*, 261912. [CrossRef]
42. Fan, Z.; Liu, Y.; Xue, S.; Rahimi, R.M.; Bahr, D.F.; Wang, H.; Zhang, X. Layer thickness dependent strain rate sensitivity of Cu/amorphous CuNb multilayer. *Appl. Phys. Lett.* **2017**, *110*, 161905. [CrossRef]
43. Misra, A.; Hirth, J.P.; Hoagland, R.G. Length-scale-dependent deformation mechanisms in incoherent metallic multilayered composites. *Acta Mater.* **2005**, *53*, 4817–4824. [CrossRef]

crystals

MDPI

Article

Quantitative Imaging of the Stress/Strain Fields and Generation of Macroscopic Cracks from Indents in Silicon

Brian K. Tanner [1], David Allen [2], Jochen Wittge [3], Andreas N. Danilewsky [4], Jorge Garagorri [5], Eider Gorostegui-Colinas [6] , M. Reyes Elizalde [5] and Patrick J. McNally [7,*]

[1] Department of Physics, Durham University, South Road, Durham DH1 3LE, UK; b.k.tanner@dur.ac.uk
[2] Department of Aerospace, Mechanical and Electronic Engineering, I.T. Carlow, Carlow R93 V960, Ireland; david.allen@itcarlow.ie
[3] Straumann GmbH, 79100 Freiburg, Germany; wittge@web.de
[4] University of Freiburg, Kristallographie, Institut für Geo- und Umweltnaturwissenschaften, 79104 Freiburg, Germany; a.danilewsky@krist.uni-freiburg.de
[5] CEIT and Tecnun (University of Navarra), Paseo de Manuel Lardizabal 15, San Sebastián 20018, Spain; jgaragorrimail@gmail.com (J.G.); relizalde@ceit.es (M.R.E.)
[6] Lortek, Arranomendi kalea 4A, Ordizia 20240, Spain; egorostegui@lortek.es
[7] School of Electronic Engineering, Dublin City University, Dublin 9, Ireland
* Correspondence: patrick.mcnally@dcu.ie; Tel.: +353-17005119

Academic Editor: Ronald W. Armstrong
Received: 10 October 2017; Accepted: 8 November 2017; Published: 14 November 2017

Abstract: The crack geometry and associated strain field around Berkovich and Vickers indents on silicon have been studied by X-ray diffraction imaging and micro-Raman spectroscopy scanning. The techniques are complementary; the Raman data come from within a few micrometres of the indentation, whereas the X-ray image probes the strain field at a distance of typically tens of micrometres. For example, Raman data provide an explanation for the central contrast feature in the X-ray images of an indent. Strain relaxation from breakout and high temperature annealing are examined and it is demonstrated that millimetre length cracks, similar to those produced by mechanical damage from misaligned handling tools, can be generated in a controlled fashion by indentation within 75 micrometres of the bevel edge of 200 mm diameter wafers.

Keywords: X-ray diffraction imaging; Raman spectroscopy; indentation geometry; plastic deformation; crack generation; plastic deformation strain imaging

1. Introduction

The work described in this paper arose from a detailed study of cracks associated with robotic handling damage in silicon wafers. Catastrophic wafer fracture [1,2] during high temperature processing is a major problem in semiconductor manufacturing with multi-million dollar associated costs on a single production line [3]. At the beginning of our research programme, it was believed that the fracture was associated with cracks introduced by mechanical damage but the location of such damage was not clear. Subsequently the origin of such failure has been shown to be cracks produced at the wafer bevel edge [4,5] due to handling tool misalignment. These cracks can be millimetres in length. Some result in high temperature wafer shattering, while some are benign. As an outcome of our project, commercial X-ray diffraction imaging tools are now available, together with associated analytical software to predict the probability of failure and hence make appropriate decisions relating to the manufacturing process.

As part of the research programme, we used nano-indentation at room temperature to generate cracks in a controlled manner and determine whether such artificially induced cracks could lead to wafer fracture during rapid thermal annealing (RTA). We used scanning electron microscopy (SEM), X-ray Diffraction Imaging (XRDI), also known as X-ray topography, and micro-Raman spectroscopy to study the detailed crack geometry and associated strain fields around Berkovich and Vickers indents. The crack geometry and associated strain fields were studied as a function of indenter load, using a Berkovich tip from 100 mN to 600 mN and a Vickers tip from there up to 5 N. As will become apparent from the results presented below, the cracks generated had high symmetry and in no case were we able to induce wafer fracture during RTA from indents generated away from the wafer edge. Only when the indentation was with a high load and within approximately 75 μm of the bevel edge could millimetre length cracks, such as those now known to induce catastrophic fracture during processing, be produced. Such indentation-generated cracks sometimes initiated catastrophic fracture during RTA. In this paper, we present results of the study of the strain fields around various indentations and explain how the low symmetry bevel edge cracks are generated.

2. Results

In all of the indentation experiments, both with Berkovich and Vickers tips, median surface-breaking cracks were observed to originate at the indenter apices (Figure 1), as was the case in the experiments of Yan et al. [6] for Vickers indents of typically 300–500 mN load on InP. Unlike the indentation of InP, however, the crack projections on the surface did not correspond to the intersection of the low surface energy planes which are {111} or {110} in the case of silicon [7]. Rotation of the Bervovich indenter apices with respect to the in-plane crystallographic directions had almost no effect on the crack geometry, the median cracks still emerging almost parallel to the projection of the indenter apices on the surface. For constant indenter load of less than 200 mN, the total crack length in the surface remained constant as the indenter was rotated.

Despite the crack extension being only a few microns beyond the indenter impression, the X-ray diffraction images (topographs) [8] were often over 100 μm in extent (Figure 2a), demonstrating that the strain field from the defective region is long range. In transmission, the images consisted of two half lobes, with the line of no contrast perpendicular to the projection of the diffraction vector, and a dark central region. The contrast differs from that in the reflection (Bragg) geometry, in which the images consist of a light central region with a complete dark circle at the edge of the contrast area [9]. Although the three images of Figure 2a are all from the same 600 mN indenter load, the image length varies significantly for the three images from 139 to 96 μm. Such a large dispersion in image size is not observed for loads below about 200 mN. On rotation of the indenter tip about the [100] axis, no change in average image dimension was observed (Figure 2b). (Note that because of the superposition of the three-fold symmetry of the indenter and the four-fold symmetry of the crystal structure, only a limited range of angle is required to cover all possible angles of rotation.)

The magnitude of the strain field close to the indenter could be imaged in detail using micro-Raman spectroscopy mapping (Figure 3). The medium range strain field is compressive with a maximum value of −0.23 GPa for a 400 mN load indent and −0.85 GPa for a 600 mN load. Close to the centre of the indent, there is tensile strain, which has a maximum value immediately below the indent. For a 400 mN indent its maximum value was 0.75 GPa, while for a 600 mN indent it was 1.2 GPa. For indenter loads above 200 mN, although there is still evidence of a three-fold symmetry in the strain field associated with the indent apex directions, the complete symmetry is lost. This behaviour is found for all indent loads above 200 mN. Annealing of the sample for 30 min at 1000 °C results in a reduction in the magnitude of the compressive strain field by a factor of about four (Figure 4). The tensile stress below the indenter tip and the quasi-three-fold symmetry is completely lost upon annealing.

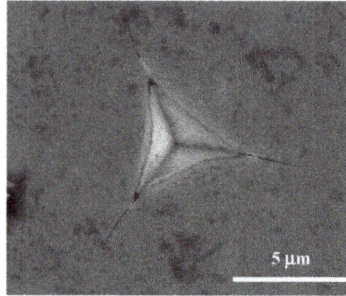

Figure 1. Scanning electron microscope image of surface-breaking cracks and associated directions from a 150 mN Berkovich indent with the indenter apex parallel to [011].

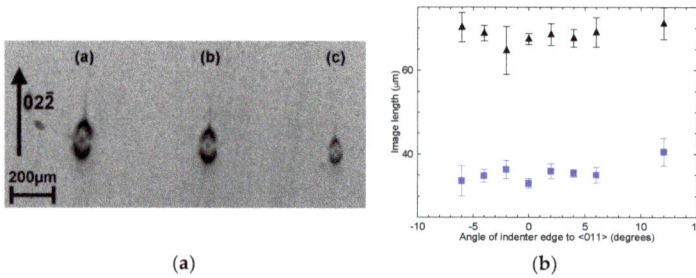

(a)

(b)

Figure 2. (a) X-ray diffraction image taken in white beam mode of 600 mN indents with the indenter tip rotated by 6° with respect to the [011] direction. (b) Extent of the X-ray diffraction contrast for 100 mN indents as a function of rotation of the Berkovich tip apex from the [011] direction. Triangles represent image extent in the direction parallel to the diffraction vector, while squares represent the extent perpendicular to the diffraction vector.

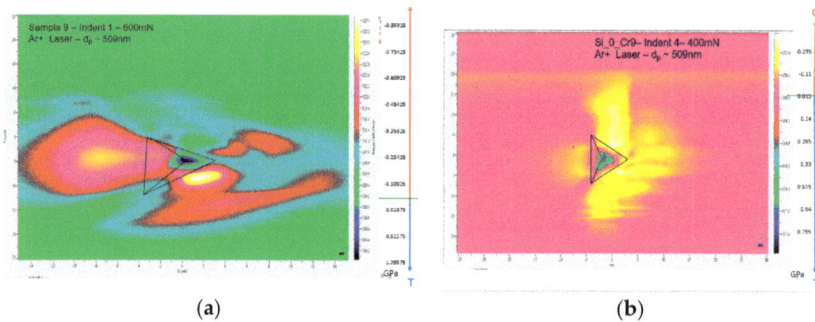

(a)

(b)

Figure 3. Two-dimensional maps of the stress as a function of position around: (a) a 600 mN; and (b) a 400 mN indent measured by micro-Raman spectroscopy profiling. The Raman spectroscopy map covers a region whose extent is 30 μm (X) by 50 μm (Y). Step size 0.5 μm. The red and blue arrows indicate whether the stress is compressive (C) or tensile (T). The highest compressive stress is indicated by the yellow regions, while the greatest tensile strain is indicated by the black/blue regions. (Figure 3a is reproduced from ref. [9]) The Raman figures and the stress figures do not correspond directly. These scans took approximately 40 h to complete so the stress figures were calculated by compensating for environmental drift using the plasma peak.

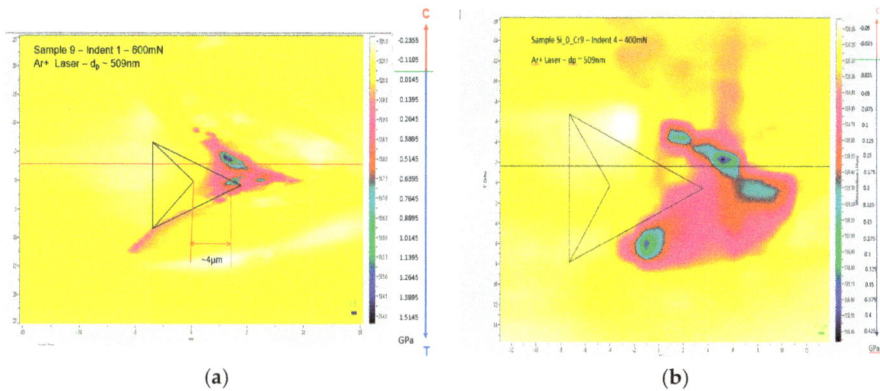

(a)   (b)

Figure 4. Micro-Raman maps of the stress around the indents shown in Figure 3 after annealing for 30 min at 1000 °C. The Raman spectroscopy maps covers regions whose extent is: 30 μm (X) by 50 μm (Y) (**a**); and 30 μm (X) by 32 μm (Y) (**b**). Step size 0.5 μm. The red and blue arrows indicate whether the stress is compressive (C) or tensile (T). The highest compressive stress is indicated by the yellow regions, while the greatest tensile strain is indicated by the black/blue regions.

The origin of the loss, in the micro-Raman image maps, of three-fold symmetry, seen in the surface-breaking crack geometry, lies in the presence of breakout above 200 mN indent load. An example of such breakout around a 500 mN indent is shown in Figure 5. There are two areas of breakout around the three edges of the indenter footprint, as is also the case for the 600 mN imprint shown in Figure 3a.

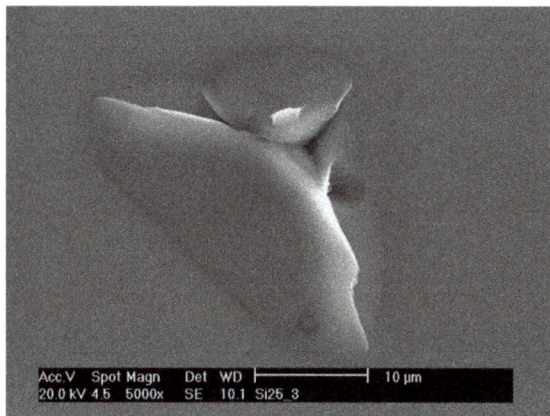

Figure 5. SEM micrograph of a 500 mN Berkovich indent, showing chipping flaws at the edges of the imprint.

Recalling that our objective was to use nanoindentation to generate macroscopic cracks which would lead to wafer fracture during subsequent high temperature processing, we continued to increase the indenter load, switching to a Vickers tip for loads above 600 mN. Due to the omnipresent breakout, we were never able to generate macroscopic length cracks, despite the symmetry breaking. We were, however, able to use the bevel edge and the breakout process to achieve this successfully. With a 5 N Vickers indent within 75 μm of the bevel at the edge of the wafer, commonly we found that two of the breakout regions propagated right through the thin bevel of the wafer as shown schematically in Figure 6a. One long crack (a of type 1) then propagated from the apex of the Vickers tip which pointed inwards from the edge, there sometimes being macroscopic cracks (b in Figure 6a) of type 2 generated from the intersection of the bevel edge with the breakout line. As shown in the white beam X-ray topographs of Figure 6b, long cracks, which again did not propagate on the low surface energy surfaces, could be generated by such an indentation technique. Here, both cracks appear of type 2, originating from the intersection of the break-out with the bevel edge. Residual stress from the indent itself can be seen between the cracks. The curved image of the wafer edge arises because of long range lattice strain associated with the indent changing the local diffraction conditions. Macroscopic crack generation was achieved at a typically 50% success level, it being sensitive to the exact distance of the indent from the bevel edge. These cracks were similar to those generated by repetitive mechanical damage from misaligned wafer handling tools.

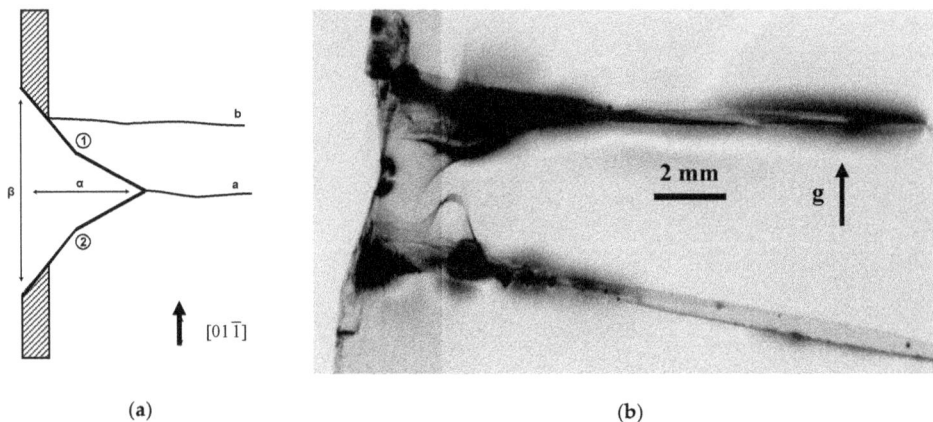

(a)   (b)

Figure 6. (**a**) Schematic diagram of the type 1 and type 2 macroscopic cracks generated by Vickers indentation within 75 μm of the bevel edge. (**b**) Composite of white beam X-ray topographs of a pair of such cracks in $02\bar{2}$ reflection. (The break-out dimensions α and β are not discussed here.)

3. Discussion

As was evident from a systematic series of experiments with increasing indenter loads, it proved impossible to generate macroscopic cracks in silicon when the indent was far from the wafer edge. In all cases, the crack lengths were limited to several micrometres, the median surface breaking cracks always being initiated at the indenter apex. Unlike conical indents [10,11], where it was found that {110} cracks were mainly introduced from the indent, indicating that fracture occurs most easily along the {110} planes among the crystallographic planes of the <001> zone, with Berkovich indents, we found no association of the crack geometry with the low surface energy planes. Indeed, ultra-fast X-ray diffraction imaging of fracture in silicon has recently shown that even when cracks are apparently following low energy {110} surfaces, there is a continual jumping between {110} and various high indexed {hkl} planes [12].

The limit on the length of crack is primarily determined by breakout associated with lateral cracks intersecting the surface. When breakout occurs, the strain is relaxed and further crack propagation does not occur. We have noted elsewhere that there is strong asterism in X-ray topographs when breakout is about to occur [9]. Lateral cracks were always found to be present even at low indenter loads when breakout did not occur. An example is shown in Figure 7a in the case of a 120 mN load indent where breakout did not occur. This is a three-dimensional reconstruction from a sequence of images taken during focused ion beam milling of the sample. Further examples can be found in reference [13]. We have shown that finite element (FE) modelling reproduces the shape of the main median crack rather well [9] when cohesive elements are included to represent the relaxation from cracking. The presence of varying length lateral cracks, explains the breaking of the symmetry in Raman strain maps of low load indents even when breakout does not occur. Without the stochastic nature of the lateral cracking process, the FE simulations always show three-fold symmetry, even to the highest loads. An example is shown in Figure 7b, which is a simulation that does not include cohesive elements. We note that the simulation shows stress concentration at the tips of the indenter, which is where the median cracks are initiated. However, in contrast to the experimental data of Figure 3, the stress at those points is tensile and a compressive strain is predicted directly below the indenter apex. The experimental Raman results of Figures 3 and 4 show tensile stresses under the indenter apex with the Ar^+ laser (where the penetration depth is 565 nm) but with the UV laser (where the penetration depth is only 9 nm), compressive stresses were observed. Development of tensile stresses implies the presence of lateral cracks and this is captured by the Raman data using the Ar^+ laser but not with the UV laser. The Raman map simulated from FE stresses should be compared with results from the UV laser as the simulated stresses are from the surface.

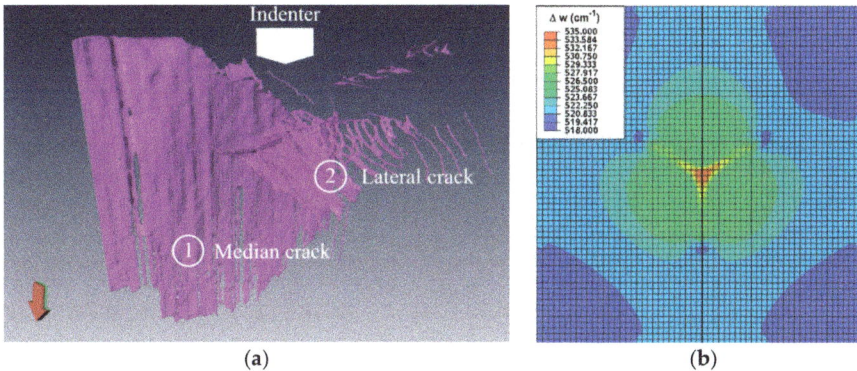

Figure 7. (**a**) 3D reconstruction of subsurface cracks corresponding to a 120 mN Berkovich indent, showing: one of the three median cracks (1); and a lateral crack (2). (**b**) Simulation of the shift of the UV Raman TO phonon line for a 150 mN indent without cohesive elements.

Due to the high strain sensitivity of X-ray diffraction imaging, the localized symmetry breaking is not reflected in the geometry of the X-ray images of the strain fields around the indents. The X-ray images form at typically 30 μm distance from the indent where the strain level is relatively low and the long range strain field is still nearly symmetric. When breakout occurs, the long range strain field falls in magnitude and the X-ray contrast begins to occur at a smaller distance from the indent. It does not, however, significantly change its symmetry (Figure 2a). The line of zero contrast, perpendicular to the diffraction vector direction, arises because the contrast only arises from displacement components in the diffraction vector direction. Along the horizontal line through the centre of the indent, the strain field is entirely in the horizontal direction; there is no component in the diffraction vector direction. There is thus no contrast along this line, giving rise to the two lobes in the image. The difference between the image widths parallel and perpendicular to the diffraction vector arises from a diffraction contrast effect. The diffraction vector lies in the incidence plane, which also includes the entrance and exit beams. So-called "direct" or "kinematical" contrast [14], such as seen here, forms at a point when the effective misorientation of the deformed lattice exceeds the perfect crystal diffraction, or Darwin, width. For deformation greater than this, the region selects a slightly different wavelength which diffracts kinematically and additional intensity appears locally in the image. However, the X-ray beams from this region have a different Bragg angle and the diffracted wave therefore travels in a different direction in space. Due to the large distance, typically 300–400 mm, between the sample and detector, the image is spread out as the propagation distance increases. No such magnification takes place in the direction perpendicular to the incidence plane (or diffraction vector) and hence there is a difference in the width in the two directions. The absence of change in the dimensions of the image as the indenter orientation is changed confirms the symmetry of this long range strain field. The magnitude of the strain field at the edge of the image is determined by the perfect crystal reflecting range. For a wavelength of 0.0506 nm and the 220 reflection, the Darwin width $\Delta\theta$ is 1.58 arc seconds, i.e., 7.6×10^{-6} radians. At the point of high symmetry parallel to the diffraction vector the lattice plane tilt is zero and the strain e is given by

$$e = \frac{\Delta d}{d} = \frac{\Delta\theta}{\tan\theta_B} \tag{1}$$

where θ_B is the Bragg angle, d is the crystal lattice spacing and Δd is the lattice displacement. Thus, the X-ray diffraction images show quantitatively that, at a distance of typically some 20 μm from the indent, the associated strain is 6×10^{-5}.

An interesting feature in all three diffraction images of Figure 2a is the central dark circular region. This does not have a line of no contrast perpendicular to the diffraction vector, indicating that the strain displacement field is not radial, as at larger distance from the indent. We have already noted that, in the micro-Raman images close to the indent (Figure 3), the strain field switches from tensile to compressive as a function of distance from the indent centre. There must therefore be a region of zero displacement around the indent and this gives rise to the circle of zero contrast between the outer dark lobes and the inner dark circular disc in the X-ray image. As the X-ray beam passes through the region of intense tensile strain, which is almost twice the magnitude of the maximum compressive strain, it must experience components of displacement in the diffraction vector direction along the central line, thus giving no line of zero contrast in this central part of the image. The central component of the image is not seen in the reflection geometry [9] due to the low penetration of the X-ray beam below the surface and hence not being scattered by the tensile strained region.

As was pointed out by Tang et al. [15], discontinuity in the unloading curve referred to as pop-out, does not signify the appearance of cracks and the 3D reconstructions from all samples studied showed the presence of lateral as well as median cracks (Figure 7a). During the loading stage, below the indenter, normal diamond cubic structure Si-I phase material is transformed to the metallic β-Sn phase Si-II. On subsequent unloading, the transformation of the Si-II phase to body centre cubic (bc8) Si-III and rhombohedral (r8) Si-XII phases is associated with the pop-out phenomenon, though the size and position of the discontinuity is related to the maximum load and the unloading rate [16]. In all indents at and above 150 mN load, below which the stress level was apparently not sufficient to transform Si-I to Si-II on loading, micro-Raman spectra taken after indentation reveal lines attributable to the Si-III and Si-XII phases. Our loading/unloading data are entirely consistent with previous work [17,18]. Further, in all samples thinned and studied by scanning transmission electron microscopy, we have observed a region of plastic deformation below the indent [9] such as has been associated [19,20] with the pop-in discontinuities which we also observed during loading.

The factor of four reduction in the maximum compressive stress and the loss of the tensile stress below the indenter point, following high temperature annealing, is associated with nucleation of dislocation loops from the indentation site at temperatures where Si is ductile [21]. As illustrated in the X-ray section topograph in Figure 8a, taken with a 15 μm wide entrance slit, the dislocation loops lie on the inclined {111} slip planes and propagate right through the 750 μm thick wafer. Depending on the profile of the temperature gradients within the annealing furnace, the loops propagate in one or both of the two orthogonal <011> directions in the wafer plane. The loops subsequently thicken into slip bands by a process of cross-slip, the early stages of which are illustrated in Figure 8b. Dislocation nucleation and propagation relieves the short range stress around the indent, imaged in the micro-Raman profiles.

An example of how very quickly the individual X-ray diffraction images of dislocations become indistinguishable during the process of slip band development is given in Figure 9. This shows two indents of the same load of 500 mN, close to the bevel edge, following an anneal for 60 s at 1000 °C.

(The varying thickness on the bevel gives rise to the almost horizontal thickness fringes at the top of the image, this being a well-known dynamical X-ray diffraction effect.) The indent A on the left, probably having experienced a stronger temperature gradient, has a well-developed slip band within which, although some dislocations can still just be resolved, most dislocation images cannot be individually distinguished. In contrast, the slip band development in the right hand image B is much less developed and the operation of a source of dislocation loops can be identified. As the loop expands, driven by the stress around the indent and the thermal gradient in the furnace, the 60° segment parallel to the interface glides out of the crystal and the screw and other 60° segments glide on the inclined {111} slip planes. The dislocation images are narrow where the dislocation intersects the exit surface of the wafer with respect to the X-ray beam and the elegant interference fringes that decorate the broadening image of the dislocation towards the X-ray entrance surface constitute the so-called intermediary image. This is a dynamical diffraction effect sometimes seen under conditions of moderate absorption, such as here. The dislocation loop originating from C, and interfering with the development of the slip band from indent B, arises from handling damage on the wafer surface. It has very similar characteristics to the effect of the indents, generating its own loop system that is developing into a slip band.

Breakout results from the intersection of the lateral cracks with the surface and exploitation of the symmetry-breaking of the bevel edge enabled us to generate millimetre length cracks into the 200 mm wafers in a controlled manner. These cracks resembled those produced by misaligned handling tools [4] and we have shown elsewhere how analysis of the ratio of the crack length to the width of the X-ray image at the crack tip provides a predictor of the probability of failure during subsequent high temperature processing [4]. The near-edge indentation technique has proven to be a satisfactory method for generating such cracks, an example of which is shown in Figure 10. This is a full X-ray diffraction image of a 200 mm wafer which was indented at both the left and right edges through the centre line of the wafer. The left hand indent did not generate a macroscopic crack, as full breakout did not occur, and the strain field in the X-ray image is localized. The right hand indent, also a 5 N Vickers indent, generated a crack of length 15 mm and the strain image extends over a similarly large radius. It is of interest to note that, because the ratio κ of the crack length (L) to the width of the X-ray image at the crack tip (D) is small, this particular crack was benign. We have shown elsewhere that, by modelling the crack as a super-dislocation, the width of the X-ray image provides a quantitative measure of the back stress which counters the opening and propagation of a crack [4]. A narrow image width D indicates a small back stress, potentially leading to an unstable crack. Using the Griffith criterion, it is straightforward to show that when κ exceeds a critical value κ_c, i.e., when

$$\kappa = L/D > \kappa_c \tag{2}$$

the crack will propagate during processing at high temperatures. Whether a particular crack is, or is not, benign depends on the temperature profile of the specific furnace and calibration is necessary to determine relevant κ_c in order to use the model predictively.

The upper crack shown in Figure 6b was similarly benign due to its low κ value, but the lower and longer of the two cracks resulted in wafer fracture. Although not shown here, the image of the tip of the lower crack is very narrow and as the crack is long, the κ value is high. Under the particular annealing conditions, $\kappa > \kappa_c$ and the wafer fractured.

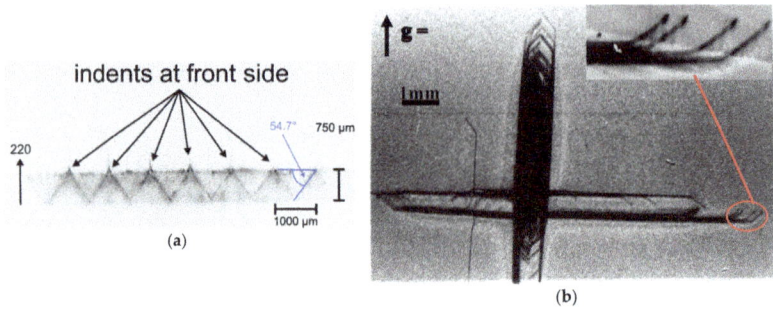

Figure 8. (**a**) X-ray section topograph of slip dislocations generated on {111} glide planes from surface indentations upon annealing. (**b**) X-ray diffraction image of the early stages of cross slip leading to thickening of the dislocation loops into a slip band and nucleation of the orthogonal slip systems.

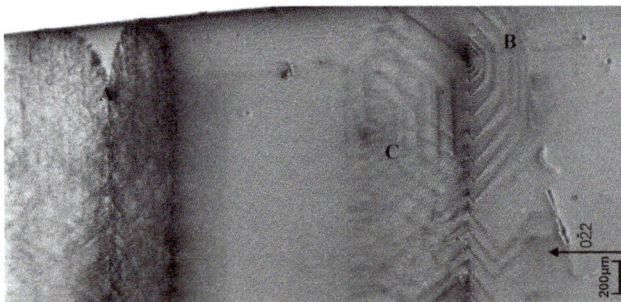

Figure 9. X-ray diffraction image of early stage slip band generation from two adjacent indents close to the bevel edge of the wafer.

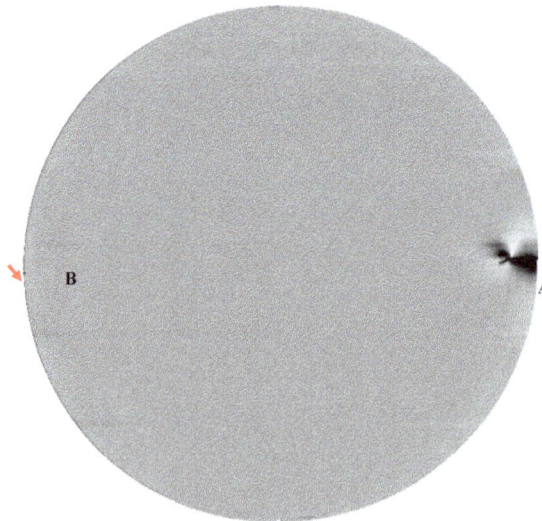

Figure 10. X-ray diffraction image, taken on a prototype of the Bruker/Jordan Valley QCTT wafer imaging tool, of two edge indents, one (**A**) of which generated a macroscopic crack in the 200 mm diameter wafer (**right**) and one (**B**, arrowed) which did not nucleate such a crack (**left**).

4. Materials and Methods

All experiments were performed on (100) oriented, integrated circuit quality, dislocation-free, 200 mm diameter silicon wafers purchased from Y Mart Inc, Palm Beach Gardens, FL, USA. Wafers were within 0.2° of (100) orientation. The nominally defect-free, double side polished, p-type wafers had resistivity below 10 ohm mm, and were of thickness 725 (±25) µm. No edge defects were visible either under optical inspection or in X-ray diffraction images of the as-received wafers, which had been packed and shipped in standard cassettes.

Two pieces of indentation equipment were used to generate controlled damage on silicon. The first was a Nanoindenter® II from Agilent (formerly Nano Instruments Inc., Oak Ridge, TN, USA), used to indent at low loads (100 mN to 600 mN), and the second was a Mitutoyo AVK-C2 hardness tester (Mitutoyo, Kawasaki, Japan) used for application of heavier loads from 500 mN up to 50 N. A Berkovich tip, was used with the Nanoindenter® II. It was a three-sided diamond pyramid with each face making an angle of 65.35° with the indenter axis, with a total included angle of 142.3°. The Mitutoyo AVK-C2 had a Vickers tip which was a four sided diamond pyramid, with a total included angle of 136°, but the same projected area-to-depth as the Berkovich indenter. All indents were performed at room temperature.

Micro-Raman spectroscopy was used to measure the strain and crystalline damage as well as high pressure phase transformations produced by the nano-indentation. The system used in this study was a Horiba Jobin-Yvon HR800® system (Horiba, Kyoto, Japan) running LabSpec 5 software (Horiba, Kyoto, Japan) and integrated with a microscope coupled confocally to an 800 mm focal length spectrometer (Figure 11a). Samples were mounted on a high precision table capable of x-y and z translation. An autofocusing system was incorporated. A diffraction grating of 1800 groves per millimetre was used to split the Raman signal into individual wavelengths, which were then directed onto a CCD detector. The HR800 uses two different CCD detectors; an air cooled Synapse CCD system and a liquid nitrogen cooled CCD3000 system used for acquisition times of greater than 120 s. Three lasers are connected to the Raman system through a periscope: a Uniphase 2014 488 nm Argon ion (Ar⁺) visible laser with a maximum output power of 20 mW, a Spectra-Physics Stablite 2017 364 nm Ar⁺ UV laser with a max. output power of 100 mW and a Kimmon Koha IK3201IR-F 325 nm Helium Cadmium (He-Cd) ultraviolet (UV) laser with a max. output power of 22.4 mW. The output light was linearly polarized.

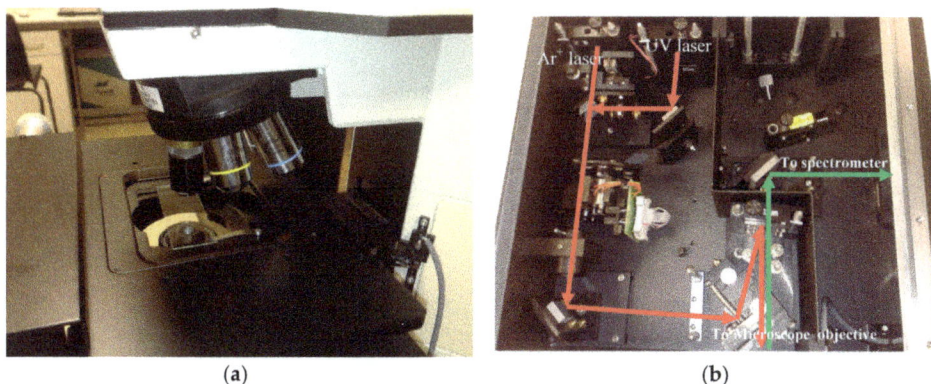

(a)   (b)

Figure 11. (a) Microscope lenses attached to the LabRam® HR800 System. The sample under test typically sits on a glass microscope slide below the objective lens. (b) Internal light paths for LabRam® HR800 System. The red line shows the path of the laser beam, while the green line shows the path of the Raman scattered beam.

For the µRS measurements, a backscattering geometry was used. In this configuration, a microscope objective is used to focus the beam onto the sample surface and the system is equipped with a number of lenses allowing different optical magnifications and can produce spot sizes as small as 1–3 µm in diameter (Figure 11a). Scattered light is collected through the same objective and passes through a beam splitter and is focused on to the entrance slit of the spectrometer which disperses the light onto one of the CCD detectors (Figure 11b). The use of a flip mirror enables the specimen under the microscope objective to be visualised using a small camera, which is particularly useful for precise positioning of the laser beam onto the area of interest.

The spectrometer was centred on the Raman line of the TO Raman phonon peak shift for unstrained silicon at 520.07 cm^{-1} [22]. There was no detectable asymmetry in the unstressed Si lineshapes. Assuming the strain in the indented silicon to be biaxial, the stress components σ_{xx} and σ_{yy} are equal and related to the spectral line shift by [22]

$$\sigma_{xx} = \sigma_{yy} = \frac{\omega_0 - \omega_1}{4} GPa \tag{3}$$

where ω_0 is the wavenumber (in cm^{-1}) of the unstrained Raman spectral line of the Si-Si transverse optical (TO) phonon and ω_1 is that of the measured Raman spectral line of the sample. Long scans required for 2D Raman maps can take from a number of hours to a number of days to complete. To compensate for temperature and laser stability, the plasma lines of the laser were recorded as fidiucal markers over the entire duration of the scan. The laser plasma lines are Rayleigh scattered and thus are insensitive to any strain in the sample [22]. The Ar$^+$ laser was used to create the maps of Figures 3 and 4.

In addition, as the laser is focused to a spot of ~1–3 µm in diameter, localised heating can arise. Any associated thermal expansion in the silicon can cause the Raman TO phonon to shift to move to a lower energy (at a rate of ~0.025 cm^{-1}/K), erroneously implying the presence of a tensile strain. However, this shift [23] is smaller than the resolution of the spectrometer used in this study, which was ~0.3 cm^{-1}.

X-ray diffraction imaging was performed in white beam mode at the TOPO-TOMO beamline at the ANKA synchrotron radiation source at Karlsruhe, Germany and at beamline B16 of the Diamond Light Source at Didcot, Oxford, U.K. At ANKA, transmission images were recorded on a CCD camera optically coupled to a macroscope and with a Lu$_3$Al$_5$O$_{12}$ scintillator. The effective pixel size of the detection system was 2.5 × 2.5 µm. At the Diamond Light Source, transmission images were recorded with a CCD camera manufactured by PCO. Imaging GmbH, Kelheim, Germany, which has 4008 × 2672 pixels and a 14 bit dynamic range. Scintillators and objective lenses were chosen such that the pixel size effective (binned) was 2 µm. Data acquisition times were between 10 and 40 s. All transmission diffraction images were taken with the 022 reflection, oriented such that there was a principal wavelength of 0.0506 nm. Further details of the ANKA instrumentation can be found in reference [24].

Scanning electron microscopy (SEM) was performed on a field emission electron microscope (LEO 1525, Leo Electron Microscopy Inc., Thornwood, NY, USA) with an in-lens secondary electron backscattering detector. A Quanta 3D dual beam FIB (focused ion beam) from FEI, Hillsboro, OR, USA, which incorporates FEGSEM (field emission gun scanning electron microscopy) and FIB cannons, was used to make microscopic observations of cracks around indents as the specimen was FIB milled down. The electron and ion beam cannons form an angle of 52° to each other. The instrument incorporated SE (secondary electron), BS (backscattered electron) and STEM (scanning transmission). From the FIB cross sections, three-dimensional reconstruction of the cracks was achieved using Amira reconstruction software. Simulation of the crack geometry was performed using the commercial software ABAQUS (version 6.8-3, Dassault Systemes, Vélizy-Villacoublay, France). The code incorporates a cohesive zone model and a simple triangular relation between stress and crack face separation was used. Conversion of stress to Raman shift was done using Equation (3).

5. Conclusions

Micro-Raman and X-ray diffraction imaging both provide maps of the strain field around indentations but on substantially different length scales. In particular, the symmetry associated with the indenter profile is lost in the X-ray images whereas the micro-Raman maps reveal in detail not only the magnitude and sense of the strain field but also its detailed symmetry. This is particularly revealing when breakout occurs. On annealing, the strain field maximum moves away from the site of the indent due to the formation of a series of dislocation loops, which are the precursors of slip bands. By exploiting the symmetry breaking of the bevelled edge of the wafer, i.e., by indentation close to the wafer edge, we have been able to generate macroscopic cracks in 200 mm wafer in a controlled, though not always predictable, manner.

Acknowledgments: Financial support was provided through the European Community FP7 STREP project SIDAM (Grant No. FP7-ICT-216382). Technical support from Kawal Sawhney, Igor Dolbnya and Andrew Malandain at beamline B16 of the Diamond Light Source and P. Vagovic, T. dos Santos Rolo and H. Schade at ANKA (Angströmquelle Karlsruhe) is gratefully acknowledged. Thanks are extended to Keith Bowen for valuable discussions and initially conceiving the SIDAM project. Funds for open access publication came from Dublin City University.

Author Contributions: Brian K. Tanner, Reyes M. Elizalde, Patrick J. McNally and Andreas N. Danilewsky collaboratively conceived and supervised the experiments within the research programme framework, which was coordinated by Brian K. Tanner; Jorge Garagorri performed most of the indentations; David Allen undertook micro-Raman analysis; Eider Gorostegui-Colinas made finite element simulations of the strain fields; Brian K. Tanner, Patrick J. McNally, Andreas N. Danilewsky, David Allen and Jochen Wittge undertook X-ray diffraction imaging experiments; and Brian K. Tanner wrote the first draft of the paper.

Conflicts of Interest: The authors declare that there are no conflicts of interest.

References

1. Chen, P.Y.; Tsai, M.H.; Yeh, W.K.; Jing, M.H.; Chang, Y. Investigation of the relationship between whole-wafer strength and control of its edge engineering. *Jpn. J. Appl. Phys.* **2009**, *48*, 126503. [CrossRef]
2. Chen, P.Y.; Tsai, M.H.; Yeh, W.K.; Jing, M.H.; Chang, Y. Relationship between wafer edge design and its ultimate mechanical strength. *Microelectron. Eng.* **2010**, *87*, 2065–2070. [CrossRef]
3. Atrash, F.; Meshi, I.; Krokhmal, A.; Ryan, P.; Wormington, M.; Sherman, D. Crystalline damage in silicon wafers and 'rare event' failure introduced by low-energy mechanical impact. *Mater. Sci. Semicond. Process.* **2017**, *63*, 40–44. [CrossRef]
4. Tanner, B.K.; Fossati, M.C.; Garagorri, J.; Elizalde, M.R.; Allen, D.; McNally, P.J.; Jacques, D.; Wittge, J.; Danilewsky, A.N. Prediction of the propagation probability of individual cracks in brittle single crystal materials. *Appl. Phys. Lett.* **2012**, *101*, 041903. [CrossRef]
5. Tanner, B.K.; Garagorri, J.; Gorostegui-Colinas, E.; Elizalde, M.R.; Bytheway, R.; McNally, P.J.; Danilewsky, A.N. The geometry of catastrophic fracture during high temperature processing of silicon. *Int. J. Fract.* **2015**, *195*, 79–85. [CrossRef]
6. Yan, J.; Tamaki, J.; Zhao, H.W.; Kuriyagawa, T. Surface and subsurface damages in nanoindentation tests of compound semiconductor InP. *J. Micromech. Microeng.* **2008**, *18*, 105018. [CrossRef]
7. Tanaka, M.; Higashida, K.; Nakashima, H.; Takagi, H.; Fujiwara, M. Orientation dependence of fracture toughness measured by indentation methods and its relation to surface energy in single crystal silicon. *Int. J. Fract.* **2006**, *139*, 383–394. [CrossRef]
8. Bowen, D.K.; Tanner, B.K. *X-ray Metrology in Semiconductor Manufacturing*; CRC Press: Boca Raton, FL, USA, 2006; pp. 279–290.
9. Tanner, B.K.; Garagorri, J.; Gorostegui-Colinas, E.; Elizalde, M.R.; Allen, D.; McNally, P.J.; Wittge, J.; Ehlers, C.; Danilewsky, A.N. X-ray asterism and the structure of cracks from indentations in silicon. *J. Appl. Cryst.* **2016**, *49*, 250–259. [CrossRef]
10. Tanaka, M.; Higashida, K.; Nakashima, H.; Takagi, H.; Fujiwara, M. Fracture toughness evaluated by indentation methods and its relation to surface energy in silicon single crystals. *Mater. Trans.* **2003**, *44*, 681–684. [CrossRef]

11. Tanaka, M.; Higashida, K.; Nakashima, H.; Takagi, H.; Fujiwara, M. Orientation dependence of fracture toughness and its relation to surface energy in Si crystals. *J. Jpn. Inst. Met.* **2004**, *68*, 787–791. [CrossRef]
12. Rack, A.; Scheel, M.; Danilewsky, A.N. Real-time direct and diffraction X-ray imaging of irregular silicon wafer breakage. *IUCrJ* **2016**, *3*, 108–114. [CrossRef] [PubMed]
13. Garagorri, J. Prediction of Critical Damage in Silicon Wafer during Rapid Thermal Processing. Ph.D. Thesis, University of Navarra, San Sebastian, Spain, 2014.
14. Bowen, D.K.; Tanner, B.K. *High Resolution X-ray Diffraction and Topography*; Taylor and Francis: London, UK, 1998; pp. 252–262.
15. Tang, Y.; Yonezu, A.; Ogasawara, N.; Chiba, N.; Chen, X. On radial crack and half-penny crack induced by Vickers indentation. *Proc. R. Soc. Lond. A* **2008**, *464*, 2967–2984. [CrossRef]
16. Chang, L.; Zhang, L.C. Deformation mechanisms at pop-out in monocrystalline silicon under nanoindentation. *Acta Mater.* **2009**, *57*, 2148–2153. [CrossRef]
17. Bradby, J.E.; Williams, J.S.; Wong-Leung, J. Transmission electron microscopy observation of deformation microstructure under spherical indentation in silicon. *Appl. Phys. Lett.* **2000**, *77*, 3749–3751. [CrossRef]
18. Gerbig, Y.B.; Stranick, S.J.; Morris, D.J.; Vaudin, M.D.; Cook, R.F. Effect of crystallographic orientation on phase transformations during indentation of silicon. *J. Mater. Res.* **2009**, *24*, 1172–1183. [CrossRef]
19. Bradby, J.E.; Williams, J.S.; Wong-Leung, J.; Swain, M.V.; Munroe, P. Mechanical deformation in Silicon by micro-indentation. *J. Mater. Res.* **2001**, *16*, 1500–1507. [CrossRef]
20. Bradby, J.E.; Williams, J.S.; Swain, M.V. In situ electrical characterization of phase transformations in Si during indentation. *Phys. Rev. B* **2003**, *67*, 085205. [CrossRef]
21. Wittge, J.; Danilewsky, A.N.; Allen, D.; McNally, P.; Li, Z.; Baumbach, T.; Gorostegui-Colinas, E.; Garagorri, J.; Elizalde, M.R.; Jacques, D.; Fossati, M.C.; et al. Dislocation sources and slip band nucleation from indents on silicon wafers. *J. Appl. Cryst.* **2010**, *43*, 1036–1039. [CrossRef]
22. De Wolf, I. Micro-Raman spectroscopy to study local mechanical stress in silicon integrated circuits. *Semicond. Sci. Technol.* **1996**, *11*, 139–154. [CrossRef]
23. Dombrowski, K.F. Micro-Raman Investigation of Mechanical Stress in Si Device Structures and Phonons in SiGe. Ph.D. Thesis, Brandenburg University of Technology, Cottbus-Senftenburg, Germany, 2000; p. 24.
24. Danilewsky, A.N.; Wittge, J.; Hess, A.; Cröll, A.; Rack, A.; Allen, D.; McNally, P.J.; dos Santos Rolo, T.; Vagovic, P.; Baumbach, T.; et al. Real-time X-ray diffraction imaging for semiconductor wafer metrology and high temperature in situ experiments. *Phys. Stat. Sol. A* **2011**, *208*, 2499–2504. [CrossRef]

Communication

A Study of Extended Defects in Surface Damaged Crystals

Claudio Ferrari [1,*], Corneliu Ghica [2] and Enzo Rotunno [1]

[1] IMEM Institute, National Research Council, 43124 Parma, Italy; enzo.rotunno@imem.cnr.it
[2] National Institute of Materials Physics, 077125 Magurele, Romania; cghica@infim.ro
* Correspondence: Claudio.Ferrari@imem.cnr.it; Tel.: +39-052-126-9223

Received: 7 November 2017; Accepted: 24 January 2018; Published: 30 January 2018

Abstract: We have analyzed by transmission electron microscopy silicon and GaAs crystals polished with sandpapers of different grain size. The surface damage induced a crystal permanent convex curvature with a radius of the order of a few meters. The curvature is due to a compressive strain generated in the damaged zone of the sample. Contrary to what was reported in the literature, the only defects detected by transmission electron microscopy were dislocations penetrating a few microns from the surface. Assuming the surface damage as a kind of continuous indentation, a simple model able to explain the observed compressive strain is given.

Keywords: curved crystals; surface damaged crystals; dislocation generation; crystal indentation

---

1. Introduction

Curved crystals may be used for the focalization of hard X- and gamma-rays through diffraction for gamma-ray astronomy [1], nuclear medicine [2], neutron beams conditioning [3], and X-ray microscopy [4]. According to X-ray diffraction theory, ideal mosaic crystals can achieve the maximum diffraction efficiency [5], intended as the integrated reflectivity of the diffraction profile. Nevertheless, the difficulty of obtaining the desired mosaic spread and mosaic structure makes the diffraction efficiency of real crystals much lower than that of ideal mosaic crystals.

A method for improving the diffraction efficiency is based on slightly bending the crystals. The curvature of lattice planes not only increases the angular interval of incident rays that are diffracted by the crystal but also enhances the diffracted peak intensity [6].

The application of a mechanical bending is not a practical method in optical systems where a large number of crystals is necessary. A possible method to produce curved crystals with permanent curvature without any external applied force is the controlled damaging of a crystal surface [7]. For instance, by surface damaging, bending with a uniform radius of curvature of 40 m of gallium arsenide, germanium, and silicon plates 2 mm in thickness could be achieved, as required for focusing gamma rays in a Laue lens for gamma ray astronomy [8,9].

Despite the empirical approach and the simplicity of the method, precise relationships have been found between the sandpaper grit size, the crystal thickness, and the induced curvature [10]:

- The curvature obtained is always convex, as seen from the damaged side of the crystal, and the same treatment induces curvatures depending on the orientation of the crystal surface and on the polarity of the crystal: a spherical curvature in (001)-oriented Si or Ge crystals and an elliptical curvature in (001)-oriented GaAs or InP crystals.
- The curvature $1/R$ is inversely proportional to the square of the sample thickness.

From these observations, it is clear that the curvature is induced by the formation of a compressive strained layer a few micrometers thick in the damaged zone [10]. To explain the formation of such a compressive strain after surface damage, different mechanisms have been proposed:

- the formation of an amorphous layer near the surface of the crystal generated by the relevant forces acting of the tips of the grains [11];
- the formation of cracks, which is the complete detachment of crystalline planes near the surface of the crystals [12,13]

To understand the formation of the compressive strain in the damaged zone, we analyzed by Transmission Electron Microscopy (TEM) several cross-sectional samples of treated crystals.

2. Experimental and Observation

2.1. Preparation of Crystals

Slices of different thicknesses of GaAs crystals grown by the liquid encapsulated Czochralski were cut from the ingots perpendicularly to the <100> growth direction, and the saw damaging was removed by a chemical etching with an HCl/HNO3 1:1 solution. Commercial GaAs samples were also used.

The surface damaging was obtained by means of a mechanical lapping process on one side of the planar samples. Two polishing machines were used: a Buehler Ecomec 4 (and a Buelher Vibromet 2 (Buehler, Uzwil, Zwizerland). Both are very versatile machines that allow one to produce different deformations by changing the grit of the sandpaper, the pressure per unit area applied on the samples, and the duration of the treatment. In the first machine, the samples were mounted on a plate with paraffin, and the plate was then positioned upside down facing the sandpaper plate. The two components rotate independently on two different axes so that the sample abrasion should be completely uniform on the entire surface of the sample plate. The sample assembling in the Vibromet 2 is similar, but in this case the sample holder is free to move on a vibrating plate covered with sandpaper. The samples were lapped for 10 min with P400 sandpapers, corresponding to an approximate grain size of 35 μm.

2.2. TEM Analysis

A TEM investigation has been conducted on a (001)-oriented GaAs sample, 500 μm thick, treated with a sandpaper P400 for 10 min which permitted to obtain a curvature radius R = 2.8 m.

The sample cross section, parallel to the (110) planes of the crystal, was mechanically polished down to ca. 300 μm. The resulting foils were ion-sputtered with a Gatan® DuoMill TM model 600 (Gatan, Pleasanton, CA, USA), to reach electron transparency. The final sample thickness was estimated to be between 70 and 120 nm, as evaluated by the extinction length of the electron beam in the GaAs lamella. A final gentle ion milling procedure was applied to the cross-section's thin lamella using a Gatan PIPS installation operated at a 3 kV accelerating voltage and 7° incidence angle.

A series of TEM micrographs recorded from neighboring areas are assembled in the panoramic image of Figure 1, which illustrates the effect of the treatment on the (001) surface, which is located on the right-hand side of the image. We did not observe defects such as cracks or inclusions in any of the observed samples but only short straight dislocation segments from the surface to ca. 3 μm in depth. We have no evidence of a formation of an amorphous phase near the surface of the sample as observed for instance in TEM specimens prepared by a focused ion beam [14,15], even if it is not possible to exclude the local formation of an amorphous layer a few nanometers thick during surface damage treatment.

The near-surface region shows a high defect density that extends for a few hundred nanometers. Below this first layer, the density of dislocations drastically decreases, and the dislocations, which appear as straight dark lines, can be observed singularly.

It is worth noting that two sets of dislocation lines can be identified in Figure 1b. The two sets are parallel to the (−111) and (1−11) planes, respectively, perpendicular to the (110) surface of the cross-section sample. Looking carefully at the dislocation segments, it appears that many dislocations cross the TEM sample, i.e., the dislocation lines are not parallel to the surface of the cross section.

Two families of parallel dislocations can be identified in Figure 1a,b. By comparing the micrograph in Figure 1b with the corresponding diffraction pattern in Figure 1c, it turns out that the habit planes of the two families of dislocations are (1−11) and (−111).

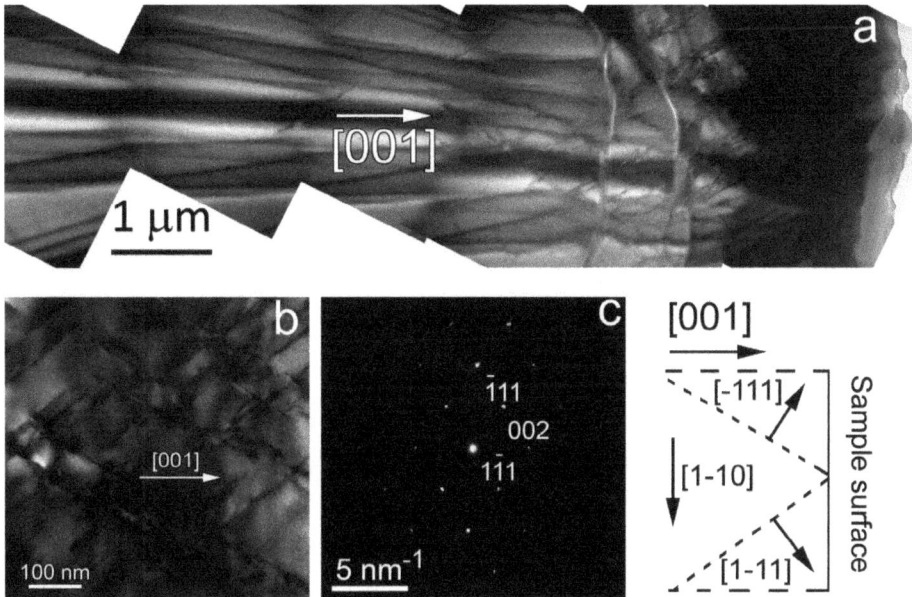

Figure 1. (**a**) Composed TEM image illustrating the material depth affected by the surface treatment and the density of dislocations along the first 7 μm below the surface. (**b**) TEM micrograph and the corresponding SAED pattern (**c**) in almost the [110] zone axis orientation from an area close to the wafer surface showing the formation of two families of dislocations included in the planes (−111) and (1−11).

By tilting the sample in such a way that only the (−111) and (1−11) reflections are excited, the two families of dislocations become respectively extinguished. It follows that, according to the $\vec{g} \cdot \vec{b} = 0$ invisibility criterion, the Burgers vectors of these dislocations are also parallel to the (−111) and (1−11) crystallographic planes, respectively.

Considering the dislocations with lines parallel to the (1−11) planes, based on the orientation-imaging conditions, it turns out that their Burgers vectors are parallel to the same planes. In the cubic structure of GaAs, an easy slip system is of the type {111}<110>, and the most common dislocations are perfect dislocations with Burgers vector a/2<110>. In the following analysis, we will assume only dislocations of this type. This may appear a critical limit of the model, but the assumption is justified by the fact that

- these dislocations have the lowest elastic energy among perfect dislocations in the face-centered-cubic (fcc) crystals;
- non-perfect dislocations are always associated to stacking faults increasing the elastic energy and limiting their mobility;
- these dislocations are typically observed in indented fcc crystals.

It follows that the Burgers vectors may have one of the following six values: a/2[110], a/2[011], a/2[−101], a/2[−1−10], a/2[0−1−1], and a/2[10−1], all of which are contained in the (1−11) plane.

A similar analysis is valid for the other set of dislocations with Burgers vectors lying in the (−111) habit plane. It follows that their Burgers vector may have one of the following six values: a/2[110], a/2[101], a/2[0−11], a/2[01−1], a/2[−1−10], and a/2[−10−1], all of which are contained in the (−111) plane.

The fact that the A and B dislocations are not extinguished simultaneously leads to the conclusion that the Burgers vector cannot be a/2[110] or a/2[−1−10]. In other words, the Burgers vector of the dislocations are not parallel to the electron beam and therefore not parallel to the wafer surface, the (001) plane. The Burgers vectors of the analyzed dislocations are one of the remaining four vectors—a/2[011], a/2[−101], a/2[0−1−1], or a/2[10−1] for the first set and a/2[101], a/2[0−11], a/2[01−1], or a/2[−10−1] for the second set. This means that all the Burgers vectors have a component $b^{||}$ parallel to the (001) surface of the processed wafer and a component $b^{\square}$ oriented perpendicular to the surface.

3. Model

To explain the compressive strain resulting from the surface grinding, we consider that the polishing by sandpaper behaves as a continuous indentation extended along the whole surface. Indentation in GaAs as in other fcc crystals results in the formation of rosettes made of dislocation loops gliding on (111) planes [16,17]. The sketch of the proposed mechanism is reported in Figure 2.

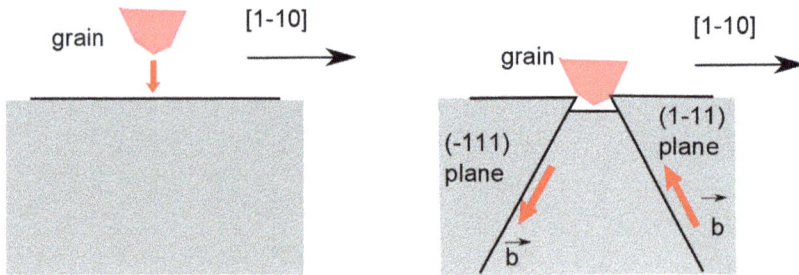

Figure 2. (**left**) Movement of the tip of a grain of the grinding paper in a direction perpendicular to the (001) surface of the crystal. (**right**) The indentation shifts part of the crystal below the original surface. The figure also shows the projection of the Burgers vector of the induced dislocations in the plane of the figure.

The indentation induces the glide of dislocations along the (1−11) planes for the set of dislocations on the right of the tip and along the (1−11) planes for the set on the left of the tip. In this illustration, we neglect for the sake of simplicity the set of dislocations that are also formed and glide on the (111) and (−1−11) planes inclined with respect to the plane of the figure. The Burgers vector components are deduced by the Burgers circuit. The Burgers vectors of the dislocations gliding on the (−111) and (1−11) planes must have opposite values of the Burgers vector component perpendicular to the surface and equivalent values of the Burgers vector component parallel to the crystal surface. According to Figure 3, the Burgers vector generated by the indentation should be

$$\vec{b_1} = \frac{a}{2}[\overline{1}0\overline{1}] \, \vec{b_2} = \frac{a}{2}[01\overline{1}] \tag{1}$$

for the (−111) glide plane and

$$\vec{b_3} = \frac{a}{2}[\overline{1}01] \, \vec{b_4} = \frac{a}{2}[011] \tag{2}$$

for the (1−11) plane. The identified Burgers vectors belong to the two sets deduced by extinction contrast in transmission electron micrographs, thus confirming the validity of the approach.

The Burgers vector have a common edge component parallel to the crystal surface, corresponding to the insertion of extra half planes form the surface of the crystal as shown in Figure 4. For a quantitative evaluation of the misfit induced by the dislocation glide, it is necessary to know the dimension of dislocation loops, a parameter not determined in the present study. A qualitative evaluation of the induced mismatch may be obtained by considering the density ρ of extra half planes as equivalent to the linear dislocation density ρ given by TEM observations. Based on Figure 1a,b, we may estimate $\rho \approx 10\ \mu m^{-1}$, leading to a misfit f value of the dislocated layer with respect to the unperturbed crystal:

$$f = \rho \cdot \frac{a}{2} \approx 8 \times 10^{-3}. \qquad (3)$$

This results in a positive misfit of the damaged zone with respect to the part of the crystal free of dislocations, which induces a convex curvature to the crystal. Assuming a damaged layer thickness $t = 2\ \mu m$ and the misfit f value given by Equation (3) from the Stoney [18] equation, we obtain

$$R = \frac{T^3}{6ft(T-t)} \approx \frac{T^2}{6ft} = 2.6\ m \qquad (4)$$

in which R is the curvature radius, T the crystal thickness, and f and t the misfit and thickness of the dislocated layer. Despite the rough approximation of the linear density of extra half planes, the result is in acceptable agreement with experimental data. The TEM analysis does not allow for the complete exclusion of the formation of an amorphous phase layer some nanometers thick at the top of the surface-treated samples. In principle, such a layer could contribute to the strain formation leading to a bending of the sample as described by Equation (4). Nevertheless, even by assuming a large mismatch between the amorphous phase and the underlying GaAs crystal, the effect of such a strained thin layer may be considered negligible with respect to the contribution originating from a 2-μm-thick dislocated layer.

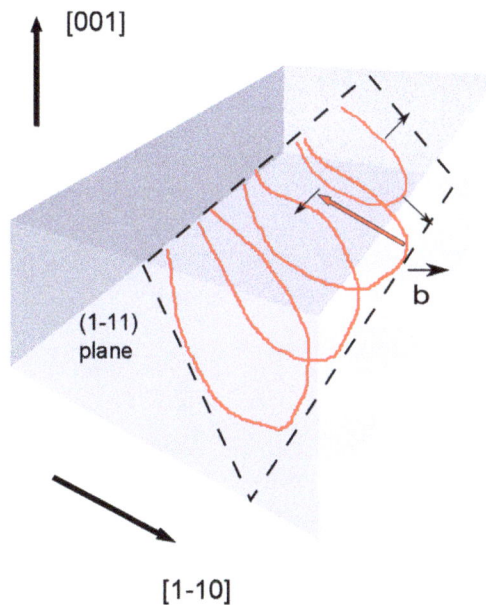

Figure 3. Loop formation and expansion due to surface damage and Burgers vector a/2[011] of a dislocation generated by indentation and gliding on the (1−11) plane.

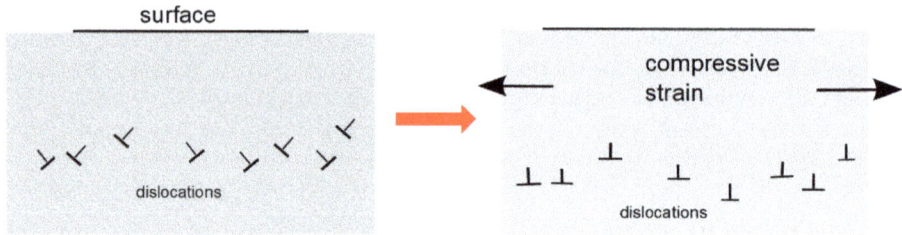

Figure 4. Sketch of the Burgers vectors of the dislocations generated by surface damaging. On the left, a randomly distributed Burgers vector is shown with a common component parallel to the surface. On the right, only the parallel component is shown. Each dislocation line corresponds to the insertion of extra half planes form the surface of the crystal, resulting in a compressive strain of the dislocated zone.

4. Conclusions

Crystals with a (001) surface damaged using sandpaper have been characterized by transmission electron microscopy in cross-sectional geometry. After the surface damaging, the samples exhibited a roughness comparable to the grit of sandpaper used, and a network of straight dislocation segments belonging to the (−111) and (1−11) glide planes inclined to the (001) surface of the crystal extended to a depth of 3 μm, comparable to the dimension of the grains of the sandpaper. No other type of defects, such as cracks or inclusions were detected.

By changing the diffraction condition, the interpretation of extinction contrast permitted the establishment that the Burgers vector of the generated dislocation are parallel to the (−111) and (1−11) glide planes.

A simple model for the formation of a compressive strain based on the dislocation glide in the damaged layer is given, in agreement with the measured sample convex curvature. Moreover, the elliptical curvature observed in surface-treated polar crystals such as GaAs and InP may be explained by the different glide velocity of dislocations on polar (111) glide planes.

We can conclude that the origin of sample curvature is the network of dislocations introduced by the damaging, considered as a continuous indentation of the crystal.

Acknowledgments: The authors acknowledge the CERIC-ERIC Consortium for the access to experimental facilities and financial support, under proposal number 20162066. CG acknowledges the financial support from the Project # PN16-480103 within the Core Program.

Author Contributions: C.F. and E.R. conceived and designed the experiments; C.G. performed the experiments; C.G. and E.R. analyzed the data; C.F. wrote the paper.

Conflicts of Interest: The authors declare no conflict of interest.

References

1. Ferrari, C.; Buffagni, E.; Bonnini, E.; Zappettini, A. X-ray diffraction efficiency of bent GaAs mosaic crystals for the LAUE project. *Proc. SPIE* **2013**, *88610D*. [CrossRef]
2. Roa, D.E.; Smither, R.K.; Zhang, X.; Nie, K.; Shie, Y.Y.; Ramsinghani, N.S.; Milone, N.; Kuo, J.V.; Redpath, J.L.; Al-Ghazi, M.S.A.L.; et al. Crystal diffraction lens for medical imaging. *Exp. Astron.* **2005**, *20*, 229. [CrossRef]
3. Courtois, P.; Bigault, T.; Andersen, K.H.; Baudin-Cavallo, J.; Ben Saïdane, K.; Berneron, M.; El-Aazzouzzi, A.; Gorny, D.; Graf, W.; Guiblain, T.; et al. Status and recent developments in diffractive neutron optics at the ILL. *Phys. B Condens. Matter* **2006**, *385*, 1271–1273. [CrossRef]
4. Borbély, A.; Kaysser-Pyzalla, A.R. X-ray diffraction microscopy: Emerging imaging techniques for nondestructive analysis of crystalline materials from the millimetre down to the nanometre scale. *J. Appl. Cryst.* **2013**, *46*, 295–296. [CrossRef]
5. Zachariasen, W.H. *Theory of X-ray Diffraction in Crystals*; Dover Pub. Inc.: New York, NY, USA, 1945.

6. Malgrange, C. X-ray Propagation in Distorted Crystals: From Dynamical to Kinematical Theory. *Cryst. Res. Technol.* **2002**, *37*, 654–662. [CrossRef]
7. Ferrari, C.; Buffagni, E.; Bonnini, E.; Korytar, D. High diffraction efficiency in crystals curved by surface damage. *J. Appl. Cryst.* **2013**, *46*, 1576–1581. [CrossRef]
8. Virgilli, E.; Frontera, F.; Rosati, P.; Bonnini, E.; Buffagni, E.; Ferrari, C.; Stephen, J.B.; Caroli, E.; Auricchio, N.; Basili, A.; et al. Focusing effect of bent GaAs crystals for γ-ray Laue lenses: Monte Carlo and experimental results. *Exp. Astron.* **2016**, *41*, 307–326. [CrossRef]
9. Liccardo, V.; Virgilli, E.; Frontera, F.; Valsan, V.; Buffagni, E.; Ferrari, C.; Bonnini, E.; Zappettini, A.; Guidi, V.; Bellucci, V.; et al. Study and characterization of bent crystals for Laue lenses. *Exp. Astron.* **2014**, *38*, 401–416. [CrossRef]
10. Buffagni, E.; Ferrari, C.; Rossi, F.; Marchini, L.; Zappettini, A. X-ray diffraction efficiency of bent GaAs mosaic crystals for the Laue project. *Opt. Eng.* **2014**, *53*, 047104.
11. Yan, J.; Asami, T.; Harada, H.; Kuriyagawa, T. Fundamental investigation of subsurface damage in single crystalline silicon caused by diamond machining. *Precis. Eng.* **2009**, *33*, 378–386. [CrossRef]
12. Davim, J.P.; Jackson, M.J. (Eds.) *Nano and Micromachining*; ISTE Ltd.: London, UK; John Wiley & Sons: Hoboken, NJ, USA, 2009; ISBN 978-1-84821-103-2.
13. Haapalinna, A.; Nevas, S.; Pähler, D. Rotational grinding of silicon wafers—Sub-surface damage inspection. *Mater. Sci. Eng. B* **2004**, *107*, 321–331. [CrossRef]
14. Holmström, E.; Kotakoski, J.; Lechner, L.; Kaiser, U.; Nordlund, K. Atomic-scale effects behind structural instabilities in Si lamellae during ion beam thinning. *AIP Adv.* **2012**, *2*. [CrossRef]
15. Korsunsky, A.M.; Guénolé, J.; Salvati, E.; Sui, T.; Mousavi, M.; Prakash, A.; Bitzek, E. Quantifying eigenstrain distributions induced by focused ion beam damage in silicon. *Mater. Lett.* **2016**, *185*, 47–49. [CrossRef]
16. White, J.E. X-ray Diffraction by Elastically Deformed Crystals. *J. Appl. Phys.* **1950**, *21*, 855. [CrossRef]
17. Bradby, J.E.; Williams, J.S.; Wong-Leung, J.; Swain, M.V.; Munroe, P. Mechanical deformation of InP and GaAs by spherical indentation. *Appl. Phys. Lett.* **2001**, *78*, 3235. [CrossRef]
18. Stoney, G.G. The Tension of Metallic Films Deposited by Electrolysis. *Proc. R. Soc.* **1909**, *82*, 172–175. [CrossRef]

crystals

MDPI

Article

Comparative Study of Phase Transformation in Single-Crystal Germanium during Single and Cyclic Nanoindentation

Koji Kosai , Hu Huangand Jiwang Yan *

Department of Mechanical Engineering, Faculty of Science and Technology, Keio University,
Yokohama 2238522, Japan; koji.kosai@keio.jp (K.K.); huanghuzy@163.com (H.H.)
* Correspondence: yan@mech.keio.ac.jp; Tel.: +81-45-566-1445

Academic Editor: Ronald W. Armstrong, Stephen M. Walley and Wayne L. Elban
Received: 30 September 2017; Accepted: 30 October 2017; Published: 1 November 2017

Abstract: Single-crystal germanium is a semiconductor material which shows complicated phase transformation under high pressure. In this study, new insight into the phase transformation of diamond-cubic germanium (dc-Ge) was attempted by controlled cyclic nanoindentation combined with Raman spectroscopic analysis. Phase transformation from dc-Ge to rhombohedral phase (r8-Ge) was experimentally confirmed for both single and cyclic nanoindentation under high loading/unloading rates. However, compared to single indentation, double cyclic indentation with a low holding load between the cycles caused more frequent phase transformation events. Double cyclic indentation caused more stress in Ge than single indentation and increased the possibility of phase transformation. With increase in the holding load, the number of phase transformation events decreased and finally became less than that under single indentation. This phenomenon was possibly caused by defect nucleation and shear accumulation during the holding process, which were promoted by a high holding load. The defect nucleation suppressed the phase transformation from dc-Ge to r8-Ge, and shear accumulation led to another phase transformation pathway, respectively. A high holding load promoted these two phenomena, and thus decreased the possibility of phase transformation from dc-Ge to r8-Ge.

Keywords: single crystal; germanium; nanoindentation; phase transformation; crystal defect; cyclic load

1. Introduction

Pressure-induced formation of various metastable polymorphs takes place in single-crystal germanium (Ge) during diamond anvil cell (DAC) and nanoindentation tests. Previous studies reported that diamond cubic Ge (dc-Ge) transforms to metallic (β-Sn)-Ge at a pressure of ~10 GPa [1–4]. When the pressure releases, this metallic phase transforms to various metastable structures depending on the experimental conditions, such as simple tetragonal phase (st12-Ge) [5–8], rhombohedral phase (r8-Ge) [6,7,9,10], body centered cubic phase (bc8-Ge) [5,6,11], as well as amorphous Ge (a-Ge) and dc-Ge as end phases [8,11]. Another phase of hexagonal diamond Ge (hd-Ge) is also confirmed by compression of a-Ge [9,12,13].

These polymorphs possess very different characteristics from their lowest energy structures, which may open up new applications in future microelectronics and photovoltaics. For this purpose, the complicated phase transformation process of Ge has been widely investigated. For example, in the DAC experiment by Nelmes et al., decompression speed was found to be critical for phase transformation, i.e., slow decompression caused phase transformation to st12-Ge while fast decompression promoted phase transformation to bc8-Ge [5]. Some other studies by nanoindentation on dc-Ge reported more complicated responses against pressure [8,10,11,14]. Slow indentation with a spherical indenter could induce no phase transformation but only plastic deformation by formation of twinning and dislocation slip [8,10,14] which prevented phase transformation by stress release. However, when using a sharp indenter tip, this kind of defect nucleation was suppressed and phase transformations were confirmed [11]. Extremely fast indentation resulted in formation of st12-Ge [8], and low temperature led to formation of r8-Ge or a-Ge [10]. Investigation of the phase transformation in a-Ge was also performed via indentation [14–17] in which phase transformations can be induced easily compared with dc-Ge because of higher resistance of a-Ge against deformation by slip and twinning. According to the results of indentation on a-Ge, two clear phase transformation pathways via (β-Sn)-Ge have been established [14]. The authors stated that the intermediate (β-Sn)-Ge may transform to two end phases depending on whether the transforming region is constrained or not. Unconstrained (β-Sn)-Ge transforms to dc-Ge with possible trace of st12-Ge while the constrained one undergoes a phase transformation to unstable r8-Ge and finally hd-Ge. They mentioned that both of these two types of phase transformation could occur in the same indent for dc-Ge. The same research team also studied phase transformation behaviors of crystalline Ge by DAC experiment [7]. They classified phase transformations of (β-Sn)-Ge on unloading depending on hydrostaticity. Namely, quasihydrostatic conditions resulted in formation of r8-Ge while the presence of shear led to nucleation of st12-Ge. Their findings were consistent with previously reported phase transformations and helpful to explain the detailed mechanisms [7,14].

Even though the basic phase transformation process of Ge has been classified using a-Ge, it is still important to study and explain dominant phase transformation behaviors of dc-Ge because of the complexity of pressure-induced behaviors [14]. For Si, which has the same crystal structure as Ge, similar pressure-induced phase transformations have been widely reported [1–4,6,18–29], and some new insights into the phase transformation behaviors were further revealed by a multi cyclic nanoindentation method [20–24]. In this current paper, nanoindentation responses of single-crystal Ge under single and double cyclic nanoindentation tests were comparatively investigated. Combined with Raman analysis of residual phases in indents, a dominant phase transformation pathway of dc-Ge under repetitive pressure was established, which may open up new insights into pressure-induced phase transformations of crystalline Ge.

2. Experimental Details

A dc-Ge (111) wafer with thickness of 0.9 mm was used for experiments. Nanoindentation tests were performed on an ENT-1100a nanoindentation instrument (Elionix Inc., Tokyo, Japan) equipped with a Berkovich diamond indenter. In this experiment, single and double cyclic nanoindentation tests were carried out according to the protocol as shown in Figure 1 and the parameters in Table 1. Firstly, the indentation load increased to 50 mN under a given loading rate, held for one second, and then decreased under the same unloading rate as loading rate. For comparison, various loading/unloading rates—10, 25, or 50 mN/s—were used. For the single mode, the indentation load returned to 0 mN and then the test finished, while for the double mode, unloading stopped at a controlled residual load (ΔP) and then held for a given load holding time (LHT). As listed in Table 1, ΔP was varied from 0 to 14 mN with LHT of 5 s. Nanoindentation tests with LHT of 20 s were also conducted under ΔP = 2 mN for comparison. After LHT, the second cycle indentation progressed with the same process as single mode from ΔP. For each experimental condition, 20 nanoindentation tests were performed to ensure the reliability of the results.

Residual phases of the all indents were measured by the NRS-3000 Raman micro-spectrometer (JASCO, Tokyo, Japan) with a 532 nm wavelength laser focused to a ~1 μm spot size. To ensure that the residual phases were minimally affected by further phase transformation at room temperature [5,11,13], the measurement of residual phases in indents was performed within three hours after indentation. In previous studies, some specific deformation responses, for example pop-in, pop-out, and elbow have been reported in indentation studies of Si [20–28] and Ge [8,10,11,14–16,23]. In this study, the effects of experimental parameters on these deformation responses were statistically analyzed as well as phase transformation process during both single and cyclic indentation.

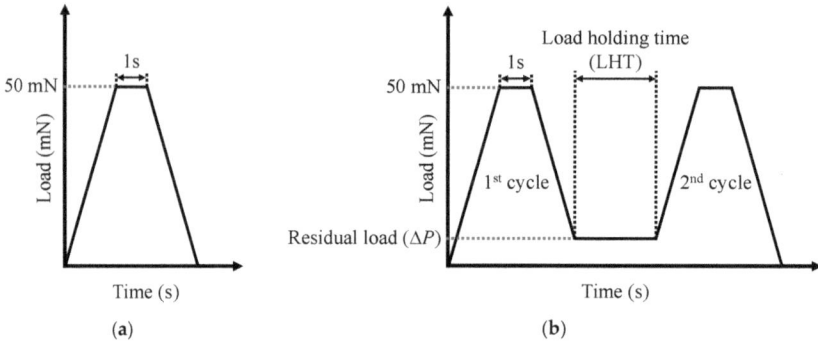

Figure 1. Nanoindentation experimental protocols: (**a**) single mode; (**b**) double cyclic mode.

Table 1. Experimental parameters for each indentation mode.

Indentation Mode	Maximum Load (mN)	Loading/Unloading Rate (mN/s)	ΔP (mN)	Length of LHT (s)
Single	50	10, 25, 50	-	-
Double cyclic			0, 2, 6, 10, 14	5 (or 20 for ΔP = 2 mN)

3. Results and Discussions

3.1. Correlation between Specific Deformation Responses during Nanoindentation Tests and Phase Transformation Behaviors

Figure 2 illustrates typical load-displacement curves obtained under single and double cyclic nanoindentation tests, with shift of the second cycle curves in double cyclic nanoindentation for clarity. For both testing modes, pop-in events marked by arrows in Figure 2 are observed during the loading process under various loading/unloading rates. This event was reported as a result of slip generation in dc-Ge sample and/or phase transformation to (β-Sn)-Ge [10,14]. In addition to the pop-in event during loading, unloading also showed some specific deformation events, i.e., pop-out and elbow as illustrated in Figure 3. The correlation between these events and phase transformation has been established for indentation of Si [23–28] and cold indentation of Ge [10]. At room temperature, ultra-fast indentation of dc-Ge also triggered these phenomena [8], but the correspondence of these events with phase transformation was not clearly addressed. Thus, the correlation between these events and phase transformation behavior will be further discussed later in this section.

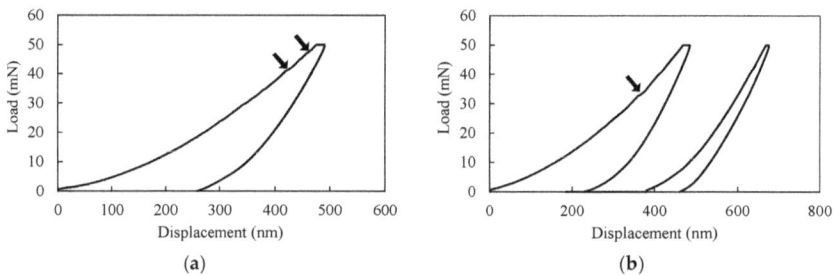

Figure 2. *Cont.*

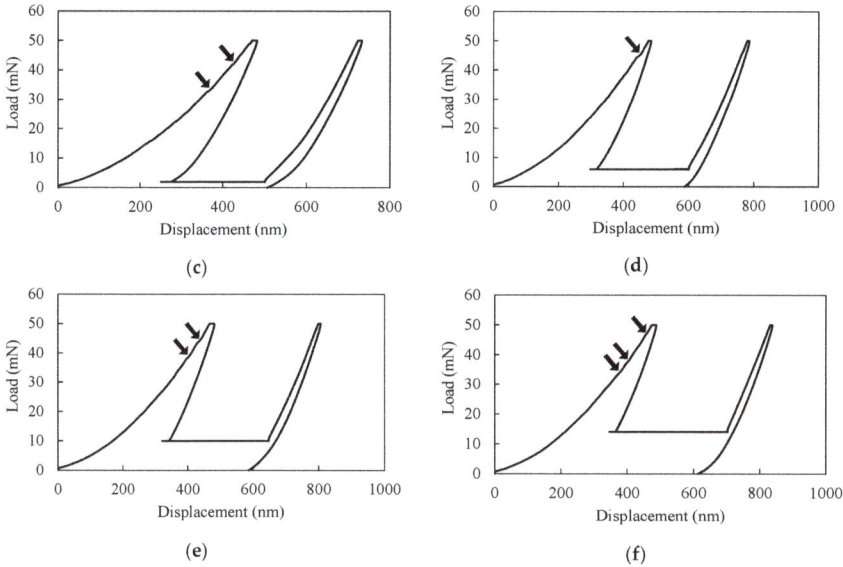

Figure 2. Load-displacement curves obtained by each mode of indentation under loading/unloading rate of 50 mN/s and LHT = 5 s: (**a**) single; (**b**) double, $\Delta P = 0$ mN (no residual load); (**c**) double, $\Delta P = 2$ mN; (**d**) double, $\Delta P = 6$ mN; (**e**) double, $\Delta P = 10$ mN; (**f**) double, $\Delta P = 14$ mN: all graphs of double cyclic indentation include artificial shift between cycles (arrows indicate pop-in events).

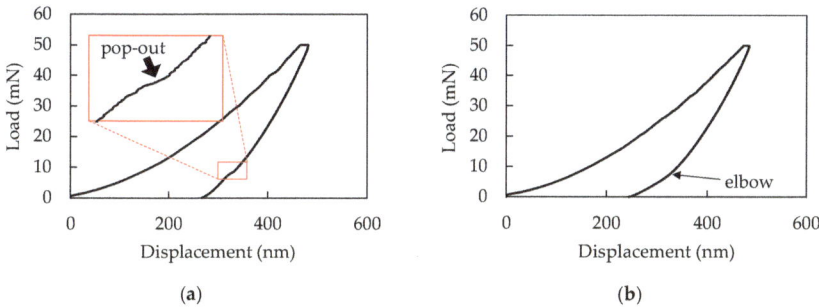

Figure 3. Specific deformation in unloading process: (**a**) pop-out; (**b**) elbow (single, loading/unloading rate of 50 mN/s).

Figure 4 illustrates Raman spectra obtained from residual indents, corresponding to the cyclic nanoindentation experiments in Figure 2. Compared with the main peak of pristine dc-Ge at 301 cm^{-1}, the corresponding peaks obtained from the indents show a little higher wavenumber as a result of compressive stress [14,23]. Furthermore, a broad component below 290 cm^{-1} corresponding to a-Ge and some peaks around 203, 225, and 247 cm^{-1} corresponding to r8-Ge phases [6,9,10,14] are observed. Bc8-Ge also shows very similar peaks to those of r8-Ge [6,11] and thus, it is difficult to distinguish these two phases from Raman data [9]. However, some recent studies indicated that the phase transformation from (β-Sn)-Ge to r8-Ge was dominant [9,14]. Accordingly, it can be concluded that the dominant transformed phase shown in Figure 4 is r8-Ge. Figure 5 presents the detection fraction of r8-Ge for every 20 indents under two loading/unloading rates of 25 and 50 mN/s with a same LHT of 5 s. For the loading/unloading rate of 10 mN/s, phase transformation was not identified by Raman detection.

This is a little different from previous research where formation of bc8-Ge was confirmed under 5 mN/s indentation [11] and it might be because of different detection resolutions for the transformed phase by Raman spectrometry. If the phase transformed region is only a part of the indented region, it may be difficult to position exactly the Raman spot on the phase transformed region with 1 μm resolution and 1 μm laser spot. However, in Figure 5, it is clear that faster indentation promoted phase transformation compared with the slower case under all conditions. This is due to an increase in loading rate, where the critical load for defect nucleation also increases, thus phase transformation becomes the dominant mechanism rather than defect nucleation in a crystalline Ge sample [8].

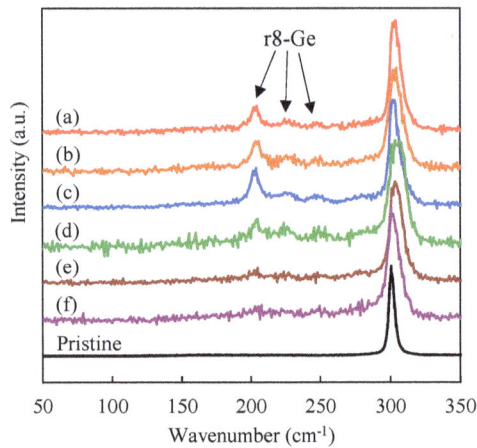

Figure 4. Raman spectra obtained from residual indents corresponding to the cyclic nanoindentation experiments in Figure 2.

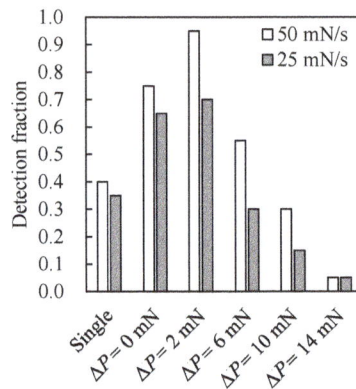

Figure 5. Detection fraction of r8-Ge under various experimental conditions (LHT = 5 s) (no detection under 10 mN/s).

Next, the correlation between the observed phase transformation to r8-Ge via (β-Sn)-Ge and the specific deformation events of pop-in, pop-out and elbow in this study will be discussed for the single indentation mode. For pop-in events, two mechanisms are suggested as the causes: slip generation in dc-Ge sample and/or phase transformation to (β-Sn)-Ge [10,14]. In this case, no correlation between r8-Ge formation and the pop-in events was confirmed, especially under the lowest loading/unloading rate of 10 mN/s which never caused phase transformation to r8-Ge in this study even though pop-in events were frequently observed. Hence, the dominant cause of the observed pop-in events is expected to be defect nucleation rather than phase transformation. There is another possibility for the cause of pop-in events by phase transformation to intermediate (β-Sn)-Ge then ending up as dc-Ge, which was reported in indentation on a-Ge [14–17]. However, it is impossible to distinguish this possibility in our experiment because the end phase is the same as the initial phase.

The above discussion indicates that the pop-in events probably do not result from the phase transformation to (β-Sn)-Ge in this case. For an elbow on unloading, amorphization was reported as the reason [10,28], but it is difficult to assess the correlation between these two phenomena in this study because of the difficulty in judging the occurrence of amorphization which only showed a weak broad band on Raman spectra in this case. On the other hand, for pop-out events, the correlation exists that all indents with r8-Ge formation are accompanied with pop-out events, although some indents show only pop-out events without the observation of phase transformation to r8-Ge. In addition, faster loading/unloading rates promoted occurrence of pop-out events (13/20 for 50 mN/s; 11/20 for 25 mN/s; 4/20 for 10 mN/s) as well as phase transformation to r8-Ge as shown in Figure 5. These facts imply that the correlation between r8 phase formation and pop-out events exist as mentioned in studies of indentation of Si [23–28] and cold indentation on dc-Ge [10]. The appearance of pop-out events alone without phase transformation in some indents might result from insufficient resolution of the Raman spectrometer. Some other possibilities may still exist, for example, phase transformation to dc-Ge from (β-Sn)-Ge also affected the occurrence of pop-out events, but additional experiments with more detailed observations are required to distinguish the cause.

3.2. Effects of Indentation Modes on Phase Transformation Behaviors

Another clear tendency for the phase transformation behavior obtained from Figure 5 is that double cyclic indentation tests with lower residual load ΔP promoted the phase transformation to r8-Ge compared with single indentation. In the double cyclic indentation, r8-Ge was confirmed. Even though the correlation between phase transformation and pop-out events for the single mode was confirmed in Section 3.2, this kind of correlation did not exist for double cyclic indentation with low ΔP. In other words, phase transformation to r8-Ge was confirmed regardless of pop-out events in the double cyclic mode. On the other hand, the detection fraction decreases with increase in ΔP and finally becomes obviously lower than single indentation. This indicates that double cyclic indentation can promote phase transformation to r8-Ge, independent of pop-out response during unloading, although phase transformation is suppressed by high residual load ΔP. The effect of residual load on the formation of r8-Ge is further confirmed from the result of different LHT experiments shown in Figure 6. Even under a small ΔP of 2 mN, longer LHT increases the influence of residual load, showing a decreased detection fraction of r8-Ge.

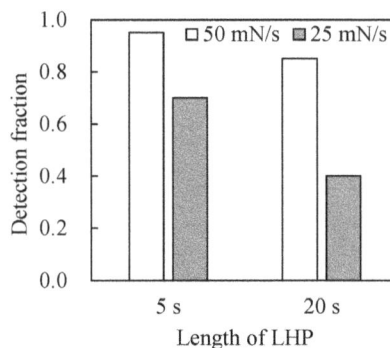

Figure 6. Detection fraction of r8-Ge under different LHT (ΔP = 2 mN, loading/unloading rate of 50 mN/s).

The mechanism for promoting the formation of r8-Ge by multi cyclic nanoindentation can be explained by referring to a previous study on Si [20]. In that study, during multi cyclic nanoindentation, it was suggested that the transformed r8-Si/bc8-Si region expanded during subsequent cycles and finally a larger r8-Si/bc8-Si region was formed compared to single indentation [20]. This mechanism can be employed to this study. Namely, compared to single indentation, double cyclic indentation with small ΔP can expand the transformed region and result in a higher detection fraction of r8-Ge. The lower detection of r8-Ge under $\Delta P = 0$ mN compared with $\Delta P = 2$ mN is possibly caused by an error of noncontact between the indenter tip and sample surface. The case for $\Delta P = 0$ mN means that the load was completely removed from the sample, and the indenter tip might leave the sample surface. In such a case, friction might occur at the second contact and cause shear in the Ge sample, which could lead to phase transformation to dc-Ge or st12-Ge [7,14,17]. It should be noted that 1 of 20 indents with a residual load of 0 mN and loading/unloading rate of 50 mN/s has peaks around 195, 250, and 280 cm^{-1} corresponding to st12-Ge. The phase transformation caused by shear does not contain r8-Ge [7,14,17], so friction during LHT might prevent the formation of r8-Ge under $\Delta P = 0$ mN compared with $\Delta P = 2$ mN even though the phase transformation to r8-Ge was still dominant compared to other higher ΔP. However, there may well be other explanations for the lower r8-Ge probability for the 0 mN LHT case, and further experiments would be needed to resolve this issue.

Although the positive effect of double cyclic indentation for promoting phase transformation is explained above by referring to the case of Si, the negative effect of increasing residual holding load should be explained in a different way from Si because a high residual load over 10 mN also promotes phase transformation for Si [21,22]. For Ge, two different mechanisms to prevent the formation of r8-Ge might be expected to occur such as nucleation of defects or phase transformation to other phases. A recent indentation study of Si with maximum load held for long times indicated that longer holding time could promote either of defect nucleation or phase transformation [29]. Similar phenomena to this previous study possibly occurred during LHT in the current study. For Ge, defect nucleation is more dominant under slow indentation of dc-Ge [8,14] which is similar to the case of LHT. So, more defects could be nucleated in non-transformed dc-Ge underneath the indenter tip during the LHT process for the case of higher ΔP. Phase transformation and defect nucleation are competitive [10,29], so expansion of the defect nucleated region means a reduction of the transformed region at the second cycle. Therefore, higher ΔP could prevent additional phase transformation during the second cycle, which could be a reason for lower detection fraction of r8-Ge under higher ΔP. On the other hand, in the region which had transformed at the first cycle, accumulation of shear pressure is expected to occur under high ΔP because of separation between phase transformation and defect nucleation regions. Under a high shear and strain condition, phase transformation to st12-Ge may became more probable [7,14,17], so the detection fraction of phase transformation to r8-Ge would decrease, even being less than that for a single indentation. However, the first explanation, that of defect nucleation during the LHT process, is expected to be more likely than st12-Ge formation.

To further verify this explanation, the indentation displacement at the beginning of the second cycle unloading and the residual displacement after the second cycle as shown in Figure 7a were statistically analyzed, and the results are presented in Figure 7b,c respectively. It is noted that even though the indentation displacement at the beginning of the second cycle unloading is almost the same with only a few nm difference as shown in Figure 7b, the residual displacement after the second cycle in Figure 7c gradually decreases with increasing ΔP except that of 0 mN which might contain some error due to re-contact as mentioned above. This is assumed to be caused by lattice volume differences. Relative volume of r8-Ge is ~10% smaller than that of dc-Ge [30]. This implies that dc-Ge with defect nucleation results in a larger difference in volume than that of the denser r8-Ge phase. Such a volume change caused by phase transformation was previously indicated in a DAC experiment [31]. This is consistent to our expectation that the dominant end phase shifted to dc-Ge with defects from r8-Ge with increasing ΔP, leading to smaller indentation displacement after the second cycle as shown in Figure 7c because of larger relative volume of dc-Ge end phase than r8-Ge.

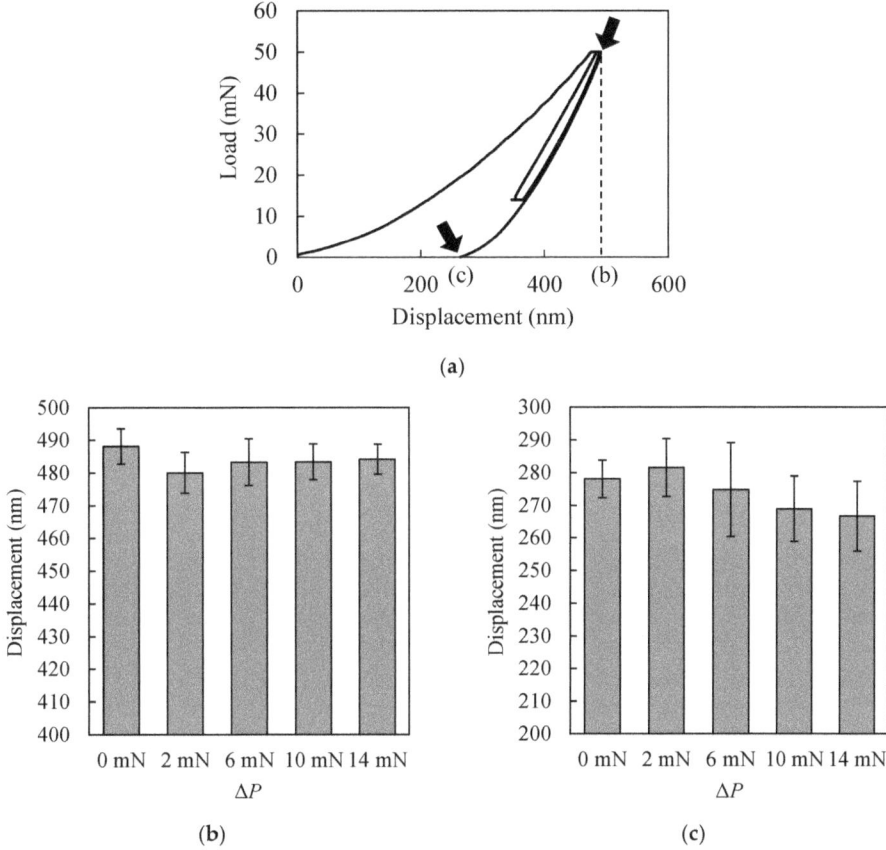

Figure 7. (**a**) An example of indentation load-displacement curve obtained under double cyclic indentation (real data without artificial shift); (**b**) the indentation displacement of the beginning of the second cycle unloading; (**c**) the indentation displacement after the second cycle.

According to the analysis mentioned above, possible phase transformation pathways during double cyclic nanoindentation corresponding to low and high residual loads are illustrated in Figure 8a,b, respectively. During loading of the first cycle, typical phase transformation to (β-Sn)-Ge [1–4] and defect nucleation in the non-transformed region [8,10,14] occurs. After unloading of the first cycle, the transformed region is a mixture of (β-Sn)-Ge and other transformed phases [5–11]. For the case of low residual load in Figure 8a, there is little defect nucleation in the surrounding region and little shear accumulation in transformed region during LHT. This enables phase transformation to r8-Ge during the second cycle and a larger transformed region than a single indentation. As a result, r8-Ge is more frequently detected as the end phase. On the other hand, high residual load causes much defect nucleation and shear accumulation as shown in Figure 8b, which limits the extension of the transformed region during the second cycle. In addition, the accumulated shear results in more defective dc-Ge than r8-Ge. These processes lead to less formation of r8-Ge.

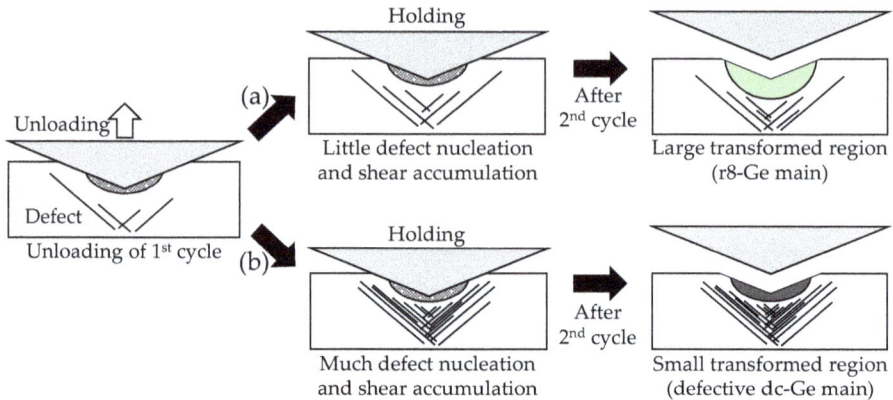

Figure 8. Possible phase transformation pathways during double cyclic nanoindentation: (**a**) case of low residual load during LHP; (**b**) case of high residual load during LHP.

4. Conclusions

Phase transformation behaviors in single-crystal Ge were comparatively investigated under single and double cyclic nanoindentation tests. The experimental results indicated that fast double cyclic indentation with low holding residual load remarkably promoted phase transformation from dc-Ge to r8-Ge regardless of the fact that deformation responses were different from single indentation. However, the occurrence of phase transformation decreased with increase in the residual load between the 1st and 2nd cycle. The possible reason is that, although fast and double cyclic indentation with low residual load expands the transformed r8-Ge region by repetitive loading and unloading, the high residual load prevents the expansion of transformed region by further defect nucleation and accumulation of shear which promote more phase transformation to another phase than r8-Ge. This study is expected to give new insights into high pressure phase transformation mechanisms in single-crystal Ge.

Acknowledgments: This research presentation is supported in part by a research assistantship of a Grant-in-Aid to the Program for Leading Graduate School for "Science for Development of Super Mature Society" from the Ministry of Education, Culture, Sport, Science, and Technology in Japan.

Author Contributions: Koji Kosai and Hu Huang conceived and designed the experiments; Koji Kosai performed the experiments; Koji Kosai and Hu Huang analyzed the data; Koji Kosai wrote the paper; Hu Huang and Jiwang Yan supervised the whole study and revised the paper.

Conflicts of Interest: The authors declare no conflict of interest.

References

1. Jamieson, J.C. Crystal Structures at High Pressures of Metallic Modifications of Silicon and Germanium. *Science* **1963**, *139*, 762–764. [CrossRef] [PubMed]
2. Minomura, S.; Drickamer, H.G. Pressure Induced Phase Transitions in Silicon, Germanium and Some III–V Compounds. *J. Phys. Chem. Solids* **1962**, *23*, 451–456. [CrossRef]
3. Gerk, A.P.; Tabor, D. Indentation Hardness and Semiconductor–Metal Transition of Germanium and Silicon. *Nature* **1978**, *271*, 732–733. [CrossRef]
4. Pharr, G.M.; Oliver, W.C.; Cook, P.F.; Kirchner, P.D.; Kroll, M.C.; Dinger, T.R.; Clarke, D.R. Electrical Resistance of Metallic Contacts on Silicon and Germanium during Indentation. *J. Mater. Res.* **1992**, *7*, 961–972. [CrossRef]
5. Nelmes, R.J.; McMahon, M.I.; Wright, N.G.; Allan, D.R.; Loveday, J.S. Stability and Crystal Structure of BC8 Germanium. *Phys. Rev. B* **1993**, *48*, 9883–9886. [CrossRef]

6.   Kailer, A.; Nickel, K.G.; Gogotsi, Y.G. Raman Microspectroscopy of Nanocrystalline and Amorphous Phases in Hardness Indentations. *J. Raman Spectrosc.* **1999**, *30*, 939–946. [CrossRef]

7.   Haberl, B.; Guthrie, M.; Malone, B.D.; Smith, J.S.; Sinogeikin, S.V.; Cohen, M.L.; Williams, J.S.; Shen, G.; Bradby, J.E. Controlled Formation of Metastable Germanium Polymorphs. *Phys. Rev. B* **2014**, *89*, 1–6. [CrossRef]

8.   Oliver, D.J.; Bradby, J.E.; Williams, J.S.; Swain, M.V.; Munroe, P. Rate-Dependent Phase Transformations in Nanoindented Germanium. *J. Appl. Phys.* **2009**, *105*, 1–3. [CrossRef]

9.   Johnson, B.C.; Haberl, B.; Deshmukh, S.; Malone, B.D.; Cohen, M.L.; McCallum, J.C.; Williams, J.S.; Bradby, J.E. Evidence for the R8 Phase of Germanium. *Phys. Rev. Lett.* **2013**, *110*, 1–5. [CrossRef] [PubMed]

10. Huston, L.Q.; Kiran, M.S.R.N.; Smillie, L.A.; Williams, J.S.; Bradby, J.E. Cold Nanoindentation of Germanium. *Appl. Phys. Lett.* **2017**, *111*, 1–4. [CrossRef]

11. Jang, J.; Lance, M.J.; Wen, S.; Pharr, G.M. Evidence for Nanoindentation-Induced Phase Transformations in Germanium. *Appl. Phys. Lett.* **2005**, *86*, 1–3. [CrossRef]

12. Xiao, S.Q.; Pirouz, P. On Diamond-Hexagonal Germanium. *J. Mater. Res.* **1992**, *7*, 1406–1412. [CrossRef]

13. Williams, J.S.; Haberl, B.; Deshmukh, S.; Johnson, B.C.; Malone, B.D.; Cohen, M.L.; Bradby, J.E. Hexagonal Germanium Formed via a Pressure-Induced Phase Transformation of Amorphous Germanium under Controlled Nanoindentation. *Phys. Status Solidi Rapid Res. Lett.* **2013**, *7*, 355–359. [CrossRef]

14. Bradby, J.E.; Williams, J.S.; Wong-Leung, J.; Swain, M.V.; Munroe, P. Nanoindentation-Induced Deformation of Ge. *Appl. Phys. Lett.* **2002**, *80*, 2651–2653. [CrossRef]

15. Deshmukh, S.; Haberl, B.; Ruffell, S.; Munroe, P.; Williams, J.S.; Bradby, J.E. Phase Transformation Pathways in Amorphous Germanium under Indentation Pressure. *J. Appl. Phys.* **2014**, *115*, 1–10. [CrossRef]

16. Oliver, D.J.; Bradby, J.E.; Ruffell, S.; Williams, J.S.; Munroe, P. Nanoindentation-Induced Phase Transformation in Relaxed and Unrelaxed Ion-Implanted Amorphous Germanium. *J. Appl. Phys.* **2009**, *106*, 1–6. [CrossRef]

17. Patriarche, G.; Le Bourhis, E.; Khayyat, M.M.O.; Chaudhri, M.M. Indentation-Induced Crystallization and Phase Transformation of Amorphous Germanium. *J. Appl. Phys.* **2004**, *96*, 1464–1468. [CrossRef]

18. Clarke, D.R.; Kroll, M.C.; Kirchner, P.D.; Cook, R.F.; Hockey, B.J. Amorphization and Conductivity of Silicon and Germanium Induced by Indentation. *Phys. Rev. Lett.* **1988**, *60*, 2156–2159. [CrossRef] [PubMed]

19. Haberl, B.; Aji, L.B.B.; Williams, J.S.; Bradby, J.E. The Indentation Hardness of Silicon Measured by Instrumented Indentation: What does It Mean? *J. Mater. Res.* **2012**, *27*, 3066–3072. [CrossRef]

20. Huang, H.; Yan, J. On the Mechanism of Secondary Pop-Out in Cyclic Nanoindentation of Single-Crystal Silicon. *J. Mater. Res.* **2015**, *30*, 1861–1868. [CrossRef]

21. Huang, H.; Yan, J. New Insights into Phase Transformations in Single Crystal Silicon by Controlled Cyclic Nanoindentation. *Scr. Mater.* **2015**, *102*, 35–38. [CrossRef]

22. Huang, H.; Yan, J. Volumetric and Timescale Analysis of Phase Transformation in Single-Crystal Silicon during Nanoindentation. *Appl. Phys. A* **2016**, *122*, 1–11. [CrossRef]

23. Gogotsi, Y.G.; Domnich, V.; Dub, S.N.; Kailer, A.; Nickel, K.G. Cyclic Nanoindentation and Raman Microspectroscopy Study of Phase Transformations in Semiconductors. *J. Mater. Res.* **2000**, *15*, 871–879. [CrossRef]

24. Jian, S.R.; Chen, G.J.; Juang, J.Y. Nanoindentation-Induced Phase Transformation in (1 1 0)-Oriented Si Single-Crystals. *Curr. Opin. Solid State Mater. Sci.* **2010**, *14*, 69–74. [CrossRef]

25. Bradby, J.E.; Williams, J.S.; Wong-Leung, J.; Swain, M.V.; Munroe, P. Transmission Electron Microscopy Observation of Deformation Microstructure under Spherical Indentation in Silicon. *Appl. Phys. Lett.* **2000**, *77*, 3749–3751. [CrossRef]

26. Haq, A.J.; Munroe, P.R. Phase Transformations in (111) Si after Spherical Indentation. *J. Mater. Res.* **2009**, *24*, 1967–1975. [CrossRef]

27. Bradby, J.E.; Williams, J.S.; Wong-Leung, J.; Swain, M.V.; Munroe, P. Mechanical Deformation in Silicon by Micro-Indentation. *J. Mater. Res.* **2001**, *16*, 1500–1507. [CrossRef]

28. Domnich, V.; Gogotsi, Y.; Dub, S. Effect of Phase Transformations on the Shape of the Unloading Curve in the Nanoindentation of Silicon. *Appl. Phys. Lett.* **2000**, *76*, 2214–2216. [CrossRef]

29. Wang, S.; Haberl, B.; Williams, J.S.; Bradby, J.E. The Influence of Hold Time on the Onset of Plastic Deformation in Silicon. *J. Appl. Phys.* **2015**, *118*, 1–6. [CrossRef]

30. Malone, B.D.; Cohen, M.L. Electronic Structure, Equation of State, and Lattice Dynamics of Low-Pressure Ge Polymorphs. *Phys. Rev. B* **2012**, *86*, 1–7. [CrossRef]
31. Menoni, C.S.; Hu, J.Z.; Spain, I.L. Germanium at High Pressures. *Rhys. Rev. B* **1986**, *34*, 362–368. [CrossRef]

crystals

MDPI

Article

Theoretical Study on Electronic, Optical Properties and Hardness of Technetium Phosphides under High Pressure

Shiquan Feng [1,*], Xuerui Cheng [1], Xinlu Cheng [2], Jinsheng Yue [3] and Junyu Li [1]

[1] The High Pressure Research Center of Science and Technology, Zhengzhou University of Light Industry, Zhengzhou 450002, China; 2014079@zzuli.edu.cn (X.C.); 2014830@zzuli.edu.cn (J.L.)
[2] Institute of Atomic and Molecular Physics, Sichuan University, Chengdu 610065, China; 49273185@163.com
[3] Luoyang Sunrui Sprcial Equipment CO., LTD, Luoyang 471003, China; yuejinsheng725@126.com
* Correspondence: fengsq2013@126.com

Received: 4 April 2017; Accepted: 15 June 2017; Published: 18 June 2017

Abstract: In this paper, the structural properties of technetium phosphides Tc_3P and TcP_4 are investigated by first principles at zero pressure and compared with the experimental values. In addition, the electronic properties of these two crystals in the pressure range of 0–40 GPa are investigated. Further, we discuss the change in the optical properties of technetium phosphides at high pressures. At the end of our study, we focus on the research of the hardness of TcP_4 at different pressures by employing a semiempirical method, and the effect of pressure on the hardness is studied. Results show that the hardness of TcP_4 increases with the increasing pressure, and the influence mechanism of pressure effect on the hardness of TcP_4 is also discussed.

Keywords: density functional theory; hardness; electronic property; optical properties

1. Introduction

Due to their excellent performance in cutting, abrasion, and drilling, exploring new types of ultrahard materials has become a subject of inherent interest in recent decades. Until now, the development of superhard materials has gone through stages of traditional superhard materials, second kinds of superhard materials, and novel superhard materials.

As the well-known hardest traditional ultrahard material, diamond is of great interest for application and basic research [1]. However, its limitations in cutting iron and other ferrous metals greatly limits its application [2]. The second types of ultrahard materials are compounds composed of light elements (B–C–N–O system) with strong and short covalent bonds. These compounds have made great progress in cutting iron and other ferrous metals, but the severe preparation conditions and high cost of these compounds limit their large-scale production and application. Novel superhard materials are transition metals combined with light elements—especially transition metal borides. These compounds have a high electron concentration due to the transition metals, and they contain short covalent bonds for the presence of boron atoms. The unique structural properties determine the high bulk moduli and hardness of transition-metal borides.

Transition-metal phosphides are a class of compounds similar to transition-metal borides, and have attracted much attention due to their wide applications in superconductivity, magnetocaloric behavior, catalytic activity, lithium intercalation capacity, and so on [3,4]. However, few investigations have been done on their hardness. Cadmium diphosphide is an important technical material as a wide-gap semiconductor. Due to its superior optical properties, CdP_2 has a wide range of application prospects in the fabrication of solar cells [5]. In addition, its large thermo-optic coefficient leads to numerous applications in thermal sensors [6]. Further, its wide band gap and anisotropic electrical properties make

CdP$_2$ a promising material in electronic engineering [7]. In addition, many investigations [8–10] have also been done on the optical and electronic properties of α-ZnP$_2$ to investigate its applications.

Neither CdP$_2$ nor ZnP$_2$ are ultrahard materials, but high pressure has an important effect on the hardness of materials. Studying the mechanism of the influence of high pressure on the hardness of materials is of great significance to improve their hardness characteristics. In our previous study, we have systematically explored the pressure effect on the elastic properties, mechanical stability, and hardness for transition-metal borides W$_2$B$_5$ [11]. Recently, we investigated the hardness properties of CdP$_2$ at high pressure, and found that the hardness of CdP$_2$ increases as the pressure is increased [12].

Technetium phosphides are a member of the transition-metal phosphides, and have seldom been studied. Using the X-ray diffraction method, Ruhl et al. [13] obtained the crystal structure of technetium phosphides for the first time in 1981. In their work, Ruhl et al. pointed out that Tc$_3$P crystallizes in the tetragonal Mn$_3$P (Fe$_3$P type) structure, and the tetragonal Tc$_3$P crystalizes in a unit cell with lattice parameters a = b = 9.568 $\pm$ 0.005 Å, c = 4.736 $\pm$ 0.003 Å, c/a = 0.4950 $\pm$ 0.0006, β = 90°, V = 433.6 $\pm$ 0.6 Å^3, and the unit cell comprises eight structure units (32 atoms); meanwhile, TcP$_4$ has an orthorhombic RePhis$_4$-type structure, and the orthorhombic cell of TcP$_4$ crystallizes in a unit cell with lattice constants a = 6.238 $\pm$ 0.001 Å, b = 9.215 $\pm$ 0.003 Å, c = 10.837 $\pm$ 0.003 Å, V = 623.0 $\pm$ 0.1 Å^3, with eight formula units (40 atoms) in the cell.

In this paper, we intend to systematically explore the structural, electronic, optical properties, and hardness of technetium phosphides on the basis of the former experimental results under high pressure. The purpose of our work is two-fold. First of all, it is to give a comprehensive and complementary investigation of the effect of pressure on the hardness of technetium phosphides. Second, it is to provide powerful guidelines for future experimental investigations; we hope that such an investigation might provide a way to explore novel ultrahard materials.

2. Computational Methods and Details

In this paper, by using the standard Kohn–Sham self-consistent density functional theory [14–17] based on SIESTA code, we study the structural, electronic, and elastic properties of two forms of technetium phosphides: Tc$_3$P and TcP$_4$ crystals. To get the calculated values, the conjugate gradient minimization method is used to relax the positions of atoms and lattice vectors of technetium phosphides at ambient and high pressures. In our MD simulation, the single-crystal diffractometer data obtained by Ruhl et al. are used as initial structure of calculations. The tetragonal Tc$_3$P crystallizes in a unit cell with lattice parameters a = b = 9.568 $\pm$ 0.005 Å, c = 4.736 $\pm$ 0.003 Å, β = 90°, and the unit cell comprises eight structural units (32 atoms); while the orthorhombic crystallizes in a unit cell with lattice parameters a = 6.238 Å, b = 9.215 Å, and c = 10.837 Å, β = 90°, and the unit cell comprises eight structural units (40 atoms). For the exchange-correlation energy, the generalized gradient approximation (GGA) designed by Perdew, Burke, and Ernzerhof (PBE) [18–20] is adopted. Additionally, the local density approximation (LDA) [21] method is also used as auxiliary calculations. At the same time, we adopt norm-conserving pseudopotentials in the form of Kleinman and Bylander [22] using Troullier and Martins' scheme to describe the valence electron interaction with the atomic core, and choose a split-valence double-ζ basis set plus polarization function (DZP) with an energy shift of 0.005 Ry as our atomic orbital basis set in all the computations.

The optical properties of a material can be described by the complex dielectric function as $\varepsilon(\omega) = \varepsilon_1(\omega) + i\varepsilon_2(\omega)$. The imaginary part of the dielectric function $\varepsilon_2(\omega)$ can be considered as detailing of the real transitions between the occupied and unoccupied states. The real part $\varepsilon_1(\omega)$ can be obtained from the imaginary part by Kramers–Kronig relationship. In addition, other optical parameters (absorption coefficient, reflectance, refractive index, and energy loss spectroscopy) can be obtained from $\varepsilon_1(\omega)$ and $\varepsilon_2(\omega)$. Here we focus on the absorption coefficient of TcP$_4$. It can be derived from the following formula [23].

$$\alpha(\omega) = \sqrt{2}\,\omega\sqrt{\sqrt{\varepsilon_1^2(\omega) + \varepsilon_2^2(\omega)} - \varepsilon_1(\omega)} \tag{1}$$

In this study, we employ the model proposed by Gao et al. [24] to calculate the theoretical Vickers hardness of TcP$_4$ at different pressures. Generally, the Vickers hardness value of a material is defined as the ratio of F/A in an experiment, where F is the pressed force applied to the diamond penetrator on the measured material in kilograms-force and A is the impression area of the resulting indentation in square millimeters. In this theoretical model, the Vickers hardness can be calculated by the following three formulae:

$$H_v = \left[\prod^{\mu} (H_v^{\mu})^{N^{\mu}} \right]^{\frac{1}{\sum N^{\mu}}} \tag{2}$$

$$H_v^{\mu} = 699 P^{\mu} \left(v_b^{\mu} \right)^{-(5/3)} \exp \left(-3005 f_m^{1.553} \right) \tag{3}$$

$$v_b^{\mu} = \frac{(d^{\mu})^3}{\sum_v (d^v)^3 (N^v/\Omega)} = \frac{(d^{\mu})^3 \Omega}{\sum_v \left[(d^v)^3 N^v \right]} \tag{4}$$

where H_v is the hardness of the calculated material, H_v^{μ} is the hardness of μ-type bond in material, N^{μ}, P^{μ}, v_b^{μ} are the total bond number, the Mulliken overlap population, and the bond volume of μ-type bond in the cell. f_m and d^{μ} are the metallicity and the bond length of μ-type bond. Ω is the cell volume. The metallicity f_m can be obtained by Formula (5),

$$f_m = \frac{n_m}{n_e} = \frac{0.026 D_F}{n_e} \tag{5}$$

where n_m is the number of electrons that can be excited at the ambient temperature, n_e is the total number of the valence electrons in a unit cell, and D_F is the electron density of states at the Fermi level.

3. Results and Discussions

3.1. Structure Properties

Combined with both GGA and LDA approximations, the calculated lattice parameters and cell volume of the crystal Tc$_3$P and TcP$_4$ at ambient pressure are obtained by the conjugate gradient (CG) minimization method. The theoretical results at ambient pressure are shown in Table 1 and are compared with the experimental ones. From Table 1, it is noted that for the tetragonal Tc$_3$P crystals, the deviations of GGA and LDA results are within 3.7% and 2.6% for the lattice parameters and cell volume, respectively. For the orthorhombic TcP$_4$ crystals, the error of the results obtained by GGA and LDA are within 4.1% and 1.0% for the lattice parameters and cell volume, respectively. So, the LDA results match the experimental values better than the GGA results, and the LDA method is a better approximation than GGA to study technetium phosphides at zero pressure.

Table 1. Calculated and experimental lattice parameters of the crystal Tc$_3$P and TcP$_4$ at zero pressure. GGA: generalized gradient approximation; LDA: local density approximation.

Lattice Parameter	Tc$_3$P			TcP$_4$		
	Theoretical Results		Experimental Values [13]	Theoretical Results		Experimental Values [13]
	GGA	LDA		GGA	LDA	
a_0 (Å)	9.490	9.376	9.568	6.341	6.280	6.238
b_0 (Å)	9.490	9.376	9.568	9.339	9.241	9.215
c_0 (Å)	4.907	4.856	4.736	10.949	10.827	10.837
V_0 (Å^3)	441.9	426.9	433.6	648.3	628.3	623.0

3.2. Electronic Properties

In this paper, the GGA approximation and LDA approximations are used to calculate the band structure of Tc$_3$P and TcP$_4$. The band structures calculated by these two methods are similar, so here we just present the GGA results in Figure 1 (only −12 to −6 eV for Tc₃P, and −8 to −2 eV for TcP₄). From Figure 1a, it can be seen that there are some points of intersection between the Fermi level and energy bands for the Tc₃P crystal. Therefore, the Tc₃P crystal can be considered as a conductor material. However, for the TcP₄ crystal in Figure 1b, there is a direct band gap of 0.91 eV at the high symmetry point G. So, the TcP₄ is a semiconductor material.

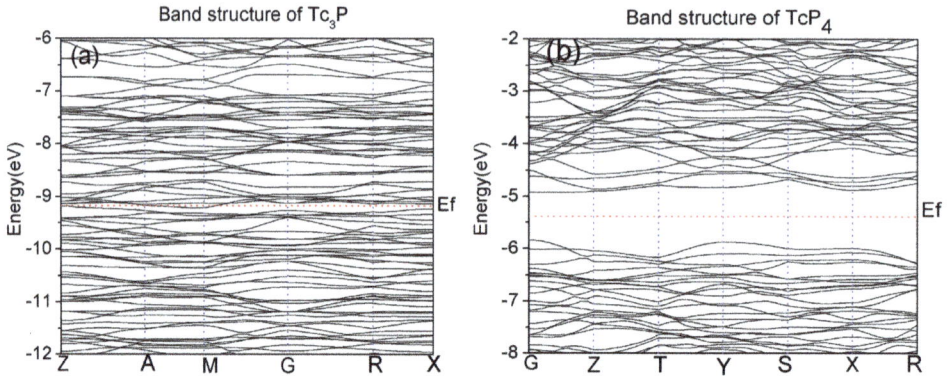

Figure 1. Electronic band structures of the (**a**) Tc₃P and (**b**) TcP₄ crystals in the vicinity of the Fermi level. The red horizontal dotted lines correspond to the Fermi level E_f.

The properties of materials have much to do with the electron configuration of molecule. So, we further calculated the total and projected density of states (TDOS and PDOS) for Tc₃P crystal to study the distribution of the electrons on different orbits to the conductibility of Tc₃P crystal.

The total and projected electronic density of states of Tc₃P with energy ranging from −16 to 0 eV is depicted in Figure 2a. At the same time, the projected density of states of some s, p, and d states of Tc and P atoms are calculated, respectively. From Figure 2a, the contribution of the Tc-4d electrons is predominant for the energy bands near the Fermi level of the Tc₃P crystal. So, the Tc-4d electrons are important for the conductibility of this crystal.

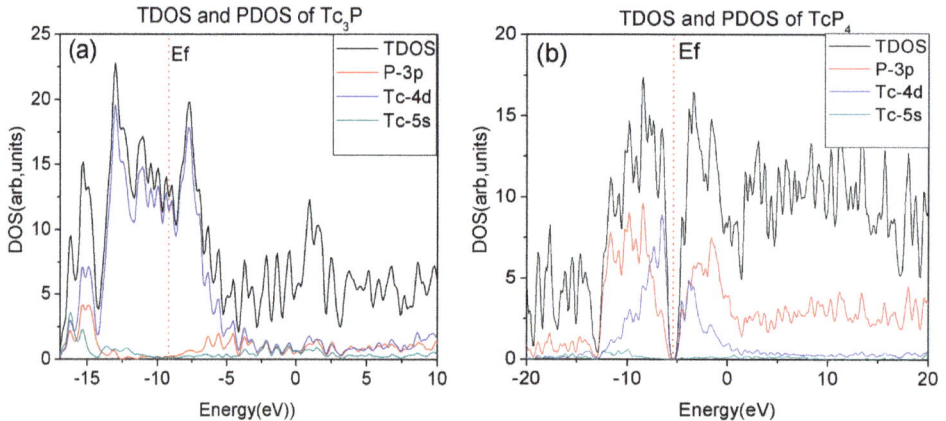

Figure 2. The total density of states (DOS) and projected DOS (PDOS) of P(3p), Tc(4d) and Tc(5s)(d) for (**a**) Tc₃P crystal in the range from −16 to 10 eV and (**b**) TcP₄ crystal in the range from −20 to 20 eV. The red vertical dotted lines correspond to the Fermi level E_f.

In addition, Figure 2b shows us the TDOS and PDOS of the TcP_4 crystal, and it can be seen that the whole band can be divided into two prominent valence-band regions and two conduction-band regions. The regions range from -20 to -13 eV, -13 to -6 eV, -6 to 1 eV, and 1 to 20 eV, respectively. The densities of states at these regions primarily consist of Tc-5s, Tc-4d, and P-3p states. From Figure 2b, we can see that the bonding peaks ranged from -20 to -13 eV and the bonding peaks between 1 and 20 eV are mainly dominated by the P-3p state. The peaks of the bonding ranged from -13 to -6 eV in the valence band and the peaks of the bonding ranged from -6 to 1 eV in the conduction band are not only predominated by the P-3p state, but also contributed from the Tc-4d state. In addition, because the electronic densities of states from -13 and -6 eV and the region range from -6 to 1 eV are located in the vicinity of the Fermi level, and the Tc-4d and P-3p states are principally for these two bands, the Tc-4d and P-3p electrons are very important for the energy bands of the TcP_4 crystal.

According to above analysis, it is noted that the conductibility of Tc_3P crystal is attributed to the Tc-4d electrons, while the distribution of the P-3p electrons weakens the conductivity of the TcP_4 crystal. As a result, the Tc_3P crystal is considered as a conductor material, while the TcP_4 crystal is a semiconductor material.

The structural and electronic properties of technetium phosphide crystals are also investigated with GGA and LDA approximations at the pressure range from 0 to 40 GPa. However, previous studies [25,26] show that the GGA method is better than LDA to study the properties of these types of crystal at a high pressure. Therefore, the following investigations are studied by GGA approximation. The cell volumes of Tc_3P and TcP_4 crystal changing with increasing pressure are shown in Figure 3a. In the pressure range between 0 and 40 GPa, it is noted that the cell volume of the Tc_3P crystal decreases smoothly with the increasing pressure. In addition, Figure 3b shows that the band gap of the TcP_4 crystal changed with the pressure. As is shown in Figure 3b, from 0 to 40 GPa, the band gap of the TcP_4 crystal increases from 0.91 to 1.30 eV. Although the band gap increases with the increasing pressure, the TcP_4 crystal is still a semiconductor under high pressure.

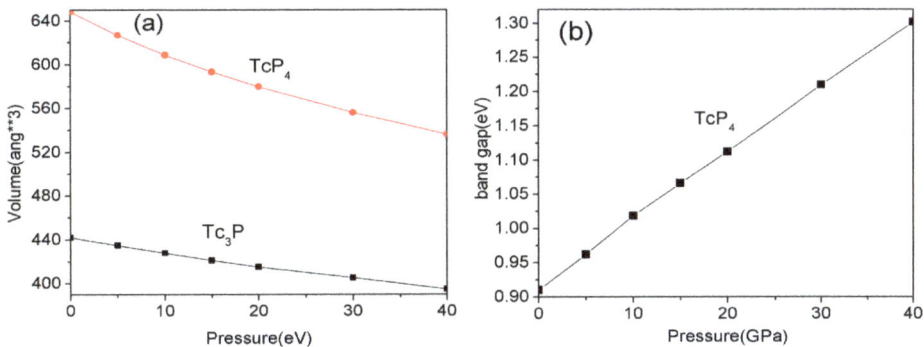

Figure 3. (a) The dependence of the cell volume on pressure for the Tc_3P and TcP_4 crystal; (b) The dependence of the band gap on pressure for the TcP_4 crystal.

3.3. Optical Properties

Commonly, transition-metal phosphides have good optical properties. The Tc3P crystal is a conductor material, while the TcP_4 is a semiconductor material with a wide direct band gap. Therefore, here we only investigate the optical properties of TcP_4. The complex dielectric function of TcP_4 at zero pressure is presented in Figure 4. In order to discuss the absorption coefficient of TcP_4 under high pressure, we computed the dielectric function of this crystal at different pressures. The imaginary part of the dielectric function and the absorption coefficient of TcP_4 crystal at different pressures are shown in Figure 5.

From Figure 5b, it is noted that a primary peak of the absorption coefficient of TcP$_4$ occurred in the wavelength region of the ultraviolet band. In addition, TcP$_4$ has a large absorption coefficient in the visible-light region, while the imaginary part of the dielectric function has a close connection with the absorption coefficient. From Figure 5a,b, we can see the change of optical properties of TcP$_4$ at high pressure. The absorption of visible and ultraviolet light for TcP$_4$ enhances as the pressure is increased. Combined with the study of the change in band gap with pressure in Section 3.2, it can be seen that the pressure effect has an important influence on the photo-catalytic properties of TcP$_4$. In the future, we can consider increasing the photo-catalytic performance of TcP$_4$ by a high-pressure method.

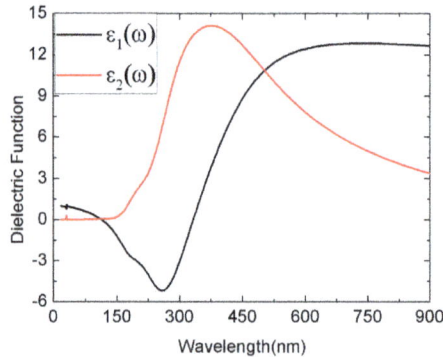

Figure 4. The complex dielectric function for the TcP$_4$ crystal; the black solid line corresponds to the real part, and the red solid line corresponds to the imaginary part.

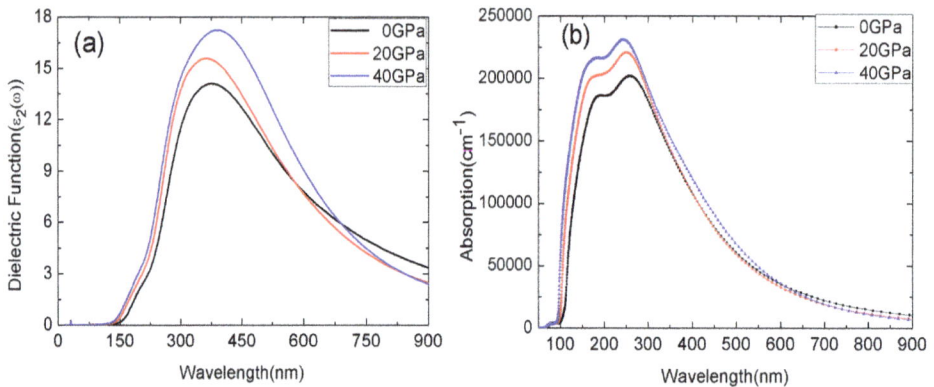

Figure 5. (**a**) The imaginary part of the dielectric function for the TcP$_4$ crystal at different pressures. (**b**) The absorption coefficient of TcP$_4$ crystal at different pressures.

3.4. Hardness

Based on previous work [27,28], it is known that the metallic bond limits the hardness of transition metal phosphides. Compared Tc$_3$P and TcP$_4$, it is not difficult to find that there are many more Tc-Tc bonds in Tc$_3$P than that of TcP$_4$, so we focus our study on TcP$_4$ in the present work. There are many theoretical models to calculate theoretical Vickers hardness. For example, Tian et al. [29] proposed the formulation of hardness for crystals, $H_v = 0.92\,k^{1.137}G^{0.708}$, where $k = G/B$, B is the bulk modulus, and G is the shear modulus. Both bulk and shear moduli are macroscopic concepts. Liu et al. [30] employed this model to calculate the hardness in pyrite-type transition-metal pernitrides. In this study, we employ the model proposed by Gao et al. [24] to calculate the theoretical Vickers hardness. This model considers the effect of metal bonds on the hardness, and presents the relationship between hardness and microscopic parameters. According to this model, the bond parameters and Vickers hardness of TcP_4 at ambient pressure are calculated and presented in Table 2. Compared with the experimental values that can be obtained, theoretical hardness obtained by Tian's and Gao's model are in better agreement with the experimental results. However, the latter can explain the origin of hardness at atomic level.

Table 2. Calculated bond parameters and theoretical hardness obtained by the models of Tian et al. and Gao et al. at ambient pressure. Experimental data are presented for comparison.

Compounds	Bond	d^μ (Å)	P^μ	Ω (Å³)	$v_b{}^\mu$ (Å³)	f_m (10^{-3})	$H_v{}^\mu$ (GPa)	$H_{v\,Tian}$	$H_{v\,Gao}$	$H_{v\,exp}$
WB_2-WB_2	B-B (1)	1.727	0.76		2.521	0		28.8	25.6 [31]	27.7 [32]
	B-B (2)	1.838	0.65		3.038	0				
	W-B (1)	2.335	0.26		6.229	1.787				
ReB_2-ReB_2	B-B	1.807	0.64		1.846	0		43.4	39.1 [31]	39.3 [32]
	Re-B	2.240	0.25		3.515	1.311				
TcP_4	P-P	2.178	0.58	621.7	5.362	0	24.68	19.7	20.1	
	P-P	2.190	0.56		5.451	0	23.19			
	P-P	2.191	0.51		5.458	0	21.07			
	P-P	2.203	0.60		5.549	0	24.12			
	P-P	2.252	0.51		5.927	0	18.36			
	Tc-P	2.342	0.52		6.667	0.929	14.39			
	Tc-P	2.356	0.33		6.787	0.929	8.86			
	Tc-P	2.358	0.54		6.804	0.929	14.44			
	Tc-P	2.377	0.53		6.970	0.929	13.62			
	Tc-P	2.425	0.57		7.401	0.929	13.25			
	Tc-P	2.523	0.30		8.335	0.929	5.72			
	Tc-Tc	3.000	0.63		14.012	0.929	5.05			

In addition, to study the effect of pressure on the hardness property of TcP_4, we computed its Vickers hardness from 0 to 40 GPa. The results are presented in Table 3. It is not difficult to see that the hardness of TcP_4 increases with the increasing pressure. By analyzing the bond parameter and bond component of TcP_4 at different pressures, we found that the hardness is most closely related with the metallicity, the Mulliken overlap population of the bonds in the crystal. Further analysis showed that there are two reasons explaining the increasing hardness at high pressure: (1) the Mulliken overlap populations of the covalent bonds (P-P bonds) and ionic bonds (Cd-P bonds) in TcP_4 increase as the pressure is increased, leading to both the hardness of P-P bonds Cd-P bonds increasing at a higher pressure; (2) the metallicity of metallic bonds in TcP_4 are weakened as the pressure increases, leading to the limit to the hardness in the crystal being weakened. In a word, due to the enhancement of covalent bonds and the weakening of metal bonds, the hardness of TcP_4 is increased at a higher pressure.

Table 3. Calculated Vickers hardness of TcP_4 at different pressures.

Compounds	0 GPa	10 GPa	20 GPa	40 GPa
Hardness (GPa)	20.14	22.35	24.88	25.74

4. Conclusions

In this paper, density functional theory is employed to investigate the structure, electronic properties, optical properties, and hardness of technetium phosphides in the pressure range of 0 to 40 GPa. The lattice parameters and the final atomic positions obtained theoretically are in excellent agreement with the experimental values at ambient pressure. In addition, the electronic properties of Tc_3P and TcP_4 crystal are also studied and presented. The results show that the Tc_3P crystal is a conductor material. However, for the TcP_4 crystal, there is a direct band gap of 0.91 eV at the high symmetry point *G*. Additionally, the band gap of the TcP_4 crystal increased from 0.91 to 1.30 eV as the pressure increased from 0 to 40 GPa. The calculated total electronic density of states and projected densities of states for the Tc_3P and TcP_4 are presented. Results show that the predominant distribution of Tc-4d to the DOS near the Fermi level leads to the classification of Tc_3P crystal as a conductor material, while the distribution of both Tc-4d and P-3p to the DOS near the Fermi level leads to the conclusion that TcP_4 crystal is a semiconductor material. What is more, the optical properties of TcP_4 are discussed, and we find the absorption of visible and ultraviolet light for TcP_4 enhances as the pressure is increased. This indicates that the high pressure method is an effective way to regulate the photo-catalytic performance of TcP_4.

At the end of our study, we research the effect of pressure on the hardness of TcP_4. Results show that both the Mulliken overlap populations of P-P bonds and Cd-P bonds in TcP_4 increase as the pressure is increased, which leads to the hardness of covalent bonds and ionic bonds increasing as the pressure is increased. Meanwhile, as the pressure increases, the metallicity weakens. This leads to the hardness of TcP_4 increasing as the pressure is increased. As the pressure rises from 0 to 40 GPa, the Vickers hardness of TcP_4 increases from 20.14 to 25.74 GPa. So, the effect of pressure plays an important role on the hardness of Tc-P crystals; it changes the hardness of the material by affecting its chemical bond parameters. This study would provide a theoretical basis for improving the hardness of transition metal phosphides and a theoretical guidance for the synthesis of novel ultrahard materials for experiment.

Acknowledgments: Supported by the National Natural Science Foundation of China (NSFC. Grant No. 11374217), the Key Scientific Research Projects of Henan Province (No. 17A140030) and the Doctoral Fund of Zhengzhou University of Light Industry (No. 2014BSJJ088 and 2015XJJZ022)

Author Contributions: Shiquan Feng conceived and designed the work, he supervised the whole work, and wrote the paper. Xuerui Cheng calculated the data, and revised part of the paper. The optical and electronic properties were analyzed by Xinlu Cheng. The hardness properties analysis was performed by Jinsheng Yue. Junyu Li revised the paper and did the work of document inquire.

Conflicts of Interest: The authors declare no conflict of interest.

References

1. Sumiya, H.; Toda, N.; Satoh, S. Mechanical Properties of Synthetic Type IIa Diamond Crystal. *Diam. Relat. Mater.* **1997**, *6*, 1841–1846. [CrossRef]
2. Komanduri, R.; Shaw, M.C. Wear of synthetic diamond when grinding ferrous materials. *Nature* **1975**, *255*, 211–213. [CrossRef]
3. Yang, S.; Liang, C.; Prins, R. Preparation and hydrotreating activity of unsupported nickel phosphide with high surface area. *J. Catal.* **2006**, *241*, 465–469. [CrossRef]
4. Muetterties, E.L.; Sauer, J.C. Catalytic Properties of Metal Phosphides: I. Qualitative Assay of Catalytic Properties. *J. Am. Chem. Soc.* **1974**, *96*, 3410–3415. [CrossRef]
5. Dmitruk, N.L.; Zuev, V.A.; Stepanova, M.A. Spectral distribution of the photoconductivity of cadmium diphosphide. *Russ. Phys. J.* **1991**, *34*, 642–644.
6. Kushnir, O.S.; Bevz, O.A.; Polovinko, I.I.; Sveleba, S.A. Temperature dependence of optical activity and circular dichroism in α-ZnP_2 crystals. *Phys. Status Solidi B* **2003**, *238*, 92–101. [CrossRef]
7. Lazarev, V.B.; Shevchenko, V.Y.; Grinberg, L. K.; Sobolev, V.V. *Semiconducting II-V Compounds*; Nauka: Moscow, Russia, 1976.
8. Morozova, V.A.; Marenkin, S.F.; Koshelev, O.G.; Trukhan, V.M. Optical absorption in monoclinic zinc diphosphide. *Inorg. Mater.* **2006**, *42*, 221. [CrossRef]
9. Sheleg, A.U.; Zaretskii, V.V. X-ray study of the commensurate—Incommensurate phase transitions in α-ZnP_2. *Phys. Status Solidi A* **1984**, *86*, 517–523. [CrossRef]
10. Slobodyanyuk, A.V.; Schaack, G. Measurement of Raman scattered intensities in media with natural or field-induced optical activity. *J. Raman Spectrosc.* **1987**, *18*, 561–568. [CrossRef]
11. Feng, S.Q.; Wang, L.L.; Jiang, X.X.; Li, H.N.; Cheng, X.L.; Su, L. High-pressure dynamic, thermodynamic properties, and hardness of CdP_2. *Chin. Phys. B* **2017**, *4*, 046301. [CrossRef]

12. Feng, S.Q.; Yang, Y.; Li, J.Y.; Jiang, X.X.; Li, H.N.; Cheng, X.L. Pressure effect on the hardness of diamond and W_2B_5: First-principle. *Mod. Phys. Lett. B* **2017**, *31*, 1750137. [CrossRef]
13. Ruehl, R.; Jeitschko, W.; Schwochau, K. Preparation and Crystal Structures of Technetium Phosphides. *J. Solid. State. Chem.* **1982**, *44*, 134–140. [CrossRef]
14. Ordejón, P.; Artacho, E.; Soler, J.M. Self-consistent order-N density-functional calculations for very large systems. *Phys. Rev. B* **1996**, *53*, R10441. [CrossRef]
15. Barth, U.; Von Hedin, L. A local exchange-correlation potential for the spin polarized case: I. *J. Phys. C. Solid State Phys.* **1972**, *5*, 1629. [CrossRef]
16. Kohn, W.; Sham, L.J. Self-Consistent Equations Including Exchange and Correlation Effects. *Phys. Rev.* **1965**, *140*, A1133. [CrossRef]
17. Strobel, R.; Maciejewski, M.; Pratsinis, S.E.; Baiker, A. Unprecedented formation of metastable monoclinic $BaCO_3$ nanoparticles. *Therm. Acta* **2006**, *445*, 23–26. [CrossRef]
18. Staroverov, V.N.; Scuseria, G.E. High-density limit of the Perdew-Burke-Ernzerhof generalized gradient approximation and related density functionals. *Phys. Rev. A* **2006**, *74*, 044501. [CrossRef]
19. Wu, Z.G.; Cohen, R.E. More accurate generalized gradient approximation for solids. *Phys. Rev. B* **2006**, *73*, 235116. [CrossRef]
20. Perdew, J.P.; Burke, K.; Ernzerhof, M. Generalized gradient approximation made simple. *Phys. Rev. Lett.* **1996**, *77*, 3865–3868. [CrossRef] [PubMed]
21. Perdew, J.P.; Zunger, A. Self-interaction correction to density-functional approximations for many-electron systems. *Phys. Rev. B* **1981**, *23*, 5075. [CrossRef]
22. Kleinman, L.; Bylander, D.M. Efficacious Form for Model Pseudopotentials. *Phys. Rev. Lett.* **1982**, *48*, 1425. [CrossRef]
23. Okoye, C.M.I. Theoretical study of the electronic structure, chemical bonding and optical properties of $KNbO_3$ in the paraelectric cubic phase. *J. Phys. Condens. Matter.* **2003**, *15*, 5945–5958. [CrossRef]
24. Gao, F. Theoretical model of intrinsic hardness. *Phys. Rev. B* **2006**, *73*, 132104. [CrossRef]
25. Fan, C.L.; Cheng, X.L.; Zhang, H. First-principles study of the structural and electronic properties of the alpha modification of zinc diphosphide. *Phys. Status Solidi B* **2009**, *246*, 77–81. [CrossRef]
26. Feng, S.Q.; Cheng, X.L. Theoretical study on electronic properties and pressure-induced phase transition in β-CdP_2. *Comput. Theor. Chem.* **2011**, *966*, 149–153. [CrossRef]
27. Feng, S.Q.; Li, X.D.; Su, L.; Li, H.N.; Yang, H.Y.; Cheng, X.L. Ab initio Study on Structural, Electronic Properties, and Hardness of Re-doped W_2B_5. *Solid. Stat. Commun.* **2016**, *245*, 60–64. [CrossRef]
28. Feng, S.Q.; Guo, F.; Li, J.Y.; Wang, Y.Q.; Zhang, L.M.; Cheng, X.L. Theoretical investigations of physical stability, electronic properties and hardness of transition-metal tungsten borides WB_x (x = 2.5, 3). *Chem. Phys. Lett.* **2015**, *635*, 205–209. [CrossRef]
29. Liu, Z.T.; Gall, Y.; Khare, S.V.D. Electronic and bonding analysis of hardness in pyrite-type transition-metal pernitrides. *Phys. Rev. B* **2014**, *90*, 134102. [CrossRef]
30. Tian, Y.J.; Xu, B.; Zhao, Z.S. Microscopic theory of hardness and design of novel superhard crystals. *Int. J. Refract. Met. H.* **2012**, *33*, 93–106. [CrossRef]
31. Zhong, M.M.; Kuang, X.Y.; Wang, Z.H.; Shao, P.; Ding, L.P.; Huang, X.F. Phase Stability, Physical Properties, and Hardness of Transition-Metal Diborides MB2 (M = Tc, W, Re, and Os): First-Principles Investigations. *J. Phys. Chem. C* **2013**, *117*, 10643. [CrossRef]
32. Gu, Q.F.; Krauss, G.; Steurer, W. Transition metal borides: Superhard versus ultra-incompressihle. *Adv. Mater.* **2008**, *20*, 3620–3626. [CrossRef]

crystals

MDPI

Review

Indentation Plasticity and Fracture Studies of Organic Crystals

Sowjanya Mannepalli and Kiran S. R. N. Mangalampalli *

SRM Research Institute, Department of Physics and Nanotechnology, SRM University, Kattankulathur, Chennai 603203, India; msowji55@gmail.com
*   Correspondence: kiranmangalampalli.k@ktr.srmuniv.ac.in

Academic Editor: Ronald W. Armstrong
Received: 23 September 2017; Accepted: 23 October 2017; Published: 27 October 2017

Abstract: This review article summarizes the recent advances in measuring and understanding the indentation-induced plastic deformation and fracture behavior of single crystals of a wide variety of organic molecules and pharmaceutical compounds. The importance of hardness measurement for molecular crystals at the nanoscale, methods and models used so far to analyze and estimate the hardness of the crystals, factors affecting the indentation hardness of organic crystals, correlation of the mechanical properties to their underlying crystal packing, and fracture toughness studies of molecular crystals are reviewed.

Keywords: nanoindentation; hardness; plasticity; organic crystals; fracture; deformation

1. Introduction

The crystals of organic molecules offer attractive physical properties that are different from the thermal, mechanical, optical, and electronic properties of conventional solids because of the presence of weak intermolecular interactions (such as van der Waals and dipole interactions, etc.) and the interplay between inter-and intramolecular degrees of freedom [1,2]. During the last decade, understanding the physics and mechanical deformation behavior of single crystals of organic molecules has become the subject of both theoretical and experimental researchers with the intention of exploring and exploiting them for various technological [2] applications such as molecular electronics [3] and pharmaceutics [4–11], etc. [12,13]. For example, in the pharmaceutical industry, the easy tableting and formulation of a drug solely depend on the mechanical properties of the bulk drug. Therefore, the establishment of structure-mechanical property relationships is key to designing and controlling the properties of molecular crystals in a more effective way [14]. In this review article, the authors have made efforts to summarize the recent advances in measuring and understanding the indentation-induced plastic deformation and fracture behavior of single crystals of a wide variety of organic molecules and pharmaceutical compounds, including the importance of hardness, H, measurement at the nanoscale, methods and models proposed and/or utilized to analyze and estimate the H of the crystals, factors affecting the H of organic crystals during small-scale testing, correlation of the mechanical properties to their underlying crystal packing, and fracture toughness studies.

1.1. Mechanical Properties

The mechanical properties of materials refer to the behavior of materials when external forces are applied. Knowledge of this area provides the basis for designing molecular solids with desirable properties and avoids failures in several engineering applications. The core concern in design to prevent structural failure is that the applied stress (force/unit area) must not exceed the strength of the crystals; otherwise, it leads to deformation or fracture failure. Deformation failure can be understood as the change in the physical dimensions or shape of crystals which cannot be recovered. When the cracking reaches to the extent that separates the crystal into two pieces, is called a fracture. So far, material failures are classified either as deformation or fracture. Deformation has been further classified as elastic and plastic upon loading. As the name indicates, elastic deformation recovers immediately upon unloading. In general, stress and strain are proportional to each other in the case of pure elastic materials. The proportionality constant, E, is the modulus of elasticity for axial loading cases. In contrast, plastic deformation is a permanent deformation process and does not recover upon unloading. Once plasticity is initiated in the material, an additional increase in stress causes further deformation, called yielding, and the beginning point of that process is known as the yield strength, σ_o. Based on the plastic deformation behavior in various materials, materials are recognized as ductile or brittle. While ductile materials are capable of sustaining large amounts of plastic deformation, brittle materials fracture without entering much into plastic deformation. While many metals exhibit ductile behavior, glasses, molecular crystals, and ceramics show brittle behavior. Materials having high values of both ultimate tensile strength, σ_u, and strain at fracture, ε_f, are recognized as tough, and these are desirable for use in structural applications. The plastic deformation that accumulates with time is termed creep [15].

1.2. Plasticity

Most real materials undergo some permanent deformation upon loading, which involves dissipation of energy. This means that the original state may be achieved by the supply of more energy, as the process is irreversible. In crystalline materials and metals, the motion of dislocations and the migration of grain boundaries are responsible for microscale level plastic deformations [16]. The theory of plasticity was initially developed by Tresca [17], who proposed yield criterion in 1864. Saint-Venant [18], Levy [19], Von Mises [20], and Hencky and Prandtl [21] have further advanced the concept of yield and plastic flow rules. Later, Prager [22] and Hill [23] developed the "classical theory" which brought many aspects into a single framework. Further developments in computational and numerical methods [24] have been developed for a better understanding of the plasticity problem in crystalline materials.

It was proposed that plastic flow occurs in molecular crystals via a slip mechanism (movement of edge dislocations) along with specific directions in the crystal [25]. Interestingly, while the dislocation climb was reported as the responsible deformation mechanism during creep for molecular crystals at elevated stresses and temperatures, edge dislocation movement under applied stress was found as the dominant deformation mechanism for plastic crystals. Since pharmaceutical industries use techniques like grinding, milling, and tableting to make tablets, the solids that deform plastically via edge dislocation movement or slip are given importance to develop a predictive approach to the yield properties of molecular crystals/compounds [26].

1.3. The Critical Resolved Shear Stress and Schmid Factor

As discussed above, slip along a crystallographic plane occurs via a dislocation motion for which a certain amount of stress is needed to overcome the resistance offered by the lattice. It is observed that the slip in a particular crystallographic plane occurs when the shear stress along the slip direction reaches a critical value on that particular plane. Therefore, this critical shear stress is related to the stress required to move dislocations across the slip plane. The yield stress (stress required to onset the plastic deformation under a tensile/compressive load) can be related to the shear stress that acts along the slip direction, as below [15]:

$$\tau = \sigma \, cos\varnothing \, cos\lambda \qquad (1)$$

where $\sigma = FA$ is the applied tensile/compressive stress. If this is applied along the long axis of the sample with cross-sectional area A (as shown in Figure 1), then the applied force along that axis is $F = \sigma A$.

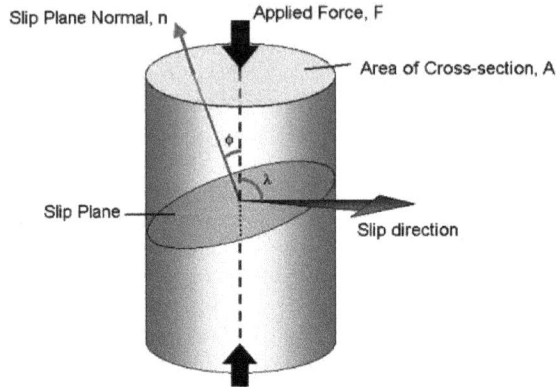

Figure 1. Schematic diagram showing slip mechanism in a single crystal under compressive loading.

The slip direction is shown in Figure 1 if the slip occurs on the plane that is shown in the schematic with plane normal n. The resolved shear stress acting parallel to the slip direction on the slip plane, τ_R, can be calculated using the following equation [15]:

$$\tau_R = \frac{resolved\ force\ acting\ on\ the\ slip\ plane}{area\ of\ slip\ plane} = \frac{Fcos\lambda}{A/cos\phi} = \frac{F}{A}cos\lambda cos\phi \qquad (2)$$

where ϕ, the angle between the force axis and the slip plane normal and λ is the angle between the force axis and slip direction. The $Fcos\lambda$ term represents the axial force that lies parallel to the slip direction. The value of τ_R at which slip occurs in a given crystal with specified density of dislocations is constant, and is known as critical resolved shear stress, τ_c. This is also known as Schmid's Law. The quantity, $cos\ \phi\ cos\ \lambda$ is called Schmid factor. Schmid's law can be written as [15]:

$$\tau_c = \sigma_y cos\phi cos\lambda \qquad (3)$$

where σ_y is known as yield stress, the stress required to cause slip on the primary slip system. There can be several slip systems in a given crystal. As the load increases, the τ_R on each slip system increases until it reaches τ_c. When the particular slip system reaches τ_c, the crystal begins to deform plastically by that slip system, hence known as the primary slip system. With the further increase of load, other slip systems may begin to operate when τ_c is reached. Schmid's law can be used to calculate the Schmid factor to estimate the primary slip system in a given crystal. The primary slip system will have the greatest Schmid factor. One can calculate the Schmid factor for every slip system in the given material to determine which slip system operates first [15].

2. Hardness Measurement Methods

Hardness testing is performed to estimate the materials' ability to resist plastic flow under applied load. However, the measurement of *H* depends on the method one chooses and is influenced by both the elastic and plastic nature of materials. Depending on the forces applied and displacements obtained, the *H* measurements can be defined as micro-, nano-, or macrohardness. While measuring macrohardness is very simple for bulk materials, thin films and microstructured materials require *H* measurement techniques at micro/nanoscale [27]. While the electromagnetic, ultrasonic, and rebound techniques are used to measure materials' hardness, indentation testing has received considerable attention because it is non-destructive and provides reliable and straightforward data. Indentation (penetration of a hard material, typically a diamond with a known geometry, into the sample) tests were first performed by Brinell [28], who used spherical balls (as shown in Figure 2a) from hardened steel ball bearings or made of cemented tungsten carbide to quantify the plastic properties of materials in 1900. Brinell's work was then followed and improved by Meyer [29] in 1908.

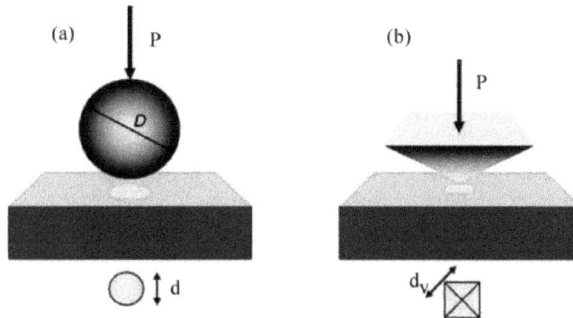

Figure 2. Indentation hardness testing methods: Schematic of (**a**) the spherical indenter (Brinell and Meyer); (**b**) diamond pyramidal indenter (Vickers). (Adapted from Reference 30, reproduced with permission).

In Meyer's work, the H was calculated using the ratio between the load (P) and the projected area (A), namely, $H = P/A$. In 1922, the Vickers [30] test was carried out using a square-based pyramid diamond indenter with a $136°$ semi-angle instead of a ball indenter, as shown in Figure 2b. The Vickers hardness was defined as the ratio between the load and the surface area of the residual impression, namely, $H_v = 1.8544\ P^2/d_v{}^2$, where d_V is the length of the diagonal of the surface area. A research paper published by Tabor [31] in 1948 advanced the understanding of indentation hardness testing, wherein he described the penetration procedure of a ball-like indenter into a material. He mentioned that, upon the application of the load, the material initially starts deforming elastically (which means that the material recovers to its original state upon removing the load) and then flows plastically (associated with work hardening mechanism after removal of the complete load). In 1951 [32], he proposed an equation relating indentation hardness and the yield stress (σ) of the material based on the theory of indentation of a rigid perfectly plastic solid, namely, $H = C\ \sigma$, where C is a constant that is dependent on the indenter geometry. Tabor furthered the understanding of the indentation response of polymers [33,34] and brittle materials [35], as well as the temperature dependence of the hardness [36] of metal oxide samples. Within a short time, the indentation technique was extended to small volume materials in the mid-1970s [37]. The indentation technique that helps in measuring the mechanical properties of small volume materials was named "nanoindentation", as the length scale of the penetration depth is usually in nanometers. In the indentation methods mentioned above, the contact area was directly measured from the residual impression area. However, in nanoindentation, since the contact area, A_c, of the residual indent is too small to measure, A_c is determined by the measured depth of penetration in nanoindentation [38].

The indentation technique became popular after 1992 with the development of a method to measure elastic modulus and hardness based on the load, P, and displacement, h, curve. Oliver and Pharr [39,40] proposed this approach in 1992. In 2007, Kucharski and Mroz [41] developed a procedure for determining stress-strain curves using cyclic spherical indentation data. Subsequently, Kruzic et al. [42] developed a method to evaluate the fracture toughness of brittle materials. Further developments, such as identifying the effects of kinematic hardening on the material response [43], the reconstruction of the axial stress-strain curve from the indentation data with conical indenter over a range of cone angles [44], and the modification of the hardness formulation within the elastic-plastic transition derived for solids, were reported by Hill [23] and Marsh [45]. Rodriguez et al. [46] found that the mechanical properties of alumina-titania nanostructured films measured using nanoindentation were higher than the conventionally measured values.

The hardness obtained by the indentation test is defined as the ratio of the maximum indentation load, P_{max}, to the contact area, of the indenter [39,40]:

$$H = \frac{P_{max}}{A_c(h_c)} \qquad (4)$$

The reader should note that the above definition of indentation hardness may deviate from the traditional hardness measurement where the area is estimated from the residual indent impression. In the latter process, the actual contact area may be underestimated if there is significant elastic recovery during indentation unloading. In general, the materials with high elastic modulus will exhibit slightly deviated values from the indentation hardness measurement.

The area of contact is a function of the indenter contact depth, h_c, and can be determined by the following expression:

$$A_c(h_c) = C_0 h_c^2 + C_1 h_c + C_2 h_c^{1/2} + C_3 h_c^{1/4} + \ldots\ldots\ldots + C_8 h_c^{1/128} \qquad (5)$$

It is important to note that only the C_0 will be used if the Berkovich indenter is assumed as a perfect tip at higher penetration depths. For the cases of imperfect tips and shallower depths, higher-order terms have to be considered, and these can be obtained from the fit of the tip area function curve for a given tip. The h_c can be estimated from the *P-h* curve, as shown in Figure 3 [39,40]:

$$h_c = h_{max} - \varepsilon \frac{P_{max}}{S} \qquad (6)$$

where h_{max} is the maximum indentation depth, and $0.75(P/S)$ denotes the extent of the elastic recovery (h_e) [39,40]. Here, the stiffness $S = dP/dh$ and ε is a constant that depends on the indenter geometry. The values of ε are 0.72, 0.75, and 1.00 for conospherical, Berkovich, and flat punch tips, respectively.

The maximum shear stress, τ_{max} [47,48], the stress required to nucleate dislocations, can be estimated using the following equation when the load-displacement curve exhibits a pop-in (i.e., a clear transition between elastic and plastic deformation):

$$\tau_{max} = 0.31 \, (6E_r^2/\pi^3 R^2)^{1/3} \, P_{max}^{1/3} \qquad (7)$$

where E_r is the reduced modulus, R is the radius of the indenter, and P_{max} represents the peak load.

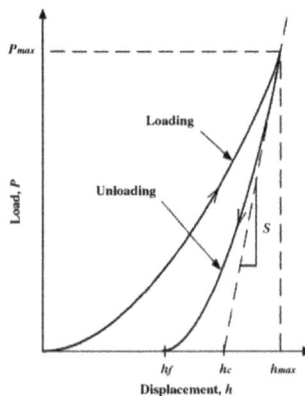

Figure 3. Schematic diagram of the indentation load-displacement curve showing important measured parameters such as peak load, P_{max}, maximum penetration depth, h_{max}, final depth after removing the load, h_f, contact depth, h_c, and the unloading stiffness, S. (Reproduced with permission from Materials Research Society, Reference 39).

2.1. Prediction of Hardness Using Crystal Morphology

Roberts and Rowe [26] developed a model to predict indentation hardness of molecular single crystals based on cohesive energy density, the weakest planes from the crystals structures, and structural parameters. The equation of hardness is given as [26]:

$$H = \left(\frac{c_1 c_2 F_a 2 N_A}{R_c^2 S_r^3 Z} \right) CED \tag{8}$$

where S_r is the slip ration, c_1 and c_2 are unit cell constants, N_A is Avogadro's number, R_c is the length of the cell, Z is the number of molecules in a unit cell, and F_a is an angular function related to α, β, and γ depending on the crystal class. CED is the Cohesive Energy Density. Since most of the organic crystals crystallize in monoclinic structure, the above equation can be re-written as [26]:

$$H = \left(\frac{bc\sin\beta 2 N_A}{a^2 S_r^3 Z} \right) CED \tag{9}$$

For a monoclinic system, $F_a = \sin\beta$; if $R_c = a$, then c_1 = b and c_2 = c. Therefore, the above equations can be utilized to predict the indentation hardness from the cohesive energy density. Two slip ratios are used, S_r of 0.7070 and 1 [26]. For orthorhombic systems, $F_a = 1$ in the above equation [26].

2.2. Factors Affecting Nanoindentation Hardness of Organic Crystals

Several factors, such as indenter calibrations, vibration during testing, indenter shape, indentation size effects [49], thermal drift [49], machine compliance [38,49], and pile-up/sink-in [38,49], etc., affect indentation hardness values severely during the testing of molecular crystals.

Thermal drift occurs during nanoindentation either due to creep (time-dependent plasticity at a constant load) within the sample caused by plastic flow or due to variation in the transducer dimensions due to temperature change-induced contraction or expansion. The latter method causes a change in the real-time penetration depth measurement under a constant P_{max} which is difficult to distinguish from the creep. However, these depth changes result in a thermal drift error on the actual penetration depth. Fisher-Cripps [38] reported that the temperature rises to 100 °C within the specimen during indentation. Although the change in the linear dimension of the specimen will be significantly smaller compared to the total penetration depth, the localized rise in temperature to 100 °C within the sample affects the viscosity and indentation hardness of the test specimen. The drift effect can be corrected by adjusting the penetration depth, if the drift rates are determined during indentation. The drift rates can be captured at the peak load. For calculating the drift rates, the data at the final unload increment can be used because creep is less likely to occur at low loads.

Another critical care to be taken during indenting molecular crystals is to provide the correct compliance value of the instrument, which is defined as the deflection in the load frame, shaft of the indenter, and sample mount. Since molecular crystals are relatively soft compared to inorganic materials, mechanical polishing of the crystals and mounting crystals in the acrylic resin is impossible. Therefore, most of the researchers use cyanoacrylate glue for firm mounting of the crystals for nanoindentation. When the load is applied, the elastic deformation in both the crystal surface and some parts of the testing machine cause an increase in the measured indentation depth that is not experienced at the indentation contact [38]. The compliance can be quantified as the ratio of the instrument's deflection to the applied load. The stiffness measured by the unloading portion of the *P-h* curve is the result of the elastic deformation behavior of both the sample and load frame. The total compliance of the machine can be obtained by adding the compliances of the specimen, indenter, and load frame. The crystal compliance can be minimized by mounting them firmly to the substrate.

To measure precise indentation hardness, finding the contact area of the indent using the residual/final penetration depth is very important. To measure such contact area, the indenter geometry should be well explored and ideally flawless, which is not common. To estimate the actual contact area of the indenter, either Atomic Force Microscope (AFM) or Scanning Electron Microscope (SEM) can be utilized, and correction can be done by dividing the real A_c of the tip. If the ratio of the real and actual A_c is greater than one, the actual indenter has a higher tip radius than the actual tip, which leads to larger contact areas at lower penetration depths [38].

Surface roughness, ρ, plays an important role in determining the mechanical behavior of materials. In general, the surface roughness should be as small as 5% of the maximum indentation depth in order to achieve reliable mechanical properties. Shibutani et al. [50] experimentally investigated the effect of ρ on pop-ins observed in nanoindentation using single-crystalline Al and found that the critical values of the load at the pop-ins are sensitive to ρ. Their results show that the first pop-in with a higher width occurs in the smoother sample at higher loads than the rough sample [38].

The determination of contact area by SEM and AFM imaging methods can go wrong if the sample surface is not aligned normal to the indenter tip. Since molecular crystals cannot be polished mechanically to obtain the flat surface, it is important to choose perfectly flat samples for nanoindentation. Otherwise, H and E_r values may be wrongly estimated because of the incorrect A_c estimation. Since it is highly impossible to obtain a 100% orthogonality condition between the sample surface and the indenter tip, one relaxation is allowed [51].

Significant pile-up and sink-in around the residual indent are observed earlier for plastically deformed materials. Such effects were found to depend on the ratio of modulus to yield stress as well as on the level of strain hardening of the sample material [38]. A relationship between the residual indentation depth and the total penetration depth also provides reliable information about these phenomena. If the ratio between the residual depth and the total depth is greater than 0.7, pile-up can be expected; otherwise, sink-in can be expected. The presence of pile-up around the residual indent causes the underestimation of the A_c and hence higher hardness [52]. When there is a pile-up around the indent, the Oliver-Pharr [39] method overestimates the H and E_r values (up to 60 and 30%, respectively) because their evaluation depends on the A_c deduced from the P-h data. Several models are available to determine the pile-up effect, such as the semi-ellipse method [53], the method put forth by Choi, Lee, and Kwon [54], and the finite element method [55]. Zhou et al. [56] estimated the pile-up free hardness (H_{actual}) and elastic modulus (E_{actual}) values using the ratio of pile-up height ($h_{pile\text{-}up}$) and contact depth (h_c). The equations are:

$$H_{actual} = H^{O\&P}\left(\frac{1 + h_{pile-up}}{h_c}\right)^{-2} \tag{10}$$

$$E_{actual} = E^{O\&P}\left(\frac{1 + h_{pile-up}}{h_c}\right)^{-1} \tag{11}$$

where $H^{O\&P}$ and $E^{O\&P}$ are the hardness and elastic modulus obtained using the Oliver-Pharr method [39,40], respectively. The dramatic increase of hardness with decreasing indenter penetration depth is known as indentation size effect (ISE). In crystalline materials, ISE was explained using the concept of geometrically necessary dislocations (GNDs), as proposed by Nix and Gao [57]. Recently, Arief et al. [58] showed, using a scanning X-ray microdiffraction technique on Cu (111) crystals at various indentation depths, that the density of GNDs increased with decreasing indentation depth. The ISE can also be due to the incorrect estimation of the A_c of the indent at shallow depths. For example, although the three-sided pyramidal Berkovich indenter is considered to be sharp and researchers use the area function of a sharp tip to estimate the contact area, the tip is not atomically sharp, as it always ends with a spherical shape. Therefore, at shallow depths, the area function of a Berkovich indenter underestimates the actual A_c to its corresponding depth, and thus measures higher H values.

3. Understanding the Plastic Behavior of Organic Crystals

Plastic bending experiments were conducted by Reddy et al. [59–61] to understand the plastic behavior of organic crystals, and they observed that molecular crystals undergo plastic deformation if and only if the intermolecular interactions strength in orthogonal directions is significantly different, and that there exists a correlation between bending and crystal packing (see Figure 4a,b). Based on that observation, molecular crystals were classified as plastic or brittle, and a model was developed for bending crystals. While the former crystals are bendable, the latter cannot be deformed plastically. Further, Reddy and Naumov [62] studied the plastic deformation mechanism in hexachlorobenzene crystals and observed changes in the unit cells parameters in the region of deformation, as shown in Figure 4c. In contrast to inorganic plastic materials, such as metals etc., no volume change including dimensions of crystals and thickness was observed following the bending of molecular crystals. However, crystals that have "cross-linked" intermolecular interactions in three orthogonal directions are hard and brittle. Recently, Sajesh et al. [63] examined the plastic bending mechanism in Dimythl Sulfone (DMS) using a new bending model that provides quantitative rationalization based on differential binding and the stacking of molecular layers in orthogonal directions.

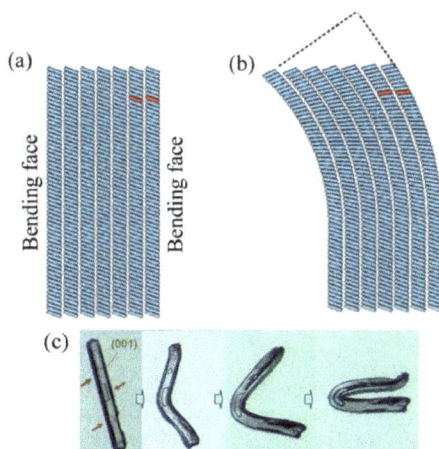

Figure 4. The plastic bending model of organic molecular crystals. (**a**) An undeformed crystal (half sectional view) where the weakest interactions represent white spaces between rows of stacks; (**b**) a bent crystal where the relative movement of the disk is highlighted in red, with pronounced deformation in interfacial angles (dashed line); and (**c**) the bending of C6Cl6 crystal. (Reproduced with permission from Royal Society of Chemistry, Reference 59).

Saha and Desiraju [64] reported a method to design hand-twisted helical crystals from plastic crystals using crystal engineering techniques (see Figure 5). The procedure was started with a 1-D plastic crystal (1,4-dibromobenzene), which was then converted to a 1-D elastic crystal (4-bromophenyl 4′-chlorobenzoate). This was achieved by introducing a molecular synthon-O-CO- in place of the supramolecular synthon Br···Br in the precursor. The 1-D elastic crystal was then modified into a 2-D elastic crystal (4-bromophenyl 4′-nitrobenzoate). These 2-D elastic crystals were then transformed into 2-D plastic crystals (4-chlorophenyl and 4-bromophenyl 4′-nitrobenzoate) with two pairs of bendable faces without slip planes by varying interaction strengths. The presence of two pairs of bendable faces which are orthogonal to each other allowed the crystals to hand twist in a helical shape [64]. This shows that prior knowledge of the structure-mechanical properties of molecular crystals such as plastic and elastic mechanical deformation are necessary to engineer the molecular crystals with desired properties.

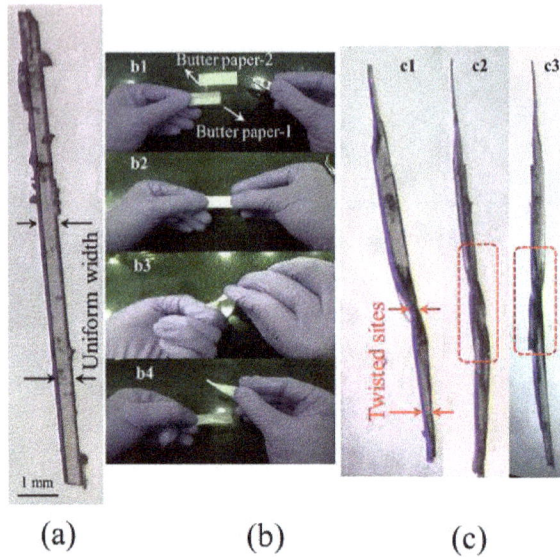

Figure 5. (**a**) As-grown crystal before performing the twist experiment. (**b1–b4**) Twisting mechanism of the butter paper encapsulated crystal by hand. (**c1–c3**) The hand-twisted helical crystal, as seen from different angles. The twisting at the middle is marked with red dotted lines. (Reproduced with permission from the American Chemical Society, Reference 64).

3.1. Indentation Hardness of Molecular Crystals

3.1.1. Cyclotrimethylenetrinitramine, (RDX) Crystals

Hagan and Chaudhri (1977) [65] measured the Vickers hardness of 24.1 kg/mm^2 for RDX single crystals between loads varied from 150 to 700 mN. Even at the smaller loads cracks have been observed. The fracture surface energy was estimated as 0.11 J/m^2 and 0.07 J/m^2 for two cleavage planes. Halfpenny et al. [66] and Chaudhri [67] measured Vickers hardness to 39 kg/mm^2 and 21 MPa, respectively, and found that the primary dislocation motion was in the (010) planes. Elban et al. [68] reported that the Vickers microhardness value varied from 310 to 380 MPa for various growth faces of RDX crystals (50-gram load). Also, Elban [69] used Knoop hardness methods to measure hardness anisotropy. Hardness varied from 170 to 700 MPa for different crystal facets. Gallagher et al. [70] utilized both microhardness and the Knoop indenter and showed crystals orientation dependency which attributed the variation in hardness for different orientations to the dominant slip system.

Roberts et al. [26] developed a model relating the indentation hardness of molecular solids to the Burgers vector's length, the cohesive energy density, the weakest plane in the crystals, and the crystals' structural parameters. The prediction of the hardness was based on the identification of the slip planes that were available in the system, and it was assumed that the primary slip plane was the weakest plane, and energetically it was the preferred slip plane. Several methods, such as attachment energy calculations, cleavage planes, and hydrogen bonding pattern information, have been used to identify the slip planes in organic crystals. It was concluded that the cleavage planes provide direct evidence for the weakest planes, which were also twinning planes, indicative of plastic deformation [26]. It is imporatnt to mention here the work carried out by Sun and Kiang [71] on the accuracy of the slip plane predictions using attachment energy calculations. They considered 14 different organic crystals that exhibited layered strcuture and predicted slip systems by their attachement energy calculations using three different current force fields, which were then compared to those identified by crystal structure visualization. They conclude that 50% of the slip/cleavage predictions were inaccurate.

Ramos et al. [72] employed nanoindentation on different faces of single crystals of RDX crystals with a conical (cone shaped tip with a rounded end) probe with the load varying from 250 μN to 10,000 μN. In general, the conical tip is used to delay the elastic-plastic transition at the shallow depths. However, in the present case, all orientations showed cracks even at very low loads with the conospherical tip. The calculated τ_{max} was within $1/15$ to $1/10$ of the shear moduli. The indentation hardness was measured between 615 and 672 MPa for different faces. In another study, Ramos et al. [73] revealed that the planes produced by the cleavage method yield at a lower applied stress but the habit planes of the as-grown crystals exhibit yield points near the theoretical shear strength. Weingarten and Sausa [74] studied nanomechanical properties of RDX crystals by the *P-h* measurements using a Berkovich diamond indenter and molecular dynamics (MD) simulations, and reported that the (210) surface was stiffer than the (001) surface.

The abovementioned experimental techniques, such as microindentation [66], nanoindentation [72], and etch-pitting [75], have revealed the (0 1 0) plane in RDX crystals to be the primary slip plane and [100] to be the cross-slip direction since it is shared by the (01 0), {021}, and {011} planes. The {011} and {021} planes are also considered as additional potential slip planes in RDX crystals. Munday et al. [76] investigated the fracture behavior of various crystal planes in RDX crystals using Rice's criterion and revealed that the (0 11), (021), (0 10), and (00 1) planes may possess active slip systems. Mathew et al. [77] carried out molecular simulation studies in order to investigate the slip asymmetry in RDX crystals. Their study revealed that: (i) the force needed to move a dislocation in RDX crystals was controlled by the mode of deformation, (ii) slip asymmetry was evident in the (010) slip plane with the lowest Peierls stress and, (iii) such asymmetry in (010) plane was caused due to steric hindrance.

Taw et al. [78] reported the mechanical properties of as-grown, conventionally processed, and sub-millimeter RDX crystals. Nanoindentation was conducted using a Berkovich tip in a low load quasi-static mode. Scanning probe microscopy images of the residual impressions showed no evidence of indentation-induced cracking. The measured mechanical properties such as elastic modulus and hardness were matched with the literature. However, the point of onset of plasticity (yield point) occurred between 0.1 and 0.7 GPa, which indicated that the powders of RDX contained a significant number of dislocation sources that were prevalent in the as-grown RDX crystals.

Liu et al. [79] performed coarse-grained MD simulations of RDX crystals to validate the limited-sample coarse-grained potential. The mechanical properties calculated with the simulations were compared with the experimental results. The deformation behavior of RDX under nanoindentation was revealed by a series of simulations that resembled the experimentally determined deformation behavior of the (100) face. Their study concluded that most of the dislocation loops were found to be parallel to the (001) plane due to the low slip threshold of the (010) [100] active slip system.

3.1.2. Hardness Anisotropy Studies in Some Organic Crystals

Joshi et al. [80] measured the microhardness of anthracene, phenanthrene, and benzoic acid single crystals on the (001) cleavage surfaces. The variation of the hardness with applied load decreased with increasing load. The σ_y (estimated from the hardness) of phenanthrene was higher compared to anthracene, which was attributed to the geometrical disposition of the molecules in the lattice, despite having similar crystal structures. However, the σ_y for benzoic acid was observed in between that of anthracene and phenanthrene, due to the occurrence of slip activity along the (010) [010] system. Nevertheless, such slip activity was absent in anthracene and phenanthrene. The authors attribute this observation to the hindrance to glide along the [100] direction and the availability of a large number of molecules in benzoic acid.

Marwaha et al. [81] measured the microhardness of different molecular crystals of anthraquinone, hexamine, and stibene along with anthracene and phenanthrene. Among the five crystals studied, hexamine belongs to the body-centered cubic structure (space group: $I\bar{4}3m$), whereas the rest belong to the monoclinic crystal structure of the space group $P2_1/a$. The indentations were performed on the (110) face of hexamine and the (001) face of all the other crystals. The study concluded that the active slip systems in monoclinic systems were $(20\bar{1})$ [010] and (100) [010] types, whereas (1 10) (1 11) and (112) (1 11) slip systems were responsible for the plastic deformation in hexamine crystals.

Sgualdino et al. [82] employed the Vickers microhardness test on (100), (001), and (110) facets of sucrose crystals and found that the microhardness correlated well with the attachment energy rather than the surface energy. Elban et al. [83] assessed the fracture behavior of (100) planes of sucrose crystals using Vickers indentation hardness testing, which was related to their attachment energy calculations. Ramos and Bahr [84] performed nanoindentation on (100) and (001) faces of sucrose crystals and reported that hardness anisotropy was not considerably high for both the orientations compared to the anisotropic nature in modulus. In fact, the hardness of the cleavage planes was greater than that of the habit planes, and it was attributed to the surface roughness of the crystals. The elastic-plastic transition point occurred at a maximum applied τ of 1 GPa, and the propagation of plastic deformation was crystal orientation-dependent, as evident from the non-uniform natured pile-up around the residual indent impressions. Previously, sucrose was considered as brittle with a limited number of slip systems, but the nanoindentation studies by Ramos and Bahr [84] revealed the inherent significant plastic deformation mechanism at the nanoscale.

Kiran et al. [85] used the instrumented nanoindentation technique to investigate the mechanical anisotropy and correlate with the intermolecular interactions in saccharin crystals. The active slip system in the saccharin crystal is (100) [011]. Both the (100) and (011) faces were indented with a Berkovich nanoindenter with an in-situ scanning probe microscopy (SPM) imaging capability. On the [100], the molecules (as centrosymmetric $NH\cdots O$ dimmers) stacked down, and make an oblique angle to the (100). Further, the molecules within stacks were stabilized through weak $\pi\cdots\pi$ interactions, and adjacent stacks were bound by the $CH\cdots O$ bonds. In contrast, in the (011) plane, stacked dimmers were arranged in a crisscross arrangement and $CH\cdots O$ bonds were arranged at $90°$ to the (011) (see Figure 6a–e). During indentation, while the loading part of the *P-h* curve of the (011) face was smooth, several distinct pop-ins were evident on the (100) face, as shown in Figure 5f. Interestingly, the first pop-in width (18 nm) was found to be the integral multiples of the interplanar spacing which was explained using the contact mechanics of a spherical indenter [85]. Further, the plastic deformation was seen on both the faces, even at a load of 0.01 mN, due to the sharp geometry of the indenter tip. While the homogeneous plastic deformation on the (011) face was attributed to the existence of several slip systems that are nearly parallel to the plane of the indentation direction, the discrete plasticity on the (100) was due to the lowest attachment energy slip planes, which act as cleavage planes and are prone to pop-ins due to their higher compressibility. The occurrence of pop-ins on (100) plane was attributed to the disruption of $CH\cdots O$ hydrogen bonds followed by an elastic compression of stacked columns through weak $\pi\cdots\pi$ interactions [1]. As a result, the columns broke away. Further, the relationship between the pop-in magnitude and the interplanar spacing was observed and rationlized the results with the aid of indentation contact mechanics. Interestingly, the pop-in magnitude measured to be the intergral multiples of the interplanar spacing. At the higher loads, anisotropic cracking was evident on the (100) planes along the corners of the sharp indenter. Furthermore, pile-up inhomogeneity around the residual indent impression indicated that plastic deformation was crystallographic orientation-dependent.

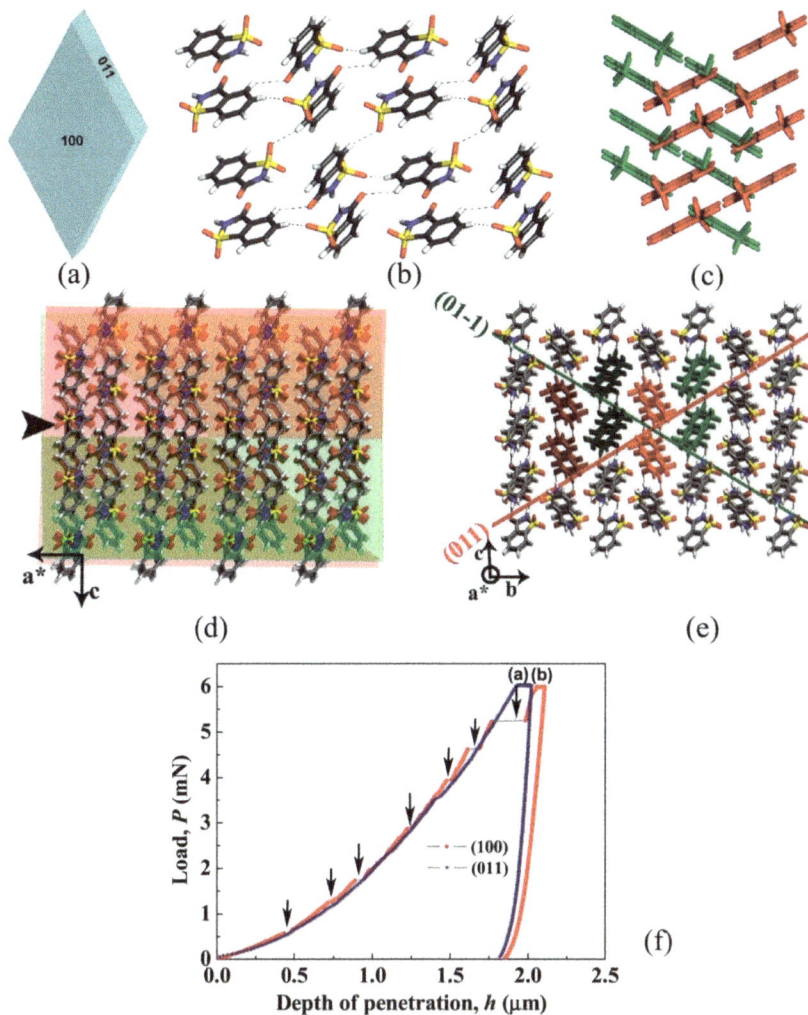

Figure 6. (a–e) Saccharin crystal with index faces, top view of intermolecular interactions in the (011) plane, oblique angle arrangement of molecules concerning the (100) plane, view of molecular arrangement along (001) plane, and the stacking of molecules in the [100] direction, respectively. The indentation direction (a*) is represented by the arrowhead. (f) Representative *P-h* curves obtained from the (100) and (011) planes. Arrows indicate pop-ins. (Reprinted with permission from the American Chemical Society, Reference 85).

Zhou et al. [86] made efforts to understand the mechanical anisotropy of 1,1-diamino-2,2-dinitroethylene (FOX-7) energetic crystals, which have wavelike π-stacks (see Figure 7a–c), using nanoindentation and density functional theory (DFT) calculations. As expected, the crystal exhibited distinct mechanical behavior from various faces upon nanoindentation, as shown in Figure 7d. While the hardness and stiffness of the (020) face were the highest, the (002) face exhibited the lowest values. Further, the (002) exhibited significant pile-up around the indent, and no cracking was observed, probably because of its soft and plastic behavior, while the (-101) and (002) faces exhibited cracks but with less pile-up. The observed mechanical anisotropy of (020) was attributed to the wavelike π stacking of FOX-7 molecules along the (020) with the support of hydrogen bonds. The uniaxial compression and shear sliding of the FOX-7 crystals calculated using DFT supported the nanoindentation results [86]. The authors conclude that the wavelike π stacking was responsible for the low impact sensitivity of the 1,1-diamino-2,2-dinitroethylene molecule, rather than the other explosives with distinct packing structures [86].

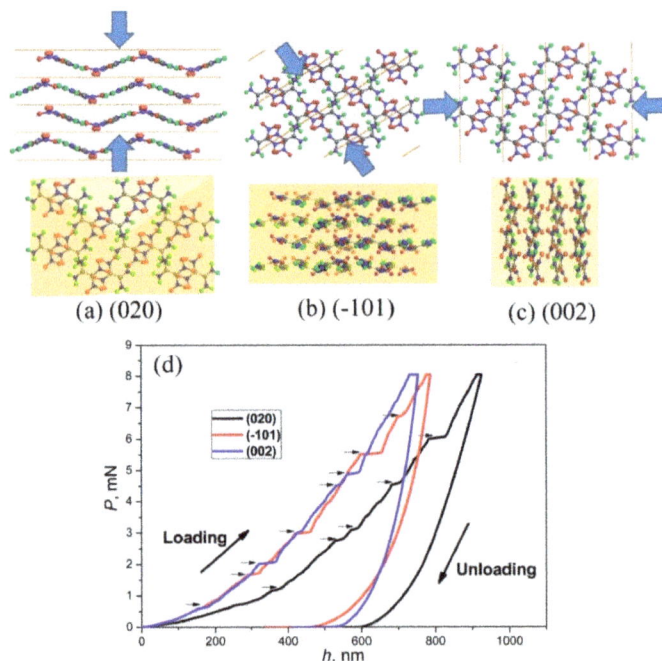

Figure 7. Molecular packing of FOX-7 crystals viewed along (top) and vertical to (bottom) the three indentation orientations of the (020) (**a**), (−101) (**b**), and (002) (**c**) faces, and (**d**) P−h curves obtained from (020), (-101), and (002) faces. The horizontal arrows represent pop-ins during loading. (Reproduced with permission from the American Chemical Society, Reference 86).

Mathew and Sewell [87] characterized the temperature dependence of the mechanical response and the very early stages of elastic/plastic deformation in 1,3,5-triamino2,4,6trinitrobenzene (TATB) by simulating nanoindentation using MD simulations. The authors used a rigid, spherical indenter and simulated the displacement-controlled nanoindentation curves on the (100), (010), and (001) planes of TATB. While the initial part of the P-h curves on the (001) basal plane follows Hertzian contact behavior (i.e., elastic), pile-up, kinking and delamination were also evident at the elastic-plastic transition. The nanoindentation hardness of the basal planes was predicted to be 1.02 ± 0.09 GPa. However, nanoindentation on the non-basal faces (100) and (010) exhibited non-Hertzian loading behavior, which was attributed to the "softening" of molecular layers due to elastic bending. In addition, the pile-up height on the non-basal planes was observed to be less significant than that on the basal plane. The anisotropic behavior observed from the basal and non-basal planes was attributed to the heating that developed during indentation, which was found to be higher for the basal plane.

Taw et al. [88] used nanoindentation to measure the elastic and plastic properties of representative as-grown sub-millimeter orthorhombic, monoclinic, and triclinic molecular crystals. So far, researchers have performed nanoindentation on relatively large crystals, but this work used as-grown small crystals for nanoindentation. The as-received molecular crystals of TATB, cyclotetramethylenetetranitramine (HMX), FOX-7, azodiaminoazoxyfurazan (ADAAF), and a trinitrotoluene and 2,4,6,8,10,12-hexanitro-2,4,6,8,10,12-hexaazaisowurtzitane cocrystal (TNT/CL-20) were indented using a Berkovich indenter in this study. The results show that the onset of plasticity on the loading part of the P-h curve occurred consistently at a τ value between 1 and 5% of the elastic modulus in all of the crystal systems studied (see Figure 8). Further, the H to E_r ratio observed in this study for different crystal systems suggested that the conventional Berkovich tips failed to generate fully self-similar plastic zones in organic crystals because the H/E_r ratio varied in the present case from 0.039 to 0.044, whereas the model for fully plastic indentations of organic crystals suggested that the H/E_r ratio should vary between 1 and 3. Therefore, the authors concluded that the indents performed by the Berkovich indenter in this study were more elastic [88].

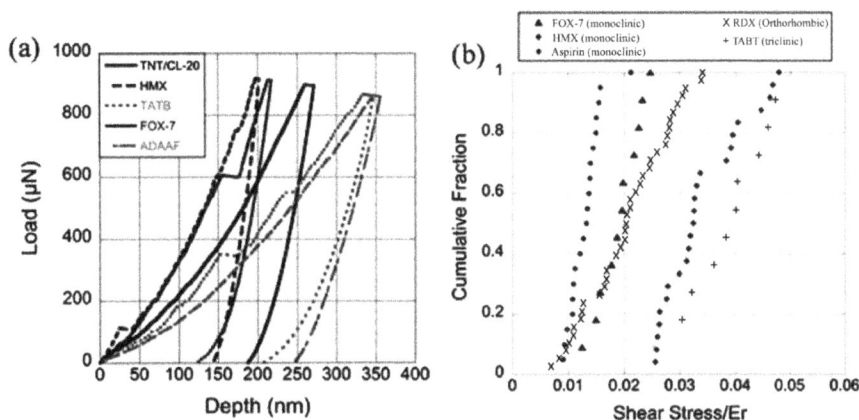

Figure 8. (a) The representative P-h curves of TNT/CL-20, TATB, HMX, FOX-7, ADAAF. (b) The τ value at a yield normalized by E_r shows that the window of plasticity is between 1 and 5% of the E_r for multiple molecular crystals. (Reproduced with permission from Materials Research Society, Reference 88).

3.1.3. Mechanical Behavior of Aspirin Polymorphs

Varughese et al. [89] utilized the nanoindentation technique to measure the mechanical properties of aspirin polymorphs and to understand the interaction characteristics and instability caused by τ. In both polymorphs, the carboxy groups formed centrosymmetric OH···O dimers and arranged as two-dimensional layers parallel to the (100) [1,89]. Although the crystal structures of the two forms are closely related, two distinct stabilizing CH···O interactions exist in form I [1,89]. They are: (1) the aromatic ring and the acetyl carbonyl group contacts, and (2) the methyl group and ester carbonyl group interactions. In contrast, in form II, the CH···O contacts between acetyl substituents of molecules related by a crystallographic 2_1 screw axis are the stabilizing interactions across the slip planes. Further, a small shift of adjacent layers parallel to one of the crystallographic axes relates form I and II structures (see Figure 9a,b) [1,89]. Nanoindentation was performed using a Berkovich tip with an end radius of 75 nm on the structurally equivalent {001} face of form I and the {102} face of form II (i.e., the indentation direction is normal to the potential slip planes but parallel to the interlayer shift direction that relates forms I and II). Upon loading, interestingly, while the loading parts of the P-h curves obtained for both the {001} of form I and the {102} of form II are smooth, pop-ins were seen on the {100} of form I, as shown in Figure 9c. The post-indent characterization on the {100} of polymorph I using scanning probe microscopy revealed a fracture along the [010] direction at higher loads, as shown in Figure 9d. Neither fracture nor pile-up was observed for either the {001} of form I or the {102} of form II. The softer nature of form II compared to form I was observed from the h_{max}. It was observed that the {001} of form I is 37% harder than the {102} of form II. The nanoindentation results justified the solid-state transformation of polymorph II to I if the samples were left under ambient conditions, but the process may be accelerateded with the application of τ.

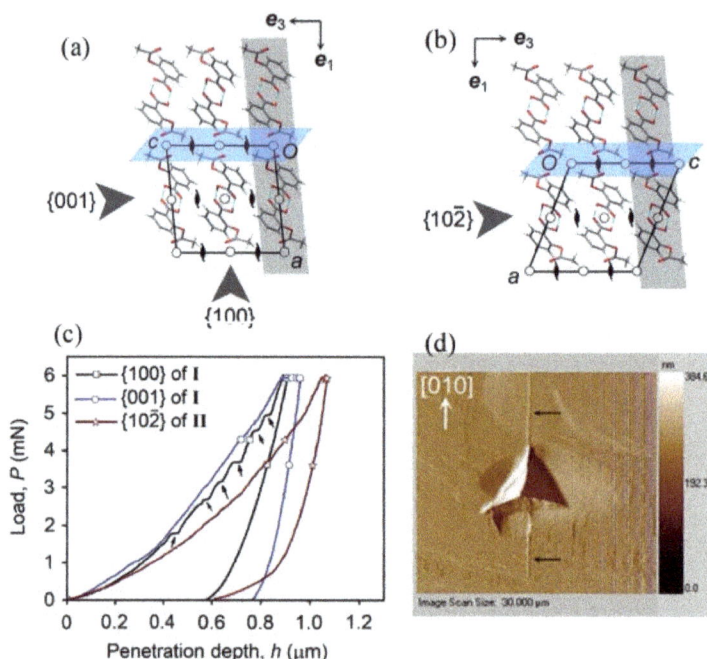

Figure 9. Crystal packing of the aspirin polymorphic forms: (**a**) form I, and (**b**) form II. The planes parallel to the {001} or {102} are highlighted by grey slabs and slip planes are colored blue. (**c**) *P-h* curves obtained from different faces of both forms, with pop-ins indicated by arrows. (**d**) Post-indent image obtained from the {100} of form I shows a crack running along the <010> at higher loads. (Reproduced with permission from the Royal Society of Chemistry, Reference 89).

Olusanmi et al. [90] investigated the anisotropic plastic deformation and fracture behavior of crystals of aspirin form I. While severe fracture was observed on the (001) plane, the (100) plane was found to be more fracture resistant, indicating that the (001) plane is the preferred cleavage plane for aspirin form I (see Figure 10a,b). The measured hardness on the (001) face was reported to be lower than the (100) face, indicating softer nature of the (001) face. Similar to the observation of Varughese et al. [89], pop-ins were noted on both the (001) and (100) faces. However, deeper and more frequent pop-ins were seen on the (100) face. While the pop-ins on the (100) face were associated with cracking, the slip mechanism was responsible for the shallow pop-ins on the (001) face. Therefore, it was suggested that the cleavage planes in aspirin crystals dominate the fracture mechanism under both quasi-static and impact loading conditions.

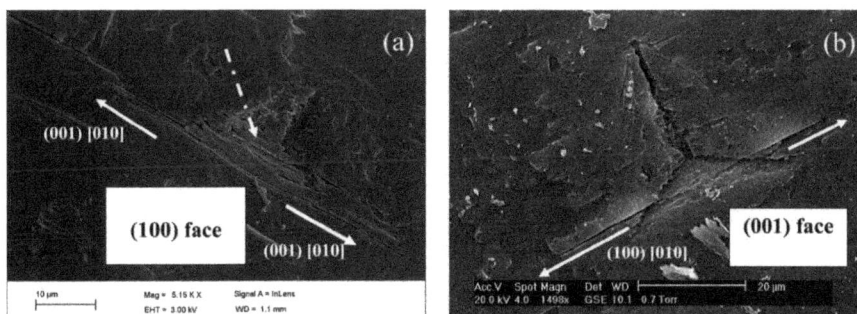

Figure 10. The scanning electron microscopic image of residual indent impression on aspirin (**a**) (100) and (**b**) (001) faces. A fracture is evident in the [010] direction. (Reproduced with permission from Elsevier, Reference 90).

3.1.4. Mechanoluminiscence Studies in Difluoroavobenzone

Krishna et al. [91] utilized load-controlled Berkovich nanoindentation on single crystals of difluoroavobenzone compounds and investigated their mechanoluminescence properties. The crystal packing of both green and cyan forms are shown in Figure 11a,b. The (100) plane of BF_2dbm (^{t}Bu)2 crystals was formed by the hydrophobic ^{t}Bu groups, which allowed for easy plastic deformation through the bending of the (001) face, which makes a 90° angle with the slip plane (100). The H value was found to be higher on the (100) plane when compared to the (001) because a slip in the (001) was easily formed, as the indentation direction was parallel to the slip plane. While the nanoindentaion measured a high hardness value for BF_2dbm(OMe)2 crystals, three-point bending experiments revealed its susceptibility to localized plastic deofrmation. This is because the (100) slip plane was parallel to the indentation direction of [010]; therefore, maximum resistance was offered against indentation penetration. In addition, the higher values of plastic and elastic properties were attributed to the strong C-H ... F interactions. The third compound, BF_2dbmOMe, exhibited lower hardness values than the second compound because the C-H ... F interactions were slightly obliquely angled to the direction of indentation. Further, the nanoindentation revealed that the crystals of the BF_2dbm(^{t}Bu)2 compound were much softer compared to both the shearing (BF_2dbm(OMe)2) and brittle (BF_2dbmOMe) type crystals. Finally, the nanoindentation results provided a rationale for the extent of the plastic deformation behavior due to the prominent mechanoluminescence in the BF_2dbm(tBu)2 compound, moderate mechanoluminescence in the BF_2dbm(OMe)2 compound, and no detectable mechanoluminescence in the BF_2dbmOMe compound, under identical conditions. The below Table 1 summarizes the plastic properties of different planes of BF_2AVB form I crystals.

Table 1. Schmid factors of various slip systems for different planes of BF_2AVB form I crystals [91].

Plane	(011)	(001)	(120)
Indentation direction	[016]	[001]	[810]
Slip plane	(010)	(010)	(010)
Slip direction	[001]	[001]	[100]
Schmid factor	0.48	0.35	0.24
Hardness (Mpa)	275 ± 12	340 ± 7	410 ± 11
2-D layers arranged (°) with respect to indentation plane	22	90	12

The above table shows that the molecular layer arrangement with respect to the (120) plane, which lies at an angle of ca. 12°, and the indenter axis, which is nearly normal to the slip plane. Therefore, pop-ins were not observed in the loading part of the *P-h* curve of the (120) (see Figure 11c). Further, the Schmid factor for the (120) planes was relatively smaller than the (011) and (001) planes; therefore, severe fracture (Figure 11d) was observed rather than slip [91].

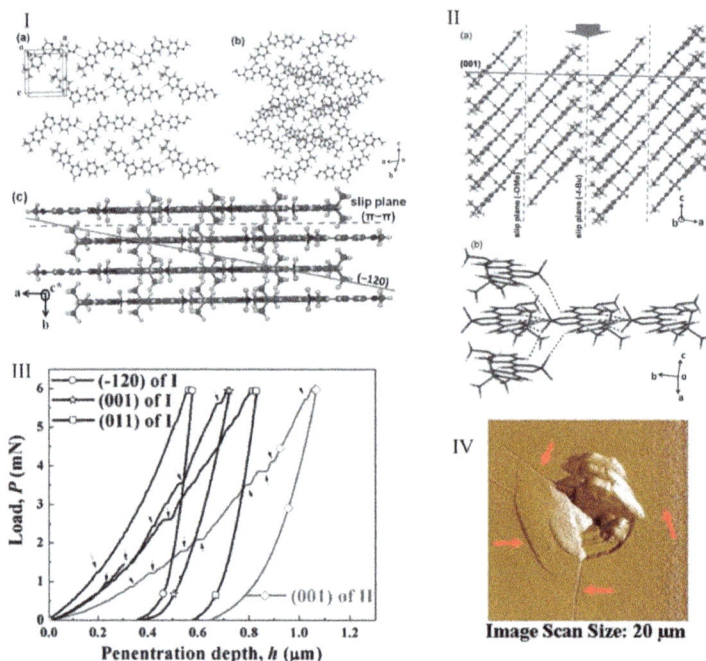

Figure 11. Crystal packing of the BF$_2$AVB, **I** (**a–c**). Green form (brittle), **II** (**a–c**). Cyan form, **III**. Representative *P-h* curves obtained from various faces ((011), (001), and (120)) of the green polymorph crystals and the (001) plane of the cyan polymorph. **IV**: The post-indent AFM image of the (120) face of the green polymorph shows cracks along the indenter corners. (Reproduced with permission from Wiley VCH, Reference 91).

3.1.5. Tuning of Hardness in Organic Crystals

Mishra et al. [92] focused on methods to tune resistance to plastic flow (hardness) in organic crystals. In the case of ductile inorganic materials such as metals etc., numerous strengthening methods (such as grain boundary strengthening in polycrystalline samples, precipitation hardening in alloy materials, and work hardening–deformation at low temperatures) were developed and understood very well. However, the methods mentioned above do not work in the case of organic crystals. Mishra et al. [92] adopted an alternative method known as solid solution strengthening or hardening to tune the resistance to plastic flow in organic crystals. In the solid solution hardening method, the target hardness is achieved by mixing solute and solvent molecules in the desired concentration. Instrumented nanoindentation was employed on a series of omeprazole polymorphs (tautomeric forms of omeprazole are shown in Figure 12a,b), and revealed that proper design of the crystals using basic crystal engineering design principles leads to improving the lattice resistance to shear sliding of the molecular layers upon application of mechanical stress, thus increasing the organic crystals' resistance to plastic flow [92]. A Berkovich tipped nanoindentation was performed on the major face {001} of five polymorphs of omeprazole. The characteristic *P-h* curves obtained from various polymorphs are shown in Figure 12c. The hardness values revealed that the hardness of polymorph V was nearly double that of polymorph I because of its layered structure and easy sliding nature during loading. However, despite having a similar layered structure in polymorphs II-V, they exhibited higher values of hardness because of the 5-methoxy group, which provides a higher resistance to shear sliding of molecular layers compared to polymorph I. The linear increase in hardness (see Figure 12d) in omeprazole polymorphic forms of I-V was attributed to the percentage of tautomer T1 in the polymorphs [92].

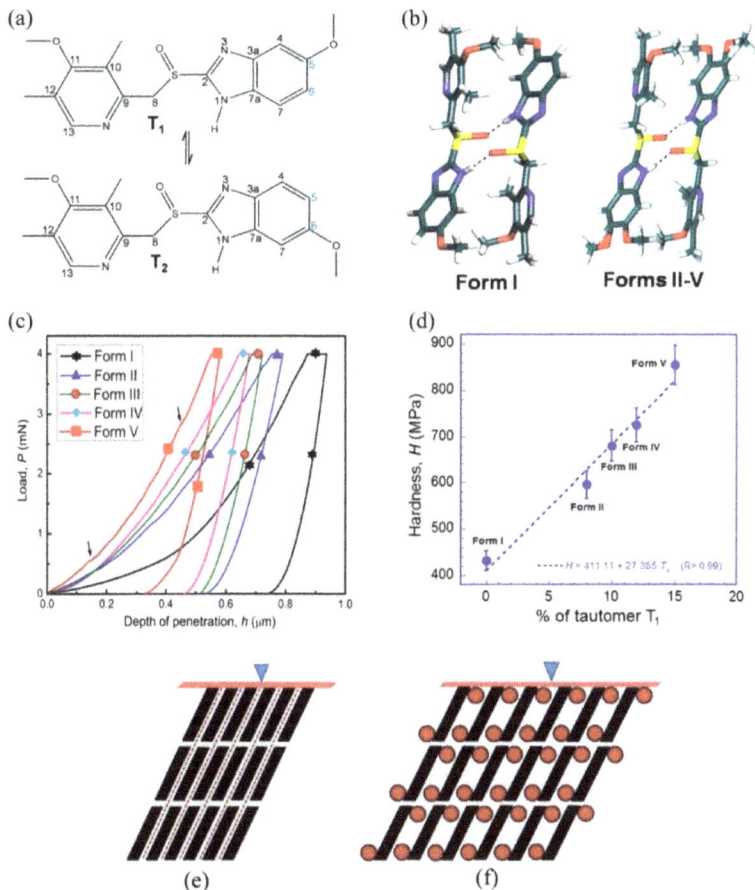

Figure 12. (a) Tautomeric forms of omeprazole. (b) All five forms of omeprazole contain centrosymmetric N-H···O=S dimers. Note the positioning of the 5- and 6-methoxy groups on the benzimidazole ring. (c) Representative *P-h* curves of all five forms of omeprazole. (d) The linear correlation between H and proportion of the 5-methoxy tautomer, T_1 in omeprazole polymorphs. (e) Schematic crystal packing of omeprazole form I (a) and forms II–V (b). The dimers of are depicted as solid parallelograms. The direction of indentation [001] is shown as a solid triangle. Slip planes of the form I are represented by red dotted lines. (f) Note that the methoxy groups are shown in solid red circles in forms II–V. (Reproduced with permission from American Chemical Society, Reference 92).

Strengthening Organic Crystals by the Co-Crystallization Approach

A three-sided pyramidal sharp nanoindentation was used by Sanphui et al. [93] to measure elastic and plastic properties and to understand the deformation behavior of voriconazole and its cocrystal and salt forms. The idea behind this study was to strengthen voriconazole, which is a highly soft material, by co-crystallizing voriconazole with both aliphatic and aromatic co-formers and forming salts with HCl and oxalic acid in different stoichiometric ratios. In fact, this method is known as the co-crystallization approach, known to alter the physicochemical and mechanical properties of Active Pharmaceutical Ingredients (APIs). [94,95]. It is evident from the literature that the mechanical properties and physiochemical properties were improved by forming cocrystals of caffeine with methyl gallate [96] ibuprofen, as well as flurbiprofenwith nicotinamide [97] and vanillin isomers with 6-chloro-2,4-dinitroaniline [98], etc. Similarly, Sanphui et al. [95] observed that the salt forms (i.e., voriconazole + HCl) were considerably stiffer (80%) and harder (58%) than voriconazole and its cocrystals. Further, the loading portions of the load-displacement curves obtained in the salt form showed pop-ins (Figure 13), indicating discontinuous plastic deformation, and their magnitude was the integer multiples of the interplanar spacing of the specific planes indented in the study. The lower hardness in voriconazole was attributed to the presence of weakly connected parallel slip planes in the indentation direction, which facilitated easy shearing or gliding for the planes during the application of mechanical pressure. However, the salt form resulted in increased hardness and stiffness because of the presence of the strong ionic interactions and hydrogen bonds in between the slip planes, which offer high resistance to the shearing of planes. Interestingly, the cocrystals exhibited in lower stiffness compared to the salt forms because of the presence of weaker non-covalent interactions in cocrystals, while salts have stronger ionic interactions.

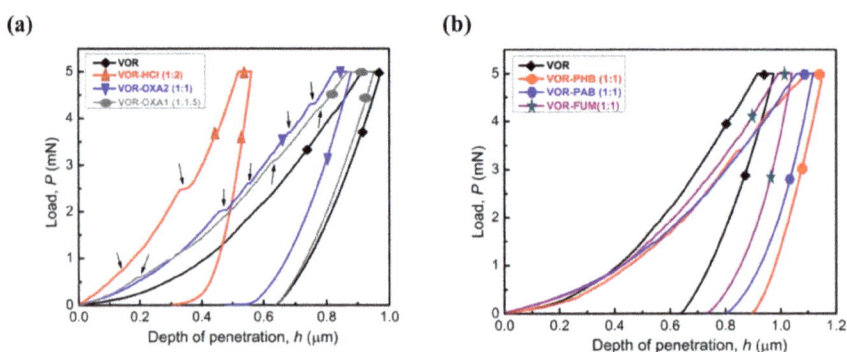

Figure 13. Representative $P-h$ curves obtained on voriconazole and its (**a**) salts and (**b**) cocrystals. (Reproduced with permission from Elsevier, Reference 93).

3.1.6. Establishing a Correlation Between Hardness and Solubility

Mishra et al. [99] investigated the correlation between the hardness and solubility of molecular crystals by performing nanoindentation on curcumin and sulfathiazole polymorphs. Among four sulfathiazole polymorphs, it was observed that polymorph I had the highest tendency to flow plastically upon mechanical pressure, and polymorph II was found to be the hardest because of the molecular layers that made a higher inclination angle with the indenter direction with respect to the different intermolecular interactions in I. In the case of curcumin, polymorph I was found to be the hardest because of the twisted molecular conformation in the crystal structure in comparison to polymorphs II and III, in which the closed cell parameters and planar structural packing allowed the polymorphs to deform plastically to a greater extent. The studies concluded that the hardness and solubility were inversely correlated in these polymorphs, as shown in Figure 14, indicating that the hardest polymorphs were less soluble and the softest were highly soluble. Further, the inverse relation suggested that the order of hardness can be utilized as a parameter to measure the solubility order in close energy-related polymorphic systems [99].

The Schmid factor values of the major faces of sulfathiazole polymorphs are summarized in Table 2 given below. The values of Schmid factor follow an inverse relation with the hardness of the crystal planes. Further, the molecular layers in forms II, III, and IV are normal to the indentation direction; therefore, pop-ins were noted in their P-h curves [99].

It is clear from the above table that the Schmid factor of form I is higher than that of forms II, III, and IV, and the inclination angle for form I is larger than that of the other forms; therefore, severe plasticity is seen in form I [99].

Table 2. The Schmid factor and plastic deformation behavior of sulfathiazole polymorphs [99].

Sulfathiazole Polymorphs	Major Face	Slip Direction	Schmid Factor	H (GPa)	Angle (Degree) between the Trace of the Molecular Layer and the Indentation Direction
Form 1	(100)	[102]	0.468	0.356 ± 0.010	145.3
Form II	(100)	[102]	0.039	1.080 ± 0.015	92.3
Form III	(100)	[001]	0.089	0.704 ± 0.018	95.1
Form IV	(10$\bar{1}$)	[001]	0.043	0.881 ± 0.012	92.5

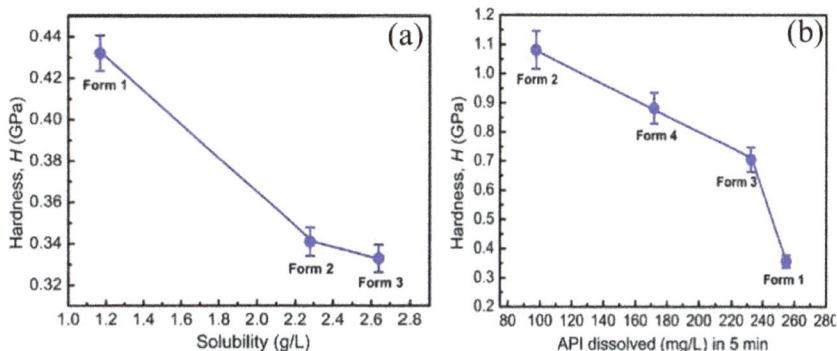

Figure 14. An inverse correlation of hardness and solubility in (a) curcumin and (b) sulfathiazole polymorphs. (Reproduced with permission from the American Chemical Society, Reference 99).

3.1.7. Indentation-Induced Plasticity in Parabens and Paracetamol

Feng and Grant [100] examined how the slip planes in parabens (such as methyl, ethyl, n-propyl, and n-butyl 4-hydroxybenzoate) influence the plastic behavior upon the application of mechanical stress. Instrumented nanoindentation was performed on the major faces of single crystal parabens of different morphologies (plate, blade, octahedral, etc.). The (111) face of methyl paraben exhibited the highest values of hardness among all the other parabens due to the absence of the slip planes. In the case of the ethyl (100) face, propyl (100) face, and butyl (002) face parabens, the slip planes facilitated severe plastic deformation upon the application of mechanical stress, which thus resulted in lower H values. Further, the highest d-spacing between the slip planes in the ethyl paraben caused increased slip activity compared to the isostructural propyl paraben; therefore, lower hardness values were observed. However, in the case of the butyl paraben, longer and more bulky alkyl chains provide resistance to the gliding of the slip planes; therefore, higher hardness compared to the ethyl and propyl parabens were observed. Duncan-Hewitt et al. [101] and Finnie et al. [102] performed indentation studies on paracetamol (acetaminophen) and revealed that the hardness varied as a function of indenter orientation with the crystallographic direction within the indentation planes.

3.1.8. Establishing the Relation between Plastic Behavior in Bulk and Single Crystals of APIs

Egart et al. [103] studied the nanomechanical properties of APIs such as nifedipine, famotidine, olanzapine, and piroxicam, in order to establish the plastic behavior correlation between bulk and single crystals. A distinct difference in the plastic behavior was observed in two polymorphs of famotidine due to the intermolecular packing under the indenter. While form I had the dense and highly cross-linked packing of molecules, form II had the slip planes that caused higher plasticity than form I. It is worth noting that lower hardness was measured for form II; though the indentation direction was normal to the slip plane and improved hardness was observed, it was still lower than that of form I. For the highly cross-linked structures, brittle behavior was observed. The authors established good correlations between bulk (Walker coefficient) and single crystal plasticity (indentation hardness) parameters. Their studies concluded that the inherent crystal deformation behavior based on the crystal packing greatly defined their compressibility and compactibility properties during tableting.

3.1.9. In Situ Nanoindentation to Study Disorders in APIs

Chen et al. [104] utilized nanoindentation along with high-resolution total scattering pair distribution function (TS-PDF) analysis coupled for mechanical property assessment and for the study of the disorders that can occur in API crystals during milling and tableting processes. While the PARP (poly (ADP-ribose) polymerase) compound was brittle in nature, sphingosine-1-phosphate receptor agonist and antagonist were plastic. The compound PARP exhibited almost four times higher hardness than the other compound. Significant fracture was evident on PARP, along with pop-ins in the loading portion of the *P-h* curve. However, the other compound exhibited pile-up, indicating severe plastic flow towards the surface along the sides of the indenter. The mechanical properties were attributed to the crystal packing in both compounds. After evaluating the mechanical properties of both compounds using nanoindentation, the crystals were milled and compacted under a variety of conditions. The resulting structural disorders during milling and compaction were then evaluated using synchrotron-based high-resolution total scattering pair distribution function (TS-PDF) analysis, and a good correlation was observed with the process conditions.

3.1.10. Strain-Rate Sensitivity Studies

Indentation strain-rate sensitivity (SRS) was examined for various organic crystals using the nanoindentation technique with a quasi-static load by Raut et al. [105]. They revealed that the plastic deformability does not depend on the rate at which the crystals are deformed. In the case of metals and alloys at room temperature, SRS is known to arise due to the lattice friction experienced by the dislocations during their glide over slip planes in response to the applied indentation stress [105]. The value of the strain-rate sensitivity index, m, ranges between 0 and 0.3 and increases with temperature, which is positive and high for metals, indicating higher resistance to the localization of plastic deformation [105]. However, the molecular crystals, which are stabilized by intermolecular interactions such as van der Waals (0.004–0.04 eV) and hydrogen bonds (0.1–0.4 eV), are considerably lower than metallic bonds (~1 eV). Therefore, the bonds break easily in organic crystals upon the application of pressure, which in turn implies that the plastic deformation does not require the movement of dislocations; slip can occur through shearing of slip planes [105]. Therefore, the m values for the organic crystals were measured close to zero and confirmed strain-rate insensitivity nature, as shown in Figure 15. Katz et al. [106] estimated the SRS using time-dependent plasticity studies via indentation creep tests on various APIs in the tablet form, and observed that the m values varied between 0.007 and 0.055, indicating that APIs are strain-rate insensitive. Though the above studies provide information regarding the role of SRS on the plastic behavior of organic crystals, they do not comment on the role of SRS in the pop-in behavior and fracture in organic crystals, which is necessary for a complete understanding of the SRS of organic crystals.

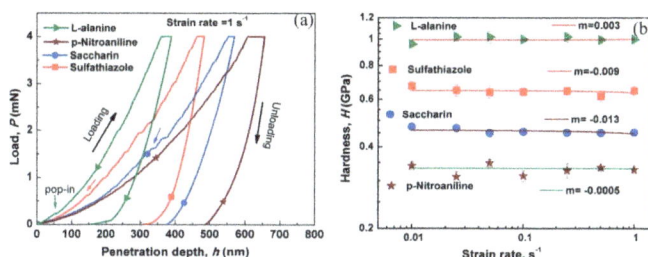

Figure 15. (**a**) Representative *P-h* curves obtained using Berkovich nanoindentation on the major faces of different organic molecular single crystals at a strain rate of $\dot{\varepsilon} = 1$ s^{-1}, and (**b**) the *m* values obtained by fitting the indentation hardness data with the flowing equation, $m = \frac{d\ln H}{d\ln\dot{\varepsilon}}$. (Reproduced with permission from Royal Society of Chemistry, Reference 105).

3.1.11. Nanoscratch Experiments: Anisotropy in Molecular Movements

Kaupp and Naimi-Jamal [107] reported the anisotropic deformation upon mechanical stress in organic crystals. They used nanoindentation and scratch experiments to demonstrate long-range anisotropic molecular movements in an organic crystal, as shown in Figure 16. Thiohydantoin crystal, which has a well-defined layered crystal with cleavage planes along the (102) planes, was indented with a cube corner tip on the (110) face. Interestingly, pile-up was observed only one side of the indenter (though it was a three-sided pyramidal tip), where the slope of the molecular monolayers matched with the indenter surface angle [107]. When nanoscratches were made at four different directions on the (110) face, four distinct molecular migration phenomena were seen (i.e., movement of molecular monolayers only to the left, only to the right, on both sides along with a pile-up in the scratch front, and the abrasion of the material) [107]. The molecular layers that were arranged in a 66° steep angle on the (110) face were greatly influenced. However, nanoscratching on the (102) cleavage plane resulted in no long-range molecular movement, but abrasion of the material was evident [107]. Further, anisotropic long-range molecular migration was detected in other compounds such as anthracene, ninhydrin, tetraphenylethylene, thiohydantoin, and thiourea, which have cleavage planes or anisotropic molecular packing, and such an effect was found to be dependent on the layer orientation arrangement and the direction of the tip movement [1].

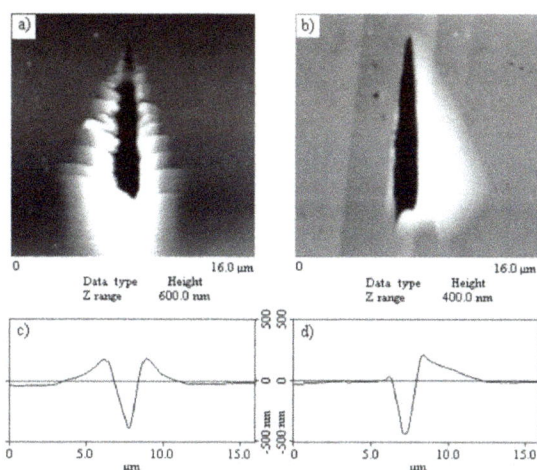

Figure 16. *Cont.*

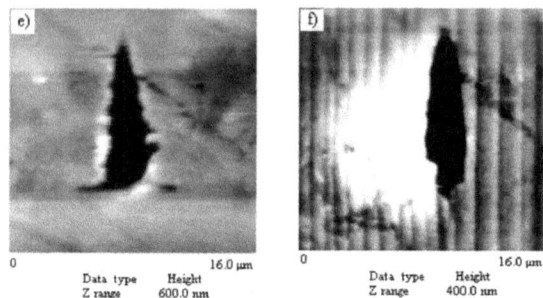

Figure 16. Atomic force microscope images of residual scratches on the (110) face of thiohydantoin crystal. (**a**) Along the skewed layers direction showing pile-up only on both sides, (**b**) scratch along the cleavage plane resulting in pile-up only on the right side, (**c**) horizontal cross-section profile at the broadest width in (**a**), (**d**) horizontal cross-section profile at the broadest width in (**b**), (**e**) scratch against the sloping of the skew layers showing no-pile up, and (**f**) scratch along the cleavage planes resulting in pile-up only on the left side. (Reproduced with permission from the Wiley, Reference 107).

Varughese et al. [108] studied the layer migration in pyrene and phenanthrene based on two charge transfer complexes of 1,2,4,5-tetracyanobenzene (TCNB) using nanoscratching experiments. A 75-nm sharp three-sided pyramidal Berkovich tip was used to scratch the major (100) and (002) faces of 1:1 TCNB–pyrene crystals, which have a layered arrangement (shown in Figure 17). While the layers mean that the plane is parallel to the (100), they make an angle of 68° tilt to (002). The large difference in the interaction characteristics and layer arrangement of both the faces resulted in significant mechanical anisotropy (H (16%) and E (21%)). Indentation on the (002) face resulted in two important plastic flow methods. They are: (1) the sliding of layers over the edge of the indenter tip and, (2) pile-up along one of the faces of the indenter. The sliding of the layers is attributed to the matching of the half angle of the indenter [108] to the molecular layers on the (002), which enables the layers to slide over the edge of the tip and creates pile-up in other orientations because of the slant arrangement of the layers [1]. The indentation scratch profile analysis and friction coefficient measurements on the (002) face revealed that the molecular migration was depndent on layer orientation and direction. As expected, the scratch along the tilt direction resulted in molecular layer migration, and such activity was noticeable on both sides of the scratch along with a small pile-up at the end of the scratch. Nevertheless, similar scratching against the tilted layered direction resulted in greater friction to the indenter movement and a significant pile-up at the end of the tip (see Figure 17 c,d). Unlike the TCNB– pyrene complex, the 1:1 TCNB–phenanthrene complex had no layer arrangement; the trimers stack down the [001], and they are approximately parallel to the (001) and perpendicular to the (020). Due to the entirely different molecular arrangements in this compound, nanoscratch experiments yield entirely different results compared to the TCNB-pyrene complex. When a scratch test was performed along the cleavage plane, limited layer migration was evident, as shown in Figure 18, which was attributed to the presence of hydrogen bonds in the interlayer region, providing high resistance to the indenter movement. Several unexpected observations were observed, such as the relation between the distance between two consecutive troughs to the multiples of interplanar spacing when scratched along the orthogonal direction.

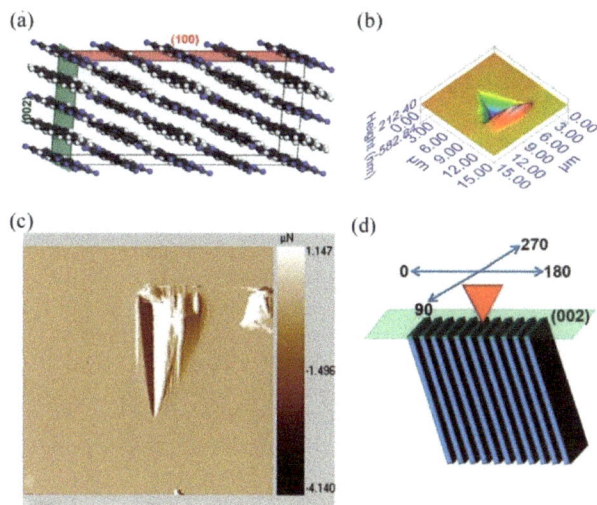

Figure 17. TCNB–pyrene complex. (**a**) Crystal packing. The major and minor faces are shown in red and green colored slabs. (**b**) Residual indent impression on the (002) showing pile-up only on one side of the indenter. (**c**) AFM scan image 25 μm in size, showing layer migration (towards the right direction only) upon nanoscratching. (**d**) Schematic representation of the indenter movement and layer arrangement. (Reproduced with permission from Wiley-VCH, Reference 108).

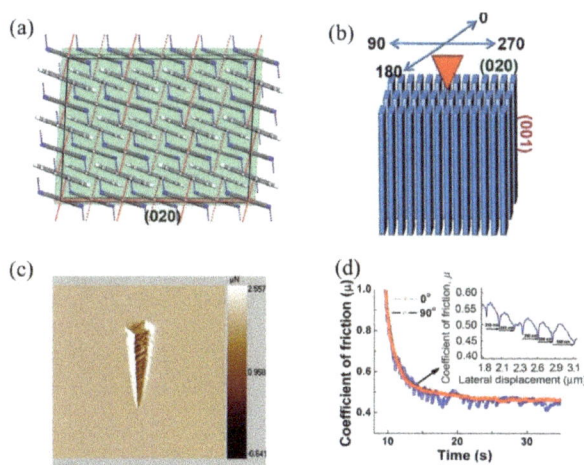

Figure 18. TCNB–phenanthrene complex. (**a**) Crystal packing. (**b**) Schematic representation of the indenter movement along different directions and layer arrangement of the (020) face. (**c**) AFM image with a scan size of 25 μm, showing layer resistance towards the indenter movement and a limited pile-up at the end of the scratch track, and (**d**) The coefficient of friction at 0° and 90° shows distinct scratching behavior. The inset shows the inhomogeneous travel of the indenter during scratching against molecular layers. (Reproduced with permission from Wiley-VCH, Reference 108).

3.1.12. AFM Nanoindentation to Study the Slip Planes

Jing et al. [109] studied the slip planes of succinic acid with a rotating sample method using AFM nanoindentation. An inhomogeneous stress field was created on the crystals using a sharp cube corner tip, which helped in activating different slip systems when the specimen was rotated. When indentation was performed on both the (001) and (010) crystal faces, the major slip planes were (010) and (111), and they were in agreement with the attachment energy calculations [109]. Interestingly, along with the predicted slip systems, several unpredicted higher index operative slips planes were also observed at different sample rotations [109]. The AFM images of residual indents on the (001) face of succinic acid, with the (010) trace being the reference, are shown in Figure 19.

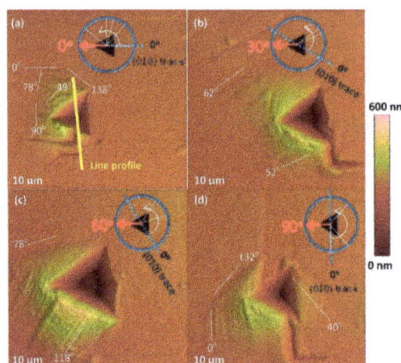

Figure 19. AFM images of residual indents on the (001) face of succinic acid, with the (010) traces being the reference. (Reproduced with permission from American Chemical Society, Reference 109).

Thakuria et al. [110] explored both the AFM and nanoindentation techniques to identify the two polymorphs of caffeine-glutaric acid cocrystals. They distinguished the variation in slip mechanism and height differences in supramolecular layers by AFM imaging and employing nanoindentation. Nanoindentation on both forms revealed that the resistance to plasticity of form I was lower than that of form II, which was attributed to the corrugation between the caffeine-glutaric acid layers that provide resistance to slide/stretch along the slip planes. The brittleness index showed that form II had a higher value than form I because of the higher ductile nature of form II. The indentation hardness data of a wide variety of organic crystals as well as their orientation dependence were measured by various groups, and are tabulated in Table 3.

Roberts and Rowe [111] proposed that the H/P_y ratio can be used to assess the mechanical deformation of materials. Here P_y indicates the yield pressure of the material. According to the H/P_y ratio, the materials can be classified as, (i) very plastic for the ratio between 1.5–2.0, (ii) brittle for H/P_y values in the range of 2.0–2.2, and (iii) plastic, if $H/P_y \geq 3$. However, Duncan-Hewitt et al. [112] proposed that a low H/E value yielded better compaction behavior of the APIs.

3.1.13. Real-Time Imaging of Indentation-Induced Structural Changes in Piroxicam

Manimunda et al. [113] coupled in situ SEM nanoindentation with Raman spectroscopy, with the intention of exploring the real-time indentation-induced structural deformation in the (011) and (011) faces of piroxicam single crystals, as shown in Figure 20. While the hardness of the (011) was 0.82 ± 0.03 GPa, the (011) face exhibited 0.64 ± 0.05 GPa. The mechanical anisotropy in both faces was attributed to the difference in the resolved shear stresses (RSS) and distinct difference in interlayer interactions along the [001] and [010] directions. In situ Raman spectroscopy at different loads on the (011) and (011) faces reveal changes in SO_2 vibrational modes and C–O stretching modes during indentation, respectively, which was attributed to the variation in interlayer and intralayer interactions in both faces. These in situ studies provided real-time information on the chemical changes and corresponding mechanical deformation behavior of piroxicam crystals.

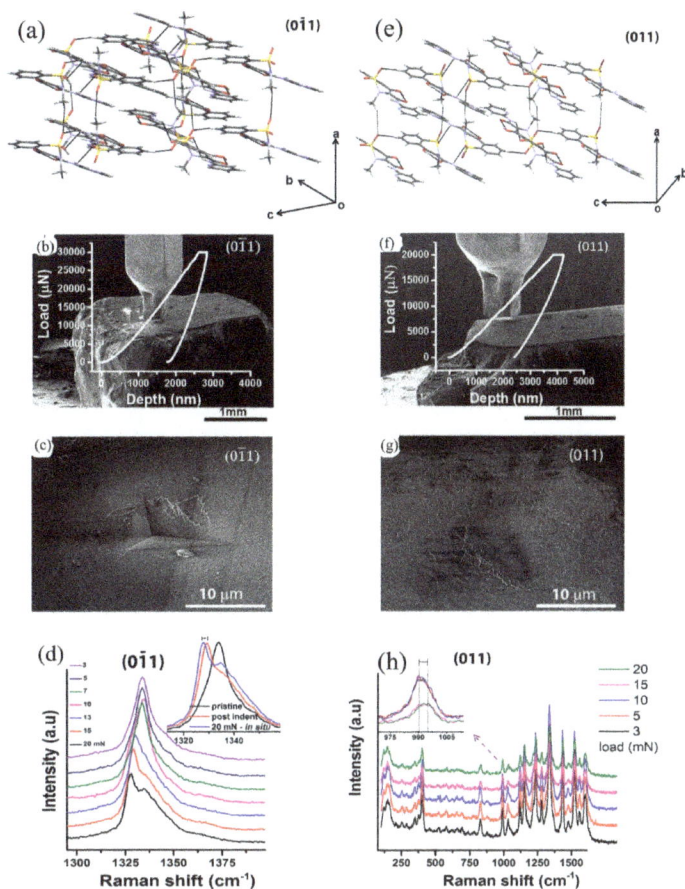

Figure 20. (**a**,**e**) Crystal packing of piroxicam molecules along the (011) and (011) faces. (**b**,**f**) In situ SEM nanoindentation on the (011) and (011) faces along with *P-h* responses. (**c**,**g**) SEM images of the residual indent impressions on the (011) and (011) indented at 30 mN load, and (**d**,**h**) Raman spectra obtained from the (011) and (011) faces at different loads. (Reproduced with Permission from Springer and Copyright Clearance Center, Reference 113).

3.1.14. Phase Transformations under Applied Load

Lie et al. [114] investigated pressure-induced amorphization in acetaminophen, sucrose, c-indomethacin, and aspirin crystals under applied load. The plastic response was calculated using a phase field dislocation dynamics theory that could predict the fraction of amorphous material formed in crystals under an applied stress. Their results showed that the volume fractions of amorphous material after the plastic deformation were quite large for c-indomethacin and sucrose, and smaller for acetaminophen and aspirin. Though there are many studies on the phase transformation behavior of various organic crystals [115] under hydrostatic compression experiments (such as the diamond anvil cell test, where a tiny amount of powdered sample is placed between diamond anvils), to the best knowledge of the authors, there is no report so far seen on indentation-induced phase transformations in organic crystals.

Table 3. The variety of organic crystals used by various research groups for nanoindentation study and their indentation hardness values.

S. No	Crystal	Crystal Face	Type of Tip Used	Hardness, H (GPa)	Reference
1	1,1-Diamino2,2-dinitroethylene	(0 2 0) (−1 0 1) (0 0 2)	Berkovich -do- -do-	0.52 ± 0.05 0.63 ± 0.02 0.67 ± 0.03	[86] [86] [86]
2	Saccharin	(1 0 0) (0 1 1) (1 0 0) (0 1 1)	Zircon Berkovich tip -do- A cube-corner indenter A cube-corner indenter	0.530 ± 3.0 0.501 ± 2.3 0.610 ± 0.01 0.550 ± 0.02	[116] [116] [85] [85]
3	L-alanine	(0 0 1) (1 0 1)	-do- -do-	0.114 ± 4.8 0.943 ± 3.1	[116] [116]
4	BF$_2$dbm(Bu)$_2$	(0 0 1)	Berkovich	0.092 ± 4.04	[98]
5	BF$_2$dbm(OMe)$_2$	(0 1 0)	Berkovich	0.264 ± 10.8	[98]
6	BF$_2$dbm(OMe)	(0 0 1)	Berkovich	0.255 ± 8.48	[98]
7	Sodium Saccharin dihydrate crystals	(0 0 1) (0 0 1) (0 1 1) (1 0 1)	Berkovich -do- -do- -do-	1.20 ± 0.04 0.78 ± 0.03 0.662 ± 0.02 0.716 ± 0.02	[117] [116] [118] [118]
8	Piroxicam form-1	(−1 0 0) (0 1 1) (0 1 −1)	-do- -do- -do-	0.56 ± 0.18 0.67 ± 0.04 0.42 ± 0.02	[2,119] [2,119] [2,119]
9	Famotidine form A	(−1 0 0) (0 0 −1)	-do- -do-	1.58 ± 0.4 1.35 ± 0.16	[2,119] [2,119]
10	Famotidine form B	(−1 0 1)	-do-	0.84 ± 0.16	[2,119]
11	Nifedipineα-form	(1 0 0)	-do-	0.71 ± 0.61	[2,119]
12	Olanzapine form 1	(1 0 0) (0 −1 −1)	-do- -do-	0.74 ± 0.04 0.72 ± 0.02	[2,119] [2,119]
13	Aspirin polymorph-1	(1 0 0) (0 0 1) (0 0 1) (1 0 0)	-do- -do- -do- -do-	0.257 ± 0.007 0.240 ± 0.008 0.10 0.12	[89] [89] [90] [90]
14	Aspirin polymorph-2	(1 0 $\bar{2}$)	-do-	0.152 ± 0.004	[89]
15	Sildenafil Citrate	—	-do-	0.52 ± 0.06	[120]
16	Voriconazole	—	-do-	0.13 ± 0.01	[120]
17	Sucrose	(1 0 0) (0 0 1) — —	-do- -do- -do- Diamond tip	1.62 ± 0.17 1.57 ± 0.07 2.3 ± 0..4 2 ± 0.5	[90] [86] [121] [122]
18	Lactose	—	Diamond tip	0.43 ± 0.08	[122]
19	Absorbic Acid	—	Diamond tip	5.6 ± 1.8	[122]
20	TATB	(0 0 1)	Berkovich	1.02 ± 0.09	[87]
21	α-RDX	(2 1 0) (2 1 0) (2 1 0) (0 2 1) (0 0 1) (0 0 1) — (0 2 1) (2 1 0) Multiple	-do- -do- -do- -do- -do- -do- -do- -do- -do- -do-	0.672 ± 0.035 0.798 ± 0.030 1.06 0.681 ± 0.033 0.615 ± 0.035 1.05 0.74 ± 0.09 0.681 0.798 0.74	[87] [87] [87] [87] [87] [87] [88] [88] [88] [88]
22	β-HMX	(0 1 0) (0 1 0)	-do- -do-	1.13 ± 0.045 0.65 ± 0.09	[87] [87]
23	HMX	— (0 1 0)	-do- -do- -do-	0.95 0.65 0.99 ± 0.06	[88] [88] [88]
24	LIM-105	(0 1 0)	-do-	0.72 ± 0.10	[87]
25	Acetaminophen	(0 1 1)	-do-	0.875 ± 0.029	[87]

Table 3. *Cont.*

S. No	Crystal	Crystal Face	Type of Tip Used	Hardness, H (GPa)	Reference
26	VOR	(1 0 0)	-do-	0.366 ± 2.8	[93]
27	VOR-HCl	(0 1 1)	-do-	0.870 ± 6.0	[93]
28	VOR-OXA1	(0 1 0)	-do-	0.426 ± 5.8	[93]
29	VOR-OXA2	(1 0 0)	-do-	0.628 ± 2.0	[93]
30	VOR-FUM	(1 0 0)	-do-	0.292 ± 3.4	[93]
31	VOR-PAB	(1 0 0)	-do-	0.264 ± 5.0	[93]
32	VOR-PHB	(1 0 0)	-do-	0.262 ± 1.6	[93]
33	Ibuprofen Lot A	—	-do-	0.6 ± 0.1	[121]
34	Ibuprofen Lot B	—	-do-	0.4 ± 0.1	[121]
35	Ibuprofen Lot C	—	-do-	0.22 ± 0.04	[121]
36	UK-370106	—	-do-	0.4 ± 0.1	[121]
37	Acetaminophen	—	-do-	1.0 ± 0.2	[121]
38	Phenacetin	—	-do-	0.9 ± 0.2	[121]
39	PHA-739521	—	-do-	1.1 ± 0.1	[121]
40	MCC	—	-do-	1.4 ± 0.3	[121]
41	Fluconazole	—	-do-	2.0 ± 0.3	[121]
42	TATB	(0 0 1) —	-do-	0.48 0.41 ± 0.04	
43	TNT/CL-20	—		0.63 ± 0.13	
44	FOX-7	(0 2 0) ($\overline{1}$ 0 1) (0 0 2) —		0.52 0.63 0.67 0.86 ± 0.08	[88]
46	ADAAF	— (2 1 0) (0 0 1)		0.23 0.672 0.615	

4. Fracture Behavior of Organic Crystals

Since organic crystals are known as brittle materials, the accurate measurement of the fracture toughness becomes challenging, because creating a sharp pre-crack is quite difficult without breaking the specimen, and notched specimens give erroneously high values [123–125]. Therefore, an alternative approach was developed to assess the fracture toughness of brittle materials by making direct measurements of cracks created using sharp probes such as Vickers, Knoop, cube-corner, and Berkovich [126–129]. The sharp tips produce high strain under the tip that leads to a fracture. Therefore cube-corner and Berkovich tips with end radii of ~75 nm and ~30 nm, respectively, are usually used to investigate the fracture behavior of brittle materials at the nanoscale. A cube-corner tip results in better fracture because the total included angle is approximately 90°, compared to the Berkovich tip, which has a total included angle of 142.3°. Then, the critical stress intensity factor (K_{IC}) is estimated using the crack length, indenting load, and the H/E ratio of the material. The fracture in molecular crystals takes place either through cleavage at certain crystallographic planes (brittle crystals) or at the maximum τ, where dislocation pile-up attains a critical density (ductile or plastic crystals) [1,90].

Lawn and colleagues [126,130] developed an expression for fracture toughness by relating the c, H, E, and P_{max}, namely:

$$K_{IC} = \xi \left(\frac{E}{H}\right)^{\frac{1}{2}} \left(\frac{P_{max}}{c^{\frac{3}{2}}}\right),$$ (12)

The value of ξ, an empirical constant that depends on the geometry of the indenter, for the cube-corner indenter and the Berkovich indenter was 0.032 and 0.016, respectively, as proposed by Harding, Oliver, and Pharr [131]. Anstis et al. [127] showed the usefulness of this relation, by studying a number of brittle materials with a wide range of fracture toughness. It is important to note that the above expression is designed for ceramic materials, where the major assumption was that K_c depends on the assumption that $P/c^{3/2}$ is constant. Taylor et al. [120] disclosed that the above assumption is also valid for pharmaceutical crystals. In their investigation, they observed that $P/c^{3/2}$ is indeed constant for pharmaceutical materials. They used a much-refined form of the equation to find out fracture toughness for organic crystals, and it is given below [120]:

$$K_c = x_v \left(\frac{a}{l}\right)^{\frac{1}{2}} \left(\frac{E}{H}\right)^{\frac{2}{3}} \left(\frac{P}{c^{\frac{3}{2}}}\right) \tag{13}$$

where x_v is the calibration constant of the indenter used, a is the indent diagonal, l is the length of a crack, and c is the crack length given by $a + l$, as shown in Figure 21.

Lawn and Marshall [132] defined the brittleness of a material as the ratio of indentation hardness to fracture toughness. According to Lawn and Marshall [132], the brittleness index (*BI*) is:

$$BI = \frac{H}{K_c} \tag{14}$$

An excellent correlation was observed between the brittleness index and milling data for pharmaceutical crystals by Taylor et al. and Olusanmi et al. [90,120]. Based on the *BI*, APIs were distinguished and classified as easy, moderate, and difficult to mill. Hence, the *BI* can be used as a means of understanding the mechanical properties of compounds early in development, and for selecting appropriate milling conditions with a minimum amount of bulk [90,120].

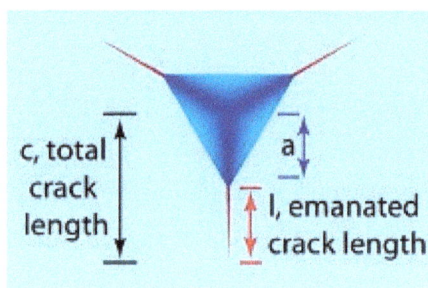

Figure 21. Schematic diagram of a cube-corner indenter and the residual indent with surface cracks emanating from the indent corners. (Reproduced with permission from Elsevier, Reference 42).

Olusanmi et al. [90] observed that the activation energies of the various plastic deformation slip systems determine the intensity of the fracture. Earlier studies have shown that the fracture behavior in organic crystals is anisotropic. Elban et al. [83] reported that the freshly cleaved sucrose crystal surfaces of the (100) have a fracture toughness of 0.055 MPa.m$^{1/2}$ at the load range of 0.15–4.9 N. However, at higher loads, the toughness value was reported to be lower by Duncan-Hewitt and Weatherly [133]. They could not observe the variation of fracture toughness for different planes, suggesting that the fracture behavior of sucrose was independent of the crystallographic direction, probably due to the isotropic nature of sucrose. Further, they observed short length cracks (compared to all other planes) on the (1 0 0) plane in the [1 1 0] direction. However, the preferred cleavage plane in sucrose is reported to be the (1 0 0), as well as the lowest attachment energy plane [134].

Duncan-Hewitt and Weatherly [134], in addition to microindentation studies by Prasad et al. [135], showed that the K_c of paracetamol was crystallographic orientation-dependent, with the lowest K_c measured on the (010) cleavage plane [90]. Also, they observed that most of the cracks formed parallel to the (010) plane when indented on different faces of paracetamol crystals. Olusanmi et al. [90] demonstrated strong fracture anisotropy in aspirin polymorph I, where cracks originated from plastically deformed regions and propagated on preferential cleavage planes. They observed that the K_c of the (001) plane was significantly lower than that of the (100), because the (001) plane is the preferred cleavage plane for aspirin polymorph I.

Varughese et al. [89] revealed interesting results on the fracture behavior of aspirin crystals. They also observed severe fracture on aspirin polymorph I on the (100), as also noted by Olusanmi et al. [90]. Since the radius of the plastic zone (r_p) in front of the crack tip is inversely proportional to the square of the hardness ($\sigma/3$), it follows that the plane with higher K_c (higher r_p) will have a lower value of hardness. Indeed, this was true in the Varughese et al. [89] case, because the (100) of form I holds the highest hardness among all other orientations and therefore severe fracture was observed. In addition, irrespective of indenter direction, fracture occurred along the <010> direction on the (100) of aspirin form I, as shown in Figure 22. The estimated K_c of the (100) of form I was 0.004 ± 0.0001 MPa m$^{0.5}$, which was much lower than the (100) and (102) of form II, indicating the high fracture-resistant behavior of aspirin form I. The brittleness index of the (100) of aspirin form I was 49×10^3 m$^{-0.5}$, which indicates that the aspirin form I crystals along the (100) were extremely brittle. Interestingly, no fracture was observed on the (001) of form I or on both faces of form II.

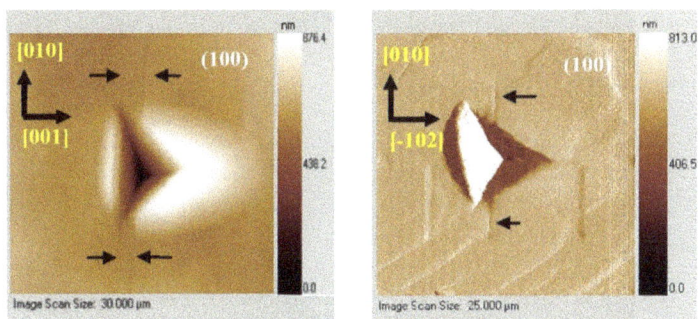

Figure 22. Irrespective of the indentation direction, the fracture is evident only in the <010> direction when indented on the {100} of aspirin polymorph I.

Kiran et al. [85] estimated the K_c of saccharin crystals to be 0.002 MPa m$^{0.5}$. Wendy et al. [112] estimated the K_c of sucrose, adipic acid, and acetaminophen, as well as that of NaCl, using the microindentation technique, and reported the K_c values to be 0.08 ± 0.001, 0.02 ± 0.005, 0.05 ± 0.006, and 0.50 ± 0.07 MPa m$^{0.5}$, respectively. The estimated brittleness index of saccharin was comparable to that of ice (2.8×10^3 m$^{-0.5}$) [85].

The K_c of the (−1 0 2) face of difluoroavobenzene (BF$_2$AVB) mechanochromic crystals was estimated [91] using a Berkovich indenter to be 0.054 ± 0.002 MPa m$^{0.5}$. The estimated BI of the BF$_2$AVB crystals was 7.3×10^3 m$^{-0.5}$, which is much higher than that of ice (2.8×10^3 m$^{-0.5}$). Since the crystals are characterized by a three-dimensional arrangement of hydrogen bonds, no significant pile-up or fracture was seen along the (001) face. However, significant pile-up along the one side of the indent and several corners and radial cracks were observed for (120). The above results suggest that the anisotropic plastic deformation and fracture behavior in organic crystals arise as a consequence of molecular packing, and that interaction strengths are determined by the crystal structure.

5. Conclusions and Outlook

It is evident from the above literature that the mechanical behavior of organic molecular crystals has gained tremendous attention over the last 10 years with the advent of small-scale mechanical testing systems such as nanoindentation and computational methods. In particular, the instrumented nanoindentation technique under quasi-static (time-independent) loading conditions was effectively used on a wide variety of molecular crystals to gain knowledge on their anisotropic mechanical behavior and their structural origins. Compared to the molecular crystals, the structure-mechanical properties of other classes of materials like ceramics, metals, polymers, polymer nanocomposites, semiconductors, and Bulk Metallic Glasses (BMGs) are well explored by performing state-of-the-art experiments, computations, simulations, and theoretical methods. Although the mechanical behavior of molecular crystals is not yet completely realized, this area of research has significant potential for applications in polymer science, crystal engineering with desired properties, pharmaceutical technology, etc. For example, the better understanding of structure-mechanical property relationships saves the pharmaceutical industries from huge economic loss. H and K_c are two important mechanical properties that determine the compressibility and tabletability of APIs. Therefore, a wide selection of materials and the proper design of molecular solids with desired physiochemical properties using basic principles of crystal engineering are required to accomplish both millability and tabletability in APIs.

Plastic properties such as H, K_c, and m of different classes of materials were collected from the literature and are represented in Figure 23. It is clear from Figure 23a that the hardness values of organic molecular crystals vary between 200 MPa and 2 GPa. However, with the wise selection of materials and the modification of intermolecular interactions using the knowledge obtained from crystal engineering concepts, the ability to resist plastic deformation can be improved in this class of materials. As shown in Figure 23b, the fracture toughness of molecular crystals is the lowest among all the other classes of materials. It is important to have materials with higher hardness, but not by sacrificing another important property, namely, fracture toughness. Therefore, more research should be focused on achieving moderate hardness as well as fracture toughness. As shown in Figure 23c, the m value for organic crystals measures closer to zero or slightly towards a negative value, indicating the strain-rate insensitivity nature of molecular crystals under an applied load.

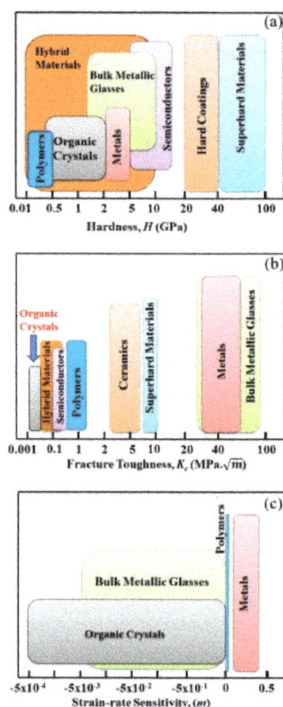

Figure 23. Schematic representation of comparison of the (**a**) hardness [2,85–90,93,98,110,113, 116–122,136–140]. (**b**) fracture toughness [85,89,90,110,112,141–147], and (**c**) strain-rate sensitivity [105,106,148–152] of organic crystals with other classes of materials.

The present authors see great progress in understanding the relationship between structure-plasticity and structure-fracture of molecular crystals using the instrumented indentation technique. However, other than quite a number of indentation studies, the theoretical and computational understanding on the deformation behavior of organic crystals is lacking. Further, studies on the fatigue, creep, and temperature dependence of molecular crystals were not explored well. These studies are important as they provide information about the time-dependent plasticity, high strain-rate dependent plasticity, and the determination of activation volumes and energies at the elastic-plastic transition points. Further, the actual cause of plastic flow in organic crystals is also not yet clearly understood. For example, in the case of inorganic crystalline materials, plasticity was explained via dislocation activity [153,154], twinning [155,156], and phase transformations [153,156] under the indenter. However, such an understanding is lacking in regard to molecular crystals because performing post-indent chacracterizations such as cross-sectional Transmission Electron Microscopy (TEM) on molecular crystals is a challenging task.

Recent advances in techniques like picoindentation, powder compaction, and high-pressure-induced spectrometry have established [136] structure–mechanical property correlations in molecular crystals and have provided new insights on the subject. As small-scale mechanical testing equipment is now getting advanced in combination with other characterization techniques and imaging capabilities, in situ Raman, in situ electrical measurement, and in situ imaging capabilities provide excellent opportunities to explore and exploit pressure-induced phase transformations, real-time monitoring of indentation-induced plasticity and fracture mechanics during loading and unloading, and electrical conductivity measurements of conductive and piezoelectric organic crystals, etc.

The modulation of mechanical properties of a molecular solid by modifying intermolecular interactions has been achieved by some researchers [64,92,93]. Therefore, future aims should work towards engineering functional molecular solids with desired physical and chemical properties with the knowledge gained from the subject of crystal engineering and structure-mechanical property correlations.

Acknowledgments: Kiran S. R. N. Mangalampalli thanks the Science and Engineering Research Board, Department of Science and Technology, Government of India for an Early Career Researcher Award (File No: ECR/2016/000827).

Author Contributions: Kiran S. R. N. Mangalampalli wrote the manuscript. Sowjanya Mannepalli helped in collecting references and writing the manuscript.

Conflicts of Interest: The authors declare no conflict of interest.

References

1. Varughese, S.; Kiran, M.S.R.N.; Ramamurty, U.; Desiraju, G.R. Nanoindentation in crystal engineering: Quantifying mechanical properties of molecular crystals. *Angew. Chem. Int. Ed.* **2013**, *52*, 2701–2712. [CrossRef] [PubMed]
2. Egart, M.; Jankovi, B.; Lah, N.; Ilic, I.; Srcic, S. Nanomechanical properties of selected single pharmaceutical crystals as a predictor of their bulk behavior. *Pharm. Res.* **2015**, *32*, 469–481. [CrossRef] [PubMed]
3. Loo, Y.L.; Someya, T.; Baldwin, K.W.; Bao, Z.; Ho, P.; Dodabalapur, A.; Katz, H.E.; Rogers, J.A. Soft, conformable electrical contacts for organic semiconductors: High-resolution plastic circuits by lamination. *Proc. Natl. Acad. Sci. USA* **2002**, *99*, 10252–10256. [CrossRef] [PubMed]
4. Kim, W.; Palilis, L.C.; Uchida, M.; Kafafi, Z.H. Efficient silole-based organic light-emitting diodes using high conductivity polymer anodes. *Chem. Mater.* **2004**, *16*, 4681–4686. [CrossRef]
5. Remenar, J.F.; Morissette, S.L.; Peterson, M.L.; Moulton, B.; Macphee, J.M.; Guzman, H.R.; Almarsson, O. Crystal engineering of novel cocrystals of a triazole drug with 1,4-dicarboxylic acids. *J. Am. Chem. Soc.* **2003**, *125*, 8456–8457. [CrossRef] [PubMed]
6. Velaga, S.P.; Basavoju, S.; Bostrom, D. Norfloxacin saccharinate-saccharin dihydrate cocrystal—A new pharmaceutical cocrystal with an organic counter ion. *J. Mol. Struct.* **2008**, *889*, 150–153. [CrossRef]

7. Kaiser, C.R.; Karlapais, C.; de Souza, M.V.N.; Wardell, J.L.; Solange, D.; Wardell, M.S.V.; Edward; Tiekink, R. Assessing the persistence of the n–h··· n hydrogen bonding leading to supramolecular chains in molecules related to the anti-malarial drug, chloroquine. *CrystEngComm* **2009**, *11*, 1133–1140. [CrossRef]
8. Schultheiss, N.; Newman, A. Pharmaceutical cocrystals and their physicochemical properties. *Cryst. Growth. Des.* **2009**, *9*, 2950–2967. [CrossRef] [PubMed]
9. Sun, C.C. Materials science tetrahedron—A useful tool for pharmaceutical research and development. *J. Pharm. Sci.* **2009**, *98*, 1671–1687. [CrossRef] [PubMed]
10. Byrn, S.R.; Pfeiffer, R.R.; Stowell, J.G. *Solid-State Chemistry of Drugs*; Ssci, Inc.: West Lafayette, IN, USA, 1999.
11. Reddy, C.M.; Krishna, G.R.; Ghosh, S. Mechanical properties of molecular crystals—Applications to crystal engineering. *CrystEngComm* **2010**, *12*, 2296–2314. [CrossRef]
12. Mørch, Y.A.; Holtan, S.; Donati, L.; Strand, B.L.; Bræk, G.S. Mechanical properties of c-5 epimerized alginates. *Biomacromolecules* **2008**, *9*, 2360–2368. [CrossRef] [PubMed]
13. Rosa, C.D.; Auriemma, F. Structural-mechanical phase diagram of isotactic polypropylene. *J. Am. Chem. Soc.* **2006**, *128*, 11024–11025. [PubMed]
14. Patel, S.; Sun, C.C. Macroindentation hardness measurement-modernization and applications. *Int. J. Pharm.* **2016**, *506*, 262–267. [CrossRef] [PubMed]
15. Courtney, T.H. *Mechanical Behaviour of Materials*, 2nd ed.; Waveland Press Inc.: Long Grove, IL, USA, 2008.
16. Rupert, T.J.; Gianola, D.S.; Gan, Y.; Hemker, K.J. Experimental observations of stress-driven grain boundary migration. *Science* **2009**, *326*, 1686–1690. [CrossRef] [PubMed]
17. Tresca, H. Memoire Sur l'ecoulement des Corps Solides Soumis a Des Fortes Pressions. *C. R. Acad. Sci. Paris* **1864**, *59*, 754–758.
18. Saint-Venant, B.D. Memoire sur l'etablissement des Equations Differentielles des Mouvements Interieurs Operes Dans Les Corps Solides Ductiles au Dela des Limites ou l'elasticite Pourrait Les Ramener a Leur Premier etat. *C. R. Acad. Sci. Paris* **1870**, *70*, 473–480.
19. Levy, M. Memoire Sur Les Equations Generales des Mouvements Interieurs des Corps Solides Ductiles au Dela Limits ou l'Elasticite Pourrait Les Rammener a LeurPpremier etat. *C. R. Acad. Sci. Paris* **1870**, *70*, 1323–1325.
20. Vonmises, R. Mechanics of solids and shells: Theories and approximations. *Göttin. Nachr. Math. Phys.* **1913**, *1*, 582–592.
21. Lin, J. *Fundamentals of Materials Modelling for Metals Processing Technologies: Theories and Applications*; Imperial College Press: London, UK, 2015.
22. Prager, W. *Introduction to the Mechanics of Continua*; Dover Publications Inc.: Mineola, NY, USA, 1961.
23. Hill, R. *The Mathematical Theory of Plasticity*; Clarendon Press: Oxford, UK, 1950.
24. Eduardo, A.; Neto, D.S.; Djordjeperic; David, R.; Owen, J. *Computational Methods for Plasticity: Theory and Applications*; Wiley Publishers: Hoboken, NY, USA, 2008.
25. Sherwood, J.N. *The Plastically Crystalline State (Orientationally-Disordered Crystals)*; Sherwood, J.N., Ed.; Wiley: Chichester, UK, 1979.
26. Roberts, R.J.; Rowe, R.C.; York, P. The relationship between indentation hardness of organic solids and their molecular structure. *J. Mater. Sci.* **1994**, *29*, 2289–2296. [CrossRef]
27. Ramamurty, U.; Jang, J. Nanoindentation for probing the mechanical behavior of molecular crystals—A review of the technique and how to use it. *CrystEngComm* **2014**, *16*, 12–23. [CrossRef]
28. Brinell, J.A. Way of determining the hardness of bodies and some applications of the same. *Tek. Tidskr.* **1900**, *5*, 69–87.
29. Meyer, E. Investigations of hardness testing and hardness. *Phys. Z.* **1908**, *9*, 66–74.
30. Hutchings, I.M. The contributions of David Tabor to the science of indentation hardness. *J. Mater. Res.* **2009**, *24*, 581–589. [CrossRef]
31. Tabor, D. A simple theory of static and dynamic hardness. *Proc. R. Soc. Lond. A* **1948**, *192*, 247–274. [CrossRef]
32. Tabor, D. *The Hardness of Metals*; Oxford University Press: Oxford, UK, 1951.
33. King, R.F.; Tabor, D. The effect of temperature on the mechanical properties and the friction of plastics. *Proc. Phys. Soc. B* **1953**, *66*, 728–736. [CrossRef]
34. Pascoe, M.W.; Tabor, D. The friction and deformation of polymers. *Proc. R. Soc. A* **1956**, *235*, 210–224. [CrossRef]

35. King, R.F.; Tabor, D. The strength properties and frictional behaviour of brittle solids. *Proc. Phys. Soc. A* **1954**, *223*, 225–238. [CrossRef]

36. Atkins, A.G.; Tabor, D. Mutual indentation hardness of single-crystal magnesium oxide at high temperatures. *J. Am. Ceram. Soc.* **1967**, *50*, 195–198. [CrossRef]

37. Liu, M. Crystal Plasticity and Experimental Studies of Nano-Indentation of Aluminium and Copper. Ph.D. Thesis, University of Wollongong, Wollongong, Australia, 2014.

38. Fischer-Cripps, A.C. A review of analysis methods for sub-micron indentation testing. *Vacuum* **2000**, *58*, 569–585. [CrossRef]

39. Oliver, W.C.; Pharr, G.M. An improved technique for determining hardness and elastic-modulus using load and displacement sensing indentation experiments. *J. Mater. Res.* **1992**, *7*, 1564–1583. [CrossRef]

40. Oliver, P. Measurement of hardness and elastic modulus by instrumented indentation: Advances in understanding and refinements to methodology. *J. Mater. Res.* **2004**, *19*, 1–20. [CrossRef]

41. Kucharski, S.; Mroz, Z. Identification of yield stress and plastic hardening parameters from a spherical indentation test. *Int. J. Mech. Sci.* **2007**, *49*, 1238–1250. [CrossRef]

42. Kruzic, J.J.; Kim, D.K.; Koester, K.J.; Ritchie, R.O. Indentation techniques for evaluating the fracture toughness of biomaterials and hard tissues. *J. Mech. Behav. Biomed. Mater.* **2009**, *2*, 384–395. [CrossRef] [PubMed]

43. Huber, N.; Tsakmakis, C. Experimental and theoretical investigation of the effect of kinematic hardening on spherical indentation. *Mech. Mater.* **1998**, *27*, 241–248. [CrossRef]

44. Masri, R.; Durban, D. Cylindrical cavity expansion in compressible mises and tresca solids. *Eur. J. Mech. A Solid.* **2007**, *26*, 712–727. [CrossRef]

45. Marsh, D.M. Plastic flow in glass. *Proc. R. Soc. A* **1964**, *279*, 420–435. [CrossRef]

46. Rodriguez, J.; Rico, A.; Otero, E.; Rainforth, W.M. Indentation properties of plasma sprayed Al_2O_3-13% TiO_2 nanocoatings. *Acta Mater.* **2009**, *57*, 3148–3156. [CrossRef]

47. Johnson, K.L. *Contact Mechanics*; Cambridge University Press: Cambridge, UK, 1985.

48. Page, T.F.; Oliver, W.C.; Mchargue, C.J. The deformation behavior of ceramic crystals subjected to very low load (nano) indentations. *J. Mater. Res.* **1992**, *7*, 450–473. [CrossRef]

49. Bennet, D.W. Multi-Scale Indentation Hardness Testing: A Correlation and Model. Doctor's Disertation, Oklahoma State University, Stillwater, OK, USA, 2008.

50. Shibutani, Y.; Koyama, A. Surface roughness effect on the displacement bursts observed in nanoindentation. *J. Mater. Res.* **2004**, *19*, 183–188. [CrossRef]

51. Agilent Technologies. *Indentation Rules of Thumb-Applications and Limits*; Agilent Technologies: Santa Clara, CA, USA, 2015.

52. Qian, L.; Li, M.; Zhou, Z.; Yang, H.; Shi, X. Comparison of nanoindentation hardness to microhardness. *Surf. Coat. Technol.* **2005**, *195*, 264–271. [CrossRef]

53. Kese, K.; Li, Z.C. Semi-Ellipse method for accounting for the pile-up contact area during nanoindentation with the berkovich indenter. *Scr. Mater.* **2006**, *55*, 699–702. [CrossRef]

54. Choi, Y.; Lee, H.S.; Kwon, D. Analysis of sharp-tip-indentation load-depth curve for contact area determination taking into account pile-up and sink-in effects. *J. Mater. Res.* **2004**, *19*, 3307–3315. [CrossRef]

55. Karthik, V.; Visweswaran, P.; Bhushan, A.; Pawaskar, D.N.; Kasiviswanathan, K.V.; Jayakumar, T.; Raj, B. Finite element analysis of spherical indentation to study pile-up/sink-in phenomena in steels and experimental validation. *Int. J. Mech. Sci.* **2012**, *54*, 74–83. [CrossRef]

56. Zhou, X.; Jiang, Z.; Wang, H.; Yu, R. Investigation on methods for dealing with pile-up errors in evaluating the mechanical properties of thin metal films at sub-micron scale on hard substrates by nanoindentation technique. *Mater. Sci. Eng. A* **2008**, *488*, 318–332. [CrossRef]

57. Nix, W.D.; Gao, H. Indentation size effects in crystalline materials: A law for strain gradient plasticity. *J. Mech. Phys. Sol.* **1998**, *46*, 411–425. [CrossRef]

58. Budiman, A.S. Indentation Size Effects in Single Crystal Cu as Revealed by Synchrotron X-ray Microdiffraction. In *Probing Crystal Plasticity at the Nanoscales*; Springer: Singapore, 2014; pp. 87–101.

59. Reddy, C.M.; Gundakaram, R.C.; Basavoju, S.; Kirchner, M.T.; Padmanabhan, K.A.; Desiraju, G.R. Structural basis for bending of organic crystals. *Chem. Commun.* **2005**, *31*, 3945–3947. [CrossRef] [PubMed]

60. Reddy, C.M.; Padmanabhan, K.A.; Desiraju, G.R. Structure-property correlations in bending and brittle organic crystals. *Cryst. Growth Des.* **2006**, *6*, 2720–2731. [CrossRef]

61. Reddy, C.M.; Kirchner, M.T.; Gundakaram, R.C.; Desiraju, G.R. Isostructurality, Polymorphism and mechanical properties of some hexahalogenated benzenes: The nature of halogen—Halogen interactions. *Chem. Eur. J.* **2006**, *12*, 2222–2234. [CrossRef] [PubMed]

62. Panda, M.K.; Ghosh, S.; Yasuda, N.; Moriwaki, T.; Mukharjee, G.D.; Reddy, C.M.; Naumov, P. Spatially resolved analysis of short-range structure perturbations in a plastically bent molecular crystal. *Nat. Chem.* **2015**, *7*, 65–72. [CrossRef] [PubMed]

63. Thomas, S.P.; Shi, M.W.; Koutsantonis, G.A.; Jayatilaka, D.; Edwards, A.J.; Spackman, M.A. The elusive structural origin of plastic bending in dimethyl sulfone crystals with quasi-isotropic crystal packing. *Angew. Chem. Int. Ed.* **2017**, *56*, 8468–8472. [CrossRef] [PubMed]

64. Saha, S.; Desiraju, G.R. Crystal engineering of hand-twisted helical crystals. *J. Am. Chem. Soc.* **2017**, *139*, 1975–1983. [CrossRef] [PubMed]

65. Hagan, J.T.; Chaudhri, M.M. Fracture surface energies of high explosives PETN and RDX. *J. Mater. Sci.* **1977**, *12*, 1055–1058. [CrossRef]

66. Halfpenny, P.J.; Roberts, K.J.; Sherwood, J.N. Dislocations in energetic materials. *J. Mater. Sci.* **1984**, *19*, 1629–1637. [CrossRef]

67. Chaudhri, M.M. The junction growth equation and its application to explosive crystals. *J. Mater. Sci. Lett.* **1984**, *3*, 565–568. [CrossRef]

68. Elban, W.L.; Hoffsommer, J.C.; Armstrong, R.W. X-ray orientation and hardness experiments on RDX crystals. *J. Mater. Sci.* **1984**, *19*, 552–566. [CrossRef]

69. Elban, W.L.; Armstrong, R.W.; Yoo, K.C. X-ray reflection topographic study of growth defect and microindentation strain fields in an RDX explosive crystal. *J. Mater. Sci.* **1989**, *24*, 1273–1280. [CrossRef]

70. Gallagher, H.G.; Miller, J.C.; Sheen, D.B.; Sherwood, J.N.; Vrcelj, R.M. Mechanical Properties of β-HMX. *Chem. Cent. J.* **2015**, *9*, 1–15. [CrossRef] [PubMed]

71. Sun, C.C.; Kiang, Y.H. On the identification of slip planes in organic crystals based on attachment energy calculation. *J. Pharm. Sci.* **2008**, *98*, 3456–3461. [CrossRef] [PubMed]

72. Ramos, K.J.; Hooks, D.E.; Bahr, D.F. Direct observation of plasticity and quantitative hardness measurements in single crystal cyclotrimethylene trinitramine by nanoindentation. *Philos. Mag.* **2009**, *89*, 2381–2402. [CrossRef]

73. Ramos, K.J.; Bahr, D.F.; Hooks, D.E. Defect and surface asperity dependent yield during contact loading of an organic molecular single crystal. *Philos. Mag.* **2011**, *91*, 1276–1285. [CrossRef]

74. Weingarten, N.S.; Sausa, R.C. Nanomechanics of RDX single crystals by force-displacement measurements and molecular dynamics simulations. *J. Phys. Chem. A* **2015**, *119*, 9338–9351. [CrossRef] [PubMed]

75. Connick, W.; May, F.G.J. Dislocation etching of cyclotrimethylene trinitramine crystals. *J. Cryst. Growth* **1969**, *5*, 65–69. [CrossRef]

76. Munday, L.B.; Solares, S.D.; Chung, P.W. Generalized stacking fault energy surfaces in the molecular crystal A-RDX. *Philos. Mag.* **2012**, *92*, 3036–3050. [CrossRef]

77. Mathew, N.; Picu, R.C. Slip asymmetry in the molecular crystal cyclotrimethylenetrinitramine. *Chem. Phys. Lett.* **2013**, *582*, 78–81. [CrossRef]

78. Taw, M.R.; Bahr, D.F. The mechanical properties of minimally processed RDX. *propellants Explos. PyrTech.* **2017**, *42*, 659–664. [CrossRef]

79. Liu, J.; Zeng, Q.; Zhang, Y.; Zhang, C. Limited-Sample coarse-grained strategy and its applications to molecular crystals: Elastic property prediction and nanoindentation simulations of 1,3,5-trinitro-1,3,5-triazinane. *J. Phys. Chem. C* **2016**, *120*, 15198–15208. [CrossRef]

80. Joshi, M.J.; Shah, B.S. On the microhardness of some molecular crystals. *Cryst. Res. Technol.* **1984**, *19*, 1107–1111. [CrossRef]

81. Marwaha, R.K.; Sha, B.S. A study on the microhardness of organic molecular solids. *Cryst. Res. Technol.* **1991**, *26*, 491–494. [CrossRef]

82. Sgualdino, G.; Vaccari, G.; Mantovani, G. Sucrose crystal hardness: A correlation with some parameters defining the growth kinetics. *Cryst. Growth* **1990**, *140*, 527–532. [CrossRef]

83. Elban, W.; Sheen, D.; Sherwood, J.N. Vickers hardness testing of sucrose single-crystals. *J. Cryst. Growth* **1994**, *137*, 304–308. [CrossRef]

84. Ramos, K.J.; Bahr, D.F. Mechanical behavior assessment of sucrose using Nanoindentation. *J. Mater. Res.* **2007**, *22*, 2037–2045. [CrossRef]

85. Kiran, M.S.R.N.; Varughese, S.; Reddy, C.M.; Ramamurty, U.; Desiraju, G.R. Mechanical anisotropy in crystalline saccharin: Nanoindentation studies. *Cryst. Growth Des.* **2010**, *10*, 4650–4655. [CrossRef]

86. Zhou, X.; Lu, Z.; Zhang, Q.; Chen, D.; Li, H.; Fudenie; Zhang, C. Mechanical anisotropy of the energetic crystal of 1,1-diamino-2,2-dinitroethylene (fox-7): A study by nanoindentation experiments and density functional theory calculations. *J. Phys. Chem. C* **2016**, *120*, 13434–13442. [CrossRef]

87. Mathew, N.; Sewell, T.D. Nanoindentation of the triclinic molecular crystal 1,3,5-triamino-2,4,6-trinitrobenzene: A molecular dynamics study. *J. Phys. Chem. C* **2016**, *120*, 8266–8277. [CrossRef]

88. Taw, M.R.; Yeager, J.D.; Hooks, D.E.; Carvajal, T.M.; Bahr, D.F. The mechanical properties of as-grown non-cubic organic molecular crystals assessed by nanoindentation. *J. Mater. Res.* **2017**, *32*, 2728–2737. [CrossRef]

89. Varughese, S.; Kiran, M.S.R.N.; Solanko, K.A.; Bond, A.D.; Ramamurty, U.; Desiraju, G.R. Interaction anisotropy and shear instability of aspirin polymorphs established by nanoindentation. *Chem. Sci.* **2011**, *2*, 2236–2242. [CrossRef]

90. Olusanmi, D.; Roberts, K.J.; Ghadiri, M.; Ding, Y. The breakage behaviour of aspirin under quasi-static indentation and single particle impact loading: Effect of crystallographic anisotropy. *Int. J. Pharm.* **2011**, *411*, 49–63. [CrossRef] [PubMed]

91. Krishna, G.R.; Kiran, M.S.R.N.; Fraser, C.L.; Ramamurty, U.; Reddy, C.M. The relationship of solid state plasticity to mechanochromic luminescence in Difluoroboron avobenzone polymorphs. *Adv. Funct. Mater.* **2013**, *23*, 1422–1430. [CrossRef]

92. Mishra, M.K.; Ramamurty, U.; Desiraju, G.R. Solid solution hardening of molecular crystals: Tautomeric polymorphs of omeprazole. *J. Am. Chem. Soc.* **2015**, *137*, 1794–1797. [CrossRef] [PubMed]

93. Sanphui, P.; Mishra, M.K.; Ramamurty, U.; Desiraju, G.R. Tuning mechanical properties of pharmaceutical crystals with multicomponent crystals: Voriconazole as a case study. *Mol. Pharm.* **2015**, *12*, 889–897. [CrossRef] [PubMed]

94. Smith, A.J.; Kavuru, P.; Wojtas, L.; Zaworotko, M.J.; Shytle, R.D. Cocrystals of quercetin with improved solubility and oral bioavailability of molecular crystals: Tautomeric polymorphs of omeprazole. *Mol. Pharm.* **2011**, *8*, 1867–1876. [CrossRef] [PubMed]

95. Sanphui, P.; Tothadi, S.; Ganguly, S.; Desiraju, G.R. Salt and cocrystals of sildenafil with dicarboxylic acids Solubility and pharmacokinetic advantage of the glutarate salt. *Mol. Pharm.* **2013**, *10*, 4687–4697. [CrossRef] [PubMed]

96. Sun, C.C.; Hou, H. Improving mechanical properties of caffeine and methyl gallate crystals by cocrystallization. *Cryst. Growth Des.* **2008**, *8*, 1575–1579. [CrossRef]

97. Chow, S.F.; Chen, M.; Shi, L.; Chow, A.H.; Sun, C.C. Simultaneously improving the mechanical properties, dissolution performance, and hygroscopicity of ibuprofen and flurbiprofen by co-crystallization with nicotinamide. *Pharm. Res.* **2012**, *29*, 1854–1865. [CrossRef] [PubMed]

98. Krishna, G.R.; Shi, L.; Bag, P.P.; Sun, C.C.; Reddy, C.M. Correlation among crystal structure, mechanical behavior, and tabletability in the co-crystals of vanillin isomers. *Cryst. Growth Des.* **2015**, *15*, 1827–1832. [CrossRef]

99. Mishra, M.K.; Sanphui, P.; Ramamurty, U.; Desiraju, G.R. Solubility-Hardness correlation in molecular crystals: Curcumin and sulfathiazole polymorphs. *Cryst. Growth Des.* **2014**, *14*, 3054–3061. [CrossRef]

100. Feng, Y.; Grant, D.J.W. Influence of crystal structure on the compaction properties of N-Alkyl 4-Hydrokxybenzoate Esters (Parabens). *Pharm. Res.* **2006**, *23*, 1608–1616. [CrossRef] [PubMed]

101. Duncan-Hewitt, W.C.; Mount, D.L.; Yu, A. Hardness anisotropy of acetaminophen crystals. *Pharm. Res.* **1994**, *11*, 616–623. [CrossRef] [PubMed]

102. Finnie, S.; Prasad, K.V.R.; Sheen, D.B.; Sherwood, J.N. Microhardness and dislocation identification studies on paracetamol single crystals. *Pharm. Res.* **2001**, *18*, 674–681. [CrossRef] [PubMed]

103. Egart, M.; Ilić, I.; Janković, B.; Lah, N.; Srčič, S. Compaction properties of crystalline pharmaceutical ingredients according to the walker model and nanomechanical attributes. *Int. J. Pharm.* **2014**, *472*, 347–355. [CrossRef] [PubMed]

104. Chen, S.; Sheikh, A.Y.; Ho, R. Pharmaceutical ingredient crystals using nanoindentation and high-resolution total scattering pair distribution function analysis. *J. Pharm. Sci.* **2014**, *103*, 3879–3890. [CrossRef] [PubMed]

105. Raut, D.; Kiran, M.S.R.N.; Mishra, M.K.; Asiri, B.M.; Ramamurty, U. On the loading rate sensitivity of plastic deformation in molecular crystals. *CrystEngComm* **2016**, *18*, 3551–3555. [CrossRef]

106. Katz, J.M.; Buckner, I.S. Characterization of strain rate sensitivity in pharmaceutical materials using indentation creep analysis. *Int. J. Pharm.* **2013**, *442*, 13–19. [CrossRef] [PubMed]

107. Kaupp, G.; Naimi-Jamal, M.R. Mechanically induced molecular migrations in molecular crystals. *CrystEngComm* **2005**, *7*, 402–410. [CrossRef]

108. Varughese, S.; Kiran, M.S.R.N.; Ramamurty, U.; Desiraju, G.R. Nanoindentation as a probe for mechanically-induced molecular migration in layered organic donar-acceptor complexes. *Chem. Asian J.* **2012**, *7*, 2118–2125. [CrossRef] [PubMed]

109. Jing, Y.; Zhang, Y.; Blendell, J.; Koslowski, M.; Carvajal, M.T. Nanoindentation method to study slip planes in molecular crystals in a systematic manner. *Cryst. Growth Des.* **2011**, *11*, 5260–5267. [CrossRef]

110. Thakuria, R.; Eddleston, M.D.; Chow, E.H.H.; Taylor, L.J.; Aldous, B.J.; Krzyzaniak, J.F.; Jonesa, W. Use of In-Situ atomic force microscopy to follow phase changes at crystal surfaces in real time. *CrystEngComm* **2013**, *52*, 10541–10544.

111. Roberts, R.J.; Rowe, R.C. The compaction of pharmaceutical and other model materials: A pragmatic approach. *Chem. Eng. Sci.* **1987**, *42*, 903–911. [CrossRef]

112. Duncan-Hewitt, W.C.; Weatherly, G.C. Evaluating the hardness, young's modulus and fracture toughness of some pharmaceutical crystals using microindentation techniques. *J. Mater. Sci. Lett.* **1989**, *8*, 1350–1352. [CrossRef]

113. Munimunda, P.; Hintsala, E.A.S.; Mishra, M.K. Mechanical anisotropy and pressure induced structural changes in piroxicam crystals probed by in-situ indentation and Raman spectroscopy. *JOM* **2017**, *69*, 57–64. [CrossRef]

114. Lei, L.; Carvajal, T.; Koslowski, M. Defect-induced solid state amorphization of molecular crystals. *J. Appl. Phys.* **2012**, *111*, 073505. [CrossRef]

115. Boldyreva, E.V.; Sowa, H.; Ahsbahs, H.; Goryainov, S.V.; Chernyshev, V.V.; Dmitriev, V.P.; Seryotkin, Y.V.; Kolesnik, E.N.; Shakhtshneider, T.P.; Ivashevskaya, S.N.; et al. Pressure-Induced phase transitions in organic molecular crystals: A combination of x-ray single-crystal and powder diffraction Raman and IR-spectroscopy. *J. Phys. Conf. Ser.* **2008**, *121*, 022023. [CrossRef]

116. Mohamed, R.M.; Mishra, M.K.; Al-Harbi, L.M.; Al-Ghamdi, M.S.; Ramamurty, U. Anisotropy in the mechanical properties of organic crystals: Temperature dependence. *RSC Adv.* **2015**, *5*, 64156–64162. [CrossRef]

117. Mishra, M.K.; Ramamurty, U.; Desiraju, G.R. Bimodal nanoindentation response of the (0 0 1) face in crystalline sodium saccharin dehydrate. *Macedonian J. Chem. Chem. Eng.* **2015**, *34*, 51–55. [CrossRef]

118. Kiran, M.S.R.N.; Varughese, S.; Ramamurty, U.; Desiraju, G.R. Effect of dehydration on the mechanical properties of sodium saccharin dihydrate probed with nanoindentation. *CrystEngComm* **2012**, *14*, 2489–2493. [CrossRef]

119. Egart, M.; Jankovi, B.; Srcic, S. Application of instrumented nanoindentation in preformulation studies of pharmaceutical active ingredients and excipients. *Acta Pharm.* **2016**, *66*, 303–330. [CrossRef] [PubMed]

120. Taylor, L.J.; Papadopoulos, D.G.; Dunn, P.J.; Bentham, A.C.; Dawson, N.J.; Mitchell, J.C.; Snowden, M.J. Predictive milling of pharmaceutical materials using nanoindentation of single crystals. *Org. Proc. Res. Dev.* **2004**, *8*, 674–679. [CrossRef]

121. Cao, X.; Morganti, M.; Hancock, B.C.; Masterson, V.M. Correlating particle hardness with powder compaction performance. *J. Pharm. Sci.* **2010**, *99*, 4307–4316. [CrossRef] [PubMed]

122. Masterson, V.M.; Cao, X. Evaluating particle hardness of pharmaceutical solids using AFM nanoindentation. *Int. J. Pharm.* **2008**, *362*, 163–171. [CrossRef] [PubMed]

123. Munz, D.; Bubsey, R.T.; Shannon, J.L., Jr. Fracture toughness determination of Al_2O_3 using fourpointbend specimens with straightthrough and chevron notches. *J. Am. Ceram. Soc.* **1980**, *63*, 300–305. [CrossRef]

124. Ritchie, R.O.; Dauskardt, R.H.; Yu, W.; Brendzel, A.M. Cyclic fatiguecrack propagation, stresscorrosion, and fracturetoughness behavior in pyrolytic carboncoated graphite for prosthetic heart valve applications. *J. Biomed. Mater. Res.* **1990**, *24*, 89–206. [CrossRef] [PubMed]

125. Fett, T.; Munz, D. Influence of narrow starter notches on the initial crack growth resistance curve of ceramics. *Arch. Appl. Mech.* **2006**, *76*, 667–679. [CrossRef]

126. Lawn, B.R.; Evans, A.G.; Marshall, D.B. Elastic/Plastic indentation damage in ceramics: The median/radial crack system. *J. Am. Ceram. Soc.* **1980**, *63*, 574–581. [CrossRef]

127. Anstis, G.R.; Chantikul, P.; Lawn, B.R.; Marshall, D.B. A critical evaluation of indentation techniques for measuring fracture toughness: I, direct crack measurements. *J. Am. Ceram. Soc.* **1981**, *64*, 533–538. [CrossRef]
128. Fett, T.; Kounga, A.B.; Rodel, J. Stresses and stress intensity factor from cod of vickers indentation cracks. *J. Mater. Sci.* **2004**, *39*, 2219–2221. [CrossRef]
129. Fett, T.; Njiwa, A.B.K.; Rodel, J. Crack opening displacements of vickers indentation cracks. *Eng. Fract. Mech.* **2005**, *72*, 647–659. [CrossRef]
130. Lawn, B. Indentation fracture: Applications in the assessment of strength of ceramics. *J. Aust. Ceram. Soc.* **1980**, *16*, 4.
131. Harding, D.S.; Oliver, W.C.; Pharr, G.M. Cracking during nanoindentation and its use in the measurement of fracture toughness. *Mater. Res. Soc. Symp. Proc.* **1995**, *356*, 663–668. [CrossRef]
132. Lawn, B.R.; Marshall, D.B. Hardness, toughness, and brittleness: An indentation analysis. *J. Am. Ceram. Soc.* **1979**, *62*, 347–350. [CrossRef]
133. Duncan-Hewitt, W.C.; Weatherly, G. Evaluating the fracture toughness of sucrose crystals using microindentation techniques. *Pharm. Res.* **1989**, *6*, 373–377. [CrossRef] [PubMed]
134. Duncan-Hewitt, W.C.; Weatherly, G. Evaluating the deformation kinetics of sucrose crystals using microindentation techniques. *Pharm. Res.* **1989**, *6*, 1060–1066. [CrossRef] [PubMed]
135. Prasad, K.V.R.; Sheen, D.B.; Sherwood, J.N. Fracture property studies of paracetamol single crystals using microindentation techniques. *Pharm. Res.* **2001**, *18*, 867–872. [CrossRef] [PubMed]
136. Mishra, M.K.; Ramamurty, U.; Desiraju, G.R. Mechanical property design of molecular crystals. *Curr. Opin. Solid State Mater. Sci.* **2016**, *20*, 361–370. [CrossRef]
137. Chung, Y.W.; Sproul, W.D. Super hard coating materials. *MRS Bull.* **2003**, *28*, 164–165. [CrossRef]
138. Mishra, K.; Verma, D.; Bysakh, S.; Pathak, L.C. Hard and soft multilayered sicn nanocoatings with high hardness and toughness. *J. Nanomater.* **2013**. [CrossRef]
139. Berding, M.A.; Sher, A.; van Shilfgaarde, M.; Chen, A.B. Fracture and Hardness Characteristics of Semiconductor Alloys. Available online: http://www.dtic.mil/dtic/tr/fulltext/u2/a204807.pdf (accessed on 23 September 2017).
140. Louzguine-Luzgin, D.V.; Louzguina-Luzgina, L.V.; Churyumov, A.Y. Mechanical properties and deformation behavior of bulk metallic glasses. *Metals* **2013**, *3*, 1–22. [CrossRef]
141. Tan, J.C.; Cheetham, A.K. Mechanical properties of hybrid inorganic-organic framework materials: Establishing fundamental structure-property relationships. *Chem. Soc. Rev.* **2011**, *40*, 1059–1080. [CrossRef] [PubMed]
142. Lee, J.; Novikov, N.V. Innovative Superhard Materials and Sustainable Coatings for Advanced Manufacturing. Available online: https://link.springer.com/book/10.1007/1-4020-3471-7?page=2 (accessed on 24 September 2017).
143. Marinescu, I.D.; Tönshoff, H.K.; Inasaki, I. *Handbook of Ceramic Grinding and Polishing*; William Andrew Publishing: Norwich, NY, USA, 2000.
144. Kutz, M. *Handbook of Materials Selection*; John Wiley and Sons: Hoboken, NJ, USA, 2002; pp. 384–424.
145. Chen, C.P. Analytical Determination of Critical Crack Size in Solar Cells. Available online: https://ntrs.nasa.gov/archive/nasa/casi.ntrs.nasa.gov/19880017352.pdf (accessed on 23 September 2017).
146. Chapter 3: Nanoindentation-Induced Mechanical Responses. Available online: https://ir.nctu.edu.tw/bitstream/11536/78079/5/180105.pdf (accessed on 23 September 2017).
147. Xu, J.; Ramamurty, U.; Ma, E. The fracture toughness of Bulk metallic glasses. *JOM* **2010**, *62*, 10–18. [CrossRef]
148. Subhash, G.; Koepper, B.J.; Chandra, A. Dynamic indentation hardness and rate sensitivity in metals. *J. Eng. Mater. Technol.* **1999**, *121*, 257–263. [CrossRef]
149. Mason, J.K.; Lund, A.C.; Schuh, C.A. Determining the Activation Energy and Volume for the onset of Plasticity during Nanoindentation. *Phys. Rev. B* **2006**, *73*, 1–14. [CrossRef]
150. Bhattacharyya, A.; Singh, G.; Prasad, K.E.; Narasimhan, R.; Ramamurty, U. On the strain rate sensitivity of plastic flow in metallic glasses mater. *Sci. Eng. A* **2015**, *625*, 245–251. [CrossRef]
151. Golbe, D.L.; Wolff, E.G. Strain-rate sensitivity index of thermoplastics. *J. Mater. Sci.* **1993**, *23*, 5986–5994.
152. Song, J.M.; Shen, Y.L.; Su, C.W.; Lai, Y.S.; Chiu, Y. Strain rate dependence on nanoindentation responses of interfacial intermetallic compounds in electronic solder joints with cu and ag substrates. *Mater. Trans.* **2009**, *50*, 1231–1234. [CrossRef]

153. Bradby, J.E.; Williams, J.S.; Wong-Leung, J.; Swain, M.V.; Munroe, P. Transmission electron microscopy observation of deformation microstructure under spherical indentation in silicon. *Appl. Phys. Lett.* **2000**, *77*, 3749–3751. [CrossRef]

154. Bradby, J.E.; Williams, J.S.; Wong-Leung, J.; Swain, M.V.; Munroe, P. Mechanical deformation of InP and GaAs by spherical indentation. *Appl. Phys. Lett.* **2001**, *78*, 3235–3237. [CrossRef]

155. Bradby, J.E.; Williams, J.S.; Wong-Leung, J.; Swain, M.V.; Munroe, P. Nanoindentation-Induced deformation of Ge. *Appl. Phys. Lett.* **2002**, *80*, 2651–2653. [CrossRef]

156. Kiran, M.S.R.N.; Tran, T.T.; Smillie, L.A.; Haberl, B.; Subianto, D.; Williams, J.S.; Bradby, J.E. Temperature-dependent mechanical deformation of silicon at the nanoscale: Phase transformation versus defect propagation. *J. Appl. Phys.* **2015**, *117*, 205901. [CrossRef]

crystals

MDPI

Article

Vickers Hardness of Diamond and cBN Single Crystals: AFM Approach

Sergey Dub [1],*, Petro Lytvyn [2], Viktor Strelchuk [2], Andrii Nikolenko [2], Yurii Stubrov [2], Igor Petrusha [1], Takashi Taniguchi [3] and Sergey Ivakhnenko [1]

[1] Institute for Superhard Materials of NASU, 2 Avtozavodskaya Str., 04074 Kyiv, Ukraine; dialab@ism.kiev.ua (I.P.); sioz@ismv13.kiev.ua (S.I.)

[2] Institute of Semiconductor Physics of NASU, 41 Nauky Pr., 03028 Kyiv, Ukraine; plyt@isp.kiev.ua (P.L.); strelch@isp.kiev.ua (V.S.); nikolenko_mail@ukr.net (A.N.); chig-ua@rambler.ru (Y.S.)

[3] National Institute for Materials Sciences, 1-1 Namiki, Tsukuba, Ibaraki 305-0044, Japan; taniguchi.takashi@nims.go.jp

*   Correspondence: Lz@ism.kiev.ua; Tel.: +380-97-278-2536

Academic Editors: Ronald W. Armstrong, Stephen M. Walley and Wayne L. Elban
Received: 23 October 2017; Accepted: 5 December 2017; Published: 12 December 2017

Abstract: Atomic force microscopy in different operation modes (topography, derivative topography, and phase contrast) was used to obtain 3D images of Vickers indents on the surface of diamond and cBN single crystals with high spatial resolution. Confocal Raman spectroscopy and Kelvin probe force microscopy were used to study the structure of the material in the indents. It was found that Vickers indents in diamond has no sharp and clear borders. However, the phase contrast operation mode of the AFM reveals a new viscoelastic phase in the indent in diamond. Raman spectroscopy and Kelvin probe force microscopy revealed that the new phase in the indent is disordered graphite, which was formed due to the pressure-induced phase transformation in the diamond during the hardness test. The projected contact area of the graphite layer in the indent allows us to measure the Vickers hardness of type-Ib synthetic diamond. In contrast to diamond, very high plasticity was observed for 0.5 N load indents on the (001) cBN single crystal face. Radial and ring cracks were absent, the shape of the indents was close to a square, and there were linear details in the indent, which looked like slip lines. The Vickers hardness of the (111) synthetic diamond and (111) and (001) cBN single crystals were determined using the AFM images and with account for the elastic deformation of the diamond Vickers indenter during the tests.

Keywords: Vickers hardness; diamond; cBN; atomic force microscopy; Raman spectroscopy

---

1. Introduction

Diamond is a difficult object for Vickers hardness testing. The small indent size, high brittleness, and high elastic recovery make the diagonals measurement difficult and not reliable under optical microscopy [1]. Moreover, the tip of a Vickers indenter is damaged after several tests. This is why the Vickers hardness of single diamond crystals remains insufficiently studied. Recently, the interest in the hardness measurement of superhard materials dramatically grew due to the emergence of novel superabrasives—nanocrystalline and nanotwinned diamonds [2,3] and cBN [4] bulk samples synthesized by the direct conversion of graphite and graphite-like BN, respectively, under very high pressure and temperature. It was reported, that the hardness of such materials is much higher than that of diamond [5] and cBN [6–9] single crystals. Contemporary imaging techniques, such as 3D optical microscopy [10], laser scanning microscopy [11], and atomic force microscopy (AFM) [1,3,12] were used to improve the precision of indent size measurements in superhard materials. However, it was also very difficult to measure the indent sizes due to the cracking within, and around, the contact sites [1,13].

The nanoindentation by a Berkovich diamond indenter is used to study the mechanical properties of superhard materials at the nanoscale. It was found that nanohardness of (111) cBN single crystals is about 61 GPa (35 mN load and 200 nm displacement) [14–17]. Moreover, nanoindentation allows one to study the onset of plasticity in nanodeformation of cBN and obtain the experimental estimates of a theoretical shear strength of (111) cBN [15,17]. Unfortunately, with respect to the nanoindentation of diamond single crystals, the contact is elastic. Thus, only elastic properties of diamonds were studied [18–21].

To overcome the above problems concerning the measuring the Vickers indents on the surface of diamond the following approach is proposed in this paper. Earlier, micro-Raman spectroscopy showed that a pressure-induced phase transformation of diamond to disordered graphite occurs during hardness tests of diamond single crystals [22,23]. For measuring the indent sizes on the surface of diamond in the above-mentioned papers [1,3,12], the ordinary AFM in topography operating mode was used, which characterized the surface relief only. In the present paper, AFM in the phase contrast mode was used to reveal the graphite layer formed in the Vickers indent in diamond and to measure its area. The (111), (001) planes of cBN single crystals and the bulk sample of nanopolycrystalline cBN were also tested for comparison.

2. Materials and Methods

Type-Ib synthetic diamond single crystals, yellow in colour, were synthesized by the spontaneous crystallization in the region of thermodynamic stability at pressures of 5.5–5.8 GPa and at temperatures of 1400–1420 °C in a toroid-type high-pressure apparatus (HPA) using iron-nickel alloy solvent with synthetic graphite MG-OSCH. The growth time was 40 min, allowing the growth of crystals 0.8–1 mm in size (Figure 1). As is known, type-Ib diamonds contain nitrogen in the amount of ~10^{19} cm^{-3} in the form of C-centers. The density of the dislocations was determined by the selective etching and was about 1.3×10^6 cm^{-2} [24].

Figure 1. (111) plane of the synthetic diamond single crystal. The arrow points to the indent which was studied by AFM and micro-Raman techniques.

cBN single crystals were synthesized in a belt-type HPA using high-pressure cells at NIMS, Tsukuba, Japan. A spontaneous nucleation and a subsequent crystal growth were implemented in the B–N–Li growth system at a pressure of 5.5 GPa and a temperature of 1500 °C for 17 h. Single crystals ~2 mm in size were synthesized. The synthesized cBN crystals have a saturated amber colour. Hardness tests were performed on (111) and (001) natural planes of cBN single crystals.

The initial high-purity graphite-like BN synthesized by CVD as dense plates 1.8–2.2 mm thick were used for producing nanopolycrystalline cBN. The necessary thermobaric conditions were created in a toroid type HPA [25]. The round plates of CVD graphite-like BN were placed into a tantalum capsule to prevent direct contact of samples with the materials of the pressure medium. The plates were separated from one another and isolated from the capsule walls using graphite interlayer of ~0.2 mm in thickness. The nanopolycrystalline cBN compact was formed as the result of the hBN→cBN direct solid phase transformation at a pressure of 8 GPa and temperature of 2300 °C for 60 s. The size of the samples was ~5 mm in diameter and ~1 mm in thickness. The microstructure of the sample was identified as close to the nanocrystalline one with grain sizes in the range from 100 to 400 nm. In all grains, there is a characteristic substructure formed by nanotwinned domains, whose thickness varies from 10 to 60 nm [17]. The finishing lapping operation was carried out using a colloidal solution based on SiO_2 with particles of 40 nm in size.

The hardness of (111) synthetic diamond and (111) and (001) cBN single crystals has been measured using a PMT-3 microhardness tester (LOMO, Leningrad, USSR) with a Vickers indenter. The indentation load was 4.91 N for synthetic diamond and nanopolycrystalline cBN samples and 1.96 N for cBN single crystals. cBN single crystals were also tested at 0.49-N load to study the initial stage of plasticity. The Vickers hardness measurements were performed on the (111) plane with diagonals of the square pyramidal indenter parallel to <110> and <112> directions (orientation A). The rotation of the indenter to 45° gives orientation B on (111) plane. The (111) cBN single crystal was tested at both orientations. The (111) diamond single crystal was tested at the orientation A only. Indent diagonal was parallel to <100> and <110> directions on the (001) cBN plane (henceforth <100> and <110> indents on the (001) cBN plane). Both Vickers hardness H_V (ratio of the applied load to lateral contact area) and Meyer hardness HM (ratio of applied load to projected contact area) were measured. Indentation fracture toughness K_{IC} (MPa m$^{0.5}$) was determined from the length of radial cracks emanating in brittle materials from Vickers indent corners [26]:

$$K_{IC} = 0.016 \left(\frac{E}{H} \right)^{0.5} \frac{P}{C^{1.5}} \tag{1}$$

where P is the applied load (N), and C is the length of radial crack measured from the indent center (m).

Three-dimensional characterisation of the indents' shapes and localisation of the indentation-induced new phases were performed by scanning probe microscopy (NanoScope IIIa Dimension 3000^{TM} microscope, former Digital Instruments, Santa Barbara, CA, USA) using atomic force microscopy (AFM) and Kelvin probe force microscopy (KPFM) techniques. The projected contact area of the indents was determined by the standard procedure for the analysis of surface elements in AFM measurements. The AFM image with an indent was levelled in a horizontal position, using the virgin surface, then the indent was sectioned at a given height and the area of this section was calculated. The height was chosen so that the edge of the indent was clearly visible and it was at the same level as the virgin surface (the pile-ups were cut off). AFM topography and viscoelastic property mappings were carried out in tapping operation mode [27] utilising conventional silicon probes with a nominal tip radius of 10 nm. Semi-quantitative mapping of the tip-surface viscoelastic interaction was realised using phase contrast imaging, where the phase lag between the excitation signal and probe response is measured [28,29]. Kelvin probe force microscopy (KPFM) differentiates the local surface areas by contact potential between the Pt/Ir tip and the surface appearing due to a difference in the tip and surface work function [30].

Micro-Raman measurements were performed at room temperature in backscattering configuration using a triple Raman spectrometer T-64000 Horiba Jobin-Yvon (Horiba Scientific, Villeneuve d'Ascq, France), equipped with an electrically-cooled CCD detector and Olympus BX41 microscope. An Ar-Kr ion laser with a wavelength of 488 nm was used for excitation. Excited radiation was focused on the sample surface with 100×/NA 0.9 optical objective, giving a laser spot diameter of about 0.6 μm. Raman mapping was performed using a piezo-driven XYZ stage with a scanning step of 100 nm. A confocal pinhole of 100 μm was placed into the focal plane of the microscope to increase the spatial resolution.

3. Results and Discussions

3.1. AFM Imaging and Micro-Raman Spectroscopy of Vickers Indents in Synthetic Ib Diamond Single Crystal

Figure 2 shows the AFM images of 4.91-N Vickers indents taken on the (111) face of synthetic Ib diamond. It can be seen from Figure 2a (topography) and Figure 2b (derivative topography) that the plasticity of the diamond is very low—there is a large number of a ring (Hertzian) cracks in the region of contact aligned along the <011> directions (Figure 2a,b). Additionally, long radial cracks emanate from the indent corners. According to Figure 2a,b a Vickers indent in diamond does not have sharp and clear boundaries and its hardness measurement is complicated. However, AFM in the phase contrast operation mode reveals the formation of a new viscoelastic phase in the contact (Figure 2c). It is known that micro-Raman spectroscopy reveals a disordered graphite in a Vickers indent in diamond [22,23]. A high pressure and high shear stress in the contact causes a phase transformation of diamond to graphite during hardness test. Probably, graphite in the indent was formed during unloading through unknown high pressure metallic phase of carbon [23] similar to that in silicon [31].

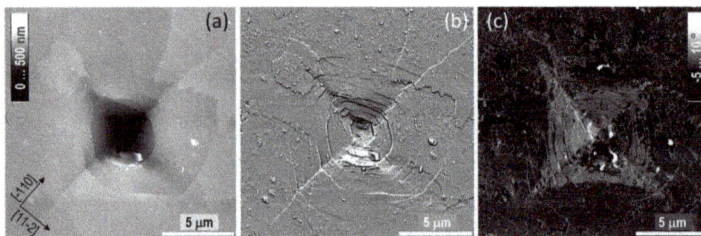

Figure 2. AFM images of a Vickers indent (5-N load) on the (111) synthetic diamond: (**a**) topography; (**b**) derivative topography; and (**c**) phase contrast.

The KPFM measurements also detects a new phase in the vicinity of the indent through an electrostatic tip-surface interaction instead of mechanical one in the previous case. Figure 3 illustrates simultaneously-captured topography and contact potential difference (CPD) over the diamond surface with the Vickers indent. The work function of the diamond could be expected within 3.9–4.2 eV [32].

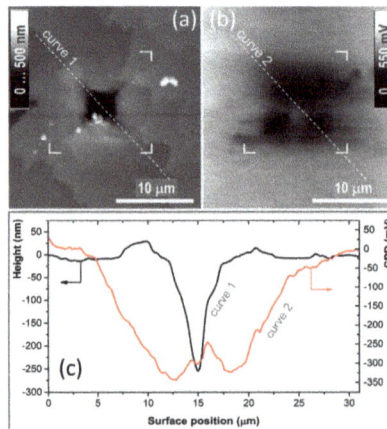

Figure 3. (**a**) Topography and (**b**) corresponding tip-surface CPD maps for the indent in diamond shown in Figure 1. (**c**) Cross-sections of maps along the same line on surface: relief profile (curve 1) and CPD profile (curve 2). Corner-like marks are shown to compare areas of low CPD and the indent area.

Taking into account the 5.0 eV work function of a thin carbon film [33] and the 5.5 eV of Pt/Ir tip, the lowering of CPD over the graphite phase in the indent should be observed (up to 0.8 eV in the ideal case). As it follows from CPD profiles (Figure 3c, curve 2) the real difference could reach a value of 350 mV. Thus, AFM and KPFM data allow us to suppose the viscoelastic phase in the region of the Vickers indent on diamond to be disordered graphite.

Confocal micro-Raman spectroscopy of the Vickers indent on the surface of (111) type-Ib synthetic diamond was applied to check the above mentioned assumption. Raman spectra taken in the indented area (Figure 4a) besides the main F_{2g} diamond phonon band at ~1334 cm^{-1} also contain broadened bands at ~1350 and ~1580 cm^{-1}, corresponding to the D and G bands of the sp^2 amorphous carbon phase, which appears due to graphitization of diamond during indentation [23]. Scanning along the indent from its centre to edge (Figure 4a) revealed non-uniform spatial distribution of the sp^2 carbon phase with its highest content in the centre of the indent and gradually decreasing to the edge, which is qualitatively demonstrated by the distribution of the relative intensity of the diamond F_{2g} band to the intensity of the graphite G-band (Figure 4b).

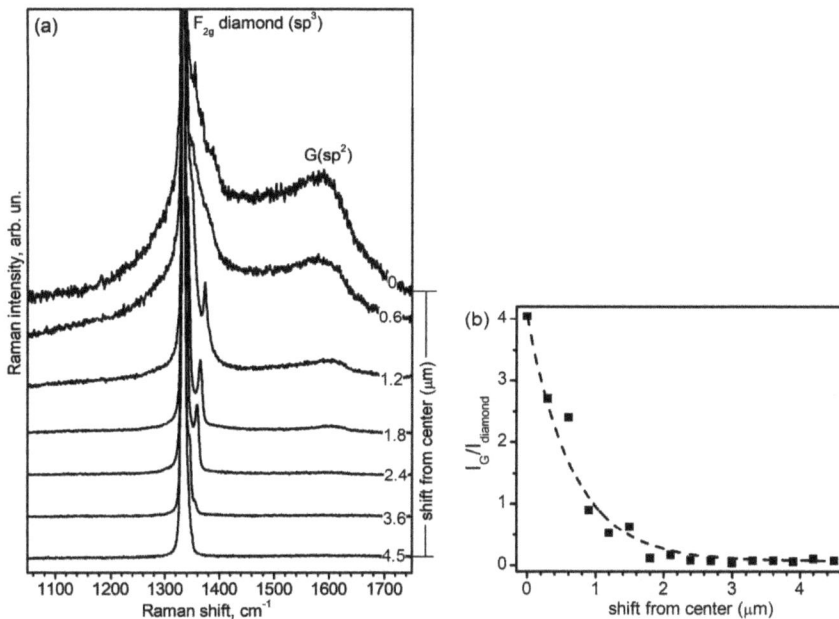

Figure 4. (a) Set of Raman spectra and (b) $I_G/I_{diamond}$ intensity ratio obtained by lateral scanning across the Vickers indent in diamond from its center to edge. The dashed line is shown to guide the eye.

It should be also noted that a series of narrow low-intensity bands with variable frequency positions were registered on the high-energy side of the diamond F_{2g} peak in the range of 1335–1390 cm^{-1} (Figure 5a). The appearance of additional bands in the Raman spectra during indentation can be associated with the formation of metastable phases and, in particular, of the hexagonal diamond (lonsdaleite) phase [34], which was predicted theoretically [35]. However, the Raman spectrum of lonsdaleite is expected to contain phonon bands in the range of 1224–1242 cm^{-1}, 1292–1303 cm^{-1}, and 1338 cm^{-1} corresponding to E_{2G}, A_{1g}, and E_{1g} vibrational modes [35,36]. Moreover, the A_{1g} mode in the region of 1292–1303 cm^{-1} should be the most intense in the spectrum, which is not observed in our case. Thus, the additional bands registered in the range of 1335–1390 cm^{-1} cannot be attributed to lonsdaleite.

On the other hand, the appearance of additional bands in the Raman spectrum of diamond on the high-frequency side of the F_{2g} band may be due to the presence of residual compressive strain that can remain in the indented area. Similar behaviour of the phonon band was observed earlier in the Raman spectra of diamond subjected to shock compressive deformations [37]. Thus, uniaxial or biaxial compressive strains can partially or completely remove the degeneracy of the triply-degenerate diamond F_{2g} band with the corresponding splitting of diamond Raman peak into two or three components with a gradual high-frequency shift of the latter with strain [38].

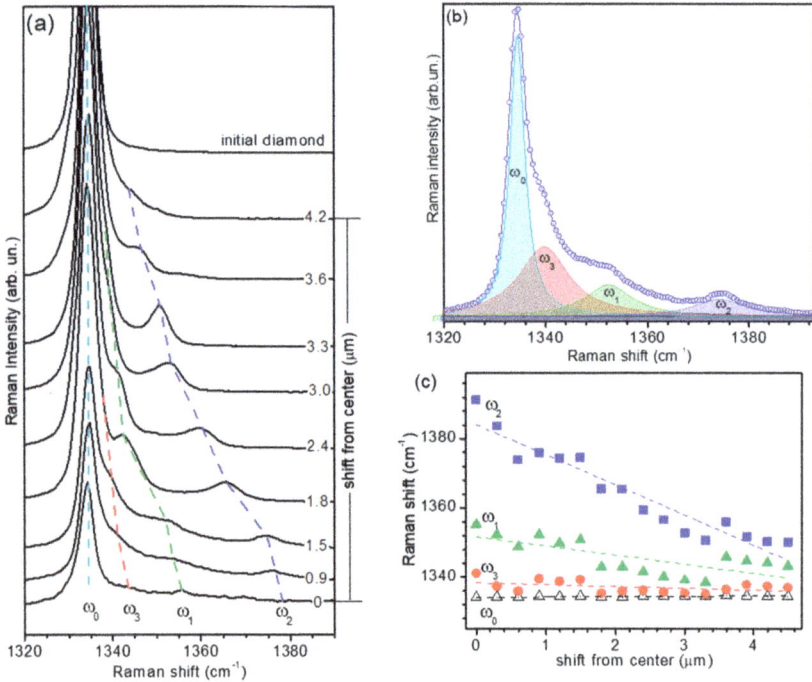

Figure 5. (a) Set of Raman spectra obtained by lateral scanning across the indent in the diamond from centre to edge; (b) Example of decomposition of typical Raman spectra taken inside the indent; (c) Frequencies of Raman modes in dependence on the spatial position from the centre of the indent.

Thus, the three registered bands in the range of 1335–1390 cm^{-1}, which are denoted as ω_1, ω_2, and ω_3 in Figure 5a,b in accordance with [38], can be presumably assigned to the components of the triplet, which is formed due to the splitting of the diamond F_{2g} band (ω_0) under uniaxial compressive strain. It should be noted that simultaneous registration in the spectrum of a more intense unsplit ω_0 diamond band indicates that both strained and unstrained regions of the diamond coexist within the area of the Raman probe. For confocal scanning at a wavelength of $\lambda_{exc} = 488.0$ nm, the lateral resolution is defined by the lateral size of the focused laser beam waist, $d_{lateral} = 0.4\lambda_{exc} \, NA^{-1} \approx 210$ nm, and the axial resolution is defined by the axial waist size, $R_{axial} = 1.4(\lambda_{exc}\mu) \, (NA)^{-2} \approx 1.8 \, \mu m$.

The gradual low-frequency shift of ω_1, ω_2, and ω_3 bands towards the diamond ω_0 band is observed at scanning from the center of the indent to its edge (Figure 5c), which apparently corresponds to the gradual strain relaxation along the studied area. A rough estimate of the strain on the basis of the data of [38] and assuming uniaxial compression in the [110] direction gives a value of the order of 20 GPa. Additionally, the Raman spectra measured near the edge of the indent revealed a weak peak (ω_4) with a frequency of about 1330 cm^{-1}, i.e., lower than the diamond ω_0 band, and which shifts slightly to the high-frequency region from 1330 to 1331 cm^{-1} when approaching the edge of the indent (Figure 6). This peak can tentatively be associated with the presence of a tensile strained diamond region near the edge of the indent, which appears due to the compressive strains inside the indent.

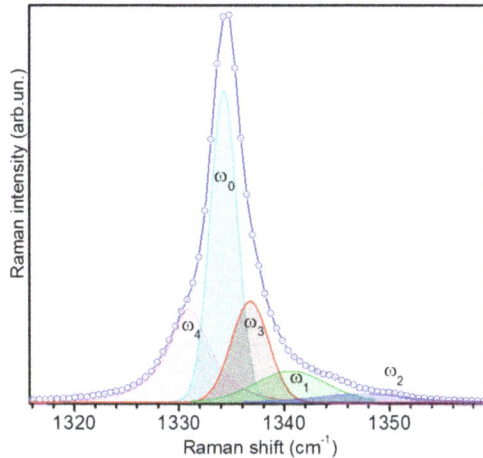

Figure 6. Example of decomposition of the Raman spectra taken near the outer edge of the Vickers indent in (111) synthetic diamond single crystal.

Thus, the new phase revealed by the AFM in the indent on the surface of diamond is graphite. Therefore, the phase-transformed contact area (60.05 × 10^6 nm^2) determined by the AFM in the phase contrast operation mode, was taken as the contact area and was used to determine the Vickers hardness of diamond.

3.2. AFM Imaging of Vickers Indents on (111) and (001) cBN Single Crystals

Figure 7 shows AFM images (1.96–N load) of A and B indents on the (111) plane of the cBN single crystals. The well-developed plastic indents are observed on the (111) plane. The shape of the Vickers indents are close to square. A long radial cracks are formed in the B indents (Figure 7a–c) on the {011}$_{90°}$ planes in the <112> directions. A short radial crack is formed along <110> direction for the A indent (Figure 7d–f) additionally to long radial cracks in <112> directions. It is well known that the {111} plane is the preferred cleavage plane of diamond. The crystalline lattice of cBN is the same as in diamond. Thus, the {111} cleavage should be expected for cBN. However, at the hardness testing of (111) cBN only {011}$_{90°}$ cleavage was observed (Figure 7). It should be noted that, for the cleavage, not only is the surface energy of the crystallographic plane important, but so too is its orientation relative to the plane being tested. For example, with the Vickers hardness test on the (001) surface of diamond the actual cleavage planes are {011}$_{90°}$ and {111}$_{54°44'}$, in this case the {011}$_{90°}$ cleavage prevails [39]. Probably, the {111}$_{90°}$ cleavage planes in cBN are possible at the testing of the (011) plane as it occurs at testing the (011) plane of diamond [40].

The AFM image of the 1.96-N load <110> indent on the (001) cBN plane in contrast to the (111) plane does not reveal radial cracks. Only the ring cracks are observed in these indents. This observation indicates a significant anisotropy of the fracture toughness in cBN single crystals. Especially interesting are the 0.49-N indents on the (001) cBN plane (Figure 8). The clear linear details along the <110> direction are observed in the <100> Vickers indent, which look like slip lines. Brookes et al. [41] annealed a room-temperature Knoop indent on the (001) plane of cBN at 900 °C and then etched the sample surface to reveal the slip lines around the indent. The orientation of the slip lines makes it possible to identify the active slip system in cBN. It was concluded that it is a {111} <110> slip system [41]. Thus, the <110> directions of the linear features in the <100> indent (Figure 8a–c) support the assumption that they are slip lines. However, further study is necessary to make a final decision. In any case, the AFM images of the Vickers indents on the (001) plane indicate unexpectedly high plasticity of cBN at room temperature as compared with diamond.

Figure 9 shows the AFM derivative topography of a 4.91-N indent on the surface of the nanopolycrystalline cBN compact. The ring cracks in the indent and around it are absent, only rather short radial cracks are observed. The shape of the indent is close to a square, and projected area of the indent is 81.40×10^6 nm^2.

Figure 7. (**a–c**) AFM images of orientations B and (**d–f**) orientation A of Vickers indents on the (111) cBN (1.96-N load): (**a,d**) topography, (**b,e**) derivative topography, and (**c,f**) phase contrast.

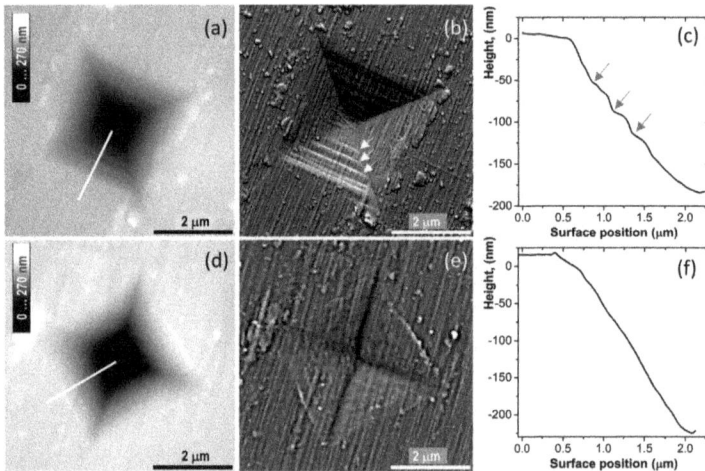

Figure 8. (**a–c**) AFM images of <100> indent and (**d–f**) <110> indent on (100) cBN plane (0.49-N load): AFM topography (**a,d**) and derivative topography (**b,e**) images; and cross-section of indents (**c,f**) along the line shown in (**a,d**). Arrows indicate pronounced steps in the <100> indent aligned to <110> direction (**b**).

Figure 9. AFM derivative topography image of Vickers indent on nanocrystalline cBN. 4.91-N load.

3.3. Vickers Hardness and Fracture Toughness of Diamond and cBN Single Crystals

The Vickers hardness H_V (GPa) is calculated as a ratio of the indentation load P (N) to the lateral indent surface using the standard equation for the Vickers geometry:

$$H_V = 2\sin\left(\frac{\varphi_i}{2}\right)\frac{P}{d^2} = 1.8544\frac{P}{d^2} \tag{2}$$

where d is the indent diagonal length (μm), and φ_i is the angle between opposite faces of the Vickers diamond indenter (136°). The following assumptions are made using this equation: (a) the shape of the indent is close to a square, and (b) the elastic deformation of a diamond indenter under the load is negligibly small (i.e., the indenter is rigid). Such assumptions are quite acceptable in hardness testing of soft and low-modulus materials. However, it is necessary to account for a change of the Vickers indenter geometry caused by the elastic deformation at the hardness testing of such high-modulus materials as diamond and cBN. We used a model proposed by B. Galanov et al. to account for the change of the Vickers indenter geometry due to elastic deformation [42]:

$$\cot\left(\frac{\varphi}{2}\right) = \cot\left(\frac{\varphi_i}{2}\right) - \frac{2HM}{E^*} \tag{3}$$

where φ is the angle between opposite faces of the Vickers diamond indenter under the load, HM is the Meyer hardness of the sample, and E^* is the reduced modulus. The effect of hardness and elastic modulus of materials on the elastic deformation of the Vickers diamond indenter are given in the Table 1.

Table 1. The effect of elastic deformation of Vickers indenter on its geometry for some materials.

Sample	E, GPa	Poisson's Ratio	HM, GPa	$\varphi/2$, °	$2\sin(\varphi/2)$
Al	72	0.35	0.36	68.04	1.8549
SiO_2	71	0.17	9	68.79	1.8646
Al_2O_3	400	0.22	22	69.94	1.8787
cBN	850	0.12	62	73.58	1.9184
Diamond	1136	0.07	100	77.17	1.9501

It can be seen from the table that the geometry of the Vickers indenter changes significantly during the penetration into the surface of hard and superhard materials. The $\varphi/2$ angle increases from 68°

(initial shape) up to 77° (a penetration into the diamond surface). However, $\sin(\varphi/2)$ changes much less than the growth of the 2HM/E* ratio (Table 1). Therefore, accounting for the elastic deformation of the Vickers diamond indenter is important only for the hardest materials. An increase of the proportionality coefficient in Equation (2) does not exceed 5% for diamond and 3.4% for cBN.

Additionally, Equation (2) supposes that the Vickers indent has the shape of a square. This is a rather rare case. The sinking of the surface around the indent (concave shape) is observed for annealed metals and the real contact area is lower as compared with a square indent with the same diagonals [43]. The situation is contrary in the case of previously-deformed metals: the formation of pile up around the indent (convex shape) takes place and the real contact area is higher than that of a square. The equivalent diagonal of the indent d_{eq} (μm) was determined to account for real projected area of the indent in diamond using an equation that connects the area of square indent with the diagonal:

$$d_{eq} = \sqrt{2A}, \tag{4}$$

where A is the projected contact area (10^6 nm²) measured by AFM. For example, according to the AFM, the projected area of the Vickers indent on the surface of the (111) diamond at 5-N load is 60,044,943 nm². From here, using Equation (4), we obtained that d_{eq} is 10.96 μm, whereas according to the AFM image (Figure 2c) the real length of diagonal is 13.06 μm. Substituting d_{eq} into Equation (2) and taking into account that $2\sin(\varphi/2)$ for the diamond sample is 1.9501 (Table 1), we finally obtained that the Vickers hardness of type-Ib (111) synthetic diamond is 79.7 GPa (4.91-N load, Table 2). Due to the concave shape of an indent on the surface of diamond, the phase-transformed contact area determined by the AFM is smaller than the indented area defined by the diagonal's length. Therefore, the H_V values for diamond single crystals in the previous publications [39,40,44] can be underestimated.

Table 2. Vickers and Meyer hardness of diamond and cBN measured taking into account the elastic deformation of the Vickers diamond indenter and the concave shape of indents.

Sample	Load P, N	Projected Contact Area A, 10^6 nm²	Equivalent Diagonal d_{eq}, μm	Vickers Hardness H_V, GPa	Meyer Hardness HM, GPa	Radial Crack Length C, μm	Fracture Toughness K_{IC}, MPa m$^{0.5}$
(111) Synthetic diamond	4.91	60.05	10.96	79.7	81.7	14.2 [44]	5.6
Nano cBN compact	4.91	81.40	12.51	60.2	61.5	11.40	7.9
(111) cBN orientation A	1.96	38.46	8.77	48.9.	51.0	10.65	3.2
(111) cBN orientation B	1.96	39.05	8.84	48.2	50.2	11.37	3.5
(001) cBN, orientation <100>	1.96	38.04	8.72	49.4	51.5	-	-
(001) cBN, orientation <110>	1.96	43.36	9.31	43.4	45.2	-	-

The Vickers hardness of the (111) plane of type-Ib synthetic diamond at 5-N load is about 80 GPa, which is appreciably lower than the Vickers hardness of natural type-Ia diamond (92 GPa [12]). This is probably the consequence of a high density of dislocations and nitrogen concentration in the type-Ib synthetic diamond sample produced by the spontaneous crystallization. This value (80 GPa) is higher than the Vickers hardness of the (011) natural diamond after etching of the indent in the KNO₃ melt (67 GPa without account for elastic deformation of Vickers indenter [45]). However, it is appreciably lower than was determined using optical microscopy for measuring the as-received indent in the (111) and (001) diamond single crystals (about 100–115 GPa for both planes [39,44]). The Vickers hardness of the (111) cBN single crystals (1.96-N load) is approximately the same for both orientations of indents on the (111) plane ~49 GPa. It is much less than the H_V of the (111) cBN data, according to the optical microscopy (62 GPa [46]). Probably, the spatial resolution of the optical microscopy is not high enough for reliable measurements of indents in diamond and cBN single crystals, especially at low applied loads. The Vickers hardness of nanopolycrystalline cBN is about 60 GPa (5-N load), which is much higher than for the (111) cBN single crystal (49 GPa, 1.96 N). A random orientation of the grains in a nanopolycrystalline sample suppresses the perfect cleavage of cBN on the {110} planes. Therefore, the fracture toughness of nanopolycrystalline cBN is also higher than for cBN single crystals, i.e., 7.9 MPa m$^{0.5}$ and 3.2 MPa m$^{0.5}$, respectively (Table 2). It follows from the tests results that grains and nanotwin boundaries in nanopolycrystalline cBN cause an increase not only in the hardness, but in the fracture toughness of materials as well.

At the present time, the ratio of the indentation load to the projected contact area is often used for hardness determination, for example, in nanoindentation. It is a Meyer hardness *HM*. The elastic deformation of the indenter does not affect the projected contact area. Since the area of pyramid base is lower than its lateral surface, the *HM* is higher than H_V. In the case of the rigid Vickers indenter, the *HM* to H_V ratio is 1.078. The Meyer hardness for diamond and cBN samples is given in Table 2.

4. Conclusions

Vickers hardness of the (111) synthetic type 1b diamond, (001) and (111) cBN single crystals, and a nanopolycrystalline cBN bulk sample was studied in this paper. Atomic force microscopy in the phase contrast operation mode, micro-Raman spectroscopy, and Kelvin probe force microscopy were used for accurate identifying the projected contact area of Vickers indents on the surface of diamond. The effect of the elastic deformation of the diamond Vickers indenter on its geometry was accounted for at the hardness measurements. It was shown that:

1. The Vickers hardness of the (111) plane of type-Ib synthetic diamond at 5-N load is about 80 GPa, which is appreciably lower than the Vickers hardness of natural-type-Ia diamond (92 GPa [12]). Probably, this is a consequence of a high density of dislocations and nitrogen concentration in the type-Ib synthetic diamond sample produced by the spontaneous crystallization.

2. The Vickers hardness of the (111) cBN single crystal at 2-N load is about 49 GPa. The appreciable anisotropy of Vickers hardness is observed on the (001) plane of cBN.

3. Grains and nanotwin boundaries in the nanopolycrystalline cBN sample resulted in the enhancement of both the Vickers hardness and fracture toughness as compared with cBN single crystals.

4. The determination of the Vickers hardness of diamond by the indent diagonals length results in the underestimated value as the concave shape of the contact under load is not taken into account.

Author Contributions: Sergey Dub, Petro Lytvyn and Viktor Strelchuk conceived the project. Igor Petrusha, Takashi Taniguchi and Sergey Ivakhnenko performed the HPHT experiments. Sergey Dub performed the hardness measurements. Viktor Strelchuk, Andrii Nikolenko and Yurii Stubrov performed the micro-Raman spectroscopy and analyzed the data. Petro Lytvyn performed the atomic force microscopy. Sergey Dub, Petro Lytvyn and Andrii Nikolenko wrote the paper. All authors discussed the results and commented on the manuscript.

Conflicts of Interest: The authors declare no conflict of interest.

References

1. Chowdhury, S.; de Barra, E.; Laugier, M.T. Hardness measurement of CVD diamond coatings on SiC substrates. *Surf. Coat. Technol.* **2005**, *193*, 200–205. [CrossRef]
2. Irifune, T.; Kurio, A.; Sakamoto, S.; Inoue, T.; Sumiya, H. Materials: Ultrahard polycrystalline diamond from graphite. *Nature* **2003**, *421*, 599–600. [CrossRef] [PubMed]
3. Sumiya, H.; Irifune, T. Indentation hardness of nano-polycrystalline diamond prepared from graphite by direct conversion. *Diam. Relat. Mater.* **2004**, *13*, 1771–1776. [CrossRef]
4. Sumiya, H.; Uesaka, S.; Satoh, S. Mechanical properties of high purity polycrystalline cBN synthesized by direct conversion sintering method. *J. Mater. Sci.* **2000**, *35*, 1181–1186. [CrossRef]
5. Huang, Q.; Yu, D.; Xu, B.; Hu, W.; Ma, Y.; Wang, Y.; Zhao, Z.; Wen, B.; He, J.; Liu, Z.; et al. Nanotwinned diamond with unprecedented hardness and stability. *Nature* **2013**, *510*, 250–253. [CrossRef] [PubMed]
6. Dubrovinskaia, N.; Solozhenko, V.L.; Miyajima, N.; Dmitriev, V.; Kurakevych, O.O.; Dubrovinsky, L. Superhard nanocomposite of dense polymorphs of boron nitride: Noncarbon material has reached diamond hardness. *Appl. Phys. Lett.* **2007**, *90*, 101912. [CrossRef]

7. Solozhenko, V.L.; Kurakevych, O.O.; Le Godec, Y. Creation of nanostuctures by extreme conditions: High-pressure synthesis of ultrahard nanocrystalline cubic boron nitride. *Adv. Mater.* **2012**, *24*, 1540–1544. [CrossRef] [PubMed]

8. Tian, Y.; Xu, B.; Yu, D.; Ma, Y.; Wang, Y.; Jiang, Y.; Hu, W.; Tang, C.; Gao, Y.; Luo, K.; et al. Ultrahard nanotwinned cubic boron nitride. *Nature* **2013**, *493*, 385–388. [CrossRef] [PubMed]

9. Nagakubo, A.; Ogi, H.; Sumiya, H.; Hirao, M. Elasticity and hardness of nano-polycrystalline boron nitrides: The apparent Hall-Petch effect. *Appl. Phys. Lett.* **2014**, *105*, 081906. [CrossRef]

10. Liu, X.; Chang, Y.-Y.; Tkachev, S.N.; Bina, C.R.; Jacobsen, S.D. Elastic and mechanical softening in boron-doped diamond. *Sci. Rep.* **2017**, *7*, 42921. [CrossRef] [PubMed]

11. Sumiya, H.; Ishida, Y.; Arimoto, K.; Harano, K. Real indentation hardness of nano-polycrystalline cBN synthesized by direct conversion sintering under HPHT. *Diam. Relat. Mater.* **2014**, *48*, 47–51. [CrossRef]

12. Xu, B.; Tian, Y. Ultrahardness: Measurement and enhancement. *J. Phys. Chem. C* **2015**, *119*, 5633–5638. [CrossRef]

13. Drory, M.D.; Dauskardt, R.H.; Kant, A.; Ritchie, R.O. Fracture of synthetic diamond. *J. Appl. Phys.* **1995**, *78*, 3083–3088. [CrossRef]

14. Solozhenko, V.L.; Dub, S.N.; Novikov, N.V. Mechanical properties of cubic BC2N, a new superhard phase. *Diam. Relat. Mater.* **2001**, *10*, 2228–2231. [CrossRef]

15. Zerr, A.; Kempf, M.; Schwarz, M.; Kroke, E.; Göken, M.; Riedel, R. Elastic moduli and hardness of cubic silicon nitride. *J. Am. Ceram. Soc.* **2002**, *85*, 86–90. [CrossRef]

16. Dub, S.N.; Petrusha, I.A. Mechanical properties of polycrystalline cBN obtained from pyrolytic gBN by direct transformation technique. *High Press. Res.* **2006**, *26*, 71–77. [CrossRef]

17. Dub, S.N.; Petrusha, I.A.; Bushlya, V.M.; Taniguchi, T.; Belous, V.A.; Tolmachova, G.N.; Andreev, A.V. Theoretical shear strength and the onset of plasticity in nanodeformation of cubic boron nitride. *J. Superhard Mater.* **2017**, *39*, 88–98. [CrossRef]

18. Richter, A.; Ries, R.; Smith, R.; Henkel, M.; Wolf, B. Nanoindentation of diamond, graphite and fullerene films. *Diam. Relat. Mater.* **2000**, *9*, 170–184. [CrossRef]

19. Sawa, T.; Tanaka, K. Nanoindentation of natural diamond. *Philos. Mag. A* **2002**, *82*, 1851–1856. [CrossRef]

20. Dubrovinskaia, N.; Dub, S.; Dubrovinsky, L. Superior wear resistance of aggregated diamond nanorods. *Nano Lett.* **2006**, *6*, 824–826. [CrossRef] [PubMed]

21. Dub, S.N.; Brazhkin, V.V.; Belous, V.A.; Tolmacheva, G.N.; Konevskii, P.V. Comparative nanoindentation of single crystals of hard and superhard oxides. *J. Superhard Mater.* **2014**, *36*, 217–230. [CrossRef]

22. Gogotsi, Y.G.; Kailer, A.; Nickel, K.G. Pressure-induced phase transformations in diamond. *J. Appl. Phys.* **1998**, *84*, 1299–1304. [CrossRef]

23. Gogotsi, Y.G.; Kailer, A.; Nickel, K.G. Transformation of diamond to graphite. *Nature* **1999**, *401*, 663–664. [CrossRef]

24. Suprun, O.M.; Ilnitskaya, G.D.; Kalenchuk, V.A.; Zanevskii, O.A.; Shevchuk, S.N.; Lysakovskii, V.V. Change of dislocations density in single crystals of various types diamonds depending on the growth temperature and rate. *Funct. Mater.* **2016**, *23*, 552–556. [CrossRef]

25. Khvostantsev, L.G.; Slesarev, V.N. Large-volume high-pressure devices for physical investigations. *UFN* **2008**, *51*, 1059–1063. [CrossRef]

26. Anstis, G.R.; Chantikul, P.; Lawn, B.R.; Marshal, D.B. A critical evaluation of indentation techniques for measuring fracture toughness: I, direct crack measurements. *J. Am. Ceram. Soc.* **1981**, *64*, 533–538. [CrossRef]

27. Zhong, Q.; Inniss, D.; Kjoller, K.; Elings, V.B. Fractured polymer/silica fiber surface studied by tapping mode atomic force microscopy. *Surf. Sci.* **1993**, *290*, L688–L692. [CrossRef]

28. Radmacher, M.; Tillmann, R.W.; Gaub, H.E. Imaging viscoelasticity by force modulation with the atomic force microscope. *Biophys. J.* **1993**, *64*, 735–742. [CrossRef]

29. Magonov, S.N.; Elings, V.; Whangbo, M.-H. Phase imaging and stiffness in tapping-mode atomic force microscopy. *Surf. Sci.* **1997**, *375*, L385–L391. [CrossRef]

30. Nonnenmacher, M.; O'Boyle, M.P.; Wickramasinghe, H.K. Kelvin probe force microscopy. *Appl. Phys. Lett.* **1991**, *58*, 2921–2923. [CrossRef]

31. Domnich, V.; Gogotsi, Y.; Dub, S. Effect of phase transformations on the shape of unloading curve in the nanoindentation of silicon. *Appl. Phys. Lett.* **2000**, *76*, 2214–2216. [CrossRef]

32. Diederich, L.; Küttel, O.M.; Aebi, P.; Schlapbach, L. Electron affinity and work function of differently oriented and doped diamond surfaces determined by photoelectron spectroscopy. *Surf. Sci.* **1998**, *418*, 219–239. [CrossRef]

33. Robrieux, B.; Faure, R.; Dussaulcy, J.P. Resistivity and electronic work function of very thin film carbon. *C. R. Acad. Sci. Ser. B* **1974**, *278*, 659–662.

34. Xu, C.; Liu, C.; Wang, H. Incipient plasticity of diamond during nanoindentation. *RSC Adv.* **2017**, *7*, 36093–36100. [CrossRef]

35. Denisov, V.N.; Mavrin, B.N.; Serebryanaya, N.R.; Dubitsky, G.A.; Aksenenkov, V.V.; Kirichenko, A.N.; Kuzmin, N.V.; Kulnitskiy, B.A.; Perezhogin, I.A.; Blank, V.D. First-principles, UV Raman, X-ray diffraction and TEM study of the structure and lattice dynamics of the diamond—Lonsdaleite system. *Diam. Relat. Mater.* **2011**, *20*, 951–953. [CrossRef]

36. Goryainov, S.V.; Likhacheva, A.Y.; Rashchenko, S.V.; Shubin, A.S.; Afanas'ev, V.P.; Pokhilenko, N.P. Raman identification of lonsdaleite in Popigai impactites. *J. Raman Spectrosc.* **2014**, *45*, 305–313. [CrossRef]

37. Boteler, J.M.; Gupta, Y.M. Shock induced splitting of the triply degenerate Raman line in diamond. *Phys. Rev. Lett.* **1993**, *71*, 3497. [CrossRef] [PubMed]

38. Boteler, J.M.; Gupta, Y.M. Raman spectra of shocked diamond single crystals. *Phys. Rev. B* **2002**, *66*, 014107. [CrossRef]

39. Novikov, N.V.; Dub, S.N. Fracture toughness of diamond single crystals. *J. Hard Mater.* **1991**, *2*, 3–11.

40. Novikov, N.V.; Dub, S.N.; Mal'nev, V.I.; Beskrovanov, V.V. Mechanical properties of diamond at 1200 °C. *Diam. Relat. Mater.* **1994**, *3*, 198–204. [CrossRef]

41. Brookes, C.A.; Hooper, R.M.; Lambert, W.A. Identification of slip systems in cubic boron nitride. *Philos. Mag. A* **1983**, *47*, L.9–L.12. [CrossRef]

42. Galanov, B.A.; Milman, Y.V.; Chugunova, S.I.; Goncharova, I.V. Investigation of mechanical properties of high-hardness materials by indentation. *J. Superhard Mater.* **1999**, *21*, 23–35.

43. Lim, Y.Y.; Chaudhri, M.M. The effect of the indenter load on the nanohardness of ductile metals: An experimental study on polycrystalline work-hardened and annealed oxygen-free copper. *Philos. Mag. A* **1999**, *79*, 2979–3000. [CrossRef]

44. Dub, S.N. Fracture Toughness Determination of Superhard Single Crystals by Indentation. Ph.D. Thesis, Institute for Superhard Materials of the Academy of Sciences of Ukraine, Kiev, Ukraine, 1 November 1984.

45. Novikov, N.V.; Dub, S.N. Hardness and fracture toughness of CVD diamond film. *Diam. Relat. Mater.* **1996**, *5*, 1026–1030. [CrossRef]

46. Novikov, N.V.; Dub, S.N.; Malnev, V.I. Microhardness and fracture toughness of cubic boron nitride single crystals. *Sov. J. Superhard Mater.* **1983**, *5*, 16–20.

![crystals logo] *crystals*

MDPI

Article

Nanoindentation of HMX and Idoxuridine to Determine Mechanical Similarity

Alexandra C. Burch [1], John D. Yeager [2] and David F. Bahr [1,*]

[1] School of Materials Engineering, Purdue University, West Lafayette, IN 47907, USA; burch12@purdue.edu
[2] Explosive Science and Shock Physics, Los Alamos National Laboratory, Los Alamos, NM 87545, USA; jyeager@lanl.gov
* Correspondence: dfbahr@purdue.edu

Academic Editors: Ronald W. Armstrong, Stephen M. Walley and Wayne L. Elban
Received: 28 September 2017; Accepted: 28 October 2017; Published: 1 November 2017

Abstract: Assessing the mechanical behavior (elastic properties, plastic properties, and fracture phenomena) of molecular crystals is often complicated by the difficulty in preparing samples. Pharmaceuticals and energetic materials in particular are often used in composite structures or tablets, where the individual grains can strongly impact the solid behavior. Nanoindentation is a convenient method to experimentally assess these properties, and it is used here to demonstrate the similarity in the mechanical properties of two distinct systems: individual crystals of the explosive cyclotetramethylene tetranitramine (HMX) and the pharmaceutical idoxuridine were tested in their as-precipitated state, and the effective average modulus and hardness (which can be orientation dependent) were determined. Both exhibit a hardness of 1.0 GPa, with an effective reduced modulus of 25 and 23 GPa for the HMX and idoxuridine, respectively. They also exhibit similar yield point behavior. This indicates idoxuridine may be a suitable mechanical surrogate (or "mock") for HMX. While the methodology to assess elastic and plastic properties was relatively insensitive to specific crystal orientation (i.e., a uniform distribution in properties was observed for all random crystals tested), the indentation-induced fracture properties appear to be much more sensitive to tip-crystal orientation, and an unloading slope analysis is used to demonstrate the need for further refinement in relating toughness to orientation in these materials with relatively complex slip systems and crystal structures.

Keywords: nanoindentation; molecular crystals; mechanical properties; hardness; elastic modulus; fracture

1. Introduction

Molecular crystals are put to a variety of different uses, including some explosives, pharmaceuticals, and foods. Despite having very different practical applications, these materials often have similar molecular structures as well as physical and mechanical properties. Often, one type of molecular crystal is used to simulate or "mock" another type of crystal for a given test or scenario. Mock materials are commonly used when the simulant allows for increased safety, lower costs, or ease of handling compared to the original material. Recently, several materials were identified as new mocks for the explosive cyclotetramethylene-tetranitramine (HMX), in terms of density and thermal stability [1]. However, the effectiveness of simulating mechanical properties with these new mocks has not been addressed. Typically, sucrose has been used to mock HMX mechanically [2] but it has a lower density and melt point, limiting the scenarios that can be simulated. In other cases, such as with pharmaceuticals, the use of a model material allows for rapid or inexpensive assessment of some processing parameters [3]. One possible way to identify materials with similar physical (elastic properties) and mechanical responses (for plastic flow and fracture) is to utilize nanoindentation to probe the properties of individual single crystals of these materials with little need for complex sample geometries.

Testing the mechanical properties of powders and small powder-like particles can be difficult both due to the small size of many of these materials in their as-formed state; testing can also be difficult with regard to forming or machining samples to meet standard mechanical test geometries. Nanoindentation resolves this limitation, as indentation can be performed on samples with lateral dimensions on the order of 10's of μms or smaller. While it is possible to assess the properties of individual grains within metallic systems (where the grains are often identified by electron backscattered diffraction [4]), it is also possible to use indentation to characterize molecular crystals such as cyclotrimethylene-trinitramine (RDX), HMX, 1,3,5-triamino-2,4,6-trinitrobenzene (TATB), sucrose, aspirin, and many more [5–10]. The small length scale of nanoindentation also enables the crystals to be tested without additional processing, whereas other techniques or larger mechanical tests (such as using a Vickers hardness test on RDX [11]) might require growing large and pristine single crystals. Nanoindentation therefore allows molecular crystals to be tested in their as-received or as-used condition. This is critical for evaluating the actual properties of the materials in application, since common techniques to grow "better" samples can alter the measured properties. For example, individual crystal growth could be enhanced using solution adjuncts which minimize nucleation, so growth rates may greatly exceed nucleation rates, which could impact the defect density within the crystals [12].

Instrumented indentation has been used to quantify a wide variety of mechanical properties including elastic modulus [13–15], which may be crystal orientation-dependent [16,17], and hardness [18,19], which is often used as a surrogate for general plastic flow behavior. Additionally, yield points during initial loading can be indicative of the onset of dislocation nucleation [20]. Finally, indentation-induced fracture [21–26] has been used by many researchers as a surrogate to quantify toughness in materials where meeting American Society for Testing and Materials (ASTM) standard specimen geometries would be challenging. Elastic modulus, hardness, and yield behavior are often measured using a Berkovich indenter probe, but more acute probes, such as the cube corner geometry, can be useful in initiating fracture.

Indentation fracture, often with a Vickers geometry and testing system [27], conventionally uses post-indentation characterization to assess if cracks are present in the sample, and then relates cracking behavior to the stress fields which drove crack propagation. Experiments have demonstrated that in some cases cracking can occur during loading of the indenter, while in other cases the cracks initiate upon unloading, as shown by Cook and Pharr [21]. Identifying a crack after the indentation is complete is typically done using optical, electron, or scanning probe microscopy, but a crack may also be identified by the load-depth curve of the indentation. Morris and co-workers showed that when a material is indented with two different indenter probes of varying acuteness, materials that have not cracked have superimposable unloading portions of the load-depth curves, whereas when the material has cracked the unloading curves will be nonsuperimposable [28]. This method of detecting fracture allows cracks to be identified even when they cannot be seen optically, for more precise identification of the onset of fracture. The need to identify fracture behavior is important in assessing materials properties, not only in determining toughness, but if modulus or hardness of two different materials are being compared it is crucial to ensure that the systems are behaving similarly. It is not appropriate to compare hardness or modulus if, in one case, a crack had formed and in the other material it had not, as the conventionally used models used to extract properties such as hardness and modulus during nanoindentation were developed for bulk materials that exhibited uniform elastic-plastic deformation and do not consider fracture.

In this paper we assess the suitability of using idoxuridine to serve as a surrogate for the mechanical response of HMX using nanoindentation for granular solids in the sub-mm regime, taking into account possible differences in material fracture behavior when comparing mechanical properties. We use two different indenter tips to conduct unloading analysis to investigate cracking and supplement this with scanning probe microscopy. We also find that using the samples as-received is sufficient for accurate data collection.

2. Results and Discussion

Hardness and elastic modulus measurements were taken using a Berkovich indenter probe at 1000 μN in both materials; the resulting load-displacement data is shown in Figure 1.

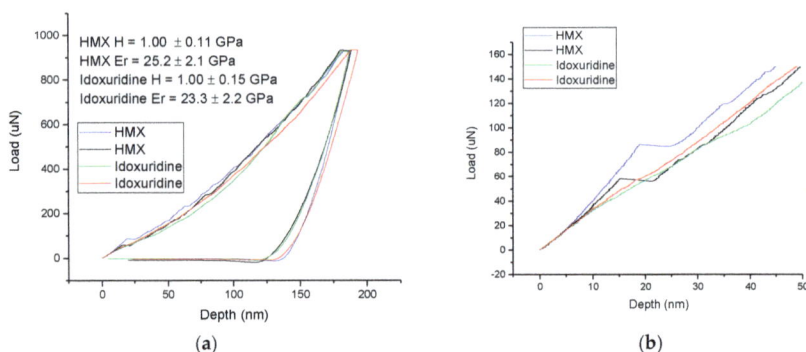

Figure 1. (**a**) Typical load-depth curves for HMX and idoxuridine, with indentations in two random crystals of each material shown. Modulus values reported in the figure are the average values of $\approx$20 indents in each material. (**b**) The same load-depth curves as in (**a**), showing only data below 150 μN. Note there is evidence of yield point behavior at loads less than 100 μN, and multiple yield events on loading, which are common in materials with limited slip systems.

For HMX, 22 indents were done on approximately 10 different crystals. Some crystals were only measured once, while others were large enough to accommodate multiple indents; all indentations were spaced at least 10 times across the residual impression diameter to ensure the pristine material was being evaluated. The average hardness measured for HMX was 1.00 ± 0.11 GPa, and the average reduced elastic modulus measured for HMX was 25.2 ± 2.1 GPa. On idoxuridine, 19 indents were done on approximately 10 crystals, with an average hardness measurement of 1.00 ± 0.15 GPa, and an average reduced elastic modulus measurement of 23.3 ± 2.2 GPa. The distribution of these measurements is shown in Figure 2. A Wilcoxon rank-sum test comparing the hardness of HMX to the hardness of idoxuridine gave a p value of 0.9167, so the hardness of the two materials is not significantly different. For comparing elastic modulus, a Wilcoxon rank-sum test gave a p value of 0.007, indicating that the elastic moduli of these materials are statistically different. However, because the elastic moduli of the two materials are within 10% of each other, they are likely similar enough to be considered "mocks" in many situations. For example, sucrose is considered a mechanical mock but has much higher elastic modulus, between 33 and 38 GPa depending on orientation [6].

As noted in Figure 1, many indentations exhibited a "pop-in" or "excursion" in the load depth curve. This is commonly considered to be indicative of the transition from elastic to plastic deformation. Yield point behavior can be quantified by the load at yield (which is proportional to the maximum applied shear stress for materials with the same elastic modulus when a common tip with a fixed radius is used). In all indentations in HMX the indentation curve exhibited a yield point, while in the case of idoxuridine only half of the indents exhibited a yield point. Figure 3 shows the difference in the distribution of load at the first yield point for all indentations performed in this study. Using a cumulative distribution plot is a convenient way to determine if different defect densities or mechanisms are being probed [29], and the relative curvature and position are indicative of an activation energy to nucleate a dislocation when one compares the only the fraction that yielded (ignoring those that exhibit no yield point). In metallic systems, higher defect densities are linked to larger numbers of indentations that do not exhibit a yield point [30], and surface preparation can shift (in load) or "tilt" in probability the cumulative fraction plot in RDX [31]. The mean load at yield for indentations that did exhibit a yield pinot for idoxuridine was 98 μN and for HMX was 93 μN, (very similar results); the maximum load exhibited (which previous studies have considered to be linked to approaching the theoretical shear stress in the crystal) are also of a similar magnitude, as one might expect for materials with similar elastic modulus values. Finally, the median of the loads which caused yield are similar for these indentations, 78.5 μN for HMX and 62.8 μN for idoxuridine. The implication here is that while the average shear stress needed to nucleate dislocations in both materials is almost identical, suggesting the nucleation phenomena is based on the same mechanism in both these samples, and the maximum shear stress is of the same order, based on the similarity of the maximum observed value in yield point, the likelihood of probing a mechanical defect (such as a pre-existing dislocation) is much higher in the idoxuridine than in the HMX for the forms of the materials tested in this study.

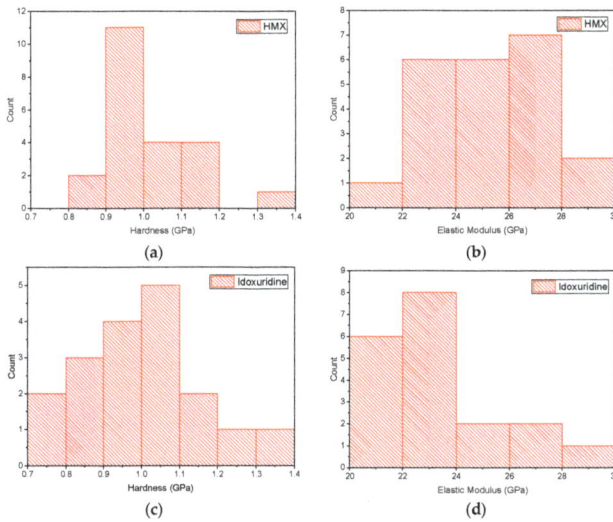

Figure 2. (**a**) HMX hardness measurement distribution. (**b**) HMX reduced elastic modulus measurement distribution. (**c**) Idoxuridine hardness measurement distribution. (**d**) Idoxuridine reduced elastic modulus measurement distribution.

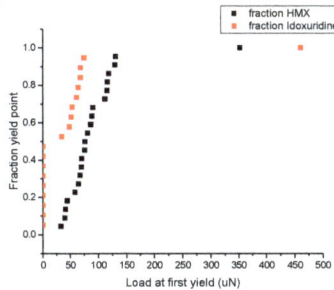

Figure 3. Cumulative fraction of yield behavior for HMX and idoxuridine. While the mean load at yield for indentations that did exhibit a distinct yield point was statistically similar (93 and 98 μN for HMX and idoxuridine, respectively), the materials exhibit different behavior in yield distribution. While all indentations in HMX showed a yield point, only half the indentations in idoxuridine showed.

One concern with assessing brittle molecular solids is the possibility of cracking. Critical loads beyond which there is fracture are often reported for many brittle materials such as RDX [24] and glass [23]. Identifying cracking in molecular crystals can be challenging. Two approaches to identifying indent-induced cracking are with post-indent microscopy and with unloading analysis [28]. Figure 4 shows scanning probe images of indents on HMX and idoxuridine with both Berkovich and cube corner indenter probes, at loads of 500 µN and 5000 µN, where cracking could not be confirmed via microscopy in any of these cases.

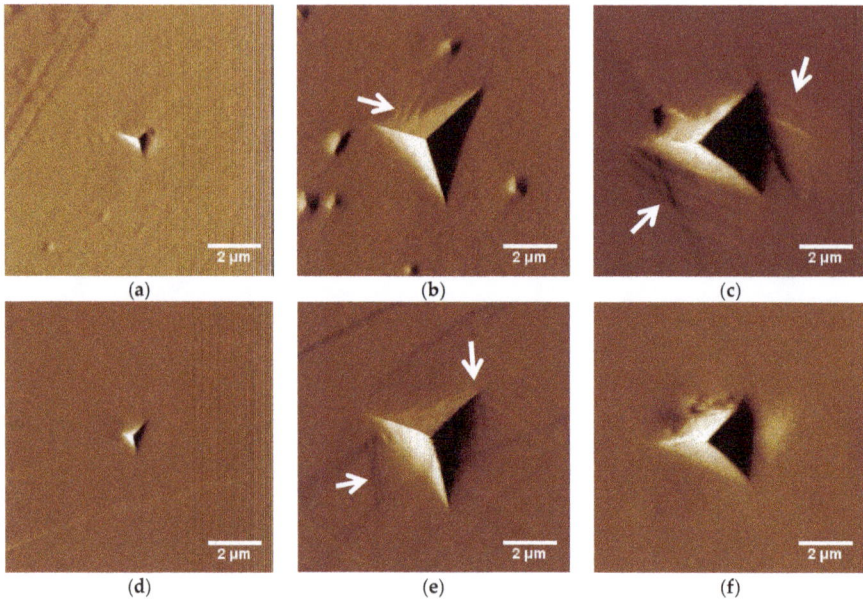

Figure 4. Scanning probe images (in deflection mode using the Hysitron Triboindenter imaging mode, which are therefore indicative of slope, not height, to accentuate small surface topography changes) of (**a**) 500 µN Berkovich indent on idoxuridine, (**b**) 5000 µN Berkovich indent on idoxuridine, (**c**) 5000 µN cube corner indent on idoxuridine, (**d**) 500 µN Berkovich indent on HMX, (**e**) 5000 µN Berkovich indent on HMX, and (**f**) 5000 µN cube corner indent on HMX. Slip steps are evident in (**b**,**c**,**e**) (noted with arrows), but it is not possible to conclusively state if cracks are present in some of the images, such as the dark band in the lower left of (**c**).

Though scanning probe images do not conclusively show surface cracks, there is still the possibility of subsurface cracking. To attempt to determine whether subsurface cracking occurred, the unloading analysis method of Morris and co-workers was used [28]. For a specific load, in this case nominal loads of 500 µN and 5000 µN, each material was indented four times with a Berkovich probe and four times with a cube corner probe, and the unloading segment of these indents were averaged, as shown in Figure 5.

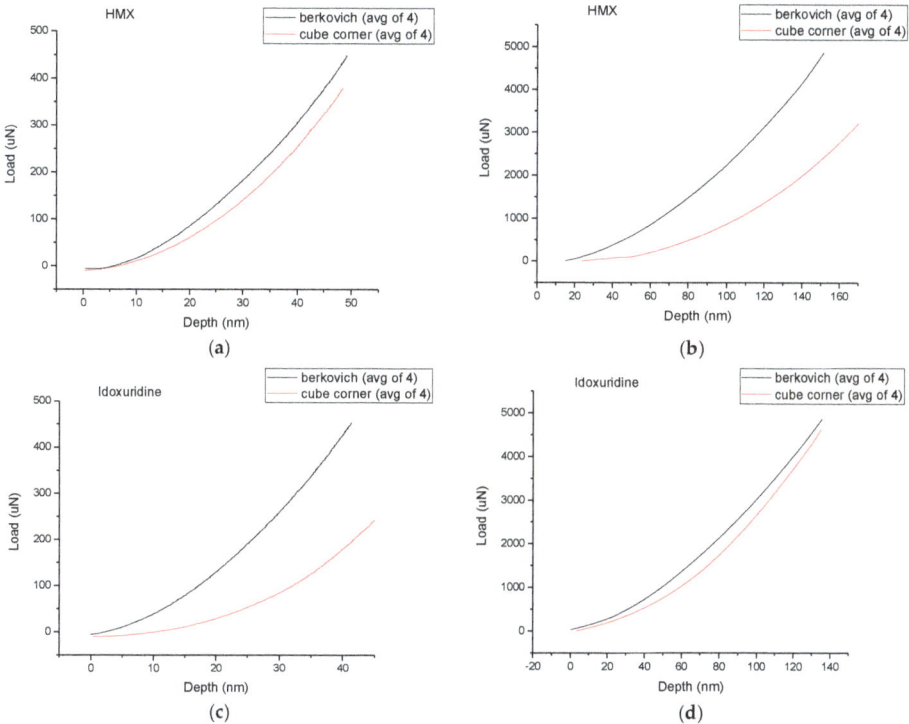

Figure 5. (**a**) HMX unloading comparison at 500 μN. (**b**) HMX unloading comparison at 5000 μN.
(**c**) Idoxuridine unloading comparison at 500 μN. (**d**) Idoxuridine unloading comparison at 5000 μN.

These indentations were carried out on random crystals with no specific orientation. Morris et al. showed that when unloading curves from the same load, using tips of different acuity was superimposable and there was no evidence of cracking, and when the unloading curves were more compliant for the sharper tip this was indicative of cracking during the indentation. While it was not possible to exactly reproduce the conditions used by Morris and coworkers in this study, the general similarity, or lack thereof, was used in this study to indicate a propensity for indentation induced fracture. For some crystals (orientations), there is no significant evidence of cracking with acute probes (Figure 5a,d show very similar unloading slopes), while other individual crystals exhibit evidence of cracking (Figure 5b,c). In all cases, there has been no direct evidence of cracking caused by a Berkovich indenter using post indent microscopy. The range of unloading behavior suggest that there is a variation between crystal faces, crystal orientations, and probe orientations in the subsequent fracture behavior that is not present in the more uniformly distributed hardness and modulus measurements. In particular, for HMX there appears to be a cracking threshold (at loads above 1 mN we consistently observed a more compliant unloading curve, indicative of fracture), while for idoxuridine some indentations at low loads show fracture with the cube corner tip (Figure 5c), while fracture doesn't appear at higher loads (Figure 5d). This variation in idoxuridine may be tied to the behavior noted in Figure 3, which indicated that the crystal to crystal variation in defect density was more significant for idoxuridine than HMX. However, without the current ability to index the individual crystals that were tested, the supposition that it is orientation-dependent cracking, and not some other crystal to crystal variation, that is leading to the less reproducible behavior in compliance on unloading is only one possible explanation, and this area of inquiry requires further study.

3. Materials and Methods

HMX and idoxuridine, shown in Figure 6, were provided by Los Alamos National Laboratory. HMX single crystals were recrystallized in acetone from Class 1 HMX produced by Holston. Idoxuridine was originally purchased from Chem-Impex International, Inc. and recrystallized in water.

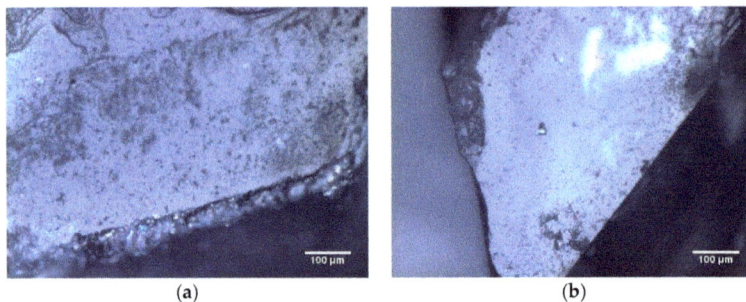

| (a) | (b) |

Figure 6. Optical micrographs of (**a**) an HMX crystal (**b**) and an idoxuridine crystal.

All samples were mounted using the technique described by Maughan et al. [32] for mounting small crystals. To briefly summarize, a flat face of a crystal was placed on an aluminum block, and a commercially available AFM "puck" (a 7 mm diameter steel disc) was suspended above the crystal by a magnet with an adhesive on the downward face of the disc. The disc was lowered such that the adhesive came in contact with the crystal, and upon inversion of the disc, the surface of the crystal was the flat face that had previously rested on the aluminum block and was therefore parallel to the AFM disc and normal to the indenter probe. Nanoindentation was performed using a Hysitron Triboindenter 950, with both Berkovich and cube corner indenter probes with tip radii of approximately 600 nm and 140 nm, respectively. All indents were quasistatic open-loop with 30 s loading, 5 s hold, and 5 s unload times; this loading profile was used in prior studies of RDX and some other organic molecular crystals [5]. The unloading curves were analyzed using the Oliver and Pharr technique; the tip had been calibrated in fused quartz and aluminum prior to indentation accounting for S (unloading stiffness), P, h, and A being load, depth, and contact area, respectively, a geometric constant γ.

$$S = \frac{dP}{dh} = \frac{2\gamma E_r \sqrt{A}}{\sqrt{\pi}} \tag{1}$$

and modulus,

$$\frac{1}{E_r} = \frac{1 - v^2}{E_i} + \frac{1 - v^2}{E_s} \tag{2}$$

where E_r is the reduced modulus, accounting for the Young's Modulus, E and Poisson's ratio, v, of the indenter tip (i) and sample (s). While E and v of the diamond indenter tip are well known (1249 GPa and 0.07, respectively), we do not know for certain the Poisson's ratio of these samples, and therefore this paper reports only E_r. Imaging (Figure 4) was carried out using the Hysitron scanning probe mode, where the tip making the indentation is used to image as a fixed load (in this case 2 μN).

4. Conclusions

We have successfully determined that idoxuridine has similar elastic and plastic mechanical properties to HMX, which can be difficult to perform tests on due to safety concerns. The similarity in hardness and elastic modulus indicate that idoxuridine can be used to test the mechanical response of composite structures typically containing HMX. When yield behavior occurs in both materials, the loads (and therefore stresses) at which dislocations are nucleated appear to be similar. However, in the as-received state, it appears idoxuridine may have a higher mechanical defect density than the HMX (for powders of the same size). Further work will be needed to quantify fracture behavior and toughness with crystal orientation and initial defect distribution. As fracture appears to be more sensitive to crystal orientation than low load hardness and the elastic modulus, the approach of sampling many randomly oriented crystals to determine average polycrystalline behavior may not be appropriate for fracture studies. Crystal orientation and the relative orientation of the indent corners to the crystal geometry mean that assessing fracture properties will have to consider both out-of-plane and relative in-plane orientation between the crystal and the indenter probe.

Acknowledgments: Funding for this work was provided by DOE/NNSA Weapons Systems Engineering Assessment Technology (WSEAT) Program and the Joint Munitions Program. A. Duque (LANL) helped with sample preparation, while M. Lewis (LANL) provided valuable direction for the project. The authors thank Bryce Tappan (LANL) for providing high quality HMX samples. Los Alamos National Laboratory, an affirmative action equal opportunity employer, is operated by Los Alamos National Security, LLC, for the National Nuclear Security Administration of the U.S. Department of Energy under contract DE-AC52-06NA25396.

Author Contributions: David F. Bahr and John D. Yeager conceived and designed the initial experiments for hardness and modulus. Alexandra C. Burch and David F. Bahr designed the experiments for indentation fracture. John D. Yeager provided the initial materials for testing, Alexandra C. Burch carried out all experimental studies, and all three authors contributed to writing the manuscript.

Conflicts of Interest: The authors declare no conflict of interest. The funding sponsors had no role in the design of the study; in the collection, analyses, or interpretation of data; in the writing of the manuscript, and in the decision to publish the results.

References

1. Yeager, J.D.; Duque, A.L.H.; Shorty, M.; Bowden, P.R.; Stull, J.A. Development of inert density mock materials for HMX. *J. Energ. Mater.* **2017**, 1–13. [CrossRef]
2. Sheffield, S.A.; Gustavsen, R.L.; Alcon, R.R. Porous HMX initiation studies—Sugar as an inert simulant. *AIP Conf. Proc.* **1998**, *429*, 575–578.
3. Buckner, I.S.; Wurster, D.E.; Aburub, A. Interpreting deformation behavior in pharmaceutical materials using multiple consolidation models and compaction energetics. *Pharm. Dev. Technol.* **2010**, *15*, 492–499. [CrossRef] [PubMed]
4. Britton, T.B.; Liang, H.; Dunne, F.P.E.; Wilkinson, A.J. The effect of crystal orientation on the indentation response of commercially pure titanium: Experiments and simulations. *Proc. Math. Phys. Eng. Sci.* **2010**, *466*, 695–719. [CrossRef]
5. Taw, M.R.; Yeager, J.D.; Hooks, D.E.; Carvajal, T.M.; Bahr, D.F. The mechanical properties of as-grown noncubic organic molecular crystals assessed by nanoindentation. *J. Mater. Res.* **2017**, *32*, 1–10. [CrossRef]
6. Ramos, K.J.; Bahr, D.F. Mechanical behavior assessment of sucrose using nanoindentation. *J. Mater. Res.* **2017**, *22*, 2037–2045. [CrossRef]
7. Ramos, K.J.; Hooks, D.E.; Bahr, D.F. Direct observation of plasticity and quantitative hardness measurements in single crystal cyclotrimethylene trinitramine by nanoindentation. *Philos. Mag.* **2009**, *89*, 2381–2402. [CrossRef]
8. Millett, J.C.F.; Bourne, N.K. The shock Hugoniot of a plastic bonded explosive and inert simulants. *J. Phys. D Appl. Phys.* **2004**, *37*, 2613–2617. [CrossRef]
9. Hudson, R.J.; Zioupos, P.; Gill, P.P. Investigating the mechanical properties of RDX crystals using nano-indentation. *Propel. Explos. Pyrotech.* **2012**, *37*, 191–197. [CrossRef]
10. Weingarten, N.S.; Sausa, R.C. Nanomechanics of RDX single crystals by force–Displacement measurements and molecular dynamics simulations. *J. Phys. Chem. A* **2015**, *119*, 9338–9351. [CrossRef] [PubMed]
11. Elban, W.L.; Armstrong, R.W.; Yoo, K.C.; Rosemeier, R.G.; Yee, R.Y. X-ray reflection topographic study of growth defect and microindentation strain fields in an RDX explosive crystal. *J. Mater. Sci.* **1989**, *24*, 1273–1280. [CrossRef]
12. Vekilov, P.G.; Rosenberger, F. Dependence of lysozyme growth kinetics on step sources and impurities. *J. Cryst. Growth* **1996**, *158*, 540–551. [CrossRef]
13. Liao, X.; Wiedmann, T.S. Measurement of process-dependent material properties of pharmaceutical solids by nanoindentation. *J. Pharm. Sci.* **2005**, *94*, 79–92. [CrossRef] [PubMed]

14. Oliver, W.C.; Pharr, G.M. An improved technique for determining harness and elastic modulus using load and displacement sensing indentation experiments. *J. Mater. Res.* **1992**, *7*, 1564–1583. [CrossRef]

15. Egart, M.; Janković, B.; Lah, N.; Ilić, I.; Srčič, S. Nanomechanical properties of selected single pharmaceutical crystals as a predictor of their bulk behavior. *Pharm. Res.* **2015**, *32*, 469–481. [CrossRef] [PubMed]

16. Vlassak, J.J.; Nix, W.D. Measuring the elastic properties of materials by means of indentation. *J. Mech. Phys. Solids* **1994**, *42*, 1223–1245. [CrossRef]

17. Kiran, M.S.R.N.; Varughese, S.; Reddy, C.M.; Ramamurty, U.; Desiraju, G.R. Mechanical anisotropy in crystalline saccharin: Nanoindentation studies. *Cryst. Growth Des.* **2010**, *10*, 4650–4655. [CrossRef]

18. Armstrong, R.W.; Bardenhagen, S.G.; Elban, W.L. Deformation-induced hot spot consequences of AP and RDX crystal hardness measurements. *Int. J. Energ. Mater. Chem. Propuls.* **2012**, *11*, 413–425. [CrossRef]

19. Sanphui, P.; Mishra, M.K.; Ramamurty, U.; Desiraju, G.R. Tuning mechanical properties of pharmaceutical crystals with multicomponent crystals: Voriconazole as a case study. *Mol. Pharm.* **2015**, *12*, 889–897. [CrossRef] [PubMed]

20. Lawrence, S.K.; Bahr, D.F.; Zbib, H.M. Crystallographic orientation and indenter radius effects on the onset of plasticity during nanoindentation. *J. Mater. Res.* **2012**, *27*, 3058–3065. [CrossRef]

21. Cook, R.F.; Pharr, G.M. Direct observation and analysis of indentation cracking in glasses and ceramics. *J. Am. Ceram. Soc.* **1990**, *73*, 787–817. [CrossRef]

22. Marshall, D.B. The compelling case for indentation as a functional exploratory and characterization tool. *J. Am. Ceram. Soc.* **2015**, *98*, 2671–2680. [CrossRef]

23. Morris, D.J.; Cook, R.F. In-Situ cube-corner indentation of soda-lime glass and fused silica. *J. Am. Ceram. Soc.* **2004**, *87*, 1494–1501. [CrossRef]

24. Yeager, J.D.; Ramos, K.J.; Singh, S.; Rutherford, M.E.; Majewski, J.; Hooks, D.E. Nanoindentation of explosive polymer composites to simulate deformation and failure. *Mater. Sci. Technol.* **2012**, *28*, 1147–1155. [CrossRef]

25. Meier, M.; John, E.; Wieckhusen, D.; Wirth, W.; Peukert, W. Influence of mechanical properties on impact fracture: Prediction of the milling behaviour of pharmaceutical powders by nanoindentation. *Powder Technol.* **2009**, *188*, 301–313. [CrossRef]

26. Olusanmi, D.; Roberts, K.J.; Ghadiri, M.; Ding, Y. The breakage behaviour of Aspirin under quasi-static indentation and single particle impact loading: Effect of crystallographic anisotropy. *Int. J. Pharm.* **2011**, *411*, 49–63. [CrossRef] [PubMed]

27. Lawn, B.; Wilshaw, R. Indentation fracture: Principles and applications. *J. Mater. Sci.* **1975**, *10*, 1049–1081. [CrossRef]

28. Morris, D.J. Instrumented Indentation Contact with Sharp Probes of Varying Acuity. *MRS Online Proc. Library Arch.* **2008**, *1049*, 111–116. [CrossRef]

29. Schuh, C.A.; Mason, J.K.; Lund, A.C. Quantitative insight into dislocation nucleation from high-temperature nanoindentation experiments. *Nat. Mater.* **2005**, *4*, 617–621. [CrossRef] [PubMed]

30. Maughan, M.R.; Bahr, D.F. Discontinuous yield behaviors under various pre-strain conditions in metals with different crystal structures. *Mater. Res. Lett.* **2016**, *4*, 83–89. [CrossRef]

31. Hooks, D.E.; Ramos, K.J.; Bahr, D.F. The effect of cracks and voids on the dynamic yield of RDX single crystals. *AIP Conf. Proc.* **2007**, *955*, 789–794.

32. Maughan, M.R.; Carvajal, M.T.; Bahr, D.F. Nanomechanical testing technique for millimeter-sized and smaller molecular crystals. *Int. J. Pharm.* **2015**, *486*, 324–330. [CrossRef] [PubMed]

![crystals logo] *crystals*

MDPI

Article

Microindentation Hardness of Protein Crystals under Controlled Relative Humidity

Takeharu Kishi [1], Ryo Suzuki [1], Chika Shigemoto [1], Hidenobu Murata [1], Kenichi Kojima [2] and Masaru Tachibana [1,*]

[1] Graduate School of Nanobioscience, Yokohama City University, Yokohama 236-0027, Japan; nano_solid@yahoo.co.jp (T.K.); n175302b@yokohama-cu.ac.jp (R.S.); n175222f@yokohama-cu.ac.jp (C.S.); hmrt@yokohama-cu.ac.jp (H.M.)

[2] Department of Education, Yokohama Soei University, Yokohama 226-0015, Japan; kkojima@soei.ac.jp

* Correspondence: tachiban@yokohama-cu.ac.jp; Tel.: +81-45-787-2307

Academic Editors: Ronald W. Armstrong, Stephen M. Walley and Wayne L. Elban
Received: 9 October 2017; Accepted: 31 October 2017; Published: 4 November 2017

Abstract: Vickers microindentation hardness of protein crystals was investigated on the (110) habit plane of tetragonal hen egg-white lysozyme crystals containing intracrystalline water at controlled relative humidity. The time evolution of the hardness of the crystals exposed to air with different humidities exhibits three stages such as the incubation, transition, and saturation stages. The hardness in the incubation stage keeps a constant value of 16 MPa, which is independent of the humidity. The incubation hardness can correspond to the intrinsic one in the wet condition. The increase of the hardness in the transition and saturation stages is well fitted with the single exponential curve, and is correlated with the reduction of water content in the crystal by the evaporation. The saturated maximum hardness also strongly depends on the water content equilibrated with the humidity. The slip traces corresponding to the $(1\bar{1}0)[110]$ slip system around the indentation marks are observed in not only incubation but also saturation stages. It is suggested that the plastic deformation in protein crystals by the indentation can be attributed to dislocation multiplication and motion inducing the slip. The indentation hardness in protein crystals is discussed in light of dislocation mechanism with Peierls stress and intracrystalline water.

Keywords: protein crystal; lysozyme crystal; indentation; hardness; dislocation; intracrystalline water; relative humidity; Peierls stress; slip

1. Introduction

The knowledge of the mechanical properties of crystals is important for the elucidation of intra-crystalline bonds and practical issues such as the limits of mechanical stability [1,2]. The mechanical properties of protein crystals is greatly affected by water content, although dislocations still play a crucial role in plastic deformation. However, our understanding of the mechanical properties of protein crystals is poor compared with those for metal and covalent crystals. The reason is that most of the classical techniques developed for studying mechanical properties of metal solid appear inapplicable due to the small size and high fragility of protein crystals. On the other hand, there are interesting studies on the mechanical response to the hydration of biological materials such as bone by using micro- and nano-indentation techniques [3–6]. Such mechanical properties in hydrated biomaterials seem to be partially similar to those in protein crystals, although they are non-crystals.

Protein crystals are composed of huge protein molecules with irregular shapes. They also contain a large amount of water with 20 to 70 vol. % [7,8]. These features are responsible for complex and weak intermolecular interactions in protein crystals. This also leads to the difficulty of protein crystallization [8]. On the other hand, it is expected that these features can lead to unique mechanical properties [9–11]. The intracrystalline water in protein crystals is qualitatively classified into two types: one is free water moving freely through the crystals and the other is bound water held around each protein molecule [12–14]. Especially, the free water can be easily evaporated when the crystals are exposed to open air. Thus the water content in the crystals is sensitive to the environmental condition such as relative humidity. The change in the water content affects the mechanical properties. Therefore, the experiments with controlled water content or relative humidity are required for not only accurate measurement but also understanding of water behavior in the crystals and the corresponding unique mechanical properties.

Most of the studies on the mechanical properties of protein crystals have been carried out for hen egg-white lysozyme (HEWL) crystals with polymorphisms such as tetragonal, orthorhombic, monoclinic, and triclinic forms. The pioneer studies on the elastic properties of cross-linked HEWL crystals had been carried out by Morozov and Morozova [15–17]. The dynamic elastic constants for native and gel-grown crystals containing sufficient intracrystalline water were measured in the ranges of MHz and GHz by the ultrasonic pulse-echo method [18–20] and the Brillouin scattering method [21–23], respectively. These measurements were carried out in the growth solution and the corresponding 98% relative humidity (% RH). Almost all elastic constants for cross-linked tetragonal (T)- [24] and orthorhombic (O)-HEWL crystals [25] containing sufficient intracrystalline water at room temperature with 98% RH were determined by the ultrasonic pulse-echo method. The value of C_{11} of the normal elastic component in T-HEWL crystals is 5.50 GPa, which is almost equal to 5.24 GPa of O-HEWL crystals. Note that these values are much lower than 12.99 GPa of the bulk modulus of hydrated lysozyme molecule [20]. On the other hand, the C_{44} of the shear component in O-HEWL crystals is 0.30 GPa, which is about a half as low as 0.68 GPa of T-HEWL crystals. The change in the shear elastic constant seems to be correlated with the water contents of 39 and 43 vol. % for native T- and O-HEWL crystals, respectively. Thus, the shear elastic constant in protein crystals is more sensitive to the water content than the normal one.

Furthermore, it was measured by the ultrasonic pulse-echo method that the normal and shear elastic constants of the T-HEWL crystals dried at 42% RH are about 2 and 4 times as large as those in the wet condition with 98% RH, respectively [26]. A similar trend depending on the relative humidity has been also observed for dynamic elastic constants measured in the range of GHz by the Brillouin scattering method [22]. These results also mean that the magnitudes of the elastic constants, especially the shear component, in protein crystals strongly depend on water content associated with relative humidity. The shear elastic constant is strongly related to the characteristics of dislocations playing a crucial role in the plastic deformation. It is therefore suggested that the plastic deformation associated with dislocations is also more sensitive to water content than elastic properties.

The studies on plastic properties of protein crystals have been carried out by using Vickers microindentation method, mainly with T-HEWL ones [27–29]. In the wet condition, the indentation marks were clearly observed on the (110) crystal plane. Slip traces were also observed around the indentation. From the analysis of the slip traces, it has been shown that the plastic deformation is controlled by the dislocation mechanism with the {110}⟨110⟩ slip system. This has been also supported by the observation of slip dislocations by X-ray topography [30,31]. The average activation energy of the dislocation motion has been also evaluated to be 0.6 eV from the measurements of the temperature dependence of the indentation hardness [28]. Furthermore, it has been found that the indentation hardness increases with the evaporation of the intracrystalline water in open air where the evaporation time dependence of the hardness has three stages such as incubation, transition, and saturation stages [29]. The maximum value of the hardness in the dried condition has been about one order of magnitude larger than that in the wet condition. Recently similar behaviors have been also observed for O-HEWL crystals [32]. However, these measurements have been carried out under ambient humidity. To clarify the hardness behavior, experiments under controlled relative humidity would be desirable.

The plastic characteristics for glucose isomerase (GI), ferritin, trypsin, and insulin crystals besides HEWL ones have been also investigated by indentation method [33] and pushing method with a glass filament [34,35]. The unique plastic behavior such as creep was observed for GI crystals [33]. Additionally, extremely high quality GI crystals were clarified by X-ray topography with dislocation images and Pendellösung fringes [36]. On the other hand, a detailed mechanical response with anisotropic properties was simulated by using a continuum-based crystal plasticity model which was calibrated with Vickers microindentation hardness data [37]. This simulation with the hardness data enabled us to deduce the critical resolved shear stress on the slip plane of the T-HEWL crystals. Therefore, it is expected that the hardness measurements under controlled relative humidity can lead to more precise plastic characteristics.

In this paper we report the indentation hardness on the (110) habit plane of the T-HEWL crystals under controlled relative humidities. The time evolution of the hardness of the crystals exposed to air with different humidities exhibits three stages such as the incubation, transition, and saturation stages. The hardness in the incubation stage keeps a constant value of 16 MPa which is independent of the humidity. The incubation hardness can correspond to the intrinsic one in the wet condition. The increase of the hardness in the transition and saturation stages is well fitted with a single exponential curve, and is correlated with the reduction of water content in the crystal by the evaporation. The saturated maximum hardness also strongly depends on the water content equilibrated with the humidity. The slip traces corresponding to the $(1\bar{1}0)[110]$ slip system around the indentation marks are observed in not only the incubation but also the saturation stages. It is suggested that the plastic deformation in the protein crystals by the indentation can be ascribed to dislocation multiplication and motion inducing the slip. The indentation hardness in the protein crystals is discussed in light of the dislocation mechanism with Peierls stress and intracrystalline water.

2. Results and Discussion

2.1. Hardness at Controlled Humidity

Figure 1 shows the time evolution of Vickers microindentation hardness on (110) habit plane of T-HEWL crystals at 296 K exposed to air with 35.9% RH. The hardness strongly depends on the exposure time to air. Note that the exposure of the crystal to air can lead to the evaporation of the intracrystalline water. The behavior of hardness exhibits three stages with increasing exposure time, as reported previously [29]. First stage is the incubation stage in which the magnitude of hardness keeps a constant value even during the water evaporation, where the indented plane is still kept in wet condition. Second stage is the transition stage in which the magnitude of hardness increases with increasing exposure time where the indented plane is partially dried. Third stage is the saturation one in which the magnitude of the hardness reaches a maximum value and almost keeps the value with increasing exposure time where the indented plane is highly dried. The maximum hardness can be controlled by the water content in the crystal equilibrated with the environmental condition such as temperature and humidity.

From data points in Figure 1, it is noted that the scatter of measured values in each stage, especially transition and saturation stages, is less than that reported elsewhere [27,29]. The low scattering of measured values is attributed to controlled relative humidity in this work. Thus, more accurate analysis of the hardness becomes possible. The value of the hardness in the incubation stage is found to be 16 MPa, as seen in Figure 1. This value is slightly lower than that reported previously [27–29]. The reason can be attributed to the high accuracy for the measurements at controlled humidity. The hardness of 16 MPa is considered to be intrinsic incubation hardness in T-HEWL crystals containing sufficient intracrystalline water, although the origin for the incubation stage is discussed later.

Furthermore, it is found that the hardness curve in transition and saturation stages is well fitted with single exponential curve given by

$$H_v = H_v^{\max} + A \exp(-k_h t), \tag{1}$$

where H_v is Vickers microhardness, k_h is rate constant for the increase of the hardness, t is exposure time, and H_v^{max} is saturated or maximum hardness. Note that a first data point with H_v of more than 20 MPa in the time evolution of the H_v, as shown in Figure 1, was defined as a starting point in the transition stage. From the fitting, we can evaluate $k_h = 0.027$ min^{-1} and $H_v^{max} = 247.6$ MPa.

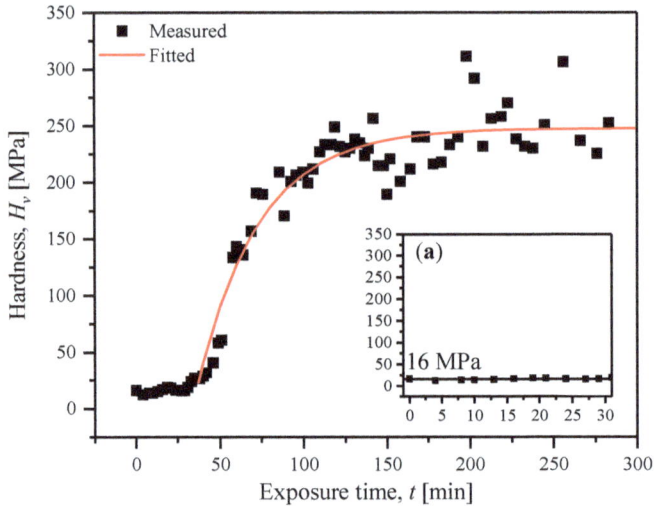

Figure 1. Time evolution of Vickers microindentation hardness on (110) habit plane of T-HEWL crystals at 296 K exposed to air with 35.9% relative humidity (RH). The hardness curve has three stages such as incubation, transition, and saturation with exposure time. The extended figure of the initial or incubation stage is shown in the inset. The fitting with single exponential curve is also drawn for the hardness curve in the transition and saturation stages.

Figure 2 shows the time evolution of Vickers microindentation hardness on (110) habit plane of T-HEWL crystals at 296 K exposed to air with different relative humidities such as 35.9, 42.1, 54.7, 73.6, and 84.0% RH, where measured crystals have different sizes of 1.6, 7.5, 1.9, 1.8, and 1.7 mm^3, respectively. All of hardness curves exhibit three stages such as incubation, transition, and saturation. At higher humidity of 84.0% RH, longer exposure time is required for the appearance of the saturation stage, as shown in Figure 2b. It should be noted that the magnitude of the hardness in the incubation stage is independent of relative humidity, and keeps a constant value, as shown in Figure 2c. This result is in good agreement with that in O-HEWL crystals reported recently [32]. Therefore, it is suggested that the hardness in the incubation stage corresponds to the intrinsic one of T-HEWL crystals with sufficient intracrystalline water in the wet condition as O-HEWL crystals.

The constant value of the hardness on (110) plane of T-HEWL crystals in the incubation stage is 16 MPa even under different humidities, as seen in Figure 2c. The value is about two times as high as the average hardness of 7.8 MPa of O-HEWL crystals reported recently [32]. Actually, the hardness value of T-HEWL crystals in the incubation stage is higher than all values of 5.7, 8.1, and 9.6 MPa on (110), (010), and (011) crystal planes of O-HEWL crystals. The high hardness can be ascribed not only to the crystal form but also to the water content with 39 vol. % in T-HEWL crystals smaller than 42 vol. % in O-HEWL crystals as mentioned above.

For transition and saturation stages, all the hardness curves are well fitted with single exponential curves, as shown in Figure 2a,b. The k_h and H_v^{max} obtained by the fitting are presented in Table 1. The rate constant, k_h, depends on the relative humidity, as shown in Table 1. The value of k_h increases with decreasing relative humidity. The value at 35.9% RH is 0.027 min^{-1}, which is about seven times as high as 0.004 min^{-1} at 84.0% RH. The high k_h for the increase of the hardness can be attributed to the high evaporation rate of the intracrystalline water under low relative humidity.

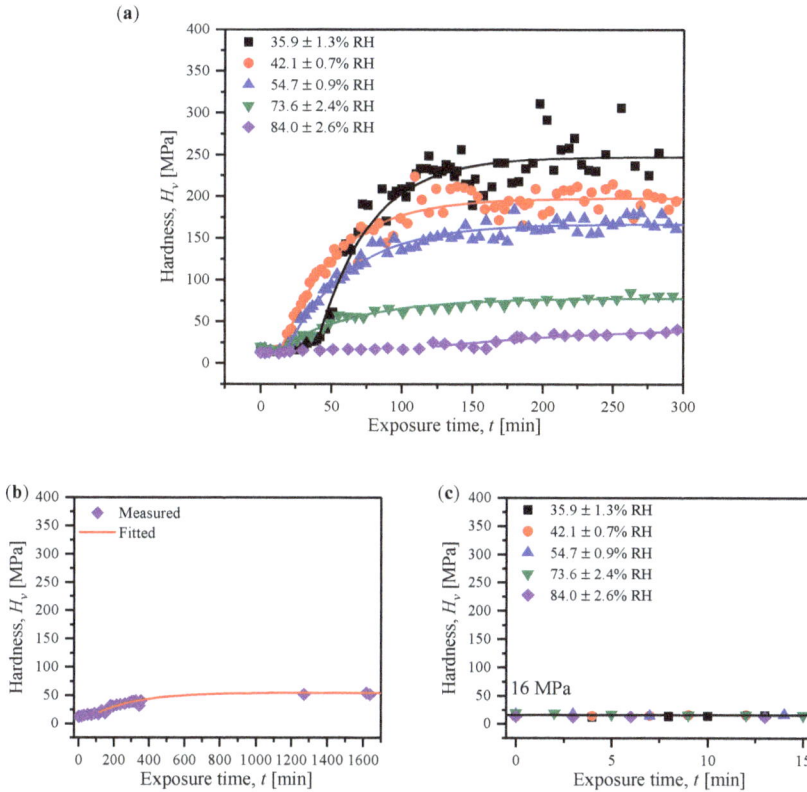

Figure 2. (**a**) Time evolution of Vickers microindentation hardness on (110) habit plane of T-HEWL crystals at 296 K exposed to air with different relative humidities such as 35.9, 42.1, 54.7, 73.6, and 84.0% RH. The hardness curve with longer exposure time at higher humidity of 84.0% RH is shown (**b**). All hardness curves in (**a,b**) have three stages such as incubation, transition, and saturation ones with exposure time. The extended figure of the incubation stages at different relative humidities in (**a**) is shown in (**c**). The fittings with single exponential curves for the hardness curves in the transition and saturation stages are also drawn in (**a,b**).

Table 1. Rate constant, k_h, for the increase of the hardness and maximum hardness, H_v^{max}, in T-HEWL crystals under different relative humidities.

Relative Humidity [% RH]	k_h [min^{-1}]	H_v^{max} [MPa]
35.9	0.027	247.6
42.1	0.026	197.8
54.7	0.022	167.2
73.6	0.018	77.8
84.0	0.004	54.7

Furthermore, it should be noted that the maximum hardness, H_v^{max}, also strongly depends on the relative humidity, as presented in Table 1. The value of H_v^{max} also increases with decreasing relative humidity. The value of H_v^{max} at 35.9% RH is 247.6 MPa, which is about 5 times as high as 54.7 MPa at 84.0% RH. The high H_v^{max} can be ascribed to the low water content by the evaporation of a large amount of intracrystalline water under low relative humidity. Thus, the H_v^{max} is controlled by the water content in the crystal equilibrated with environmental conditions such as relative humidity. Additionally, the hardness at 100% RH is extrapolated from a fitted curve for the humidity dependence of the maximum hardness, as shown in Figure 3. The extrapolated value is 7 MPa, which is even lower than 16 MPa in the incubation stage related to the wet condition, as mentioned above. The low value of the extrapolated hardness can correspond to real one of T-HEWL crystals in the solution at 100% RH.

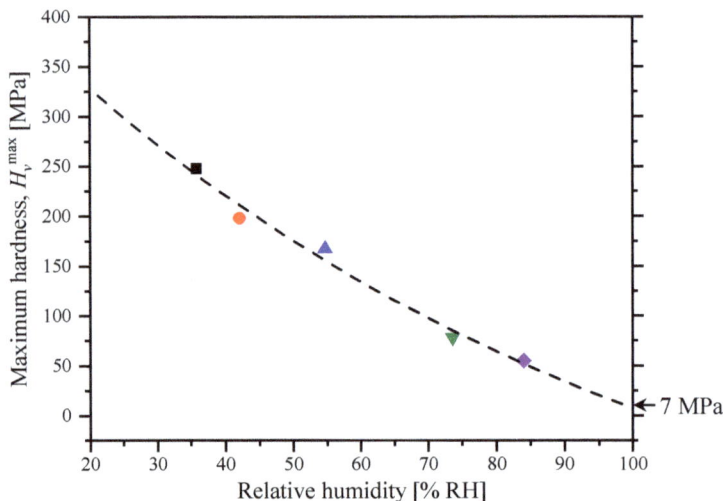

Figure 3. Humidity dependence of the maximum hardness, H_v^{max}, obtained by the fitting with single exponential curve for the hardness curves in Figure 2a.

2.2. Evaporation of Intracrystalline Water

In order to know the behavior of water evaporation, the change in crystal weight of T-HEWL crystals exposed to different relative humidities was measured by using a thermogravimetric analyzer in which the humidity is controlled by the flow ratio of wet and dry N_2 gases. Figure 4a shows time evolutions of crystal weights of T-HEWL crystals at 296 K exposed to N_2 gas with different relative humidities such as 39.1, 55.7, 74.3, and 92.9% RH. Note that the crystal weight at $t = 0$ corresponds to the sum of weights of intrinsic crystal with sufficient intracrystalline water and excess water around the crystal. As seen in Figure 4a, the crystal weight monotonically decreases with time evolution. The decay curves at 55.7, 74.3, and 92.9% RH are well fitted with single exponential curves given by

$$W = W_0 + A\exp(-k_{w1}t), \tag{2}$$

where W is the relative weight to the initial one at $t = 0$, k_{w1} is first rate constant for the reduction of weight, i.e., the evaporation of water, and t is the exposure time to N_2 gas with controlled humidity. The typical fitting with single exponential curve for the measured decay curve at 92.9% RH is shown in Figure 4b. The well-fitting by single exponential curve means that there is no significant change in the evaporation rate for the intracrystalline water and common water around the crystal. Namely, the characteristic of the evaporation of crystalline water, probably free water, is similar to that of common water. Additionally, the single exponential fitting is in good agreement with that in the hardness curves in transition and saturation stages, as shown in Figure 2. Thus, it is suggested that the change in the hardness in transition and saturation stages can be strongly correlated with the behavior of the evaporation of intracrystalline water, probably free water. The rate constant, k_{w1}, for the reduction of crystal weight, i.e., the evaporation of intracrystalline water, is estimated by the fitting for the decay curve. The values of k_{w1} for the decay curves at different humidities in Figure 4 are presented in Table 2.

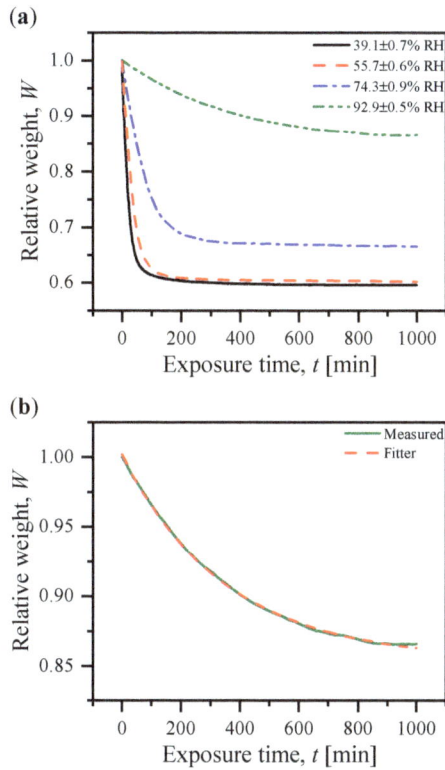

Figure 4. (**a**) Time evolution of the crystal weight of T-HEWL crystals at 296 K exposed to N_2 gas with different relative humidities such as 39.1, 55.7, 74.3, and 92.9% RH. The typical fitting with single exponential curve for the decay curve at 92.9% RH is shown in (**b**).

Table 2. Rate constants of k_{w1} and k_{w2} for the reduction of crystal weights at different relative humidities.

Relative Humidity [%]	k_{w1} [min^{-1}]	k_{w2} [min^{-1}]
39.1	0.049	0.006
55.7	0.029	-
74.3	0.014	-
92.9	0.003	-

On the other hand, the decay curve at the lowest humidity of 39.1% RH is well fitted not with single but with two exponential curves given by

$$W = W_0 + A_1 \exp(-k_{w1}t) + A_2 \exp(-k_{w2}t), \qquad (3)$$

where k_{w2} is second rate constant for the reduction of crystal weight and A_1 and A_2 are the ratios of two kinds of evaporation processes with first and second rate constants, respectively. The clear difference in the fitting accuracy with single and two exponential curves is confirmed in Figure 5a,b. The well-fitting with two exponential curves means that the decay curve contains two kinds of evaporation processes with fast and slow rate constants. The two rate constants are also presented in Table 2. From comparing the values of rate constants in Table 2, fast evaporation process corresponding to k_{w1} is observed in all humidities. As mentioned previously, there are two kinds of intracrystalline waters such as free water and bound water in protein crystals. It is therefore considered that the fast component, k_{w1}, is related to the evaporation of free water which can be easily evaporated through the crystal.

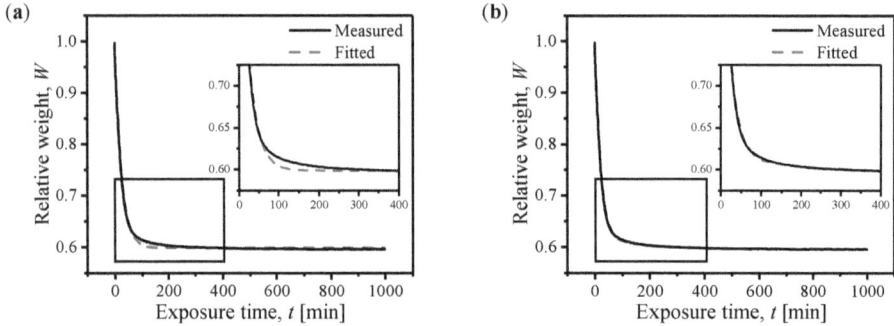

Figure 5. The decay curve of the crystal weight of T-HEWL crystals at 39.1% RH in Figure 4 (a) and the corresponding fitting curves with single (**a**) and two exponential curves (**b**).

Figure 6 shows the fast rate constant, k_{w1}, for the reduction of crystal weight, i.e., the evaporation of free water, as a function of the relative humidity. The k_{w1} increases with a decrease in the relative humidity. This means that the low humidity leads to the increase of the reduction rate of crystal weight, i.e., the evaporation rate of free water. The value of k_{w1} at lowest humidity of 39.1% RH is 0.049 min^{-1}, which is larger by more than one order compared with 0.003 min^{-1} at highest humidity of 92.9% RH. This trend depending on the humidity is similar to the behavior of k_h for the increase of the hardness with decreasing humidity, as shown in Figure 6. On the other hand, the values of k_{w1} for the reduction of crystal weight, especially for low humidities, are higher than those of k_h for the increase of the hardness, as seen in Figure 6. The discrepancy in the values might be due to the difference in humidity-control systems with dry-wet and N_2 gas flow used in the hardness and crystal weight measurements, respectively.

The second, i.e., slow, rate constant, k_{w2}, of evaporation processes is observed at only lowest humidity of 39.1% RH. The slow component, k_{w2}, can be related to the evaporation of bound water around each protein molecule. Strictly bound water around protein molecules forms hydration layers [12,14,38]. The water in the outer layer is loosely bound to the protein compared with the inner layer. The loosely bound water can be evaporated at low humidity, although it is strongly bound with the protein compared with the free water. On the other hand, as shown in Figure 2, the hardness curve is well-fitted with single exponential one even under the lowest humidity. These results imply that the evaporation of the loosely bound water gives no significant effect on the behavior of the crystal hardness in this work.

As mentioned so far, the monotonical reduction of crystal weight, i.e., the evaporation of intracrystalline water, can explain the change in the hardness in the transition and saturation stages, whereas it cannot be simply correlated with a constant value of the hardness in the incubation stage. Now, let us consider the mechanism for the incubation. The intracrystalline water, mainly free water, is monotonically evaporated. On the other hand, the hardness first keeps a constant value in the incubation stage, as mentioned above. This means that the surface region corresponding to the indentation depth in the incubation stage is kept at the wet condition, although the intracrystalline water is monotonically evaporated. According to the studies on the drying mechanism in porous materials [39–41], the evaporation of water at the surface is followed by the flow of interior water to the surface. Similar process can occur in protein crystals with free water. When the evaporation rate of the surface water is equal to the diffusion rate of interior water to the surface, the crystal surface is always kept in the wet condition. Namely, the water content at the surface is kept at nearly constant, although the intracrystalline water is monotonically evaporated. Such equilibrium of evaporation and diffusion rates can be kept in high water content so that a constant value of the hardness at the wet condition appears as the incubation stage.

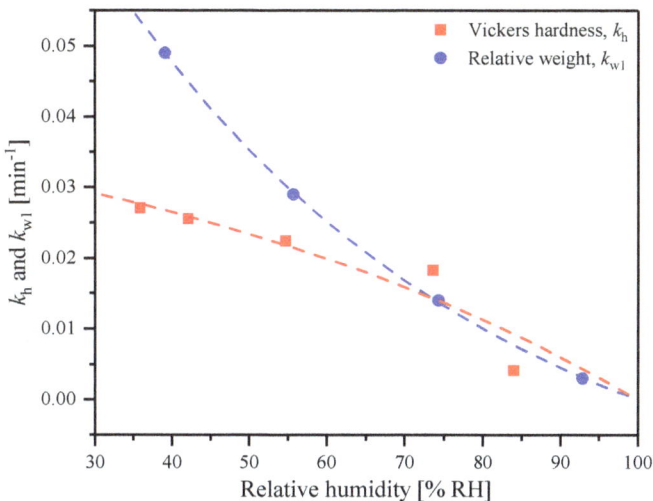

Figure 6. Comparison of rate constants of k_h and k_{w1} for the increase of Vickers hardness and the reduction of crystal weight, respectively, as a function of the relative humidity.

Further reduction of water content leads to the decrease of the water evaporation and diffusion rates. Especially, the diffusion rate of interior water to the surface is more reduced compared with the evaporation demand after a critical water content in the crystal [42]. This reduction of diffusion rate leads to the drying at the indentation surface. As a result, the transition stage with the increase of the hardness appears with the drying. Thus, the constant hardness in the incubation stage can be explained based on the drying process in porous materials [39–41]. This also means that the drying mechanism of protein crystals with free water is similar to that of porous materials. Additionally, according to the drying process in porous materials [39,40], the rapid drying before the incubation stage occurs, although it is actually difficult to measure it. Thus, real hardness of protein crystals in the solution becomes smaller than 16 MPa in the incubation stage in Figure 2. This is consistent with the small hardness of 7 MPa at 100% RH extrapolated from the fitted curve for the humidity dependence of the H_v^{max} in Figure 3.

Such drying behavior in protein crystals affects the intermolecular interaction, e.g., lattice constant and elastic constant. The change in the intermolecular interaction greatly influences the dislocation mechanism, playing a crucial role in the plastic deformation.

2.3. Dislocations and Peierls Stress

Figure 7 shows indentation marks formed on (110) planes in the three stages. As seen in Figure 7a, the slip traces indicated by arrows around the indentation mark are clearly observed in the incubation stage related to the wet condition, as reported previously [27–29]. It is suggested that plastic deformation brought about by indentation mainly results from dislocation multiplication and motion, inducing the slip in the crystal. On the other hand, no clear slip trace around the indentation mark has been observed in the saturation stage related to the dried condition so far. This might be attributed to small plastic deformation corresponding to small indentation mark due to the high hardness in the saturation stage. In this work, the indentions with higher loads were also applied in the saturation stage. As a result, the clear slip traces around the larger indentation marks were sometimes observed even in the saturation stage related to dried condition as seen in Figure 7d. These results suggest that dislocation multiplication and motion can occur for the plastic deformation in all stages. However, actually, it is still difficult to observe the slip traces around the indentation marks even by high load indentation. This might be related to the poor crystal quality in the saturation stage, i.e., dried condition.

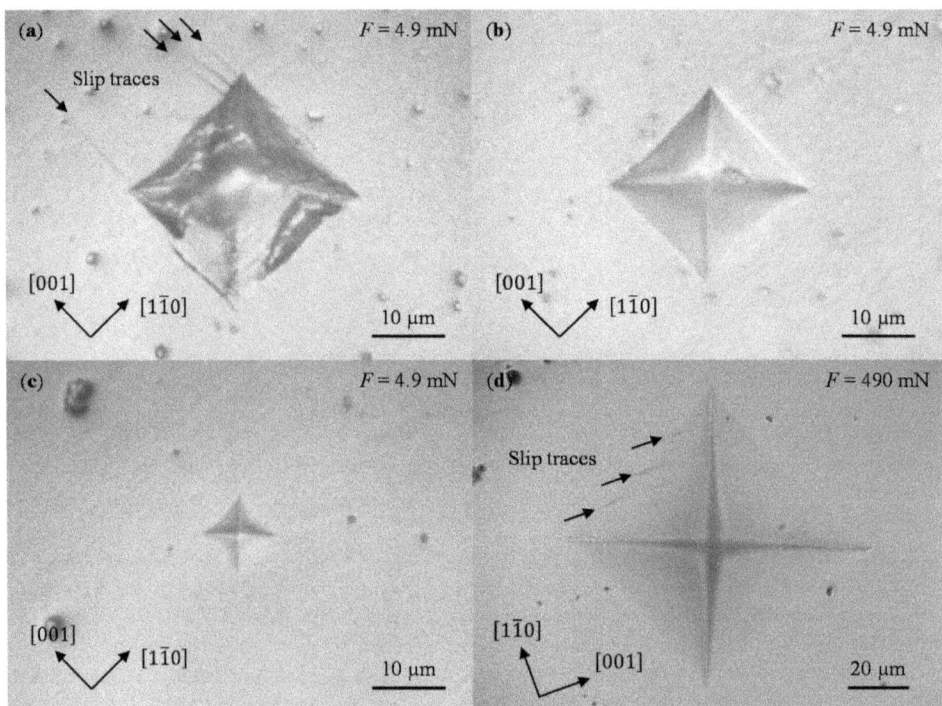

Figure 7. The morphologies around the indentation marks formed by the indentations on (110) habit planes of T-HEWL crystals in (**a**) incubation, (**b**) transition, and (**c,d**) saturation stages at 64% RH. Note that a load of 4.9 mN (0.5 g weight) was used in (**a–c**), whereas a higher load of 490 mN (50 g weight) was employed in (**d**). The slip traces are indicated by arrows in (**a,d**).

The surface morphology inside the indentation mark in the incubation stage is rough compared with the smooth surface in the transition and saturation stages, as seen in Figure 7. Additionally, the edges of the indentation mark in the incubation stage are partially disturbed in contrast to the sharp edges in another stages. Such roughness of the indentation mark can be ascribed to the pull-out effect due to the adhesion depending on the hydration by the indenter [43–45]. The correction of the adhesion on protein crystals by the indenter would be required for more accurate analysis of the indentation hardness.

The directions of all slip traces indicated by arrows in Figure 7a,d are parallel to $\langle 001 \rangle$, as reported previously [27–29]. According to the dislocation self-energy in previous papers [28], $\{110\}\langle 001 \rangle$

(b = 3.79 nm) and $\{110\}\langle 110 \rangle$ (b = 11.1 nm) are suggested as possible slip systems corresponding to $\langle 001 \rangle$ slip traces, where b is the magnitude of Burgers vector. However, the main slip system of $(1\bar{1}0)\langle 001 \rangle$ does not appear on the (110) surface, since the (110) plane contains the $\langle 001 \rangle$ axis. On the other hand, when the secondary slip system of $(1\bar{1}0)[110]$ is active, the slip traces of $\langle 001 \rangle$ directions can be observed on the (110) plane. Thus, the slip traces observed near the indentation marks on the (110) surface, as seen in Figure 7a,d, correspond to secondary slip systems of $(1\bar{1}0)[110]$.

Finally let us consider the mechanism of plastic deformation and hardness in protein crystals by indentation. The applied stress due to the indenter is concentrated in the indentation region and rapidly decreases away from it. When the indenter contacts the specimen surface, dislocations are generated beneath the indenter where the stress is very high. Then, the generated dislocations are able to move away from the indented region and thus T-HEWL crystals can deform plastically. Generally speaking, it is difficult to describe quantitatively the hardness value of crystals in terms of dislocation mechanism because the stress distribution around an indentation is very complicated. Peierls stress required to make a dislocation move in the crystal is estimated, although it cannot be directly related to the hardness. To evaluate the Peierls stress, we use a simple form of classic Peierls stress [46,47] given by

$$\sigma = \frac{2G}{1 - \nu} \exp\left(\frac{-2\pi d}{b(1 - \nu)}\right), \tag{4}$$

where G is the shear modulus, ν is Poisson's ratio, d is the distance between slip planes, and b is the magnitude of Burgers vector. In this work, the Peierls stresses are evaluated for $(1\bar{1}0)[110]$ (b = 11.1 nm) slip system experimentally observed in both wet and dried T-HEWL crystals in the incubation and saturation stages, respectively, as seen in Figure 7. The Peierls stress of wet or hydrated T-HEWL crystals at 98% RH is estimated to be 10.7 MPa with G = 0.70 GPa, ν = 0.42, d = 5.59 nm, and b = 11.1 nm where those values used in the calculation are experimental ones obtained from the measurements of sound velocities and X-ray diffractions of hydrated T-HEWL crystals at 98% RH, reported previously [24]. On the other hand, the Peierls stress of dried or dehydrated T-HEWL crystals at 42% RH is evaluated to be 57.2 MPa with G = 2.64 GPa, ν = 0.37, d = 5.23 nm, and b = 10.5 nm. Note that those values used in the calculation are also experimental ones obtained from the measurements of sound velocities and X-ray diffractions of dehydrated T-HEWL crystals at 42% RH, reported previously [26]. The value of Peierls stress at 42% RH is 57.2 MPa, which is about six times as high as 10.7 MPa at 98% RH. This trend depending on the humidity is in good agreement with the increase of one order of the hardness experimentally observed at 42% RH, as shown in Figure 2. Additionally, the values of the Peierls stress are similar order to 16 and 198 MPa in the incubation and saturation stages at 42% RH, respectively, as shown in Figure 2. Thus, it is suggested that the hardness in protein crystals can be comparably correlated with Peierls stress based on simple model as typical metal and covalent crystals.

3. Materials and Methods

Three times crystallized HEWL (Wako Pure Chemical Industries, Ltd., Osaka, Japan) was used without further purification. T-HEWL crystals ($P4_32_12$, $a = b$ = 7.91 nm, c = 3.79 nm, Z = 8) were grown by means of a salt-concentration gradient method at 296 K in test tubes held vertically and using $NiCl_2$ as a precipitant [48]. Large crystals up to a size of 5 mm were grown over two weeks. Almost all the crystals had habit plane such as $\{110\}$ and $\{101\}$. In this experiment, T-HEWL crystals with (110) habit plane of approximately 2×2 mm^2 were used for the measurements of Vickers hardness and crystal weight.

The Vickers hardness, H_v, was measured by using a microindentation testing machine (HM-221, Mitutoyo Co., Kawasaki, Japan). In order to measure the hardness at controlled relative humidity, the testing machine was covered with a simplified plastic chamber with 12.1×10^{-3} m^3. The relative humidity in the chamber is controlled by using water, silica-gel (Wako Pure Chemical Industries, Ltd., Osaka, Japan), and humidity control agents (DRY WET, Toshin Chemicals Co., Tokyo, Japan).

The lower humidity of 35.9% RH was controlled by using the silica-gel (190 g). The middle humidity of 42.1 and 54.7% RH is realized by using the DRY WET (40 g). The higher humidities of 73.6 and 84.0% RH were reached by using water (300 mL). Note that the controlled humidity in the chamber slightly depended on the outside humidity, since the simplified chamber had a little leak from the outside. The time evolution of the hardness was measured at 296 K in air with the controlled relative humidities. Just after the crystal is transferred from solution on the indentation stage in open air, the crystal plane is covered with solution droplet. In that situation, it is difficult to indent the crystal plane and/or observe the indentation marks. The clear indentation marks are confirmed after a few minutes with the evaporation of water. That time when the first indentation mark is observed is defined as $t = 0$ of exposure time to air.

The indentation was carried out on (110) habit planes of T-HEWL crystals. The indenter, with a load of 4.9 mN (0.5 g weight) and 490 mN (50 g weight), was pulled down to the crystal plane at a velocity of 0.01 mms^{-1}. The contact period of the indenter with the plane was 5 s, which is hold time at maximum load. The indentation marks were observed by using an optical microscope with a magnification of 100. In this experiment, the length of the diagonal of the indentation mark was approximately 20 μm in the incubation stage. The distance between the indentation marks and crystal edges was 50 μm at least. The separation of the indentation marks is more than the same length of indentation marks at least. It is difficult to separate the indentation marks with longer length since the area of the indentation marks is limited due to the small crystal. A standard block of hardness (HV700, Yamamoto Scientific Tool Laboratory, Funabashi, Japan) was used for calibration of the microindentation testing machine. The H_v was determined with equation $1.854(Fd^{-2})$, where F (N) and d (mm) are the load and average length of the diagonal of the indentation mark, respectively. Note that the hardness, as evaluated above, would include the error of 10% at least assuming the error of 1 μm in the measured value of the diagonal of $d = 20$ μm.

The weight measurement was carried out at 296 K by using a thermogravimetric analyzer (STA7000, Hitachi High-Technologies Co., Tokyo, Japan). In this analyzer, the relative humidity was controlled by a gas mixture of dry and wet N$_2$ gases with controlled water vapor. The flow rates of dry and wet N$_2$ gases were 200 and 100 mL/min, respectively.

4. Conclusions

We have shown the indentation hardness of T-HEWL crystals with intracrystalline water under controlled relative humidities. The hardness strongly depends on the water content in the crystals associated with the evaporation and humidity. The evaporation process is similar to that in porous materials. The slip traces related to dislocations multiplication and motion are clearly observed around the indentation marks. The hardness and plastic deformation in protein crystals by the indentation can be explained by the dislocation mechanism with Peierls stress and the change in the water content. The knowledge of such a dehydration process on the hardness of protein crystals is useful for the elucidation of not only the fundamental interest but also various applications and practical issues such as the handing of the protein crystals, e.g., substrate or drug binding, heavy-atom compound binding, and cryoprotectant soaks.

Acknowledgments: This work was supported in part by KAKENHI Grant-in-Aid for Scientific Research (C) (No. 25420694 and 16K06708).

Author Contributions: Takeharu Kishi, Hidenobu Murata, and Masaru Tachibana conceived and designed the experiments; Takeharu Kishi, Ryo Suzuki, and Chika Shigemoto performed the experiments; all authors analyzed the data; Takeharu Kishi, Ryo Suzuki, Chika Shigemoto, Kenichi Kojima, and Masaru Tachibana wrote the paper.

Conflicts of Interest: The authors declare no conflict of interest.

References

1. Meyers, M.A.; Chawla, K.K. *Mechanical Behavior of Materials*, 2nd ed.; Cambridge University Press: Cambridge, UK, 2008; ISBN 9780521866750.
2. Gilman, J.J. *Chemistry and Physics of Mechanical Hardness*, 1st ed.; John Wiley and Sons: Hoboken, NJ, USA, 2009; ISBN 9780470226520.
3. Bembey, A.K.; Oyen, M.L.; Bushby, A.J.; Boyde, A. Viscoelastic properties of bone as a function of hydration state determined by nanoindentation. *Philos. Mag.* **2006**, *86*, 5691–5703. [CrossRef]
4. Bembey, A.K.; Bushby, A.J.; Boyde, A.; Ferguson, V.L.; Oyen, M.L. Hydration effects on the micro-mechanical properties of bone. *J. Mater. Res.* **2006**, *21*, 1962–1968. [CrossRef]
5. Oyen, M.L. Poroelastic nanoindentation responses of hydrated bone. *J. Mater. Res.* **2008**, *23*, 1307–1314. [CrossRef]
6. Oyen, M.L. Nanoindentation of hydrated materials and tissues. *Curr. Opin. Solid. State Mater. Sci.* **2015**, *19*, 317–323. [CrossRef]
7. McPherson, A. *Crystallization of Biological Macromolecules*; Cold Spring Harbor Laboratory Press: New York, NY, USA, 1999; ISBN 9780879695279.
8. Matthews, B.W. Solvent content of protein crystals. *J. Mol. Biol.* **1968**, *33*, 491–497. [CrossRef]
9. Vilenchik, L.Z.; Griffith, J.P.; Clair, N.S.; Navia, M.A.; Margolin, A.L. Protein crystals as novel microporous materials. *J. Am. Chem. Soc.* **1998**, *120*, 4290–4294. [CrossRef]
10. Margolin, A.L.; Navia, M.A. Protein crystals as novel catalytic materials. *Angew. Chem. Int. Ed.* **2001**, *40*, 2204–2222. [CrossRef]
11. Abe, S.; Ueno, T. Design of protein crystals in the development of solid biomaterials. *RSC Adv.* **2015**, *5*, 21366–21375. [CrossRef]
12. Otting, G.; Liepinish, E.; Wuthrich, K. Protein hydration in aqueous solution. *Science* **1991**, *254*, 974–980. [CrossRef] [PubMed]
13. Morozov, V.N.; Kachalova, G.S.; Evtodienko, V.U.; Lanina, N.F.; Morozova, T.Y. Permeability of lysozyme tetragonal crystals to water. *Eur. Biophys. J.* **1995**, *24*, 93–98. [CrossRef]
14. Jones, M.J.; Ulrich, J. Are different protein crystal modifications polymorphs? A discussion. *Chem. Eng. Technol.* **2010**, *33*, 1571–1576. [CrossRef]
15. Morozov, V.N.; Morozova, T.Y. Viscoelastic properties of protein crystals: Triclinic crystals of hen egg white lysozyme in different conditions. *Biopolymers* **1981**, *20*, 451–467. [CrossRef] [PubMed]
16. Morozov, V.N.; Morozova, T.Y.; Kachalova, G.S.; Myachin, E.T. Interpretation of water desorption isotherms of lysozyme. *Int. J. Biol. Macromol.* **1988**, *10*, 329–336. [CrossRef]
17. Zenchenko, T.A.; Pozharskii, E.V.; Morozov, V.N. A magnetic micromethod to measure Young's modulus of protein crystals and other polymer materials. *J. Biochem. Biophys. Methods* **1996**, *33*, 207–215. [CrossRef]
18. Tachibana, M.; Kojima, K.; Ikuyama, R.; Kobayashi, Y.; Ataka, M. Sound velocity and dynamic elastic constants of lysozyme single crystals. *Chem. Phys. Lett.* **2000**, *332*, 259–264. [CrossRef]
19. Tachibana, M.; Kojima, K.; Ikuyama, R.; Kobayashi, Y.; Ataka, M. Erratum to: Sound velocity and dynamic elastic constants of lysozyme single crystals. *Chem. Phys. Lett.* **2002**, *354*, 360. [CrossRef]
20. Tachibana, M.; Koizumi, H.; Kojima, K. Effect of intracrystalline water on longitudinal sound velocity in tetragonal hen-egg-white lysozyme crystals. *Phys. Rev. E* **2004**, *69*, 051921. [CrossRef] [PubMed]
21. Caylor, C.L.; Speziale, S.; Kriminski, S.; Duffy, T. Measuring the elastic properties of protein crystals by Brillouin scattering. *J. Cryst. Growth* **2001**, *232*, 498–501. [CrossRef]
22. Speziale, S.; Jiang, F.; Caylor, C.L.; Kriminski, S.; Zha, C.S.; Thorne, R.E.; Duffy, T.S. Sound velocity and elasticity of tetragonal lysozyme crystals by Brillouin spectroscopy. *Biophys. J.* **2003**, *85*, 3202–3213. [CrossRef]
23. Hashimoto, E.; Aoki, Y.; Seshimo, Y.; Sasanuma, K.; Ike, Y.; Kojima, S. Dehydration process of protein crystals by micro-brillouin scattering. *Jpn. J. Appl. Phys.* **2008**, *47*, 3839–3842. [CrossRef]
24. Koizumi, H.; Tachibana, M.; Kojima, K. Elastic constants in tetragonal hen egg-white lysozyme crystals containing large amount of water. *Phys. Rev. E* **2009**, *79*, 061917. [CrossRef] [PubMed]
25. Kitajima, N.; Tsukashima, S.; Fujii, D.; Tachibana, M.; Koizumi, H.; Wako, K.; Kojima, K. Elastic constants in orthorhombic hen egg-white lysozyme crystals. *Phys. Rev. E* **2014**, *89*, 012714. [CrossRef] [PubMed]

26. Koizumi, H.; Tachibana, M.; Kojima, K. Observation of all the components of elastic constants using tetragonal hen egg-white lysozyme crystals dehydrated at 42% relative humidity. *Phys. Rev. E* **2006**, *73*. [CrossRef] [PubMed]

27. Tachibana, M.; Kobayashi, Y.; Shimazu, T.; Ataka, M.; Kojima, K. Growth and mechanical properties of lysozyme crystals. *J. Cryst. Growth* **1999**, *198*, 661–664. [CrossRef]

28. Koizumi, H.; Tachibana, M.; Kawamoto, H.; Kojima, K. Temperature dependence of microhardness of tetragonal hen-egg-white lysozyme single crystals. *Philos. Mag.* **2004**, *84*, 2961–2968. [CrossRef]

29. Koizumi, H.; Kawamoto, H.; Tachibana, M.; Kojima, K. Effect of intracrystalline water on micro-Vickers hardness in tetragonal hen egg-white lysozyme single crystals. *J. Phys. D* **2008**, *41*, 074019. [CrossRef]

30. Tachibana, M.; Koizumi, H.; Izumi, K.; Kajiwara, K.; Kojima, K. Identification of dislocations in large tetragonal hen egg-white lysozyme crystals by synchrotron white-beam topography. *J. Synchrotron Radiat.* **2003**, *10*, 416–420. [CrossRef] [PubMed]

31. Mukobayashi, Y.; Kitajima, N.; Yamamoto, Y.; Kajiwara, K.; Sugiyama, H.; Hirano, K.; Kojima, K.; Tachibana, M. Observation of dislocations in hen egg-white lysozyme crystals by synchrotron monochromatic-beam X-ray topography. *Phys. Stat. Sol.* **2009**, *206*, 1825–1828. [CrossRef]

32. Suzuki, R.; Kishi, T.; Tsukashima, S.; Tachibana, M.; Wako, K.; Kojima, K. Hardness and slip systems of orthorhombic hen egg-white lysozyme crystals. *Philos. Mag.* **2016**, *96*, 2930–2942. [CrossRef]

33. Tait, S.; White, E.T.; Litster, J.D. Mechanical characterization of protein crystals. *Part. Part. Syst. Charact.* **2008**, *25*, 266–276. [CrossRef]

34. Nanev, C.N.; Dimitrov, I.; Tsekova, D. Adhesion of protein crystals: Measurement of the detachment force. *Cryst. Res. Technol.* **2006**, *41*, 505–509. [CrossRef]

35. Nanev, C.N. Brittleness of protein crystals. *Cryst. Res. Technol.* **2012**, *47*, 922–927. [CrossRef]

36. Suzuki, R.; Koizumi, H.; Kojima, K.; Fukuyama, S.; Arai, Y.; Tsukamoto, K.; Suzuki, Y.; Tachibana, M. Characterization of grown-in dislocations in high-quality glucose isomerase crystals by synchrotron monochromatic-beam X-ray topography. *J. Cryst. Growth* **2017**, *468*, 299–304. [CrossRef]

37. Zamiri, A.; De, S. Modeling the mechanical response of tetragonal lysozyme crystals. *Langmuir* **2010**, *26*, 4251–4257. [CrossRef] [PubMed]

38. Pal, S.K.; Zewail, A.H. Dynamics of water in biological recognition. *Chem. Rev.* **2004**, *104*, 2099–2123. [CrossRef] [PubMed]

39. Bray, Y.L.; Prat, M. Three-dimensional pore network simulation of drying in capillary porous media. *Int. J. Heat Mass Tran.* **1999**, *42*, 4207–4224. [CrossRef]

40. Yiotis, A.G.; Tsimpanogiannis, I.N.; Stubos, A.K.; Yortsos, Y.C. Pore-network study of the characteristic periods in the drying of porous materials. *J. Colloid. Interface Sci.* **2006**, *297*, 738–748. [CrossRef] [PubMed]

41. Lehmann, P.; Assouline, S.; Or, D. Characteristic lengths affecting evaporative drying of porous media. *Phys. Rev. E* **2008**, *77*. [CrossRef] [PubMed]

42. Yiotis, A.G.; Salin, D.; Tajer, E.S.; Yortsos, Y.C. Drying in porous media with gravity-stabilized fronts: Experimental results. *Phys. Rev. E* **2012**, *86*. [CrossRef] [PubMed]

43. Ebenstein, D.M.; Pruitt, L.A. Nanoindentation of biological materials. *Nano Today* **2006**, *1*, 26–33. [CrossRef]

44. Ferguson, V.L. Deformation partitioning provides insight into elastic, plastic, and viscous contributions to bone materials behavior. *J. Mech. Behav. Biomed.* **2009**, *2*, 364–374. [CrossRef] [PubMed]

45. Oyen, M.L. Nanoindentation of biological and biomimetic materials. *Exp. Tech.* **2013**, *37*, 73–87. [CrossRef]

46. Hirth, J.P.; Lothe, J. *Theory of Dislocations*, 2nd ed.; Wiley: Hoboken, NY, USA, 1982; ISBN 9780894646171.

47. Hull, D.; Bacon, D.J. *Introduction to Dislocations*, 5th ed.; Butterworth-Heinemann: Oxford, UK, 2011; ISBN 9780080966724.

48. Tachibana, M.; Kojima, K. Growth, Defects and mechanical properties of protein single crystals. *Curr. Top. Cryst. Growth Res.* **2002**, *6*, 35–49.

crystals

MDPI

Article

Mechanical Anisotropy in Austenitic NiMnGa Alloy: Nanoindentation Studies

Ashwin Jayaraman [1,2], M. S. R. N. Kiran [1,3],* and Upadrasta Ramamurty [1]

[1] Department of Materials Engineering, Indian Institute of Science, Bangalore 560012, India;
    ashwin.jayaraman@gmail.com (A.J.); ramu@materials.iisc.ernet.in (U.R.)
[2] Harvard John A. Paulson School of Engineering and Applied Sciences, Cambridge, MA 02138, USA
[3] SRM Research Institute and Department of Physics and Nanotechnology, SRM University, Kattankulathur, Chennai 603203, India
* Correspondence: kiranmangalampalli.k@ktr.srmuniv.ac.in

Academic Editor: Ronald W. Armstrong
Received: 23 June 2017; Accepted: 15 August 2017; Published: 17 August 2017

Abstract: Mechanical anisotropy in an austenitic ferromagnetic shape memory alloy (SMA), $Ni_{50}Mn_{26.25}Ga_{23.75}$, is investigated along (010), ($\bar{1}$20), ($\bar{1}2\bar{1}$), ($23\bar{1}$) and (232) using nanoindentation. While (010) exhibits the highest reduced modulus, E_r, and hardness, H, (232) shows the lowest amongst the grain orientations examined in this study. The significant elastic anisotropy measured is attributed to differences in planar packing density and number of in-plane Ni–Mn and Ni–Ga bonds, whereas the plastic anisotropy is due to the differences in the onset of slip, which is rationalized by recourse to Schmid factor calculations. This would help determine the grain orientations in austenitic NiMnGa which exhibit better mechanical properties for SMA applications such as improving vibration damping characteristics of the alloy.

Keywords: NiMnGa; mechanical anisotropy; nanoindentation; hardness; modulus

1. Introduction

Ni-Mn-Ga ferromagnetic shape memory alloy (FSMA) is one of the most promising materials for possible applications in magnetic actuation and sensor applications as well as structural and damping applications due to high recoverable strains (above 10%) [1,2].

However, a major impediment to using them, especially in structural applications, is their extreme brittleness, which has been attributed to fracture along the low strength grain boundaries that is, in turn, is a result of the high directionality of the bonds in the ordered structure which breaks down at the grain boundaries. Also, their coarse-grained microstructure combined with large mechanical anisotropy makes the alloys susceptible to intergranular cracking [3,4]. Some quantitative data regarding stiffness constants of different orientations (single crystals) and studies on composition dependence on the mechanical perperties of NiMnGa alloys using nanoindentation technque are present in the literature [5–10], but to our knowledge, quantitative investigation of mechanical anisotropy across individual orientations in an as-cast polycrystal of room temperature austenitic $Ni_{50}Mn_{26.25}Ga_{23.75}$ is not reported yet. Nanoindentation technique, because of its ability to probe mechanical properties of relatively small volume materials, allows for measuring mechanical properties along various crystallographic directions and hence estimates the anisotropy [11–17]. Some of the current authors have extensively utilized nanoindentation technique to characterize organic, pharmaceutical and metal-organic framework systems to correlate molecular-level properties such as interaction characteristics, crystal packing, and the inherent anisotropy with micro/macroscopic events [18–29]. This has been attempted in this work wherein the elastic and plastic properties of grains in an austenitic $Ni_{50}Mn_{26.25}Ga_{23.75}$, which are oriented in different crystallographic directions, are evaluated by employing the nanoindentation technique. This would help determine the grain orientations in austenitic NiMnGa which exhibit better mechanical properties for SMA applications such as improving vibration damping characteristics of the alloy, thus in better engineering of SMAs for various structural applications.

2. Experimental

Polycrystalline $Ni_{50}Mn_{26.25}Ga_{23.75}$ ingots were manufactured by vacuum arc melting technique using 99.8% pure powders. Note that this composition is off-stoichiometric as Ni_2MnGa is the stoichiometric alloy. The average grain size in this alloy was ~500 µm, as observed by using optical microscopy. Differential scanning calorimetry (Figure S1 in Supplementary Information) shows that the austenitic start (A_s) and finish (A_f) temperatures for this alloy are 266.1 and 276.7 K, respectively, whereas the martensitic start (M_s) and finish (M_f) temperatures are 265.2 and 253.9 K, respectively. Thus, the alloy is in the fully austenitic state at room temperature (~298 K) at which the nanoindentation experiments were performed. To identify the crystallographic orientations of different grains in the microstructure, electron backscattered diffraction (EBSD) was performed using a field-emission scanning electron microscope (FEI Nova NanoLab 200, FEI Company, Hillsboro, OR, USA) on an electro-polished sample. (Following are the optimum electropolishing conditions: voltage = 9 V, temperature = 243 K, time = 20 s, electrolyte = 20% perchloric acid +80% methanol, cathode = pure Ti). The step size given for EBSD scans was 10 µm. A confidence index of ~0.25 indicates the high reliability of the data. Point analysis in EBSD (was used to characterize the orientation of each grain in the sample. The elemental compositional analysis in different grains has been performed using electron probe micro-analysis (EPMA).

Nanoindentation experiments were performed on the five different grains with crystallographic orientations of (010), ($\bar{1}$20), ($\bar{1}$2$\bar{1}$), (23$\bar{1}$) and (232) (See Table 1 for complete crystal orientation details) using the Triboindenter (Hysitron Corp., Minneapolis, MN, USA) which has a coupled in-situ imaging capability. These orientations were chosen since they were the primary ones with large grain size located in the central region of the sample. The nanoindenter was fitted with a Berkovich diamond tip with a tip radius of ~75 nm. The loading and unloading rates during the nanoindentation experiments were 0.9 mN/s and the load was paused for 10 s at the peak load of 9 mN. A minimum of 10 indentations were performed on each grain. They were always located in the central region of the grains (and sufficiently far away from the grain boundaries) so that the grain boundaries do not influence the measured mechanical properties. The coarse grain size of the alloy is particularly beneficial in this aspect. The EPMA results reveal negligible compositional variation across the grains, confirming that the elastic and plastic anisotropies observed in the present study are not due to compositional variation. The images of the residual indent impressions were captured immediately on unloading with the same indenter tip, now functioning as the stylus. The load, *P*, vs. depth of penetration, *h*, curves were analyzed using the Oliver-Pharr (O-P) method [17] to determine E_r and H of the particular grain orientation. The E_r was determined using the equation.

$$E_r = \frac{\sqrt{\pi}}{2} \frac{\beta S}{\sqrt{A}} \tag{1}$$

where S is the stiffness of the test material, which was obtained from the initial unloading slope by evaluating the maximum load and the maximum depth, i.e., $S = dP/dh$. β is a shape constant that depends on the geometry of the indenter and is 1.034 for the Berkovich tip.

Table 1. The crystallographic orientations, corresponding color schemes, and Euler angles obtained from EBSD on austenitic $Ni_{50}Mn_{26.25}Ga_{23.75}$ prior to nanoindentation.

Color Scheme	Φ_1 (Degrees)	Φ (Degrees)	Φ_2 (Degrees)	Plane
Purple	138.8	65.1	203.9	$3.45°$ from $(\bar{1}21)$
Cream	277.2	82.1	333.2	$6.4°$ from $(\bar{1}20)$
Orange	254.8	99.5	179.4	$8.1°$ from (010)
Blue	47.3	58.3	32.8	$2.9°$ from (232)
Pink	63	104	33.4	$5.3°$ from $(23\bar{1})$

The indentation modulus, E_M of the individual grains was obtained using the following equation [15,16]:

$$\frac{1}{E_r} = \left(\frac{1-v^2}{E}\right)_{Indenter} + \left(\frac{1}{E_M}\right)_{Sample} \tag{2}$$

where v and E are Poisson's ratio and elastic modulus, respectively. The indenter properties used in this study are $E_i = 1140$ GPa, and Poisson's ratio for the indenter is $v_i = 0.07$.

3. Results and Discussion

Powder X-ray diffractometry (Figure S2 in Supplementary Information) shows crystalline peaks corresponding to *bcc* structure indicating a Heusler cubic superlattice with the L_{21} order with a lattice parameter of 0.58 nm. Figure 1a shows a combined EBSD scan image (obtained from two different regions of the sample where nanoindentations were performed) whereas Figure 1b shows the corresponding inverse pole figure map with the color scheme used. The boundaries between two color coded regions indicate the approximate location of the grain boundary. The orientation imaging microscopy (OIM) software was used to analyze the data and obtain the Euler angles of the different orientations and pertinent *(hkl)* planes, which are indicated on Figure 1b and listed in Table 1. (Note that the planes mentioned are low index equivalents of the high index planes obtained from EBSD scan analysis).

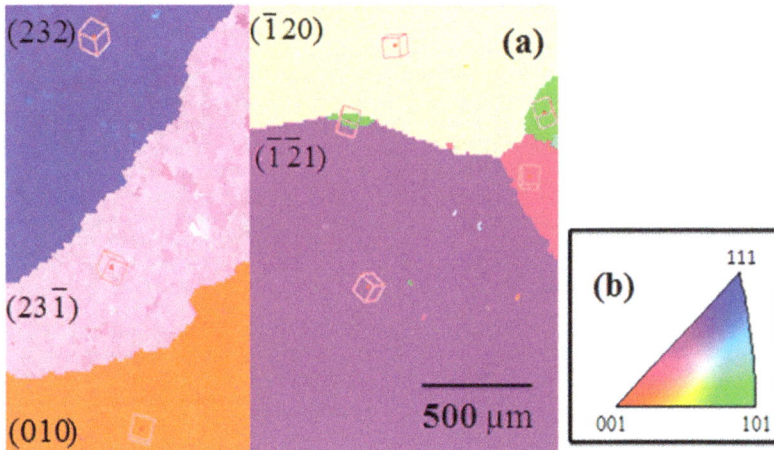

Figure 1. (a) Orientation Imaging Microscopy (OIM) scans of individual grains in room temperature austenitic $Ni_{50}Mn_{26.25}Ga_{23.75}$ obtained using electron backscattered diffraction (EBSD), (b) corresponding color coded inverse pole figure map.

Representative *P-h* curves obtained on different grains are shown in Figure 2a and the average E_r and *H* values extracted from these are listed in Table 2. The *P-h* curves are smooth with no evidence of pop-ins in the loading part of the curves, suggesting that dislocation activity not happened in sudden bursts. No kinks (or pop-outs) were observed in the unloading curves either, implying that the material underneath the indenter might not have undergone sudden phase transformation. The Differential Scanning Calorimetry (DSC) shows that the Curie temperature (T_c) of paramagnetic to ferromagnetic transformation of $Ni_{50}Mn_{26.25}Ga_{23.75}$ is 376 K. The studied alloy undergoes transformation to 5 M/5 fold modulated martensite at M_s since $T_c > M_s$. The maximum percentage strains (ε) due to stress induced martensite under uniaxial compressive loading of different orientations have been calculated using the shape strain matrix of the austenite to 5 M martensite transformation [30]. Corresponding room temperature uniaxial compressive stresses required for austenite to martensite transformation, σ_T, have been computed using the modified Clausius-Clapeyron equation [31] as follows.

$$\sigma_T = \left(\frac{dH \times \rho}{M_s \times \varepsilon}\right) \times (R_T - M_s) \qquad (3)$$

where *dH* is the enthalpy change on transformation obtained from DSC curve (4.3 J/g), ρ is the density of alloy (8.13 g/cc), R_T is room temperature (~298 K). The ε values along <100>, <120>, <121> and <232> of the austenite form are found to be 3.868, 3.175, 3.782, and 3.718 % with corresponding σ_T values being 109.2, 133, 111.7 and 113.6 MPa, respectively. These stresses are smaller, by more than an order of magnitude than the corresponding *H*. Also, *H* and σ_T are not correlated. These observations suggest that the measured anisotropy in *H* is not a reflection of the anisotropy in σ_T.

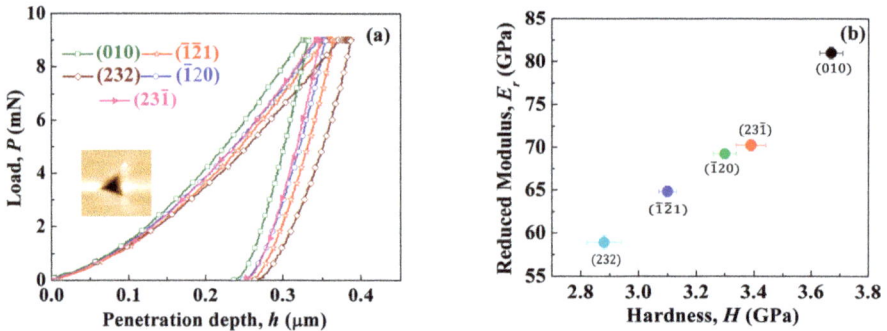

Figure 2. (**a**) Representative *P-h* curves of nanoindentation along (010), $(\bar{1}20)$, $(\bar{1}\bar{2}1)$, $(23\bar{1})$ and (232). (Inset shows AFM image of residual indent impression after immediate unloading), (**b**) variation of E_r, with *H*.

The O-P method, used for extracting *E* and *H* from the *P-h* curves, can give inaccurate values if there is significant pile-up or sink-in due to plastic flow underneath the indenter. However, the AFM images of the indentation imprints (a representative one is shown in the inset of Figure 2) do not show any such features. Further, they do not give any evidence of formation of slip lines or martensitic twin variants along the edges or corners of the indenter, which for example was reported in the case of nanoindentation of individual grains of $Cu_{83.1}Al_{13}Ni_{3.9}$ SMA under similar loading-unloading conditions [32].

Data presented in Table 2 shows that (010) is the stiffest and also the hardest whereas (232) is the most compliant and softest amongst the crystallographic planes studied in the work. The extent of elastic and plastic anisotropies is significant, with 37.5% and 27.6% differences in *E* and *H* values of (232) and (010). Kumar et al. [14] have mapped the anisotropic indentation modulus in different cubic materials and showed that the maximum variation in modulus across orientations in highly anisotropic materials like Pb, Th and Ni_3Al is ~14%. Vlassak and Nix [15,16] have measured the elastic anisotropy factor, defined as the ratio of the highest E_M to the lowest measured amongst various crystallographic directions, using nanoindentation on different metals. The data reported by them is listed in Table 3 along with that obtained in the present study on NiMnGa. While metals like W and Al are nearly-isotropic, brass shows maximum anisotropy with a factor of 1.25. The anisotropy factor for the FSMA examined in this work is 1.31, much larger than that of brass, clearly highlighting the fact that NiMnGa is highly anisotropic.

Table 2. Slip System, corresponding Schmid factors and calculated mechanical properties on nanoindentation along different crystallographic directions.

Direction	Slip System	Maximum Schmid Factor	h_{max} (nm)	Hardness, H (GPa)	Reduced Modulus, E_r (GPa)
[010]		0.09	326	3.7 ± 0.04	81 ± 0.7
[23$\bar{1}$]		0.34	344	3.39 ± 0.05	69.2 ± 0.7
[$\bar{1}$20]	{$\bar{1}$10} <001>	0.36	346	3.3 ± 0.04	70.2 ± 0.7
[$\bar{1}\bar{2}$1]		0.45	360	3.2 ± 0.03	64.8 ± 0.6
[232]		0.49	377	2.9 ± 0.06	58.9 ± 0.8

Table 3. Comparison between elastic anisotropy factors on nanoindentation of different materials [15] and the austenitic NiMnGa studied. * denotes that the values presented in the table are "Indentation modulus, E_M" calculated from the E_r, Poisson's ratio and indenter modulus information and using Vlassak and Nix model [15] in order to compare the NiMnGa anisotropic factor with the materials listed in Ref. [15].

Material	$E_{Mhighest}$ (GPa)	$E_{Mlowest}$ (GPa)	Elastic Anisotropy Factor ($E_{Mhighest}/E_{Mlowest}$)	Space Group
W	439	438	1.002	$Im\bar{3}m$
Al	79	77	1.025	$Fm\bar{3}m$
Cu	137	124	1.104	$Fm\bar{3}m$
Brass	130	104	1.250	$I4\bar{3}m$
Ni-Mn-Ga	86.68 *	65.83 *	1.31	$Fm\bar{3}m$

As shown in Figure 2b, cross-plotting of the E_r and H values obtained for various crystallographic planes shows that the planes that are stiffest are also the hardest. Sometimes, such a correlation could be an experimental artifact, a result of significant pile-up or sink-in during indentation. The possible reasons for the observed anisotropy in mechanical properties are discussed below.

The elastic modulus of material primarily depends on two factors: the bonding characteristics and the structure of the material. In the NiMnGa alloys, both the Ni–Mn and Ni–Ga bonds are metallic in nature and have the same bond length (2.527 Å). Therefore, it is reasonable to expect that differences in bonding characteristics cause no significant anisotropy. Then, the possible reason could be significant differences in planar packing densities along different orientations. Projections of the (010) and (232) planes for stoichiometric NiMnGa are shown in Figure 3a,b respectively. It is seen that the former is more densely packed with a planar packing density of 0.1195 atoms/Å^2. This translates into a much greater resistance to bond stretching by elastic deformation in the [010] direction of indentation compared to [232].

Next, we focus on the hardness anisotropy. To gain insight into the plastic deformation processes under nanoindentation, *a priori* knowledge of the possible slip systems and the Schmid factors (SFs) for them are essential. In general, the possible slip systems in L2$_1$ structures are {$1\bar{1}$0} <111>, {11$\bar{2}$} <111>, {$1\bar{1}$0}<110> and {$1\bar{1}$0}<001> [33,34]. Slip is expected to happen earlier on orientations with higher estimated SF. The reader should note that Schmid's law is for a uniaxial stress; however, the stress under indenter is heterogeneous. Nevertheless, estimation of SFs provides insights into the slip mechanism under the indenter. Recently, some authors have tried to define a Schmid factor for indentation in a better way [35]. However, in the present case it is seen that $\{1\bar{1}0\}<001>$ could be the favored slip system since SF calculations for slip along all possible slip plane- direction combinations of $\{1\bar{1}0\}<001>$ reinstate the hardness trends seen. The maximum values of the SFs estimated for different crystallographic orientations along which indentations are performed, with the slip system being $\{1\bar{1}0\}<001>$ are listed in Table 2. It is seen that higher the estimated SF, lower is the measured H.

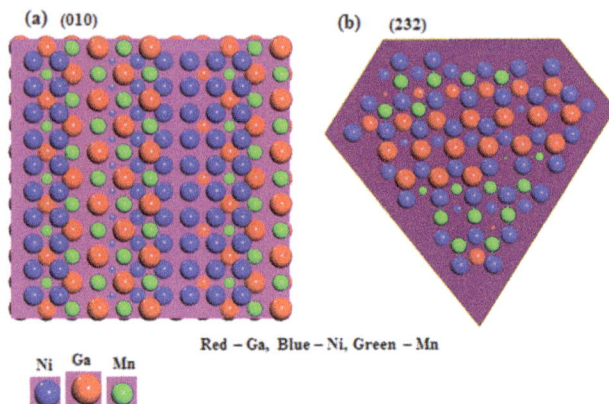

Figure 3. Planar Projections of (**a**) (010) and (**b**) (232) in NiMnGa, normal to which nanoindentations are performed along [010] and [232] directions, respectively.

The aforementioned can be argued qualitatively as well. The <001>, <110> and <111> are possible slip directions in NiMnGa. Now, slip along <110> is difficult since it is not densely packed as compared to the other two directions. Thus, slip along it can only be activated at elevated temperatures. The <001> direction in NiMnGa has neighboring Mn and Ga atoms which are not bonded to each other but are coordinated to Ni atoms in the adjacent (001) plane (See-Figure 4). Unlike <001>, the <111> is seen to have a bonded repetitive chain of -Ga-Ni-Mn-Ni-Ga- atoms. All these atoms are bonded to pertinent atoms in the adjacent plane. Our hypothesis is that the chain of bonded atoms in <111>, along with multiple bonds with the adjacent plane, makes it comparatively more difficult for slip compared to <001> (See Figure 5). It is evident from the theoretical and experimental literature that for fcc [36] and bcc [37] crystals, (100) is the most complaint and (111) is the strongest. To check our hypothesis, we have evaluated <110> and <001>-orientations on another austenitic NiMnGa sample whose texture shows <110> grain orientation on the surface using nanoindentation in which <110> has a higher reduced modulus (83 GPa) than <001> (81 GPa) confirming <001> is likely to be the favorable slip direction. It is also concluded that slip along $\{1\bar{1}0\}<001>$ could have a significantly low lattice friction stress allowing easy shear of atomic planes in slip direction. However, this hypothesis has to be experimentally verified through transmission electron microscopy of plastically deformed samples of this FSMA.

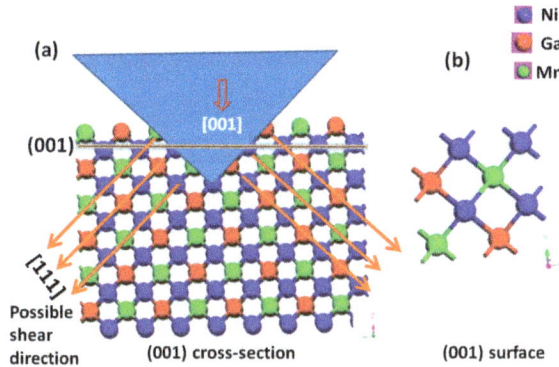

Figure 4. (**a**) Cross-section of (001) atomic arrangements underneath the plane being indented for indentations along [001] and, (**b**) planar projections of (001) in NiMnGa, normal to which nanoindentations are performed.

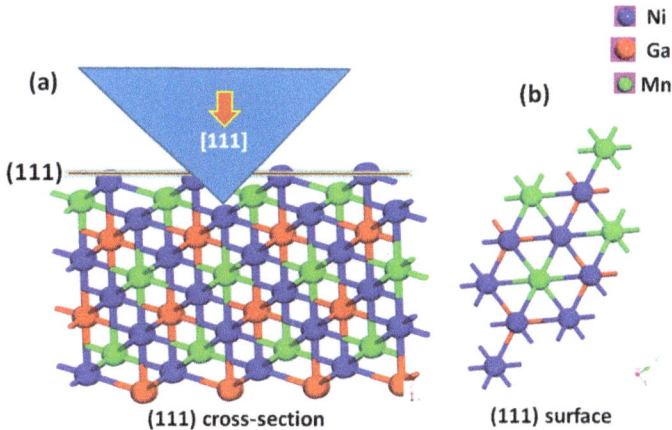

Figure 5. (**a**) Cross-section of (111) atomic arrangements underneath the plane being indented for indentations along [111] and, (**b**) planar projections of (111) in NiMnGa, normal to which nanoindentations are performed.

The cross-sections underneath the plane being indented are shown in Figure 6a,b for (232) and (010) respectively. It is seen that shearing of the planes along the [001] direction shown in Figure 6a is a distinct possibility on indenting along (232). On the other hand, along [010], no such particular direction of shear seems apparent. This corroborates our results according to which (232) is the softest and (010) is the hardest under nanoindentation amongst the crystallographic orientations studied here.

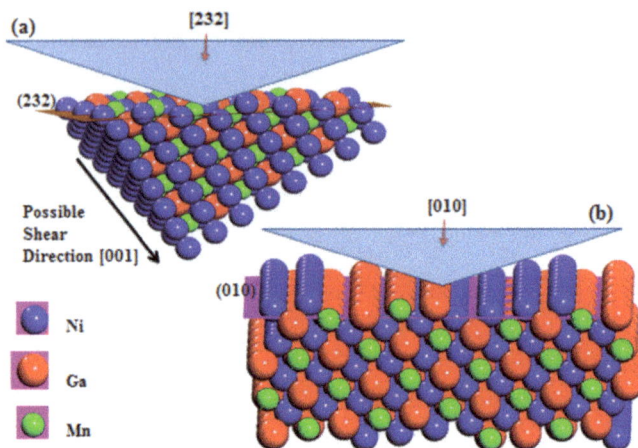

Figure 6. Cross-sections of atomic arrangements underneath the plane being indented for indentations along (**a**) [232] and (**b**) [010].

4. Conclusions

In summary, we have examined the orientation dependence of mechanical properties austenitic polycrystalline $Ni_{50}Mn_{26.25}Ga_{23.75}$ sample by nanoindentation technique. The experimental finding reveal (010) to be the hardest and stiffest while (232) is the least hard and most compliant amongst the crystallographic planes studied in the work. The anisotropy in hardness on indenting different orientations is attributed to the differing Schmid factors for slip propagation on pertinent slip plane. The significant variation in reduced modulus across orientations follows the same trend as hardness and is attributed to variation in crystal packing and atomic density. Having prior knowledge about the hardest and stiffest grains, one can texture these alloys in the processing stage itself to improve the mechanical properties and hence, enhance their structural applicability.

Supplementary Materials: The following are available online at http://www.mdpi.com/2073-4352/7/8/254/s1, Figure S1: Differential Scanning Calorimetry results obtained from the austenitic composition of NiMnGa studied in our work, Figure S2: X-Ray Diffractometry results obtained from the martensitic form of the austenitic NiMnGa studied in our work clearly showing the presence of indexed 7M martensite peaks, Table S1: Calculated Schmid Factor for all the slip systems possible in the NiMnGa Heusler structure with respect to the indentation stress direction, Table S2: Maximum Schmid factors for the 5 different family of slip systems possible in NiMnGa superstructure with respect to the indentation stress directions, Table S3: Slip System, corresponding Schmid factors and calculated mechanical properties on nanoindentation along different crystallographic directions.

Acknowledgments: M.S.R.N.K. thanks the Science and Engineering Research Board, Department of Science and Technology, Govt. of India for an Early Career Researcher Award (File No: ECR/2016/000827). U.R. acknowledges the Govt. of India for a J. C. Bose National Fellowship.

Author Contributions: U.R. and M.S.R.N.K. conceived and designed the experiments; A.J. and M.S.R.N.K. performed the experiments; A.J. and M.S.R.N.K. analyzed the data; A.J., M.S.R.N.K. and U.R. wrote the paper.

Conflicts of Interest: The authors declare no conflict of interest.

References

1. Pasquale, M. Mechanical sensors and actuators. *Sens. Actuators A Phys.* **2003**, *106*, 142–148. [CrossRef]
2. Suorsa, I.; Tellinen, J.; Ullakko, K.; Pagounis, E. Voltage generation induced by mechanical straining in magnetic shape memory materials. *J. Appl. Phys.* **2004**, *95*, 8054–8058. [CrossRef]
3. Ham-Su, R.; Healey, J.P.; Underhill, R.S.; Farrell, S.P.; Cheng, L.M.; Hyatt, C.V. Fabrication of magnetic shape memory alloy/polymer composites. *Proc. SPIE* **2005**, *5761*, 490–500.

4. Chen, F.; Cai, W.; Zhao, L.; Zheng, Y.F. Mechanical Properties and Fracture Analysis of Mn-Rich Ni-Mn-Ga Polycrystalline Alloys. *Key Eng. Mater.* **2006**, *325*, 691–694. [CrossRef]
5. Manosa, L.; Comas, A.G.; Obrado, E.; Planes, A.; Chernenko, V.A.; Kokorin, V.V.; Cesari, E. Anomalies related to the TA 2-phonon-mode condensation in the Heusler Ni_2MnGa alloy. *Phys. Rev. B* **1997**, *55*, 11068–11071. [CrossRef]
6. Wuttig, M.; Liu, L.; Tsuchiya, K.; James, R.D. Occurrence of ferromagnetic shape memory alloys. *J. Appl. Phys.* **2000**, *87*, 4707–4711. [CrossRef]
7. Worgull, J.; Petti, E.; Trivisonno, J. Behavior of the elastic properties near an intermediate phase transition in Ni_2MnGa. *Phys. Rev. B* **1996**, *54*, 15695–15699. [CrossRef]
8. MacLaren, J.M. Role of alloying on the shape memory effect in Ni_2MnGa. *J. Appl. Phys.* **2002**, *91*, 7801–7803. [CrossRef]
9. Zhou, L.; Giri, A.; Cho, K.; Sohn, Y. Mechanical anomaly observed in Ni-Mn-Ga alloys by nanoindentation. *Acta Mater.* **2016**, *118*, 54–63. [CrossRef]
10. Jakob, A.M.; Müller, M.; Rauschenbach, B.; Mayr, S.G. Nanoscale mechanical surface properties of single crystalline martensitic Ni-Mn-Ga ferromagnetic shape memory alloys. *New J. Phys.* **2012**, *14*, 033029. [CrossRef]
11. Kiran, M.S.R.N.; Varughese, S.; Reddy, C.M.; Ramamurty, U.; Desiraju, G.R. Mechanical anisotropy in crystalline saccharin: Nanoindentation studies. *Cryst. Growth Des.* **2010**, *10*, 4650–4655. [CrossRef]
12. Viswanath, B.; Raghavan, R.; Ramamurty, U.; Ravishankar, N. Mechanical properties and anisotropy in hydroxyapatite single crystals. *Scripta Mater.* **2007**, *57*, 361–364. [CrossRef]
13. Gollapudi, S.; Azeem, M.A.; Tewari, A.; Ramamurty, U. Orientation dependence of the indentation impression morphology in a Mg alloy. *Scripta Mater.* **2011**, *64*, 189–192. [CrossRef]
14. Kumar, A.; Rabe, U.; Hirsekorn, S.; Arnold, W. Elasticity mapping of precipitates in polycrystalline materials using atomic force acoustic microscopy. *Appl. Phys. Lett.* **2008**, *92*, 183106. [CrossRef]
15. Vlassak, J.J.; Nix, W.D. Measuring the elastic properties of anisotropic materials by means of indentation experiments. *J. Mech. Phys. Solids* **1994**, *42*, 1223–1245. [CrossRef]
16. Vlassak, J.J.; Nix, W.D. Indentation modulus of elastically anisotropic half spaces. *Philos. Mag. A* **1993**, *67*, 1045–1056. [CrossRef]
17. Oliver, W.C.; Pharr, G.M. An improved technique for determining hardness and elastic modulus using load and displacement sensing indentation experiments. *J. Mater. Res.* **1992**, *7*, 1564–1583. [CrossRef]
18. Varughese, S.; Kiran, M.S.R.N.; Ramamurty, U.; Desiraju, G.R. Nanoindentation in crystal engineering: Quantifying mechanical properties of molecular crystals. *Angew. Chem. Int. Ed.* **2013**, *52*, 2701–2712. [CrossRef] [PubMed]
19. Kiran, M.S.R.N.; Varughese, S.; Ramamurty, U.; Desiraju, G.R. Effect of dehydration on the mechanical properties of sodium saccharin dihydrate probed with nanoindentation. *CrystEngComm* **2012**, *14*, 2489–2493. [CrossRef]
20. Mishra, M.K.; Varughese, S.; Ramamurty, U.; Desiraju, G.R. Odd–even effect in the elastic modulii of α, ω-alkanedicarboxylic acids. *J. Am. Chem. Soc.* **2013**, *135*, 8121–8124. [CrossRef] [PubMed]
21. Mishra, M.K.; Desiraju, G.R.; Ramamurty, U.; Bond, A.D. Studying microstructure in molecular crystals with nanoindentation: Intergrowth polymorphism in Felodipine. *Angew. Chem. Int. Ed.* **2014**, *53*, 13102–13105. [CrossRef]
22. Sanphui, P.; Mishra, M.K.; Ramamurty, U.; Desiraju, G.R. Tuning mechanical properties of pharmaceutical crystals with multicomponent crystals: Voriconazole as a case study. *Mol. Pharm.* **2015**, *12*, 889–897. [CrossRef]
23. Krishna, G.R.; Kiran, M.S.R.N.; Fraser, C.L.; Ramamurty, U.; Reddy, C.M. The Relationship of Solid-State Plasticity to Mechanochromic Luminescence in Difluoroboron Avobenzone Polymorphs. *Adv. Funct. Mater.* **2013**, *23*, 1422–1430. [CrossRef]
24. Ghosh, S.; Mondal, A.; Kiran, M.S.R.N.; Ramamurty, U.; Reddy, C.M. The role of weak interactions in the phase transition and distinct mechanical behavior of two structurally similar caffeine co-crystal polymorphs studied by nanoindentation. *Cryst. Growth Des.* **2013**, *13*, 4435–4441. [CrossRef]
25. Spencer, E.C.; Kiran, M.S.R.N.; Li, W.; Ramamurty, U.; Ross, N.L.; Cheetham, A.K. Pressure-Induced Bond Rearrangement and Reversible Phase Transformation in a Metal–Organic Framework. *Angew. Chem. Int. Ed.* **2014**, *53*, 5583–5586. [CrossRef]

26. Li, W.; Kiran, M.S.R.N.; Manson, J.L.; Schlueter, J.A.; Thirumurugan, A.; Ramamurty, U.; Cheetham, A.K. Mechanical properties of a metal–organic framework containing hydrogen-bonded bifluoride linkers. *Chem. Comm.* **2013**, *49*, 4471–4473. [CrossRef]

27. Li, W.; Barton, P.T.; Kiran, M.S.R.N.; Burwood, R.P.; Ramamurty, U.; Cheetham, A.K. Magnetic and Mechanical Anisotropy in a Manganese 2-Methylsuccinate Framework Structure. *Chemistry* **2011**, *17*, 12429–12436. [CrossRef]

28. Varughese, S.; Kiran, M.S.R.N.; Ramamurty, U.; Desiraju, G.R. Nanoindentation as a Probe for Mechanically-Induced Molecular Migration in Layered Organic Donor–Acceptor Complexes. *Chem. Asian J.* **2012**, *7*, 2118–2125. [CrossRef]

29. Raut, D.; Kiran, M.S.R.N.; Mishra, M.K.; Asiri, A.M.; Ramamurty, U. On the loading rate sensitivity of plastic deformation in molecular crystals. *CrystEngComm* **2016**, *18*, 3551–3555. [CrossRef]

30. Sontakke, P.; Gupta, A.; Hiwarkar, V.; Krishnan, M.; Samajdar, I. Self-Accommodating Microstructure and Intervariant Interfaces of 5M and NM Martensites in Off-Stoichiometric Ni2MnGa Alloys. In *International Conference on Martensitic Transformations (ICOMAT)*; John Wiley & Sons, Inc.: Hoboken, NJ, USA, 2009.

31. Otsuka, K.; Wayman, C.M. *Shape Memory Materials*; Cambridge University Press: Cambridge, UK, 1999.

32. Crone, W.C.; Brock, H.; Creuziger, A. Nanoindentation and microindentation of CuAlNi shape memory alloy. *Exp. Mech.* **2007**, *147*, 133–142. [CrossRef]

33. Yoo, M.H.; Horton, J.A.; Liu, C.T. *Micromechanisms of Deformation and Fracture in Ordered Intermetallic Alloys: 1, Strengthening Mechanisms*; [Ni/sub 3/Al and CuZn]; Oak Ridge National Lab.: Oak Ridge, TN, USA, 1988.

34. Yamaguchi, M.; Umakoshi, Y. Deformation of single crystals of the L2 1 ordered Ag2MgZn. *J. Mater. Sci.* **1980**, *15*, 2448–2454. [CrossRef]

35. Li, T.L.; Gao, Y.F.; Bei, H.; George, E.P. Indentation Schmid factor and orientation dependence of nanoindentation pop-in behavior of NiAl single crystals. *J. Phys. Sol.* **2011**, *59*, 1147–1162. [CrossRef]

36. Haušild, P.; Materna, A.; Nohava, J. Characterization of anisotropy in hardness and indentation modulus by nanoindentation. *Metallogr. Microstruct. Anal.* **2014**, *3*, 5–10. [CrossRef]

37. Patel, D.K.; Al-Harbi, H.F.; Kalidindi, S.R. Extracting single-crystal elastic constants from polycrystalline samples using spherical nanoindentation and orientation measurements. *Acta Mater.* **2014**, *79*, 108–116. [CrossRef]

crystals

MDPI

Article

Local Stress States and Microstructural Damage Response Associated with Deformation Twins in Hexagonal Close Packed Metals

Indranil Basu [1,*], Herman Fidder [1,2], Václav Ocelík [1] and Jeff Th.M de Hosson [1,*]

[1] Department of Applied Physics, Zernike Institute for Advanced Materials and Materials Innovation Institute, University of Groningen, 9747AG Groningen, The Netherlands; h.fidder@rug.nl (H.F.); v.ocelik@rug.nl (V.O.)

[2] Department of Mechanical Engineering, Cape Peninsula University of Technology, Cape Town 7535, South Africa

* Correspondence: i.basu@rug.nl (I.B.); j.t.m.de.hosson@rug.nl (J.T.d.H)

Received: 17 November 2017; Accepted: 18 December 2017; Published: 21 December 2017

Abstract: The current work implements a correlative microscopy method utilizing electron back scatter diffraction, focused ion beam and digital image correlation to accurately determine spatially resolved stress profiles in the vicinity of grain/twin boundaries and tensile deformation twin tips in commercially pure titanium. Measured local stress gradients were in good agreement with local misorientation values. The role of dislocation-boundary interactions on the buildup of local stress gradients is elucidated. Stress gradients across the twin-parent interface were compressive in nature with a maximum stress magnitude at the twin boundary. Stress profiles near certain grain boundaries initially display a local stress minimum, followed by a typically observed "one over square root of distance" variation, as was first postulated by Eshelby, Frank and Nabarro. The observed trends allude to local stress relaxation mechanisms very close to the grain boundaries. Stress states in front of twin tips showed tensile stress gradients, whereas the stress state inside the twin underwent a sign reversal. The findings highlight the important role of deformation twins and their corresponding interaction with grain boundaries on damage nucleation in metals.

Keywords: titanium; digital image correlation; detwinning; twin-grain boundary interactions; plastic deformation

1. Introduction

The mechanical behavior of metals, is to a large extent, influenced by the changes in intrinsic length scales and the interaction between different microstructural features associated with them. A microstructure can be essentially defined as the overall arrangement of crystallites/grains and material defects (point defects, dislocations and grain/twin boundaries). Depending upon the volume of the probed region these features can considerably vary both topologically as well as dimensionally. Since the local stress state in a material is directly proportional to the density and spatial configuration of defects, this also means that internal/residual stresses can strongly vary across different length scales i.e., macro-, meso- and microscopic dimensions.

2. Background

Plastic deformation in metals is primarily carried out by the creation and motion of linear defects viz. dislocations. In polycrystalline materials with diverse grain orientations, the interfaces between differing crystallite orientations can present themselves as severe obstacles to dislocation motion. The resultant interaction between line defects and such grain boundaries often gives rise to complex geometrical configurations of stored dislocations that are associated with long-range elastic stress fields.

These stored dislocations are termed as geometrically necessary dislocations (GNDs), since they ensure the geometrical compatibility of deformed grains across the grain boundaries. Superposition of such stress fields invariably results in a strong spatial heterogeneity in local stress states. Needless to say, the variation in local GND density levels directly influences the distribution of microscopic residual stresses in the grain and grain boundaries. Internal stresses are also influenced by mutual interaction of line defects leading to local entanglements of dislocations i.e., forest dislocations or statistically stored dislocations (SSDs), but their numeric contribution in comparison with GNDs progressively diminishes with increasing applied strains due to saturation in dislocation densities of the former for strains above ~0.1–0.2 [1]. On the other hand, GND dislocation associated with grain boundaries continue to increase linearly with applied shear strain [1], thus acting as the primary contributors to strain hardening. Such correlation between GND density levels and local stress gradients has been utilized in the past to explain local hardening phenomenon due to dislocation pile-up at grain boundaries. Eshelby et al. [2] showed analytically that a dislocation pile-up ahead of an insurmountable obstacle such as a grain boundary would result in a stress gradient that varies as "one over the square root" of the distance from the obstacle. In fact, the stress field in front of the spearhead of the dislocation pile-up resembles the stress field of a crack singularity in a linear elastic medium. Subsequent experimental observations by Hall [3,4] and Petch [5] independently re-established such a behavior in metals as the well-known Hall-Petch effect, wherein the mechanical strength of the material increases with a decreasing intrinsic length scale of the grain size.

The dislocation configuration near a grain boundary strongly determines the degree of pile-up and corresponding local stress concentration. Depending on the crystallography of the grain boundary certain slip/twin systems may find conjugate systems in the neighboring grain that facilitate complete or partial strain transfer. It must be mentioned here that like slip deformation, twins are described by their twinning plane and the direction of shear. A theoretical estimate of the feasibility of slip/twin transmission can be captured by the strain transfer parameter [6,7], expressed as,

$$m' = (n_1 \cdot n_2) \cdot (b_1 \cdot b_2), \tag{1}$$

where n_1 and n_2 are the normalized intersection lines common to the slip/twinning planes and the boundary plane, and b_1 and b_2 are the normalized slip/twinning shear directions in the pile-up and emission grains. The value of m' provides a measure of the probability for possible transmissivity of a slip or twinning dislocation across the grain boundary. Maximization of strain transfer parameter m' abates dislocation pile-up and promotes easier strain transfer across the grain boundary.

Precise measurement of residual stresses at different length scales is extremely vital in order to acquire a fundamental understanding of damage mechanisms in present day structural materials. Despite the availability of diverse techniques for estimating internal stresses such as, using hard X-rays, hole drilling, contour method, slitting and ring coring, very few methodologies allow estimation of stresses up to micron and sub-micron scale resolution. Diffraction techniques utilizing convergent beam or nano-beam, hard X-rays from synchrotron are some of the available methods that can resolve stress at inter/intra-granular level [8], but the availability of such facilities are scarce [9]. Lately, the measurement of local scale stresses can be performed by high resolution electron back scatter diffraction (HR-EBSD) wherein Kikuchi patterns from reference (un-deformed) and deformed states are used to measure the residual displacements, and subsequently calculate the local elastic strain and stress state [10–14]. However, the method is limited to 2-dimensional investigations wherein only subsurface information is obtained [8]. With the advent of dual-beam focused ion beam (FIB) field emission gun microscopes, measurement of residual stresses with simultaneous sub-micron lateral and depth resolution in a semi-automated and robust way [9] is made possible. In this way both the subsurface and bulk deformation contribution on internal stress build up is accounted for during stress quantification. The methodology utilizes correlative imaging and milling to remove material and estimate the local stress relaxation in the neighborhood. Depending upon the milling geometry, either multiple stress components of the whole stress tensor can be determined or spatial stress gradients along one stress component is evaluated. Digital image correlation (DIC) is utilized to determine relaxation induced displacements in the vicinity of the milled region.

In previous studies [15,16] the authors introduced a site-specific technique utilizing electron back scattered diffraction (EBSD) and FIB-DIC linear slit milling to accurately determine spatially resolved stress profiles in the vicinity of grain boundaries in commercially pure titanium. The investigations in the vicinity of different grain boundaries in commercially pure titanium revealed the appearance of a local stress minimum just next to the boundary. This was followed by a Hall-Petch type monotonic stress decrement to a steady state regime. Correlations with the GND density and local misorientation data further validated the observed trends (c.f. Figure 1a,b). The results further showed that the width over which the stresses relaxed in the vicinity of the boundary was strongly dependent upon the obstacle strength of the grain boundary i.e., the Hall-Petch coefficient, k_{HP} (c.f. Figure 1c). The observed stress drop was justified by a local change in elastic stress fields arising from dislocation-dislocation and dislocation-grain boundary interactions that may lead to a relative depletion of dislocation densities in the vicinity of specific grain boundaries. It was shown that the stress fields due to dislocation-grain boundary interactions are long range in nature and can be of the order of 10^{-3} G even at distances $\approx 10^4 \cdot b_d$ (where, G is the shear modulus and b_d is the Burgers vector for active dislocation slip) from the grain boundary plane.

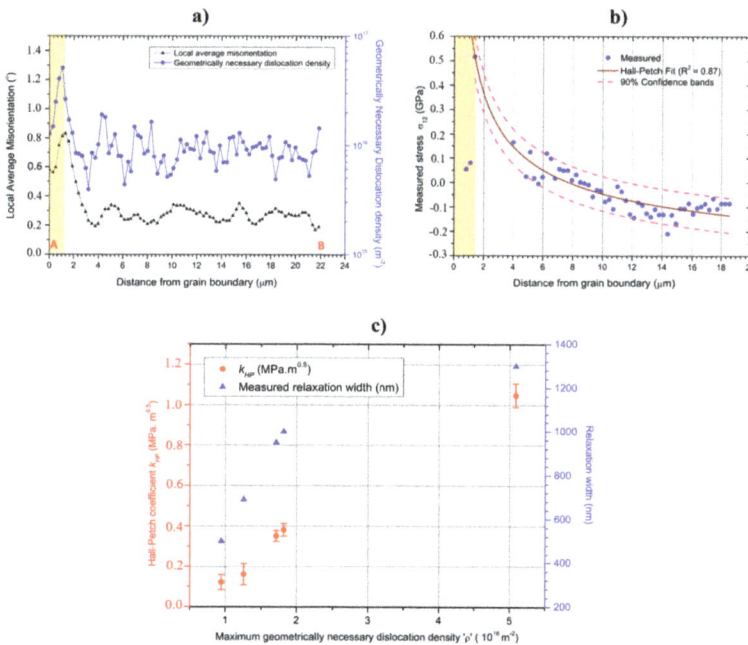

Figure 1. (a) Local misorientation and stored dislocation densities in the vicinity of the grain boundary; (b) corresponding residual stress profile as a function of distance from the grain boundary; (c) variation of k_{HP} and relaxation width with the peak GND density measured in the vicinity of different grain boundaries (adapted from ref. [15]).

Hexagonal crystal' structures, due to their low symmetry, often exhibit anisotropic deformation behavior unlike face centered cubic metals with cubic symmetry. This typically arises from a lack of easily available slip deformation modes along the c-axis and the role of mechanical twinning in accommodating strain out of the basal plane [17–20]. Due to its inherent nature, deformation twinning strongly impacts crystallographic texture evolution as well as the grain scale stress evolution [21,22].

While plastic slip is spatially more homogeneous and lattice strain evolution is gradual, the onset of twinning involves sudden reorientation of a part of the crystal associated with significant lattice strain and localized shear. The localized lattice rotation during twinning leads to the creation of twin boundaries. These newly formed boundaries can dynamically refine the grain size and lead to significant latent hardening inside the twins on subsequent straining [23,24]. Furthermore, dislocations, grain-boundaries and twins can mutually interact. Such complex local plastic response often gives rise to significant stress heterogeneities in the vicinity of twins. The significance of understanding such localized stress fields is in determining their associated effects either on accommodating deformation close to crack tips or on nucleation and propagation of cracks and thus their tendency to limit ductility [21,22]. Unfortunately, due to the sudden nature of twin formation that comprises nucleation and propagation, capturing load partitioning between the twin and untwinned parent experimentally is still extremely important.

The current work therefore aims to extends the aforementioned combination of EBSD and FIB-DIC methodology to quantify stress fields arising from the interaction of twin boundaries with parent grains and grain boundaries in commercially pure titanium. The observed trends are subsequently discussed with respect to underlying physical processes and the subsequent impact of deformation twinning on fracture behavior of titanium is acknowledged.

3. Experimental Methods

Commercially available grade II titanium was subjected to room temperature in situ four-point bending test inside a Tescan Lyra dual beam (FEG-SEM/FIB) scanning electron microscope (Brno, Czech Republic). The initial microstructure comprised of coarse grains with a mean grain size of ~100 μm. Prior to mechanical testing, bending specimens were prepared for EBSD measurements using conventional metallographic techniques [25]. Specimens were strained to a final surface true strain, $\varepsilon = 0.18$. Microstructural characterization was performed by means of EBSD, thereby extracting both topographical and orientation information about the individual grains. A step size of 0.3 μm and hexagonal type of grid was used for the measurements. The acquired raw EBSD data was subsequently analyzed using EDAX-TSL OIM^{TM} Analysis 7.3 (software and MTEX open source Matlab toolbox [26]. Slip traces in individual grains were imaged using in situ scanning electron microscopy (SEM) (Tescan Lyra, Brno, Czech Republic). The orientation of the grain boundary plane was determined by milling into the region containing the boundary using focused ion beam and examining the grain boundary trace along the milled cross section. All observations were made on the tensile surface of the bent specimen, with the direction of viewing parallel to the surface normal, hereinafter referred to as the A3 sample axis.

Residual Stress Measurement by FIB-DIC Slit Milling

The protocol followed during the measurement of residual stress starts with the acquisition of a scanning electron microscopy (SEM) image of the area to be analyzed. After recording the first image, a slit is milled on the surface. Then, a second image of the same area is taken. From the comparison of these two SEM images recorded before and after stress release by DIC the displacements are obtained. These displacements are compared with those obtained by the analytical solution for an isotropic elastic material, and the value of residual stress is obtained from the slope of the fitting [27].

Figure 2 shows the geometry of the slit used in our experiments including its dimensions: a length L, a width w and a depth a. The evaluated displacements of the surface, u_x, are normal to the plane of the slit. The origin of coordinates is placed at the center of the slit. Considering the geometry, in plane displacements U_{dir} can be related to residual stress σ_{dir} in the same direction by an analytical expression (see Equation (2)) such that [27–29]:

$$U_{dir} = \frac{2.243}{E'}\sigma_{dir}\int_0^a cos\theta\left(1 + \frac{sin^2\theta}{2(1-v)}\right) \times (1.12 + 0.18 \cdot sech(tan\theta))dz, \qquad (2)$$

where E' is $E/(1 - v^2)$, E is the Young's modulus, v is the Poisson's ratio, θ is $arctan(d/a)$, a is the depth of the slit and d is the distance from the slit; *dir* represents x or y directions. The displacements caused by the stress release depend on the slit depth a and are directly proportional to the σ/E ratio. Moreover, the extraction of the residual stress requires the a priori knowledge of the elastic properties of the material under study (i.e., Young's modulus and Poisson's ratio). In the present study, the adopted rectangular slit geometry allows measurement of only one stress component that is aligned laterally to the longitudinal direction of the slit.

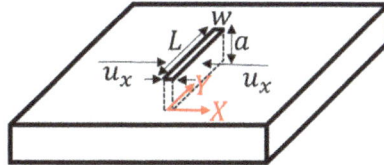

Figure 2. Schematic of slit introduced by FIB milling.

SEM images acquired before and after FIB milling are processed using a commercial digital image correlation software GOM Correlate v. 2016 and subsequent displacements lateral to the slit are recorded for each facet (group of pixels) position.

Figure 3 shows an example of a typically obtained DIC contour map of relative displacements normal to the slit corresponding to the local stress release due to milling. The scale bar of the image is in the range of tens of nanometers. Each color means that a group of pixels is displaced over the respective number of nanometers. When studying the displacements normal to the plane of the slit, the displacement is to the right (red color) or to the left (blue). Consequently, data arrays comprising of facet ID, coordinates and relative displacements are exported for post processing and residual stress determination. A Matlab based script is utilized to empirically determine the residual stress values from the experimentally measured displacements, as per Equation (2).

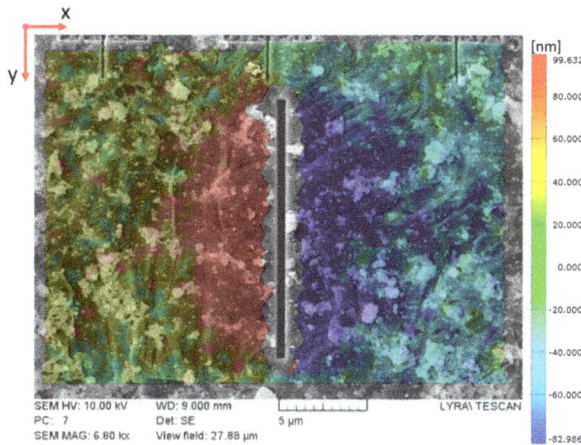

Figure 3. An arbitrary example illustrating the displacement field map from DIC analysis in GOM; image acquired over a field of view of ~28 μm, the color bar in the right side shows the magnitude of recorded surface displacements due to stress relaxation by FIB milling along direction x on both sides of the slit. The values range between −100 to +100 nm.

In the current work, linear slits, oriented normal to the grain/twin boundary trace, with widths between 0.3–0.5 μm, depths from 1.5–2.5 μm, and lengths varying from 15–25 μm (depending on the twin/grain size), were milled inside individual grains/twins showing pile-up, classified on the basis of measured local lattice misorientation values near the grain/twin boundary. For each slit, multiple SEM images of resolution 768 × 768 pixels were acquired at high magnifications (field of view of 10–15 μm) to ensure a high spatial resolution of measured displacement field. In order to obtain statistically sufficient data points, DIC was performed using a facet size of 19 × 19 pixels with a step width of 16 pixels. Yttria-stabilized Zirconia (YSZ) nano-particles were used for surface decoration to obtain optimum image contrast for high accuracy DIC analysis.

To quantify displacements in the range of nanometers, the precision of measurements at reduced length scales at high magnifications is critical. The accuracy of the measured displacement field from DIC, $u_{x,y}$ depends upon the image pixel size along x and y directions, $N_{x,y}$ and sub-pixel shift resolution parameter k, given by the expression [28,30]:

$$u_{x,y} = k \cdot (N_{x,y}). \tag{3}$$

Under favorable imaging conditions the value of k achievable from the DIC algorithm is 0.01, which amounts to a precision of 1×10^{-5} [30,31]. It must be mentioned here that the sub-pixel shift resolution is a major criterion for stress measurement with high spatial resolution. Due to the characteristics of the above expression, the measurement accuracy can be progressively improved with smaller imaging sizes i.e., view fields. In realistic cases, including imaging related drift inaccuracies, the sub-pixel shift resolution varies from 0.01 to 0.1. For instance, in the current work the pixel sizes range between 13–18 nm (for the above defined view fields and image resolution) leading to the resultant sensitivity of the DIC software of each measured displacement being in the range of 0.1 nm to 1.8 nm. In the present work, SEM imaging parameters were optimized as per ref. [30] to minimize experimental drift related inaccuracies.

Stress distributions at sub-microscopic length scales need not always be homogeneous i.e., constant stress all throughout the material, but may considerably vary spatially. In such cases, stress determination by simplistic averaging of all displacements along the slit length, misrepresents the actual stress state of the material. A multiple fitting approach [15,29], wherein a stress value is obtained for displacements corresponding to each row was implemented to account for spatially heterogeneous stress states. All stress calculations were made using orientation dependent elastic modulus values, extracted from ref. [32], wherein the angle between the indentation loading axis and the crystal c-axis was varied to obtain an angular dependence of the elastic modulus with respect to c-axis orientation.

4. Results and Discussion

4.1. Stress Gradients across Twin-Parent Interface

Figure 4 illustrates a representative case of coherent twin-parent interface in titanium. Figure 4a shows the inverse pole figure (IPF) map of the highlighted twin boundary in the inset image. The viewing axis corresponds to the A3 direction. The twinned region and the parent grain are labelled as 'T$_{grain}$' and 'P$_{grain}$', respectively. While the c-axis of twinned grain is oriented parallel to the viewing axis, the parent grain was oriented such that its c-axis was aligned with the tensile axis. Figure 4a also shows the orientation of the milled slit lying between points A and B. Figure 4b,c represent the kernel average misorientation (KAM) and local average misorientation (LAM) mappings. The KAM physically describes the average misorientation spread between a reference pixel and its nearest neighbor pixels for a defined kernel size. The LAM angle corresponds to the misorientation averaged over all nearest neighbor pairs within a kernel. Both LAM and KAM values were calculated for the 2nd nearest neighbor with a threshold value of 2° [33]. $\{10\bar{1}2\}$ Tension twin boundaries are shown in red in Figure 4b,c. The characterization of twins in the EBSD maps (c.f. Figure 4b,c) was done on the basis of the characteristic misorientation angle of 85.03° about the $\langle 11\bar{2}0 \rangle$ rotation axis (given in minimum angle-axis pair), which corresponds to $\{10\bar{1}2\}\langle 10\bar{1}1 \rangle$. tension twin. A maximum angular deviation of $\pm 6°$ was considered [34,35]. The KAM and LAM maps indicate signs of stress concentration at the interface.

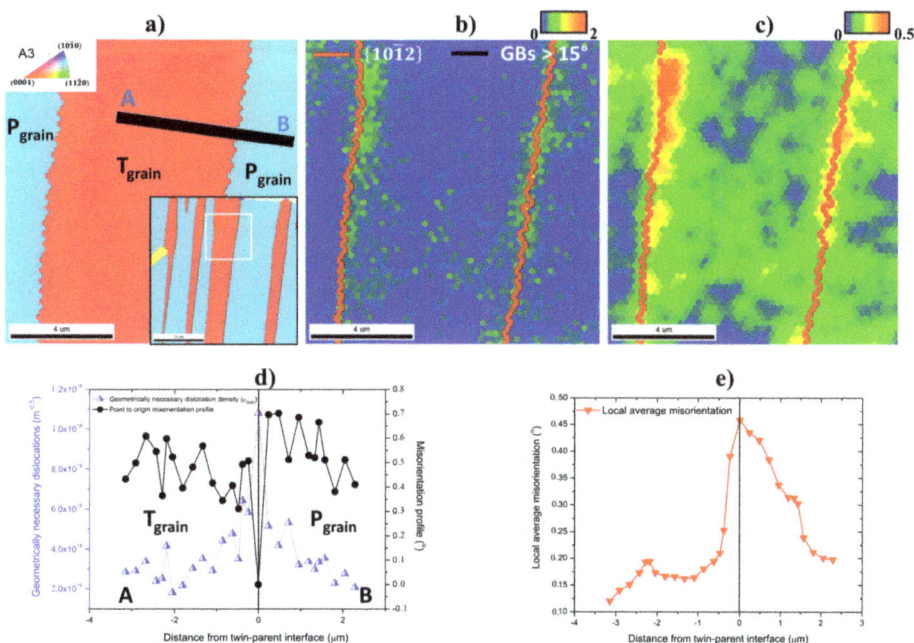

Figure 4. (a) Inverse pole figure (IPF) map of twin lamellae intersecting a grain boundary; inset image shows a low magnification IPF map enclosing the area of interest highlighted in black square. Slit orientation between points A and B is shown schematically; parent and twin grains are labelled as P_{grain} and T_{grain}; (b) KAM and (c) LAM maps ($\{10\bar{1}2\}\langle 10\bar{1}1 \rangle$ tension twin boundaries shown in red and high angle grain boundaries shown in black); (d) GND density and misorientation profile (with respect to the grain boundary) inside twin and neighbor grain; and (e) LAM profile between points A and B.

Figure 4d shows the misorientation profile with respect to the interface along the twin grain and parent grain. The degree of misorientation between the twin center and the interface is of the order of ~0.6°, whereas between the boundary and parent grain interior is ~0.7°. GND density values are additionally shown as a function of distance from the twin-parent interface. GND density (ρ_{GND}) values from EBSD data were calculated using the strain gradient approach [36,37], given by the expression:

$$\rho_{GND} = \frac{2\theta}{n\lambda|b_d|} \tag{4}$$

where θ is the experimentally measured KAM value, λ is the step size, n is the number of nearest neighbors averaged in the KAM calculation, and b_d is the Burgers vector corresponding to the active slip system in the grain. It must be noted that the GND values obtained from Equation (4) provide a lower bound estimate as they can only account for contributions from the non-paired edge dislocation segments and dislocation walls (since both lead to an effective unclosed burgers circuit, thereby contributing to the measured local misorientation). Figure 4e displays the LAM values with respect to the distance from the twin-parent interface. The maximum LAM value is recorded at the boundary as 0.45°. Both the GND and LAM values decrease monotonically on moving away from the twin boundary. The excellent agreement between the LAM and GND values is not surprising since both values are derived from the measured local misorientation.

Figure 5a shows the interaction of $(10\bar{1}0)$ prismatic slip bands with the twin-parent interface shown in Figure 4. The corresponding local misorientation gradient due to slip accumulation at the interface is shown in the inset LAM map. Figure 5b,c represents the FIB-DIC analysis and corresponding stress measurements for the region shown in Figure 4. Figure 5b shows the spatial orientation of the milled slit with respect to the twin boundary. The white dots correspond to the YSZ particles used for surface decoration. The twin boundary plane orientation is highlighted by the yellow parallelogram, with the twin plane normal defined as T_n. The measured stress values were resolved along the twin plane to obtain the shear component acting on the twinning plane along the twin boundary trace, designated as $\sigma_{1,Twinplane}$. Figure 5c displays the variation of $\sigma_{1,Twinplane}$ as a function of normal distance from the twin boundary i.e., along direction 2. The stress values all throughout the twin and parent remain compressive in nature, with a continuous transition across the interface. The magnitude of the stress is highest at the interface reaching up to a value of ~ -180 MPa, subsequently dropping to -80 MPa near the twin center. On the other hand, the stress values in the parent grain interior stabilize at ~ -110 MPa. On comparing the stress gradients with Figure 4d,e the agreement seems excellent, thereby indicating that the observed stress fluctuations indeed confer to the actual stress state in the twin and parent grains.

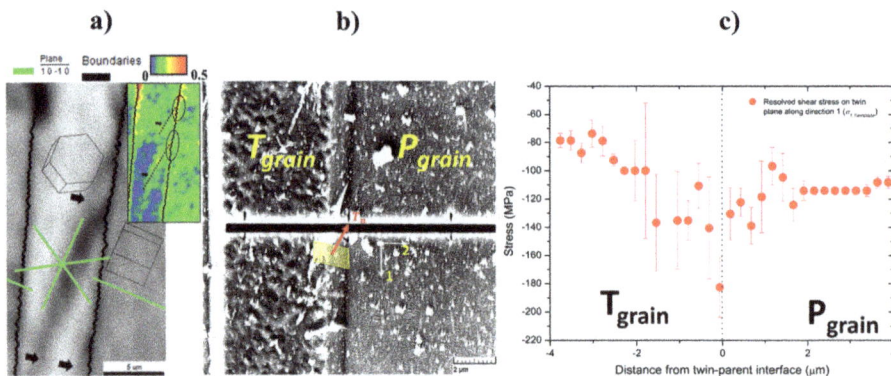

Figure 5. (a) Image quality map showing instances of $(10\bar{1}0)$ prismatic slip bands interacting with twin parent interface; inset image shows corresponding local misorientation gradient along the identified slip trace; (b) SEM image of decorated twin-parent interface from Figure 4a; the twin boundary plane orientation is shown in yellow with twin plane normal T_n. Slit coordinate system labelled by directions '1' and '2'; (c) corresponding measured resolved shear stress component ($\sigma_{1,Twinplane}$) along line AB.

The observations indicate the presence of compressive residual stress states acting parallel to the twin boundary plane. In order to understand the implications of such a stress state it is important to delve into the mechanistic of migration of coherent twin-parent interfaces. The lateral broadening of a twin typically involves shear coupled migration of twin boundaries, whereby the normal translation of the boundaries is simultaneously accompanied by shearing of the parent grain. The magnitude of the theoretical twin shear S^t is characteristic for the twin type and in the case of $\{10\bar{1}2\}\langle10\bar{1}1\rangle$ tension twins is given as: $S^t = \frac{\left(\frac{c}{a}\right)^2 - 3}{\left(\frac{c}{a}\right)\sqrt{3}}$. For titanium with $c/a = 1.587$, $S^t = 0.171$.

Twin boundary motion typically involves the glide of twinning dislocations/zonal dislocations, which are defined as regions wherein non-homogeneous shear at the twin matrix interfaces is accomplished at the expense of pure atomic shuffling in multilayer twin lamellae (c.f. Figure 6a). The mechanism of lateral thickening is demonstrated in Figure 6a. During tensile loading, the stress component acting along the twin plane $\tau_{applied}$ typically drives the motion of twinning dislocations along the twin plane, resulting in shearing of the adjacent parent region and simultaneous thickening of the twin by a value Δh. This process repeats itself as long as the applied stress is sufficient to move the twinning dislocations and sustain the thickening process. In the unloaded state however, the presence of a compressive residual stress state generates a negative shear that may promote the motion of twinning dislocations in a direction opposite to that in case for externally applied stress (c.f. Figure 6b). Such a scenario typically indicates the favorability of the twin lamellae to undergo thinning during unloading and also disappear when applied load is reversed. This further explains why lower applied stresses are required to activate de-twinning as compared to twin nucleation during cyclic loading behavior, since the already present internal compressive stresses act as an additive stress to the applied load [38,39]. Indeed, it must be noted that the considerations of anisotropy in twin boundary motion in the longitudinal and transverse directions, due to the former being shear dominated and the latter primarily driven by atomic shuffling, is also crucial to accurately understand the de-twinning phenomenon. In the forthcoming section, it will be also shown how the stress state at the twin tips further contribute to the aforementioned de-twinning effects, often observed in hexagonal close packed metals [40].

a) During Loading   b) Unloaded state

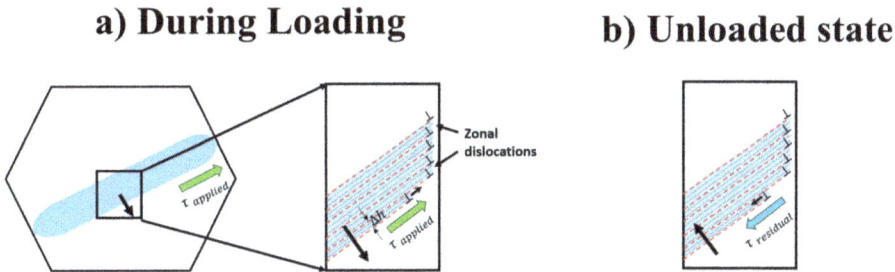

Figure 6. (a) Schematic showing twin boundary migration leading to twin broadening under applied stress; the stress component parallel to the twin boundary drives the motion of zonal dislocations that result in simultaneous shearing and twin thickening i.e., shear coupled twin boundary motion (b) In unloaded state, compressive residual stress states can exert negative shear forces and subsequently favor opposite movement of twinning dislocations and concurrent twin shrinkage i.e., de-twinning.

4.2. Stress Gradients Arising from Twin-Grain Boundary Interactions

Figure 7 illustrates an instance of twin-grain boundary interaction wherein $\{10\bar{1}2\}$ tension twin lamellae impinge and are subsequently being blocked at the grain boundary. Figure 7a represents the inverse pole figure map of the twin-grain boundary intersection zone, wherein two tension twins of the same variant meet the grain boundary (marked as GB). The selected area corresponds to the magnified view of the highlighted region in the inset image. The inset image indicates that the investigated twin could either arise in the blue grain by means of propagation of the dark pink twin in the neighboring green grain across the grain boundary or due to simultaneous nucleation of twins at the grain boundary, which subsequently propagate inside both green and blue grains. As in Figure 4a, the parent grain is denoted as 'P_{grain}', whereas the twinned domains are labelled as 'T_{grain}' and the neighboring grain is represented as 'N_{grain}'. A schematic of the orientation of the milled slit from point A to B is additionally shown in Figure 7a. Figure 7b,c represent the KAM and LAM maps of the same region, with twin boundaries highlighted in red and grain boundaries shown in black. Grain boundaries were designated by a lower threshold of 15° in Figure 7b,c. A grain boundary map color coded with respect to the ease of twin transmission is presented in Figure 7d. The values at the grain boundary of interest indicate a poor probability of twin transfer, evident by an $m\prime$ value of 0.3 (c.f. Equation (1)). Grain and twin orientations are further depicted by the spatial orientation of the hexagonal crystals. Traces corresponding to the twin plane (in orange) and twin shear directions (in blue) are also plotted in Figure 7d. Figure 7e shows the distribution of GND values as well as the misorientation profile with respect to the grain boundary along line AB. Figure 7f indicates the LAM distribution along line AB. Figure 7e,f indicate a peak lattice distortion and dislocation density in the vicinity of the grain boundary that decays to lower values on moving left (towards twin interior) or right (into the neighboring grain). Lattice rotations inside the twin near the twin-grain boundary interface are significantly high and drop drastically, within a width of 1 μm away from the grain boundary. On the other hand, the strain gradient in the neighboring grain shows a less sharp decrease. The GND and LAM values in the neighbor grain indicate a local minimum in the vicinity of the grain boundary, within a width of 440 nm.

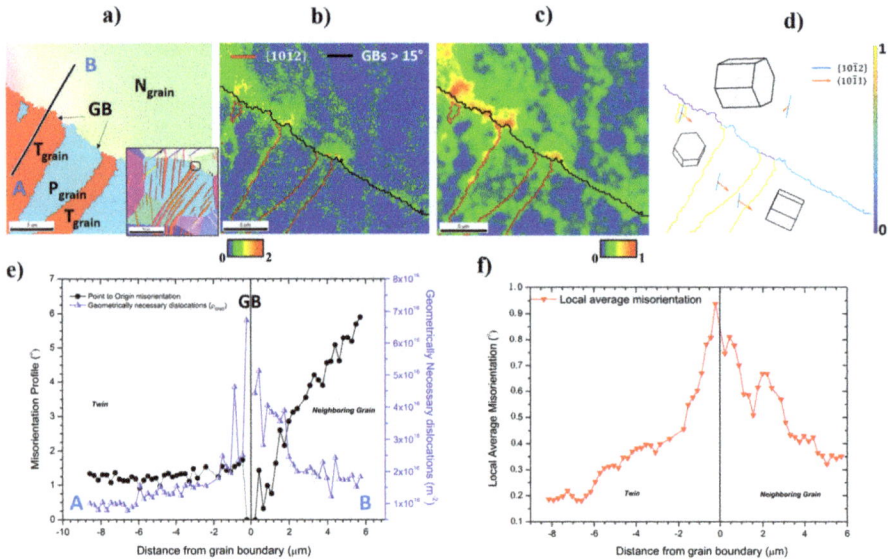

Figure 7. (**a**) Inverse pole figure (IPF) map of twin lamellae intersecting a grain boundary; inset image shows a low magnification IPF map enclosing the area of interest highlighted in black square. Slit orientation between points A and B is shown schematically; parent, twin and neighbor grains are labelled as P_{grain}, T_{grain} and N_{grain}; (**b**) KAM, (**c**) LAM and (**d**) m/ maps corresponding to twin transmission across grain boundary ($\{10\bar{1}2\}\langle10\bar{1}1\rangle$ tension twin boundaries shown in red and high angle grain boundaries shown in black); (**e**) GND density and misorientation profile (with respect to the grain boundary) inside twin and neighbor grain and (**f**) LAM profile between points A and B.

Figure 8 presents the stress measurements obtained from the FIB-DIC slit milling technique. Figure 8a presents the image quality map of the region shown in Figure 7a, with a schematic of the slit. The twins, parent and neighbor grains are labelled as T_{grain}, P_{grain} and N_{grain} respectively. Figure 8b corresponds to an SEM image captured post slit milling of the region highlighted in Figure 8a. The twin boundary plane orientation is depicted and the corresponding inclination of the twinning plane normal with respect to the sample surface is measured as 137°. The twinning shear direction, S_d is aligned with the normal to the slit wall, represented by axis 1. The longitudinal axis of the slit is labelled as 2. Since the measured stress component from slit milling corresponds to σ_{11}, the corresponding stress component lying on the twinning plane and in the direction of twinning shear can be described as $\sigma_{1R} = \sigma_{11} * cos47°$. Figure 8c shows the orientation of the grain boundary plane, marked by red arrow, to be perpendicular to the longitudinal axis of the slit, labelled as '2'. Figure 8d represents the measured profile of resolved stress on the twin plane along the twinning shear direction, σ_{1R} along line AB.

The stress gradients inside the twin domain reveal high compressive stresses at the twin tip, reaching values up to ~−170 MPa. On moving inwards, the stress values decrease considerably reaching values in the less negative/low positive range. On the other hand, the stress profile in the neighboring grain registers tensile stresses as high as ~50 MPa near the grain boundary that subsequently drop to very low positive values on moving away from the grain boundary. Agreeing with the trends seen in Figure 7e,f, the stress values in the neighbor grain show a local minimum near the grain boundary.

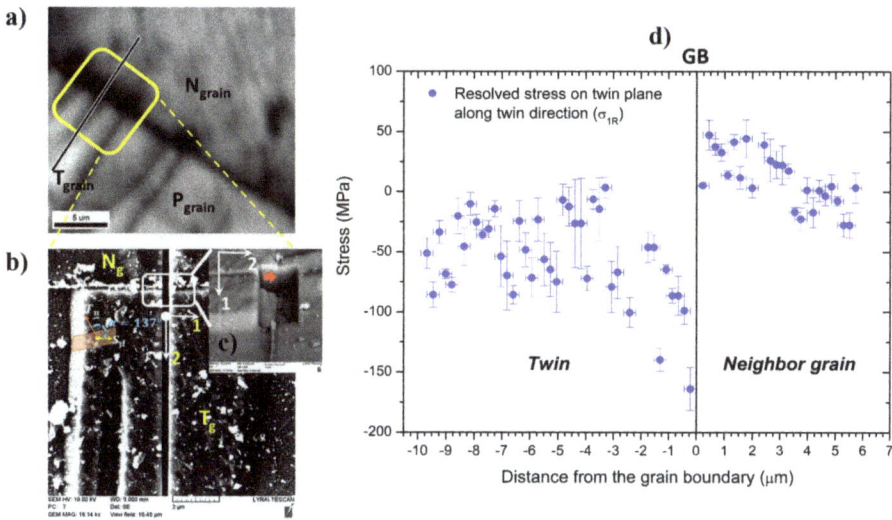

Figure 8. (**a**) Image quality map of region shown in Figure 7a with schematic of milled slit; (**b**) SEM image of decorated twin-grain boundary intersection from Figure 7a; twin boundary plane orientation is shown in orange with twin plane normal T_n and twin shear direction S_d. Slit coordinate system labelled by directions '1' and '2', measured angle between twin boundary normal and axis '1' (normal to the slit wall) is displayed as 137°; (**c**) the orientation of the grain boundary plane (marked by red arrow) with respect to the slit length aligned normal to the wall of the milled trench; (**d**) corresponding measured resolved shear stress component (along the twinning plane in the direction of twinning shear) along line AB.

Residual stress profiles measured inside the twin and the neighbor grain indicate a sign reversal in stress states across the twin-grain boundary interface. Previous studies employing simulations also reported similar observations of sign reversal of stress states in twin and neighboring grains [41–44]. In general, grain orientations in low symmetry hexagonal crystals can be classified into crystallographically 'soft' or 'hard' orientations, depending on whether they are initially favorably oriented for strain accommodation along the <a>-axis (i.e., basal or prismatic slip) or not. $\{10\bar{1}2\}$ tension twinning typically results when the local stress along the c-axis of a crystallographically 'hard' parent orientation is tensile in nature. Under an externally applied stress, the soft orientations are typically the first to yield while the hard grain orientations, being elastically stiffer, undergo elastic straining due to lack of slip accommodation. This mechanism continues until a threshold stress is reached, whereby twinning is able to activate in the crystallographic hard grain orientations. Twinning in hexagonal crystals proceeds primarily via 3 stages: nucleation, propagation and lateral growth of the twin lamella. Nucleation involves formation of viable twin nuclei, a few atoms thick, preferably at grain boundaries associated with high localized stress concentration. It has been suggested that the twin nucleation mechanism is triggered by the interaction of grain boundary dislocations and stress driven slip dislocations [40]. Typically, the nucleation process is governed by two main factors, that is, the local resolved shear stress along the twinning shear direction on the twin plane and the ease of accommodation of twin associated shear in the neighbor grain (either by twinning or dislocation slip). The accommodation strains imposed on the neighboring grains by twinning can be readily calculated from the twinning shear by rotating its displacement tensor into the crystallographic reference frame of the neighboring grain.

Following nucleation, the twin propagates along the longitudinal direction, elongating in shape (Figure 9a). The green arrows in Figure 9a,b indicate the direction of the applied stress in the current study. Propagation refers to the process of the twin front moving, by means of glide of twinning dislocations, into the bulk of the grain and eventually terminating on encountering an obstacle such as grain boundary. The orange arrow in Figure 9b indicates the direction of the resolved shear stress component that drives twin propagation. Hereafter, further stress increase triggers lateral growth of the twins leading to their thickening (Figure 9b). The mechanism of twin thickening is already described in Section 4.1 (c.f. Figure 6a).

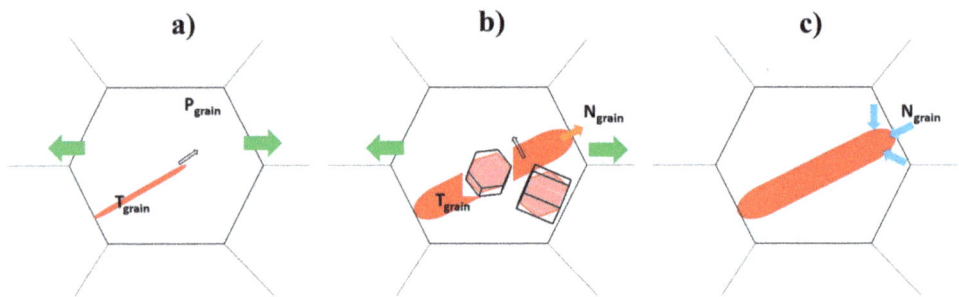

Figure 9. (**a**) Schematic showing twin propagation in a parent grain, applied stress in the present study is depicted by green arrows, direction of propagation shown by black open arrow; (**b**) twin growth leads to lateral thickening under external stress once the twin hits the grain boundary, the forward stress component driving motion of twinning front (shown in blue arrow) is accommodated in the neighboring grain; (**c**) direction of back stresses acting on the twin in the investigated region during unloading that resist further twin growth.

While the part of the parent undergoing twinning undergoes significant stress relaxation, the untwinned parts of the parent grain, as well as the neighboring grains devoid of twinning show a significant increase in the internal stresses on further straining. Additionally, owing to the nearly 90° crystallographic reorientation of the twinned volume, the twins also assume a plastically hard orientation in terms of both slip as well as $\{10\bar{1}2\}$ tension twinning (Figure 9b), leading to build up of large compressive internal stresses.

During unloading it is expected that the large internal stresses stored in the untwinned parent grain and the neighboring grains would impose considerable back stress on the twin (schematically illustrated by the blue arrows in Figure 9c), which explains the observed compressive stress state inside the twin domain (Figure 8d). The values of stresses near the twin tip in the present study reached values of -170 MPa. Comparing these values and the ones obtained in Section 4.1 to the typical critical resolved shear stresses for twinning in pure titanium that is around 125 MPa [45], indicates that the reaction stresses at the twin tips as well as along the twin boundary plane are significant enough to trigger mechanisms such as de-twinning in these regions. Furthermore, the high back stresses near the tips and low values at the mid region of the twin also comply with the typically seen lamellar twin morphology with converging tips (higher resistance to lateral growth) and a relatively thick mid-section (easier thickening under external stress).

The tensile stress gradient observed in the neighbor grain (Figure 8d) most likely arises from the forward stress component (Figure 9b) driving twin propagation (owing to the directional nature of twinning) and countering the aforementioned back stress. When the twin hits the grain boundary this positive stress in front of the twin tip is plastically accommodated in the neighboring grain, either by slip or twinning. In the present case absence of twins and a steady change in misorientation observed in the neighbor grain (c.f. Figure 7a–d) typically indicates slip induced strain accommodation. Similar to the observations in reference [15], the stress profile indicated a local minimum close to the boundary (Figure 8d). This characteristic is attributed to the role of stress fields arising from superposition of twinning dislocations and grain boundary dislocations, which in turn influence the pile-up configuration.

4.3. Implications for Macroscopic Damage Performance and Fatigue Behavior of Hexagonal Materials

The implications of the observations in the current work are significant in terms of understanding the role of twins on the fracture behavior in titanium. The findings indicate that stress development inside twins is significantly impacted by neighbor grain deformation as well as the plastic response of the untwinned parent. The values presented in the current study also highlight the role of twins in crack nucleation in the adjacent grain, especially at the twin tips. Furthermore, compressive back stresses acting parallel to the twin-parent interface in the unloaded state, along with the reaction forces at the twin tips explain the frequently observed dynamic microstructural changes induced due to internal back stresses during unloading via mechanisms such as, de-twinning or re-twinning [39,40]. The measured values also indicate that stresses can be significantly large to easily drive such reverse migration of twin boundaries, thereby corroborating the frequently observed behavior of twins disappearing during cyclical loading experiments.

The quantitative estimates of local stress profiles near twin boundaries provides an in depth understanding of microscale stress evolution, which is essential for designing microstructures that can enhance bulk scale mechanical performance.

5. Conclusions

A novel correlative technique utilizing EBSD and a FIB-DIC method for obtaining site specific microstructural and local stress information is presented. Stress gradients due to dislocation pile-up at pure titanium twin boundaries and at twin-grain boundary intersections are quantified. The following conclusions are drawn:

1. Stress gradients across the tension twin-parent interface were compressive in nature, with the maximum stresses recorded at the twin boundary. A resolved stress of ~ -180 MPa acting along the twin boundary is reported. The results indicate that the in-built stresses are significant enough to promote reverse migration or de-twinning during reverse loading.
2. Stress profiles at twin grain boundary intersections show a sign reversal, being compressive inside the twin and tensile in the neighboring grain. The results provide a quantitative measure of back stresses exerted on the twin in unloaded condition (which reach values as high as ~ -170 MPa near the twin tips) and stress gradients originating in the neighbor grain due to the interaction of twinning dislocations and a grain boundary.
3. The stress values at the twin tips and in the twin center also highlight the role of local stresses in defining the typically observed lamellar morphology of twins with wider mid-sections and converging tips.
4. The observations in the current work highlight the contribution of residual stresses associated with deformation twinning in hexagonal close packed metals in predicting their damage behavior.

Acknowledgments: This research was carried out under project number T61.1.14545 in the framework of the Research Program of the Materials Innovation Institute (M2i) (www.m2i.nl). The authors would like to acknowledge T.B. Britton for his valuable inputs in significantly improving the manuscript.

Author Contributions: Indranil Basu, Václav Ocelík and Herman Fidder conceived and designed the experiments. Indranil Basu and Herman Fidder performed the experiments. Indranil Basu, Herman Fidder, Václav Ocelík, and Jeff Th.M de Hosson contributed to the data analysis and scientific interpretation of the work. Indranil Basu drafted the article. Indranil Basu, Václav Ocelík and Jeff Th.M de Hosson made critical revisions to the the article.

Conflicts of Interest: The authors declare no conflict of interest.

References

1. Nes, E.; Marthinsen, K. Modeling the evolution in microstructure and properties during plastic deformation of f.c.c.-metals and alloys—An approach towards a unified model. *Mater. Sci. Eng. A* **2002**, *322*, 176–193. [CrossRef]
2. Eshelby, J.D.; Frank, F.C.; Nabarro, F.R.N. XLI. The equilibrium of linear arrays of dislocations. *Lond. Edinb. Dublin Philos. Mag. J. Sci.* **1951**, *42*, 351–364. [CrossRef]
3. Hall, E.O. The Deformation and Ageing of Mild Steel: III Discussion of Results. *Proc. Phys. Soc. Sect. B* **1951**, *64*, 747. [CrossRef]
4. Hall, E.O. Variation of Hardness of Metals with Grain Size. *Nature* **1954**, *173*, 948–949. [CrossRef]
5. Petch, N.J. The Cleavage Strength of Polycrystals. *J. Iron Steel Inst.* **1953**, *174*, 25–28.
6. Shen, Z.; Wagoner, R.H.; Clark, W.A.T. Dislocation and grain boundary interactions in metals. *Acta Metall.* **1988**, *36*, 3231–3242. [CrossRef]
7. Clark, W.A.T.; Wagoner, R.H.; Shen, Z.Y.; Lee, T.C.; Robertson, I.M.; Birnbaum, H.K. On the criteria for slip transmission across interfaces in polycrystals. *Scr. Metall. Mater.* **1992**, *26*, 203–206. [CrossRef]
8. Guo, Y.; Collins, D.M.; Tarleton, E.; Hofmann, F.; Tischler, J.; Liu, W.; Xu, R.; Wilkinson, A.J.; Britton, T.B. Measurements of stress fields near a grain boundary: Exploring blocked arrays of dislocations in 3D. *Acta Mater.* **2015**, *96*, 229–236. [CrossRef]
9. Winiarski, B.; Withers, P.J. Novel implementations of relaxation methods for measuring residual stresses at the micron scale. *J. Strain Anal. Eng. Des.* **2015**, *50*, 412–425. [CrossRef]
10. Britton, T.B.; Wilkinson, A.J. Measurement of residual elastic strain and lattice rotations with high resolution electron backscatter diffraction. *Ultramicroscopy* **2011**, *111*, 1395–1404. [CrossRef] [PubMed]
11. Britton, T.B.; Wilkinson, A.J. Stress fields and geometrically necessary dislocation density distributions near the head of a blocked slip band. *Acta Mater.* **2012**, *60*, 5773–5782. [CrossRef]
12. Guo, Y.; Britton, T.B.; Wilkinson, A.J. Slip band–grain boundary interactions in commercial-purity titanium. *Acta Mater.* **2014**, *76*, 1–12. [CrossRef]
13. Jiang, J.; Britton, T.B.; Wilkinson, A.J. Mapping type III intragranular residual stress distributions in deformed copper polycrystals. *Acta Mater.* **2013**, *61*, 5895–5904. [CrossRef]
14. Guo, Y.; Abdolvand, H.; Britton, T.B.; Wilkinson, A.J. Growth of {11-22} twins in titanium: A combined experimental and modelling investigation of the local state of deformation. *Acta Mater.* **2017**, *126*, 221–235. [CrossRef]
15. Basu, I.; Ocelík, V.; De Hosson, J.T.M. Measurement of spatial stress gradients near grain boundaries. *Scr. Mater.* **2017**, *136*, 11–14. [CrossRef]
16. Basu, I.; Ocelík, V.; De Hosson, J.T.M. Experimental determination and theoretical analysis of local residual stress at grain scale. *WIT Trans. Eng. Sci.* **2017**, *116*, 3–14.
17. Hirsch, J.; Al-Samman, T. Superior light metals by texture engineering: Optimized aluminum and magnesium alloys for automotive applications. *Acta Mater.* **2013**, *61*, 818–843. [CrossRef]
18. Basu, I.; Gottstein, G.; Zander, B.D. Recrystallization Mechanisms in Wrought Magnesium Alloys Containing Rare-Earth Elements. Bachelor's Dissertation, RWTH Aachen University, Aachen, Germany, December 2016.
19. Basu, I.; Al Samman, T.; Gottstein, G. Recrystallization and Grain Growth Related Texture and Microstructure Evolution in Two Rolled Magnesium Rare-Earth Alloys. *Mater. Sci. Forum* **2013**, *765*, 527–531. [CrossRef]
20. Basu, I.; Al-Samman, T. Triggering rare earth texture modification in magnesium alloys by addition of zinc and zirconium. *Acta Mater.* **2014**, *67*, 116–133. [CrossRef]
21. Yoo, M.H. Slip, twinning, and fracture in hexagonal close-packed metals. *Metall. Trans. A* **1981**, *12*, 409–418. [CrossRef]
22. Yoo, M.H.; Lee, J.K. Deformation twinning in h.c.p. metals and alloys. *Philos. Mag. A* **1991**, *63*, 987–1000. [CrossRef]
23. El Kadiri, H.; Oppedal, A.L. A crystal plasticity theory for latent hardening by glide twinning through dislocation transmutation and twin accommodation effects. *J. Mech. Phys. Solids* **2010**, *58*, 613–624. [CrossRef]

24. Qiao, H.; Guo, X.Q.; Oppedal, A.L.; El Kadiri, H.; Wu, P.D.; Agnew, S.R. Twin-induced hardening in extruded Mg alloy AM30. *Mater. Sci. Eng. A* **2017**, *687*, 17–27. [CrossRef]
25. Taylor, B.; Weidmann, E. *Metallographic Preparation of Titanium, Struers Application Notes*; Struers: Ballerup, Denmark, 2008.
26. Hielscher, R.; Schaeben, H. A novel pole figure inversion method: specification of the MTEX algorithm. *J. Appl. Crystallogr.* **2008**, *41*, 1024–1037. [CrossRef]
27. Kang, K.J.; Yao, N.; He, M.Y.; Evans, A.G. A method for in situ measurement of the residual stress in thin films by using the focused ion beam. *Thin Solid Films* **2003**, *443*, 71–77. [CrossRef]
28. Sabaté, N.; Vogel, D.; Gollhardt, A.; Marcos, J.; Gràcia, I.; Cané, C.; Michel, B. Digital image correlation of nanoscale deformation fields for local stress measurement in thin films. *Nanotechnology* **2006**, *17*, 5264. [CrossRef]
29. Mansilla, C.; Martínez-Martínez, D.; Ocelík, V.; De Hosson, J.T.M. On the determination of local residual stress gradients by the slit milling method. *J. Mater. Sci.* **2015**, *50*, 3646–3655. [CrossRef]
30. Mansilla, C.; Ocelík, V.; Hosson, J.T.M.D. A New Methodology to Analyze Instabilities in SEM Imaging. *Microsc. Microanal.* **2014**, *20*, 1625–1637. [CrossRef] [PubMed]
31. Nelson, D.V.; Makino, A.; Schmidt, T. Residual Stress Determination Using Hole Drilling and 3D Image Correlation. *Exp. Mech.* **2006**, *46*, 31–38. [CrossRef]
32. Britton, T.B.; Liang, H.; Dunne, F.P.E.; Wilkinson, A.J. The effect of crystal orientation on the indentation response of commercially pure titanium: experiments and simulations. *Proc. R. Soc. Lond. Math. Phys. Eng. Sci.* **2010**, *466*, 695–719. [CrossRef]
33. Calcagnotto, M.; Ponge, D.; Demir, E.; Raabe, D. Orientation gradients and geometrically necessary dislocations in ultrafine grained dual-phase steels studied by 2D and 3D EBSD. *Mater. Sci. Eng. A* **2010**, *527*, 2738–2746. [CrossRef]
34. Drouven, C.; Basu, I.; Al-Samman, T.; Korte-Kerzel, S. Twinning effects in deformed and annealed magnesium-neodymium alloys. *Mater. Sci. Eng. A* **2015**, *647*, 91–104. [CrossRef]
35. Basu, I.; Al-Samman, T. Competitive twinning behavior in magnesium and its impact on recrystallization and texture formation. *Mater. Sci. Eng. A* **2017**, *707*, 232–244. [CrossRef]
36. Kubin, L.P.; Mortensen, A. Geometrically necessary dislocations and strain-gradient plasticity: A few critical issues. *Scr. Mater.* **2003**, *48*, 119–125. [CrossRef]
37. Konijnenberg, P.J.; Zaefferer, S.; Raabe, D. Assessment of geometrically necessary dislocation levels derived by 3D EBSD. *Acta Mater.* **2015**, *99*, 402–414. [CrossRef]
38. Qiao, H.; Agnew, S.R.; Wu, P.D. Modeling twinning and detwinning behavior of Mg alloy ZK60A during monotonic and cyclic loading. *Int. J. Plast.* **2015**, *65*, 61–84. [CrossRef]
39. Wu, L.; Agnew, S.R.; Brown, D.W.; Stoica, G.M.; Clausen, B.; Jain, A.; Fielden, D.E.; Liaw, P.K. Internal stress relaxation and load redistribution during the twinning–detwinning-dominated cyclic deformation of a wrought magnesium alloy, ZK60A. *Acta Mater.* **2008**, *56*, 3699–3707. [CrossRef]
40. Basu, I.; Al-Samman, T. Twin recrystallization mechanisms in magnesium-rare earth alloys. *Acta Mater.* **2015**, *96*, 111–132. [CrossRef]
41. Arul Kumar, M.; Kanjarla, A.K.; Niezgoda, S.R.; Lebensohn, R.A.; Tomé, C.N. Numerical study of the stress state of a deformation twin in magnesium. *Acta Mater.* **2015**, *84*, 349–358. [CrossRef]
42. Arul Kumar, M.; Beyerlein, I.J.; McCabe, R.J.; Tomé, C.N. Grain neighbour effects on twin transmission in hexagonal close-packed materials. *Nat. Commun.* **2016**, *7*. [CrossRef] [PubMed]
43. Knezevic, M.; Daymond, M.R.; Beyerlein, I.J. Modeling discrete twin lamellae in a microstructural framework. *Scr. Mater.* **2016**, *121*, 84–88. [CrossRef]
44. Zhang, R.Y.; Daymond, M.R.; Holt, R.A. A finite element model of deformation twinning in zirconium. *Mater. Sci. Eng. A* **2008**, *473*, 139–146. [CrossRef]
45. Wu, X.; Kalidindi, S.R.; Necker, C.; Salem, A.A. Prediction of crystallographic texture evolution and anisotropic stress-strain curves during large plastic strains in high purity α-titanium using a Taylor-type crystal plasticity model. *Acta Mater.* **2007**, *55*, 423–432. [CrossRef]

crystals

MDPI

Article

A New Method for Evaluating the Indentation Toughness of Hardmetals

Prem C. Jindal

Independent Scholar, 615 Westchester Drive, Greensburg, PA 15601, USA; jindalprem@yahoo.com

Received: 2 March 2018; Accepted: 25 April 2018; Published: 3 May 2018

Abstract: This paper proposes a new method of evaluating the indentation toughness of hardmetals using the length of Palmqvist cracks (C) and Vickers indentation diagonal size (d_i). Indentation load "P" is divided into two parts: P_i for plastic indentation size and P_c for Palmqvist cracks. Pi depends upon the square of the indentation size (d_i^2) and Pc depends upon ($C^{3/2}$). The new method produces a very good linear relationship between the calculated indentation toughness values and the standard conventional linear elastic fracture mechanics toughness values with the same cemented carbide materials for a large number of standard Kennametal grades for both straight WC-Co carbide grades and grades containing cubic carbides. The new method also works on WC-Co hardmetal data selected from recently published literature. The technique compares the indentation toughness values of WC-Co materials before and after vacuum annealing at high temperature. The indentation toughness values of annealed carbide samples were lower than for un-annealed WC-Co hardmetals.

Keywords: WC-CO cemented carbide materials; Vickers hardness; Palmqvist indentation cracks; indentation toughness; linear elastic fracture mechanics toughness, K_{IC}, G_{IC}

1. Introduction

WC-Co based cemented carbide materials, also known as hardmetals, with and without the addition of cubic carbides such as TiC, TaC, and NbC to the base material, are extensively used in metalcutting, mining, metalforming, and other speciality wear-resistant applications. Many hardmetal components rely on material hardness, and while a number of application-relevant properties, such as strength, elastic modulus, and hardness are easy to measure, the conventional linear elastic fracture mechanics approach for measuring the fracture toughness, critical energy release rate (G_{IC}), and critical stress intensity factor (K_{IC}) requires considerable effort. Specifically, the pre-cracking of specimens has remained a serious obstacle.

The Palmqvist indentation cracking test is sometimes used for the characterization of the toughness of cemented carbides [1]. The test provides a measure of the indentation crack resistance of a brittle material from the length of cracks induced with a Vickers diamond hardness impression and applied load as per Equation (1)

$$W = P/C \tag{1}$$

where W is the Palmqvist indentation toughness, P is the indentation load on a Vickers diamond indenter, and C is the sum of the four Palmqvist cracks lengths, $(C_1 + C_3) + (C_2 + C_4)$ emanating from the four corners of the indentation after the load has been removed. Crack length $C_1 + C_3$ is measured along one indentation diagonal length and $C_2 + C_4$ is measured along the other indentation diagonal length [2]. It is to be noted that W has the unit of kg/mm, similar to G_{IC} in linear elastic fracture mechanics formulation.

Palmqvist cracks geometrically different from half-penny cracks are essentially confined to the specimen surface and therefore surface preparation is extremely important and critical for the evaluation of indentation toughness. Exner [2] further examined the issue of specimen surface preparation techniques such as diamond polishing of the ground specimen so that the deformed binder phase layer near the surface and surface residual compressive stress observed in the WC phase are minimized, and further recommended a high-temperature (1000–1100 °C) vacuum annealing procedure after diamond-polishing procedures so that reproducible Palmqvist cracks are generated at each indentation load.

2. Indentation Toughness versus Linear Elastic Fracture Mechanics Toughness

In recent years, considerable efforts have been directed at relating indentation toughness W or equivalent K values with conventional linear elastic fracture mechanics (K_{IC}) or equivalent G_{IC} values for the same cemented carbide materials. Niihara [3] and Warren and Matzke [4] independently suggested the relationship in Equation (2)

$$K_{IC} = b(H \cdot W)^{1/2} \tag{2}$$

The above relationship is based upon the formation of half-penny cracks which have not been observed in cemented carbide materials. In the above equation, "b" is a non-dimensional constant dependent on the ratio of Young's modulus"E" and Vickers hardness "H" in Niihara's analysis. The value of constant in Warren and Matzke's analysis is unspecified. These investigators collected a large body of experimental data on WC-Co hardmetals and showed good linear correspondence with Equation (2) for K_{IC} values up to ~17 MPa·m$^{1/2}$. The latest model is that of Shetty and colleagues [5], who used a wedge loaded crack as a fracture mechanical analogue to the situation in Palmqvist cracks and showed that K_{IC} can be evaluated as Equation (3)

$$K_{IC} = 0.0889(H \cdot W)^{1/2} \tag{3}$$

Shetty's model has become an accepted model for evaluating the indentation toughness of hardmetals and is being used extensively by the carbide industry for that purpose [6] The indentation toughness values have a good linear relationship with K_{IC} values determined by the conventional linear elastic fracture mechanics procedures for values up to ~20 MPa·m$^{1/2}$ but the linear relationship breaks down for carbide materials with very high toughness values. The reason for this discrepancy is that Palmqvist cracks are extremely small compared with indentation diagonal size, so that ratio of $C/2d_i$ is extremely small, at much less than 1. In that case indentation toughness values are very large compared with K_{IC} values. This paper proposes a new method to address this problem.

3. The New Approach for Evaluating the Indentation Toughness

Two effects are observed whenever a flat and properly polished specimen of a cemented carbide material is indented with a Vickers indenter with load "P". One can observe Vickers plastic indentation with size "d_1" and "d_2" along with Palmqvist cracks emanating from the four corners of the Vickers indentation. The size of the average indentation diagonal $d_i = (d_1 + d_2)/2$ and lengths of cracks depend upon the mechanical properties (plastic deformation and toughness properties of a given carbide material which in turn depend upon the chemical composition of WC-Co, WC grain size, and the average thickness of the binder phase). Sometimes the indentation load has to be sufficiently large to induce Palmqvist cracks on all four corners of the Vickers indentation in very-high toughness cemented carbide materials.

The technical approach adopted here is as follows:

One can divide indentation load "P" into two components, P_i and P_c. Pi is responsible for causing average indentation "d_i" and P_c for causing Palmqvist cracks $C = C_1 + C_2 + C_3 + C_4$. One can write the Equation (4) as

$$P = P_i + P_c \tag{4}$$

It is well-known that P_i is proportional to the square of the average indentation "d_i". Therefore, Equation (5) can be written as

$$P_i = X_i \cdot d_i^2 \tag{5}$$

Also, P_c is proportional to $C^{3/2}$ and therefore Equation (6) is as follows:

$$P_c = X_c \cdot C^{3/2} \tag{6}$$

Therefore, one can write the two equations into Equation (7)

$$P = X_i \cdot d_i^2 + X_c \cdot C^{3/2} \tag{7}$$

Now, if the indentation load is in "kg", the indentation size is in mm and C is in mm. Then, "X_i" is described in kg/mm²) and "X_c" is in kg/mm$^{3/2}$. Assuming that $X_i = 1$; and $X_c = 1$, this leads to Equation (8)

$$P = d_i^2 + C^{3/2} \tag{8}$$

One can combine these equations and arrive at Equations (9) and (10)

$$K_m = P_c / C^{3/2} \tag{9}$$

$$W_m = P_c / C \tag{10}$$

Therefore, one can calculate "K_m" and "W_m" by using the P_c and C from the measured values of indentation size and total lengths of Palmqvist cracks. It should be understood that K_m and W_m are different from conventional indentation toughness "W", as is mentioned in Equation (1).

4. Results and Discussions

The results are presented in three sections as follows.

4.1. Application to Kennametal Cemented Carbide Grades

Detailed investigations [7] were undertaken in early 1980s on the Palmqvist toughness and the linear elastic fracture mechanics toughness (K_{IC}) of a large number of commercially available Kennametal carbide grades covering metal cutting, mining, metal forming, and specialty grades. Metalcutting grades contained fair amounts of cubic carbides such as TiC, NbC etc., whereas others were essentially straight WC-Co grades with less than 0.5% cubic carbides. The properly polished samples were indented at various indentation loads varying from 30 to 120 kg. Three measurements were conducted at each load for indentation size and Palmqvist crack measurements. Considerable variation in Palmqvist crack lengths was noted even within a single indentation from one corner to the opposite corner. Linear elastic fracture mechanics measurements (K_{IC}) were also conducted from the same batch of carbide samples using the Terra Tek procedure [8]. Indentation toughness "K_m" was calculated at 100 kg indentation load and compared with the average value of K_{IC}. Figure 1 shows the K_m versus K_{IC}. The linear agreement between K_m and K_{IC} is quite reasonable across the whole range of carbide materials.

4.2. Application to Recently Published Crack Length and Vickers Hardness Data

Recently, Seikh and colleagues published a paper [9] measuring the indentation toughness and K_{IC} values on a large number of straight WC-Co cemented carbide samples using an indentation load of 30 kg for both Vickers hardness and Palmqvist crack measurements. Ten measurements were performed for each WC-Co material for a total of eight different carbide materials. K_{IC} measurements were also conducted for all of the eight carbide materials. The sum of Palmqvist crack lengths "C" was calculated from the given data and indentation toughness (K_m) was calculated for each carbide material. Figure 2 shows the plot of indentation toughness "K_m" versus "K_{IC}" for all of the samples. The linear agreement between K_m versus K_{IC} is excellent across the whole range of carbide materials. This shows the clear difference indentation toughness between the method adopted in this approach versus the previous methods, as detailed in Section 2 of this paper.

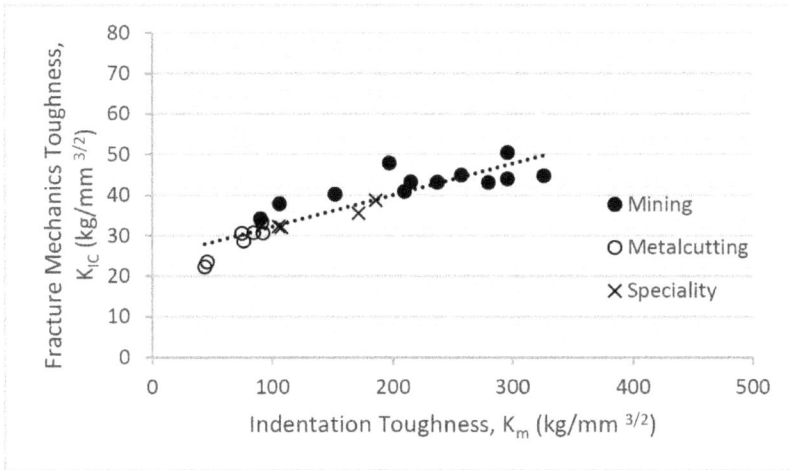

Figure 1. Indentation toughness (K_m) versus fracture mechanics toughness (K_{IC}) for Kennametal grades.

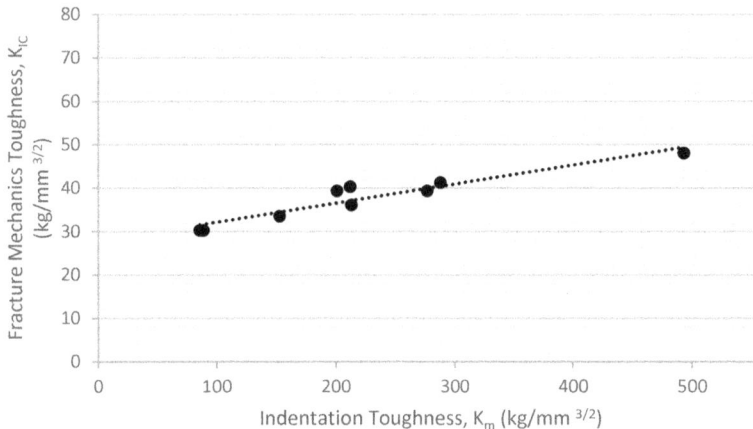

Figure 2. Indentation toughness (K_m) versus fracture mechanics toughness (K_{IC}), for WC-Co hardmetals.

4.3. Effect of Vacuum Annealing on Indentation Toughness of Carbide Materials

Exner et al. [10] conducted Palmqvist crack measurements on a number of straight WC-Co carbide grades, which were vacuum annealed at 1100 °C before Palmqvist crack lengths were carried out at indentation loads of 30, 45, 60, 100, and 150 kg. It was not possible to compare indentation toughness before and after vacuum annealing in that work because no crack measurements were conducted on the as-sintered un-annealed samples. However, it was possible to compare the results with the published data of Seikh and colleagues [9], who performed extensive Palmqvist crack measurements at an indentation load of 30 kg. Therefore, indentation toughness was calculated on a few vacuum annealed WC-Co samples at an indentation load of 30 kg and the indentation toughness results were compared with the indentation toughness data taken from the work of Seikh and colleagues [9]. The results are summarized in Tables 1 and 2.

Table 1. Data from Exner et al. [10] on Co Vol %; Vickers hardness and toughness.

SP# (Co Vol %)	Vickers Hardness	K_m	W_m
5.1	1705	69	50.9
10.1	1603	104	66
14.8	1390	175	103

Table 2. Data from Seikh et al. [9] on Co Vol %; Vickers hardness and toughness.

SP# (Co Vol %)	Vickers Hardness	K_m	W_m
4.2	1782	89	60
7.5	1748	86	59
10	1591	200	97.3
15.6	1483	213	108

One can note that both K_m and W_m values are higher for the as-sintered WC-Co materials (samples from Seikh et al. [9]) as compared with the vacuum-annealed carbide materials (samples from Exner et al. [10]) for essentially similar WC-Co compositions, in spite of the fact that un-annealed specimens have higher Vickers hardness values. In general, toughness is inversely proportional to Vickers hardness for these hardmetal materials. This result indicates that vacuum annealing reduces the indentation toughness of WC-Co carbide materials.

This result is completely unexpected and contradicts the results of various investigators [10,11] who compared vacuum annealed indentation toughness values with K_{IC} and G_{IC} values, which were generally measured on un-annealed as-sintered carbide samples assuming explicitly that vacuum annealing of WC-Co material should not reduce or degrade any mechanical properties of the as-sintered carbide materials. This is probably based on the fact that Vickers hardness does not change after annealing. To the best of our knowledge, uniaxial yield stress and K_{IC} measurements have not been conducted on high-temperature vacuum-annealed WC-Co materials and reported in the open published literature.

The work of Pickens and Gurland [12] is worth mentioning to explore this issue further. These authors evaluated the K_{IC} and G_{IC} of a large number of WC-Co materials with varying volume fraction of cobalt, WC grain sizes, and cobalt-based binder phase layer thickness, and proposed Equation (11) to explain the results:

$$G_{IC} = a \cdot \sigma_y \cdot l \tag{11}$$

where "a" is a constant, σ_y is the in-situ yield stress of the binder phase, and "l" is the average thickness of the binder phase.

Vacuum annealing at (1000–1100 °C) is not expected to change the value of binder phase thickness. Also, it has been observed during routine X-ray diffraction of the polished carbide samples that the major cobalt-based binder phase XRD peak becomes sharper and of higher intensity for the annealed sample than that of the un-annealed as-sintered polished sample, which is broad and of low intensity. This observation indicates that in situ yield stress of the binder phase (σ_y) has decreased, resulting in a lower G_{IC} value after annealing. This result is consistent with the lower indentation toughness of annealed samples compared with un-annealed samples as shown in our results.

This result is also consistent with lower transverse rupture strength of CVD-coated carbide samples routinely observed in CVD-coated samples as compared with uncoated polished samples.

It has also been well established that carbide materials coated with CVD coatings (multi-layer TiCN/TiC/Al$_2$O$_3$) have performed poorly in metal cutting machining operations, especially for rotating tools (interrupted cutting operations such as milling applications) relative to high quality ion-plated PVD TiN, TiCN, and TiAlN coatings, even though CVD coatings have higher abrasive wear resistance (hot hardness) and also higher crater wear resistance (chemical inertness) than PVD TiN, TiCN, and TiAlN coatings. The primary reason is that CVD coatings routinely deposited at high temperatures (~1050–1250 °C) reduce the toughness of the base carbide materials. PVD coatings are generally deposited at around ~500 °C and do not degrade the transverse rupture strength of the base material.

5. Conclusions

1. A new method of evaluating the indentation toughness of hardmetals has been proposed.
2. The new measured indentation toughness values provide very good linear agreement with K_{IC} values measured by conventional linear elastic fracture mechanics procedures.
3. Vacuum annealing of as-sintered cemented carbide materials at 1000–1100 °C lowers the indentation toughness of cemented carbide materials.

Acknowledgments: I would like to dedicate this paper to Ronald W. Armstrong for being an inspirational teacher and guide. I would like to thank Kennametal for allowing me to publish portions of the work done while under their employment. I give special thanks to Binky Sargent for the graphics in this paper.

Conflicts of Interest: The author declares no conflict of interest.

References

1. Palmqvist, S. Method att Bestamma Seghten hos Spread Material. *Sarskit Hardmettaler. Jernkont. Ann.* **1957**, *141*, 300–307.
2. Exner, H.E. The Influence of Sample Preparation on Palmqvist's Method for Toughness Testing of Cemented Carbides. *Trans. TMS AIME* **1969**, *245*, 677–683.
3. Niihara, K. Indentation Fracture Toughness of Brittle Materials for Palmqvist Cracks. *J. Mater. Sci. Lett.* **1983**, *2*, 2. [CrossRef]
4. Warren, R.; Matzke, H. Indentation Testing of a Broad Range of Cemented. In Proceedings of the International Conference on The Science of Hard Materials, Jackson, WY, USA, 23–28 August 1981; Viswanadham, R.K., Rowcliffe, D.J., Gurland, J., Eds.; Plenum Press: New York, NY, USA, 1983; p. 563.
5. Shetty, D.K.; Wright, I.G.; Mincer, P.N.; Clauer, A.H. Indentation fracture of WC-Co Cermets. *J. Matrt. Sci.* **1985**, *20*, 1873–1882. [CrossRef]
6. Hardness Toughness Tests. *VAMAS Interlaboratory Exercise*; Report No.48; NPL: Middlesex, UK, 2005.
7. *Kennametal Technology Report*; Technical Report; Kennametal Inc.: Latrobe, PA, USA, 1983.
8. Barker, L.M. A Simplified Method for Measuring Plane Strain Fracture Toughness. *Eng. Fract. Mech.* **1977**, *9*, 361–364. [CrossRef]
9. Seikh, S.; M'Saoubi, R.; Flasar, P.; Schwind, M.; Persson, T.; Yang, J.; Llanes, L. Fracture Toughness of Cemented Carbides: Testing Method and Microstructural effects. *Int. J. Ref. Met. Hard Mater.* **2015**, *49*, 153–160. [CrossRef]
10. Exner, E.L.; Pickens, J.; Gurland, J. A Comparison of Indentation Crack Resistance and Fracture Toughness of Five WC-Co alloys. *TMS AIME* **1978**, *9*, 736–738. [CrossRef]
11. Spiegler, R.; Schmauder, S.; Sigl, L. Fracture Toughness Evaluation of WC-Co Alloys by Indentation Testing. *J. Hard Mater.* **1990**, *1*, 147.
12. Pickens, J.R.; Gurland, J. The Fracture Toughness of WC-Co Alloys Measured on Single-edge Notched Beam Specimens Precracked by Electron Discharge Machining. *Mater. Sci. Eng.* **1978**, *33*, 135–142. [CrossRef]